Springer Series in Synergetics

Springer Series in Synergetics
Founding Editor: H. Haken

The Springer Series in Synergetics was founded by Herman Haken in 1977. Since then, the series has evolved into a substantial reference library for the quantitative, theoretical and methodological foundations of the science of complex systems.

Through many enduring classic texts, such as Haken's *Synergetics and Information and Self-Organization*, Gardiner's *Handbook of Stochastic Methods*, Risken's *The Fokker Planck-Equation* or Haake's *Quantum Signatures of Chaos*, the series has made, and continues to make, important contributions to shaping the foundations of the field.

The series publishes monographs and graduate-level textbooks of broad and general interest, with a pronounced emphasis on the physico-mathematical approach.

Len Pismen

Patterns and Interfaces in Dissipative Dynamics

Revised and Extended, Now also Covering Patterns of Active Matter

Second Edition

Springer

Len Pismen
Israel Institute of Technology
Technion
Haifa, Israel

ISSN 0172-7389 ISSN 2198-333X (electronic)
Springer Series in Synergetics
ISBN 978-3-031-29578-2 ISBN 978-3-031-29579-9 (eBook)
https://doi.org/10.1007/978-3-031-29579-9

This Springer imprint is published by the registered company Springer Nature Switzerland AG
The registered company address is: Gewerbestrasse 11, 6330 Cham, Switzerland

Foreword

Nature is full of marvels, be they in the inanimate world or in the realm of life. Quite often their beauty and attractive power come from the regularity of some sort of "pattern," or – more subtly – from some obvious or hidden irregularity (as the French poet Verlaine said, "always favour the uneven"). Obvious natural patterns are found in the stripes of sea-shells or on the fur of mammals, such as our domestic cats. Stones such as malachites or agathes also show finely crafted layers of matching or contrasting colors. A less obvious case of "pattern" is seen in the opals, whose opalescence reflects wonderfully regular arrays of spheres of silica. Civilizations of the Far East highly praised these miraculous examples of beauty in minerals.

The way things and living beings are made or have been made began to attract the interest of scientists of the Classical Age, particularly after Newton in the query 31 at the end of his Opticks claimed that the microscopic world, made of atoms, follows the same kind of laws as the Universe at large. The first instance of a "regular pattern" produced artificially was by the spontaneous modulation of the surface of a liquid vibrated vertically. It was first believed that this Faraday instability was a kind of linear resonance of the capillary waves at the frequency of the excitation. It was Rayleigh who first explained the striking fact that actually the dominant frequency of the surface oscillations is half that of the exciting acceleration, a result that is not obvious. Rayleigh's theory gave rise to a whole field of research, the study of ordinary differential equations with time periodic coefficients. Once transformed into maps of the circle this class of problems has been a major source of mathematical progress over the years. The next example of an artificial pattern in experiments was the famous (and first very poorly understood) Bénard cells. Henri Bénard, one of the first great experimentalists in fluid mechanics, showed that a thin layer of whale oil heated from below set itself spontaneously in motion in a remarkably regular array of hexagonal cells. Bénard himself did explain this phenomenon in a rather confused way and another 50 years passed before Scriven and Sterling showed that the instability is due to the Marangoni effect, that is the temperature dependence of the surface tension. It took another 10 years to see the explanation of the hexagonal pattern by Palm, who proved that hexagons are the generic structure to expect beyond a bifurcation yielding wavy patterns with an arbitrary orientation in a plane. The Rayleigh–Bénard instability due to buoyancy (instead of the Marangoni effect) of a fluid layer heated from below does show rolls, because of a particular, nongeneric symmetry.

Over the years an almost infinite number of "pattern-forming" bifurcations in physical systems have been discovered, as well as many kinds of instabilities. The famous Brusselator explained the oscillating chemical reactions in the well-known Belousov–Zhabotinsky system. The Turing instability shows up in driven systems with widely different diffusion coefficients. Quite often the scientists responsible for the discoveries and/or the theory of pattern-forming systems saw in a more or less explicit way a relationship between their observations and the structures of living beings. Such a rather close connection surely exists for patterns on sea-shells that can be explained by simple reaction–diffusion equations without any recourse to a detailed "program" written in the DNA of the snails.

May be less noticed than discoveries in the experimental field, there has been a tremendous progress in our rational understanding of what I would call loosely the qualitative behavior of solutions of partial differential equations and/or ordinary differential equations. A very crucial idea, first developed by Poincaré and by Andronov, is the one of "robustness." Loosely speaking, it allows us to describe a dynamical system (= the equation(s) of motion of the system we consider) independently of their detailed form. This is of course a "topological theory" in the deepest sense, as it amounts to consider that two systems are "the same" if one can deform one smoothly into the other. Except at "bifurcation" value this holds true for almost all values of the parameters into the equation(s) of motion: changing a bit a parameter value changes only numerical values typical of the solutions, not their structure.

The study of the qualitative properties of dynamical systems has experienced an explosive growth in the late 1970s and early 1980s. This is very nicely presented in the present book. The mathematicians managed later on to prove many results that had only been guessed by physicists in this field. Emboldened by the success of this notion of "structural stability," some scientists, Len Pismen among them, tried to deal in the same way with the pattern-forming systems. This amounted to summarizing the properties of the system at hand into a set of equations as simple as possible, but with all the relevant symmetries. This method traces back its origin in Poincaré's normal forms; near a bifurcation point, one can reduce the complexity of the equations by keeping as variables a few time-dependent amplitudes only, with simple polynomial nonlinearities. This reduction of the complexity is, after all, what we do spontaneously when we see a set of rolls in aligned clouds or chaos and turbulence in a cloud painted by Tiepolo or Corot.

The question is posed by this reduction: Can we obtain a picture of the phenomena general enough to yield the most obvious features of the solutions we would like to describe, although the mathematics remain simple enough to be amenable to our methods of analysis? Of course, such a reductionist view is at the heart of our physical description of the world. To take an example, the Navier–Stokes equations of a viscous fluid reduce all the complex motion of an almost infinite number of atoms and molecules to the knowledge of the value of the shear viscosity of this fluid and, eventually, to the pressure–density relation.

Len Pismen starts from quite general equations, kind of generalized reaction–diffusion equations for an arbitrary number of chemical species. He shows masterfully how to reduce them to simpler forms near bifurcation points (the famous

Newell–Withehead–Segel amplitude equation, the Kuramoto–Sivashinsky equation, the Ginzburg–Landau equation to name just a few). He then follows this general scheme of reduction, but with very little "hubris." Reading his text, I am amazed by the conciseness and efficiency of his presentation. Only a few lines of text and some very clever mathematical steps bring all the necessary light in very messy questions, in the great tradition of scientists such as Zel'dovich, Kolmogorov, and Landau. He balances very finely the logical step, the development of a new intuition, with the help of classical analysis at its best. This text has the double interest to introduce the reader to a whole fascinating new field, still in a stage of fast development, and to show how to sort out quite complex issues in an efficient way. Besides its obvious pedagogical aspects, this monument of scholarship is also the very first one to cover in a fully unified way an extremely wide range of topics. It reaches quite often the cutting edge of present-day research. For instance, it gives a very thorough review of everything known about the spirals in excitable media, including the instabilities of the center of the spiral that may oscillate, meander, and so on. Although the amplitude equations have been known for a number of years, this is – as far as I can tell – the first time all their properties are discussed in a unified text.

As the title shows, part of the book is devoted to fronts and interfaces. This is done in Chap. 2, which is very welcome as it puts in a single framework topics such as the Maxwell–van der Waals theory of liquid–vapor interface, the growth of a stable pattern in an unstable medium (the so-called KPP problem), and other "phase field models" that have been lately very thoroughly studied.

The amplitude equation section (Chap. 4) is an exposition by a master of the field. Everything is covered in a smooth way, including the very tricky question of the motion of defects, where the distortion brought to the phase field by the motion is analyzed in depth. The latest progress in pinned fronts, grain boundaries, etc. is well covered. The advantage of a single book is most evident here, because the discussion relies on a rather delicate analysis of bifurcations of trajectories in a 3D space.

The final chapter is about wave patterns and spirals, with all their instabilities, interactions, etc. This book follows another one by the same author, on vortices in nonlinear fields, and the two bring a unified view of an extremely active and interesting field of modern research.

Paris, France *Yves Pomeau*

Preface to the First Edition

This book involved considerable rewriting and reshuffling, despite the fact that I thought when I started it that I knew the material well. As usual, it was difficult to decide what to include and what not. A monograph should not imitate a journal review; it is too easy to send readers to find their own way in the thick of the woods. However, the narrative may expand without limit when attempting to be self-contained. The final compromise may sometimes be subjective and imperfect.

Few appreciated the hard labor of Procrustes, who had to fit all kinds of strangers into his infamous bed. The author finds himself in a somewhat similar position, with an additional constraint on the use of violent means and an obligation to keep his customers alive and well, while resizing them to common logic and notation. This is not easy even when processing one's own work of yesteryear. One wishes to give up and write another paper instead, but this becomes impossible after a critical point has been passed and the invested effort has grown too heavy to discard. The result is before the reader.

Patterns and interfaces live in space; I could not avoid, nevertheless, to include a long chapter on dynamical systems, which live in time only, but provide both technical means and examples to follow for more robust spatial structures. Ardent students of dynamical systems may find it almost sacrilegious to see a variety of bifurcations and endless abyss of chaos crowded into a few dozen pages, still shared with the dynamics of *regular* patterns (which imitate far simpler systems by forfeiting their imperfections and natural variability). These are, of course, pure essentials that are needed to understand further material.

The core of the book lies in various applications of perturbation theory, which attempts to tackle nonlinear spatiotemporal structures that, at first sight, are likely to break our modest analytical tools. A modern paradox is that these structures are often easier to simulate numerically than to carve analytically. Simulation is, however, not a substitute for understanding, and one hopes that analytical skills will not fade away with the change of generations.

Just imagine a feeble knight-errant, armed only with a paper sword – and, perhaps, a laptop with symbolic computation software: can he dare to overcome a horde of nonlinear beasts? There are, essentially, two tricks known from ancient times: separate them or try to surprise them while they are still waking up. The first method is called *scale separation* and is the principal tool in Chaps. 2 and 3, where we study first fronts, and then structures built up by their combinations in reaction–diffusion and related equations. The second method, called *amplitude expansion*, deriving and analyzing universal equations near symmetry-breaking transitions, prevails in Chaps. 4 and 5.

Reaction–diffusion systems, with their plethora of chemical and biological applications and common but versatile structure, provide the bulk of material of the book. Other common pattern-forming systems rooted in fluid mechanics and nonlinear optics are not considered explicitly, but they converge to the same universal equations of amplitude and phase dynamics.

I am indebted to many colleagues and friends of our "nonlinear community", which has forged during the late decades of 20th century our basic understanding of pattern formation far from equilibrium. The new century, that seems to be interested, in its youth, in the particular more than in the universal, may still find this knowledge useful for building up future biomorphic technologies based on bottom-up self-organization rather than top-down manufacturing.

Technion – Israel Institute of Technology
Haifa, November 2005

L.M. Pismen

Preface to the Second Edition

The age of a "feeble knight-errant" featured in the first preface has passed. A successful modern scientist is armed with grants and computational software and leads a troop of students and postdocs. I heard an old man, still publishing meagerly cited papers, observing cynically: "They color the grey hair of nonlinear science using color printers". Nevertheless, the core of nonlinear analysis developed late in the last century remains a part of the golden fund of science, supporting and widening the scope of specific applications. Human and artificial intelligence differ in one crucial aspect: we are able not only to simulate but to understand, and analytical skills may not fade away with the change of generations.

The contents of this edition is extended, but the guiding principle remains the same: I am trying to present, whenever possible, all derivations on these pages in detail and in a common style. This book is not a review: the references are kept to a bare minimum (which still spreads out to a long list), and I try to keep the descriptions self-sustained rather than sending the reader to look for details elsewhere, in a forest of diverse notations. An excessive number of self-citations could be excused by the affinity of their style and content to the approach of this book. Most illustrations, except pictures from experiments and simulations, are original, and those from the first edition are redrawn using contemporary graphics. As if to confirm the suspicions of the cynical observer, this edition contains color illustrations, while the first one was totally black-and-white.

The scope of pattern-forming phenomena is very wide. The emphasis in this edition still rests on presenting general tools of nonlinear analysis, but the important new element is the extension of dissipative dynamics to its opposite in the novel field of "active matter" – a term that was still unknown at the time I was working on the first edition. In order to keep the volume within reasonable limits, I avoid overabundant applications to specific chemical and hydrodynamic processes, as well as to rotating, magnetic, electrostatic, and optical systems (even though I worked on some of them myself), and all kinds of latter-day studies (largely numerical) of the various modifications of standard pattern-forming models.

The first chapter has been completely rewritten. I cut down the material on dynamical systems lacking spatial dependence and excluded chaotic dynamics, as these topics are treated in more detail in a wide range of specialized literature, including my recent textbook (Pismen, 2021a). Instead, I added here the basic analysis of thermo-convective instabilities in order to ease the emphasis on reaction–diffusion systems. Other chapters, centered on multiscale perturbation methods, remain closer to the first edition, but, besides thorough editing, some new material, both original (especially in the third chapter) and based on relevant recent publications, has been added throughout.

The new chapter on active matter turned out to be the longest and, naturally, took most time to write. I do not enter here into particular biological and biomorphic applications, which I had the pleasure to describe on a qualitative and informal level in my recent semi-popular book (Pismen, 2021b), welcomed by a full-page review in Physics Today (Shankar, 2002). Here the emphasis is different: to underline the applications of the established methods of nonlinear science to this new field. So, in some instances, I had to rework the derivations to comply with the style and the standard of this book. This material necessitated an introduction to media with orientational order and their affinity with non-equilibrium patterns, which remain the core of this book.

I have benefited from discussing different topics, mostly online, as it goes these days, with Igor Aranson, Markus Bär, Jean-François Joanny, Alexander Nepomnyashchy, Harald Pleiner, Boris Rubinstein, and Uwe Thiele.

Technion – Israel Institute of Technology
Haifa, January 2023

L.M. Pismen

Contents

Introduction

Quest for Complexity

The human eye is bored by monotonous landscapes. We are depressed by gray skies, unending plains under the wing of an airliner, featureless urban sprawl, geometric canvasses in empty museum halls. Having evolved to meet the challenges of a complex world, we enjoy complexity. A stormy sea entertains us by a spectacle of *spatiotemporal* complexity. It is imitated by entertainers overwhelming us with the information deluge of blinking images in video clips. We do also enjoy, however, the calm of *spatial* complexity: a mountain landscape, a forest, a meadow, a mosaic.

Complexity is particularly appealing when it arises spontaneously, rather than by design. This is where nature and industry differ most: nature starts from a simple form and builds complexity by evolution and differentiation, while industry manufactures objects according to a preconceived form. As products of modern technology become increasingly complex, and the size of details goes down to the micro- and nanoscales, industrial methods may become biomorphic as well. They would then make use of natural processes of spontaneous structure formation to grow complex entities in the way they are nurtured by nature itself.

Our aim is to understand how relatively simple nonlinear systems obeying symmetric equations may generate a dazzling variety of forms. We shall be mostly concerned with *distributed* systems dependent on spatial coordinates as well as on time. The standard physical setting for our study is an extended (ideally, infinite) homogeneous isotropic system, and the description will invariably be on a macroscopic (mean field, or hydrodynamic) level. A much simpler *dynamical system* arises when spatial dependence is suppressed. Dynamical systems may still exhibit complex temporal behavior, and serve as an indispensable tool for understanding more complex spatially distributed systems when a few dominant modes are sufficient for an adequate approximate description of the spatiotemporal dynamics.

L. Pismen, *Patterns and Interfaces in Dissipative Dynamics*, Springer Series in Synergetics,
https://doi.org/10.1007/978-3-031-29579-9_1

Field Variables

The level of detail on which the physical system is described is always a compromise between faithfulness and transparency, between available tools and desired results. The physicist may be satisfied by the simplest description that catches the essence of the phenomena, while the engineer strives for quantitative precision, often illusory. Either one sees (or prefers not to see) that the quest is open-ended, and any problem is solvable only in an imperfect way, on a level that shifts with time, or with changing circumstances and aims.

The level on which we operate here is one of macroscopic classical theory. The physical system is described by a certain set of *mean field* variables, which characterize the thermodynamic state of the system. The list of variables may include temperature, density or pressure, concentrations of chemical species, as well as internal characteristics of material, such as molecular orientation, magnetization, electric polarization, etc. It might be convenient to organize these variables in an n-dimensional array, which should completely determine the state of the physical system in question. The mean field description is sufficient in very large ensembles, under conditions when fluctuations are negligible. In some cases, fluctuations may be reintroduced phenomenologically through random inputs to mean-field equations.

The same set of thermodynamic variables also serves to characterize a non-equilibrium system. All these variables should be then defined *locally* and are, generally, functions of spatial variables and time. The aim of the mean field theory is to formulate and solve *evolution equations* describing the dynamics of state variables. Since non-equilibrium processes usually involve motion and mass transport, it might be necessary to complement the set of thermodynamic variables by the velocity or deformation field. The set of evolution equations will then include coupled hydrodynamic equations for flow velocity and transport equations for thermodynamic variables in fluids and equations of elasticity theory in solids.

The mean-field description of a spatially inhomogeneous and temporally evolving system is, in some way, self-contradictory. Averaged state variables and flow velocities can be defined in a *macroscopic* volume only, which is formally at odds with the notion of an *infinitesimal* volume used in derivation of continuum equations. The contradiction is resolved by the requirement that all state variables should vary on spatial and temporal scales far exceeding all relevant "microscopic" scales, such as intermolecular distances or free path length and molecular relaxation times. This requirement is, unfortunately, too stringent in many situations, especially near interphase boundaries or at particular values of parameters (in the vicinity of critical points) when the scale of fluctuations reaches macroscopic dimensions. The crossover from microscopic to mean field description is one of "internal frontiers" of modern physics, where problems are challenging and answers are few and incomplete.

Even as the mean-field description is oversimplified from the fundamental point of view, it may be, at the same time, too complex and technical. A more transparent theory may elucidate the most important qualitative features of the system under study. For this purpose, Landau has coined the notion of "order parameter". The order parameter may distinguish between alternative equilibria, which correspond to different phases. When the phases coexist, the order parameter changes continuously across the interphase boundary. Thus, we should speak of an *order parameter field* (with some regret for misuse of the word "parameter" that commonly denotes a fixed coefficient rather than a field variable). It may be either one of the physical fields or a fictitious field that lumps the effect of all changes accompanying the phase transition. The free energy should be expressed through the order parameter and its derivatives in such a way that its minima correspond to stable phases.

The hierarchy of simplified descriptions ascending to larger scales and more complex phenomena is a common structure of knowledge. Even the order parameter field may be subject to further averaging and coarsening on the next hierarchical level if it develops its own elaborate spatio-temporal structure. Composite entities play a role of elementary objects in diverse applications, from dynamics of granular media to motion of galactic clusters. In biology and social sciences, a mean field theory may operate with still more complex and coarse-grained variables, such as population densities, while applying equations and techniques not much different from those of chemical physics.

Conservative Systems

The earliest dynamical models originated in classical mechanics. These models contain one or more *integrals of motion*, which are conserved in the course of evolution. Systems of this kind, called *conservative* (Hamiltonian), may exhibit persistent motion without relaxing to a stationary equilibrium state. Until the end of 19th century, classical mechanics was preoccupied with *integrable* systems, which can be separated into an ensemble of oscillators and display *periodic* or *quasiperiodic* behavior, depending on whether the oscillation periods are commensurate or not. Henri Poincaré was the first to realize that even such a simple system as two interconnected pendula may already exhibit far more complex aperiodic dynamics. However, it took another half a century or more for the notion of *deterministic chaos* to establish itself in the scientific community.

It was quantum mechanics that destroyed the belief in classical determinism. The latter's ultimate expression was the famous maxim by Laplace: a "supreme intelligence" capable of solving dynamical equations for all particles in the universe and knowing all initial conditions could predict the future in its entirety. Quantum mechanics showed this to be impossible, introducing chance into physics. Many scientists and philosophers, including the great Einstein, who detested God playing dice, regretted this. The paradox is that throwing a dice, the very symbol of a game of chance, follows the laws of classical mechanics. Its determinism is illusory, as precision in fixing the initial conditions needed for predicting the future state rapidly increases with the growing length of the prediction interval. Quantum mechanics with its indeterminism actually smoothes out chaotic classical dynamics, so that prediction of averaged expectations becomes easier in quantum theory.

The advent of computers played a crucial role in understanding complex chaotic dynamics, or at least in making the notion of deterministic chaos familiar to the wider scientific community and general public. Before the computer age, research was concentrated on a tiny and non-representative minority of mechanical systems described by analytically solvable models, and it took the imagination of men of genius to understand the complexity of behavior that is transparent nowadays to a lay person observing a trajectory on a computer screen.

An extension of conservative dynamical systems is *field theory*, studying the conservative dynamics of spatially extended systems with an infinite number of degrees of freedom. Nonlinear fields described by nonlinear partial differential equations possess the richest mathematical structure and are capable of generating an infinite variety of spatiotemporal structures, both ordered and chaotic. The first and perhaps most widely studied nonlinear field theory was classical hydrodynamics of ideal fluids. The phenomenon of *turbulence* was first studied in a fluid-mechanical context, and only later extended to a wider class of distributed systems displaying spatiotemporal chaos. A fascinating analytical theory has been developed to describe the behavior of a special class of completely integrable nonlinear partial differential equations that possess an infinite number of integrals of motion. The most prominent feature of these models is the existence of particle-like localized solutions, called *solitons*, which are able to preserve their identity in interactions and collisions. Completely integrable field equations are as non-generic as completely integrable dynamical systems. Yet, they often give the only opportunity to study nonlinear structures analytically, and the results of this study, besides their mathematical beauty, are useful for the analysis of more realistic systems where the conservation laws are weakly broken, so that integrals of motion do not remain strictly constant but slowly evolve in time.

Dissipative Systems

As conservation laws are woven into the canvas of physics, *dissipative* (nonconservative) models always appear as approximate or coarse-grained models describing the averaged behavior of complex systems. From a formal point of view, conservative systems are non-generic: a randomly generated set of equations will have no conservation laws. Dissipation may appear in a mechanical system as an additional term taking account of friction, which is, in fact, the result of disordered motion of a great number of microscopic particles lumped by a macroscopic model characterized by few degrees of freedom. Mean field models of thermodynamic or chemical origin are always dissipative, as they describe averaged dynamics characterized by a small number of thermodynamic variables. In fluid mechanics, dissipation is brought in by viscosity and diffusion, stemming, like friction, from disordered motion on a microscopic level.

Dissipative mean field models are also usually derived from mass, energy, and momentum conservation laws, but contain sources or sinks that destroy conservation of the respective integrals. By the second law of thermodynamics, such a system has to relax to a state of thermodynamic equilibrium. Since thermodynamic *free energy* decreases in the course of evolution, any system of equations modeling a closed system should possess an integral $\mathcal{F}$ monotonically decreasing with time[1].

[1] In some models, existence of a monotonically decreasing integral, called a *Lyapunov functional* can be established *ad hoc* even when not supported by thermodynamics

This restricts dynamics in a severe way, but the outcome of evolution is not unique when multiple equilibria are present.

Equilibria may differ *qualitatively* by their *internal* characteristics. For example, a solid state may possess different crystalline structures. At melting, the order may disappear not completely, leading instead to a liquid crystal, losing translational ordering but retaining a certain degree of orientational order. Such a liquid may undergo further phase transitions to decreasingly ordered states. Any equilibrium state can be characterized by a certain value of free energy, and so, generically, there will be a single state of absolute equilibrium with the lowest value of $\mathcal{F}$, while the other equilibria will be *metastable*. Evolution to the state of absolute equilibrium may be, however, hindered by high activation barriers, so that various metastable equilibria may persist indefinitely.

We shall mostly deal here with *open* macroscopic dissipative systems exchanging mass and energy with the environment. Besides dissipation in the common sense of the word – loss of energy, momentum, etc. – open systems may be "antidissipative" due to either external pumping or internal sources of energy. The latter is characteristic of *active* matter, which is attracting particular attention in biological applications. Behavior of such systems, constrained neither by conservation laws nor by the boring laws of thermodynamics, is particularly rich and interesting; most pattern-forming systems we shall study belong to this class.

Historical and Bibliographical Notes

The study of dynamics and pattern formation far from equilibrium (a major branch of "nonlinear science") is now a mature discipline that has gone through a common sequence of development stages. The following is an incomplete view from the perspective of this book.

Prehistory: Faraday instability (Faraday, 1831) of the free surface of a vibrated fluid layer was the first example of spontaneous pattern formation, but it had to wait for more than a century for theoretical explanation.

Dawn: the study of particular phenomena, lacking general connections (1900s–1960s). Spontaneous symmetry breaking had been known in fluid mechanics since the experiments by Bénard (1900) and their forceful, though factually incorrect, explanation by Rayleigh (1916). In chemical and biological applications, an understanding of this phenomenon had to wait another half a century till the famous work of Turing (1952), far weaker analytically, though philosophically charged. Chemical oscillations were considered taboo because of thermodynamical misconceptions well into the 1960s, while *electro*chemical oscillations and waves had been known since the turn of the century (Ostwald, 1900). Landau (1944) treated turbulence as a field theory with a very large number of degrees of freedom, while wide implications of chaotic behavior in simple mechanical systems, known already to Poincaré, were not understood. A paradigmatic model of chaotic flow by Lorenz (1963) remained unnoticed for 15 years.

Development: general theory comes of age (late 1960s to early 1980s). The synergetics of Haken (1983) and the self-organization principle of the Brussels school

(Nicolis and Prigogine, 1977) offered paradigms of dynamics far from equilibrium. The Belousov–Zhabotinsky reaction (Zhabotinskii, 1974) was explored as a spectacular example of chemical oscillations, prompting theoretical studies of chemical waves and diffusion fronts (Ortoleva and Ross, 1974; Kuramoto and Tsuzuki, 1976; Fife, 1979, and many others). The nonlinear theory of convective patterns was firmly established (Busse, 1978; Normand, Pomeau and Velarde, 1977) and tested in precision experiments (Ahlers, 1974). Further input came from combustion theory (Zeldovich, 1985; Buckmaster and Ludford, 1983; Sivashinsky, 1977) and the theory of solidification (Langer, 1980). Simple pattern-forming models – Brusselator, Kuramoto–Sivashinsky, Swift–Hohenberg equations – were put forward and explored.

Summit (mid-1980s to mid-1990s): *acme*. Common techniques were applied to nonequilibrium structures of different physical origin. Space-dependent amplitude equations and phase dynamics led to understanding of the genesis and behavior of realistic patterns, constrained by boundaries and blemished by defects. Chemical patterns were studied under controlled conditions (Ouyang and Swinney, 1991), and the theory of spiral waves, prompted by chemical and physiological applications, attracted great attention. The NATO Special Program on Chaos, Order and Patterns did much to unify physicists' approach and diffuse the knowledge, and it might not be by accident that its end coincided with the publication of the review by Cross and Hohenberg (1993), which remains to this day the most comprehensive survey of the field. Other monographs, emphasizing different aspects of nonequilibrium dynamics and pattern formation were published by Manneville (1990), Mikhailov (1991), Nicolis (1995), and Walgraef (1997), while the books by Bergé et al (1984) and Ott (1993) concentrated on problems of chaos and turbulence. Specific problems of patterns and fronts in reaction–diffusion equations were treated in (very dissimilar) books by Murray (1989) and Fife (1988) – one geared to biological applications and the other to combustion problems.

Maturity (turn of the 21st century): a more detailed look at complexity. Attention returned to specific applications; among them, nonlinear optics and studies of granular media came to the forefront. Forcing and control of patterns, either enhancing or suppressing the complexity of behavior, were studied in detail. As the humble laptop turned into a supercomputer, more fascinating patterns, the envy of abstract expressionists, were generated by model equations of increasing complexity.

Towards biology (current trend): With time, attention is shifting to more complex systems, and analytical methods are giving way to numerics. The most actively pursued novel direction is the study of *active matter*. Even this term was not yet in use at the time of the first edition of this book, while it had appeared in about 1500 papers listed by Web of Science by the end of 2022, including 115 reviews, most prominently, Ramaswamy (2010), Marchetti et al (2013), Doostmohammadi et al (2018), and Gompper et al (2020). It is largely motivated by biological applications, as well as by emerging biomorphic technologies. Soft matter driven and modified by internal sources of energy is able to display complex dynamic behavior. Although it is rarely amenable to analytical theory, the basic tools of nonlinear analysis remain indispensable for guiding numerical simulations and interpreting their results.

1 Genesis of Patterns

1.1 Basic Models

1.1.1 Representative Models of Open Systems

An *open* system is sustained in a nonequilibrium state by external fluxes. Such a system, generally, can be assigned no energy functional; its dynamics is more interesting, since it is not restricted by *a priori* rules. There is, however, a restriction imposed on the *geometry* of an open system. While a closed system may be isotropic in three dimensions, the direction of external fluxes sets a preferred axis in an open system; we shall tentatively call this direction "vertical." The system may retain isotropy in the normal "horizontal" plane, provided neither the properties of the system itself nor the strength of external fluxes depend on spatial coordinates explicitly and the influence of lateral boundaries can be discarded. A suitable geometric setting is an infinite plane normal to the direction of external fluxes (Fig. 1.1).

The system may be truly two-dimensional (Fig. 1.1a) if the process is taking place on a surface and all relevant field variables are surface densities. Otherwise, the system would possess some "vertical structure" in the direction of fluxes (Fig. 1.1b) but retain isotropy in horizontal directions. Then the "trivial" state of a distributed nonequilibrium system is uniform in the plane normal to the direction of external fluxes, but dependent on the "vertical" coordinate, as in Fig. 1.1. This is necessarily so in the presence of fluid flow when hydrodynamic equations should be solved together with transport equations to determine a vertical structure satisfying appropriate boundary conditions. Other systems with vertical structure include directional solidification, light propagation in nonlinear media, and orientation textures in liquid crystals. The set of variables $\mathbf{u}$ describing systems of this kind may include vector as well as scalar fields, and the applicable equations will typically include differential terms, in addition to algebraic nonlinearities. The most general form is an array of nonlinear operators of the type

$$\mathcal{H}(\mathbf{u}, \partial_t \mathbf{u}, \partial_z \mathbf{u}, \nabla \mathbf{u}, \ldots) = 0, \tag{1.1.1}$$

L. Pismen, *Patterns and Interfaces in Dissipative Dynamics*, Springer Series in Synergetics,
https://doi.org/10.1007/978-3-031-29579-9_2

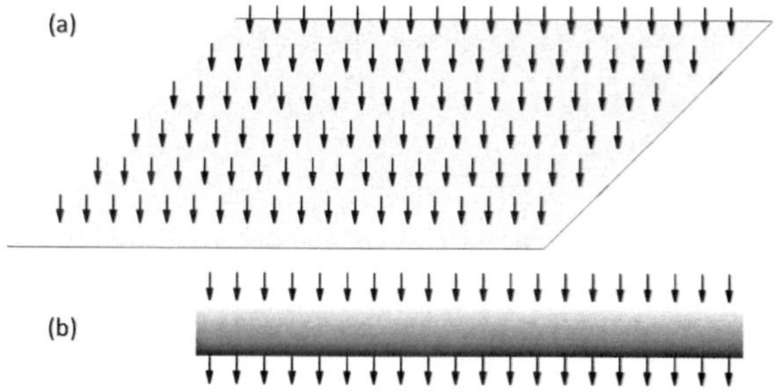

Fig. 1.1. An open system isotropic in two dimensions. (a) A truly two-dimensional system. (b) A cut through a system with vertical structure (shown symbolically by varied shading). The arrows indicate the direction of external fluxes

where we distinguish between the derivatives in the "vertical" direction ∂_z and the derivatives in "horizontal" directions, expressed by the 2D gradient operator ∇.

A vertical structure can be often eliminated when it remains almost constant in the horizontal plane. More precisely, the reduction is possible when the characteristic horizontal scale is much larger than the characteristic vertical scale, say, $\partial_z = O(1)$, $\nabla = O(\epsilon)$, $\epsilon \ll 1$. Another possibility is a slow modulation of a structure periodic in the horizontal direction. This involves multiscale expansion of the gradient operator $\nabla = \nabla_0 + \epsilon\nabla_1$. Equations obtained as a result of elimination of a vertical or periodic structure have, as a rule, a relatively simple form, which is determined by symmetries of the underlying system. Equations of identical structure may emerge therefore in different physical applications. Such representative equations can be commonly obtained by expanding more complex underlying equations of different origin in a small parameter, and therefore are most suitable for study of typical dynamic phenomena in nonlinear systems. A substantial part of the material of this book is based on studying such equations.

The following equations for a single scalar field $u(\mathbf{x})$ can be written simply by requiring spatial isotropy as well as a specified symmetry to transformations of u, and restricting to a certain order in u and ∇; they can also be written by scaling both u and ∇, as indicated below in brackets, and retaining the lowest order terms:

- Cubic KPP equation (Kolmogorov et al, 1937) – second order in ∇, third order in u, inversion symmetry:

$$\partial_t u = \nabla^2 u + u(1 - u^2) \quad [\nabla \sim O(\epsilon), \partial_t \sim O(\epsilon^2)]. \tag{1.1.2}$$

- Swift–Hohenberg (SH) equation (Swift and Hohenberg, 1977) – fourth order in ∇, third order in u, inversion symmetry:

$$\partial_t u = \mu u - (q^2 + \nabla^2)^2 u - u^3 \quad [\nabla \sim q \sim O(\sqrt{\epsilon}), \mu \sim \partial_t \sim O(\epsilon^2)]. \tag{1.1.3}$$

- Burgers equation (Burgers, 1948) – second order in ∇, second order in u, symmetric to its translations (dependent on derivatives of u only):

$$\partial_t u = \nabla^2 u + |\nabla u|^2 \quad [u \sim O(1), \ \nabla \sim O(\epsilon) \quad \partial_t \sim O(\epsilon^2)]. \tag{1.1.4}$$

- Kuramoto–Sivashinsky (KS) equation (Kuramoto and Tsuzuki, 1976; Sivashinsky, 1977) – fourth order in ∇, second order in u, symmetric to its translations:

$$\partial_t u = -\epsilon \nabla^2 u - (\nabla^2)^2 u + |\nabla u|^2 \quad [u \sim O(\epsilon),\ \nabla \sim O(\sqrt{\epsilon}) \quad \partial_t \sim O(\epsilon^3)]. \tag{1.1.5}$$

All these equations are written in a rescaled form with the minimum number of parameters and with signs chosen in a way ensuring stability to short-scale perturbations and preventing runaway to infinity. The KPP equation is the simplest model of spontaneous symmetry breaking. The SH equation is the simplest pattern-forming system. Its linear part is designed to produce a dispersion relation with a maximum at a finite wavenumber q; the height of the maximum is controlled by the parameter μ. The cubic term serves to saturate the instability developing at $\mu > 0$. The Burgers ant KS equations model phase dynamics.

Similar equations can be written for a *complex* scalar variable. Equations of this kind often appear as *envelopes* of periodic structures, and will be studied in detail in Chaps. 4 and 5. The lowest-order form analogous to (1.1.2) is the complex Ginzburg–Landau (CGL) equation

$$\partial_t u = \eta \nabla^2 u + u(\mu - \nu |u|^2). \tag{1.1.6}$$

The coefficients η, μ, ν are generally complex, but the particular cases when they are either all real or all imaginary are important in their own right. In both cases, all coefficients can be reduced to unity by rescaling; the resulting parameterless forms are called, respectively, the real Ginzburg–Landau (RGL) equation

$$\partial_t u = \nabla^2 u + u(1 - |u|^2), \tag{1.1.7}$$

and the nonlinear Schrödinger (NLS) equation

$$-\mathrm{i}\partial_t u = \nabla^2 u + u(1 - |u|^2). \tag{1.1.8}$$

Equations (1.1.2), (1.1.3), and (1.1.7) describe *gradient* dynamics, and are derivable from an appropriate energy functional $\mathcal{F}$:

$$\partial_t u = -\delta \mathcal{F}/\delta u, \quad \mathcal{F} = \int \mathcal{L}(u, \nabla u)\, \mathrm{d}\mathbf{x}. \tag{1.1.9}$$

Since these equations are applicable to spatially isotropic systems, the gradient operator in the Lagrangian $\mathcal{L}$ should be expressed through the rotationally invariant combination $|\nabla u|^2$, e.g., in the simplest case of (1.1.2) we have

$$\mathcal{L} = \tfrac{1}{2}|\nabla u|^2 + \tfrac{1}{4}(1 - |u|^2)^2. \tag{1.1.10}$$

The RGL equation is actually a shorthand for two real equations of u and its complex conjugate $\overline{u}$, so that the explicit form of (1.1.9) is

$$\partial_t u = -\delta \mathcal{F}/\delta \overline{u}, \quad \mathcal{L} = \nabla u \cdot \nabla \overline{u} - u\overline{u} + \tfrac{1}{2}(u\overline{u})^2. \tag{1.1.11}$$

The essential property of gradient dynamics is that it ensures a *dissipative* temporal evolution. Indeed, the energy integral must decrease until a minimum is reached:

$$\frac{\mathrm{d}\mathcal{F}}{\mathrm{d}t} = \int \frac{\delta \mathcal{F}}{\delta u}\frac{\partial u}{\partial t}\,\mathrm{d}\mathbf{x} = -\int \left(\frac{\delta \mathcal{F}}{\delta u}\right)^2 \mathrm{d}\mathbf{x} \leqslant 0. \tag{1.1.12}$$

The gradient dynamics is retained for the RGL equation since

$$\frac{\mathrm{d}\mathcal{F}}{\mathrm{d}t} = \int \left(\frac{\delta \mathcal{F}}{\delta u}\frac{\partial u}{\partial t} + \frac{\delta \mathcal{F}}{\delta \overline{u}}\frac{\partial \overline{u}}{\partial t}\right) \mathrm{d}\mathbf{x} = -\int \left|\frac{\delta \mathcal{F}}{\delta u}\right|^2 \mathrm{d}\mathbf{x} \leqslant 0. \tag{1.1.13}$$

This is a particular case of the Onsager reciprocal relations which require the matrix $\delta\mathcal{F}/\delta u_i$ to be Hermitian to ensure monotonic relaxation to equilibrium (more on this in Sect. 1.6.3). The gradient property may be just approximate and appear in reduced equations even when the original equations describing an open system lack gradient structure.

1.1.2 Spectral Decompositions

Spatially extended nonequilibrium systems, especially those lacking gradient structure, are capable, in principle, of exhibiting extremely complex dynamics. With a great number of degrees of freedom present, one could expect turbulent behavior of unfathomable complexity to be a norm. This, however, does not necessarily happen. More often than not, the latent complexity of behavior is not realized, and ordered or weakly distorted patterns prevail. This suggests that a dynamical system involving a limited number of variables might be sufficient to faithfully describe the dynamic behavior in lieu of a full partial differential equation (PDE) system, at least approximately and within a certain range of parameters.

The choice of such a set of variables is by no means obvious. In the absence of other hints, one can expand the variables in a Fourier series, say, $\mathbf{u}(\mathbf{x},t) = \sum_j \mathbf{U}_j(\mathbf{k})\mathrm{e}^{\mathrm{i}(\mathbf{k}\cdot\mathbf{x}-\omega_j t)}$. The dependence of the frequency ω_j on the wave vector $\mathbf{k}$ in each spectral band labeled by the index j is defined by a dispersion relation $\omega(\mathbf{k})$. The vectors $\mathbf{U}_j$ are complex, unless both $\mathbf{k}_j$ and ω_j vanish. In an open system of the kind sketched in Fig. 1.1, the spatial expansion is two-dimensional (2D), but when a vertical structure is present, $\mathbf{U}_j$ are functions of z depending on $\mathbf{k},\omega$ as parameters.

When such an expansion is used in the original PDE, different modes couple through nonlinear terms, and, generally, no finite number of modes would ever suffice, as, given two modes with wave vectors $\mathbf{k}_a$, $\mathbf{k}_b$, modes $\mathbf{k}_a \pm \mathbf{k}_b$ will be generated. The expansion can be, however, truncated, leaving only a small number of modes and neglecting the rest. The resulting ODE system with polynomial functions $\mathbf{F}(\mathbf{U})$ dependent on amplitudes of chosen modes may look like "chemical" systems derived from the mass action law. Of course, one can never be sure which, if any, features of the original system are preserved by such a bold action, but, with good luck and insight, it may bring success. In this way, Lorenz (1963) constructed a three-variable dynamical system by retaining three modes of the Navier–Stokes and convective diffusion equations describing Rayleigh–Bénard convection. The resulting system is not actually faithful to the original, but it turned (after some 15 years of obscurity) into a celebrated paradigm of chaotic dynamics.

A more systematic (though empirical) approach to spectral truncation is offered by the Karhunen–Loewe method used, in particular, for compressing images and discerning "coherent structures" in turbulent flow (Sirovich and Everson, 1992). This method employs statistical analysis of measured or computed data to find out an optimal combination of modes sufficient to reproduce the data in a faithful but economic way. This combination would, of course, change when the underlying system is modified or driven to a different regime.

It is far more appealing to be able to discern relevant modes analytically. The key is given by *scale separation*: some modes may evolve on far longer characteristic times than the others do. Fast modes relax to their quasistationary values dependent on the current values of slow modes, which remain virtually unchanged during this relaxation period. Following this, slow modes evolve on a longer time scale, while fast modes follow them quasistationarily, being bound by quasiequilibrium algebraic relations – much in the same way as rapidly reacting chemical species reaching a partial equilibrium. This is expressed by Haken's (1983) *slaving* principle: "fast modes are slaved by slow modes."

Clearly, only slow modes are relevant for long-time evolution. Therefore, the original PDE system, however complex it might be, can be reduced to a set of equations for the amplitudes of slow modes. Their general form can be predicted on the basis of scaling and symmetry considerations. Exact derivations connecting their parameters with parameters of the underlying system are possible when the amplitudes of slow modes are small, so that the state of the system deviates only slightly from a known "trivial" homogeneous stationary state. Amplitude equations derived through this approach can reproduce dynamics of the original underlying system faithfully only in a limited range of parameters and in a limited region of the phase space, since slow modes would, generally, be modified or cease to be slow when either parameters or values of dependent variables change. Nevertheless, they may reproduce dynamics correctly in a qualitative sense even outside the domain where their derivation can be justified formally.

1.1.3 Linear Analysis

A comprehensive study of a distributed (space-dependent) system of a general form (1.1.1), may start in a parametric domain where the dynamic behavior is most simple. The system may possess a *trivial* solution that retains all symmetries of the equations, or a homogeneous stationary state (HSS). Close to thermodynamic equilibrium, when driving fluxes are weak, this state is expected to be stable and serve as an attractor of the system, so that it is approached asymptotically starting from any initial conditions in its vicinity. If this equilibrium is unique, it is a global attractor; otherwise, it would have a finite attraction basin. When the system is driven farther from thermodynamic equilibrium, its behavior may become far more complicated. The trivial solution may become unstable, and the system may tend asymptotically to attractors of different nature: oscillatory, inhomogeneous (patterned), or chaotic (turbulent) states.

The first step in a search for nontrivial states is therefore to find the locus of a *primary bifurcation* where the trivial state first becomes unstable. This transition is indicated by local linear analysis in the vicinity of the HSS. In a general case, the linearization of (1.1.1) can be written as

$$\mathbf{L}(\partial_t, \partial_z, \nabla)\phi(\mathbf{x}, z, t) = 0. \tag{1.1.14}$$

where $\phi = \mathbf{u} - \mathbf{u}_0$ is a deviation from some known solution $\mathbf{u} = \mathbf{u}_0$. In the geometry of Fig. 1.1, both $\mathbf{x}$ and ∇ are defined in 2D, while the "vertical" coordinate z is singled out. If the trivial solution is stationary and homogeneous, the operator $\mathbf{L}$ does not contain explicit time and coordinate dependence (except, perhaps, on the "vertical" coordinate z). Then the general solution of (1.1.14) is presented as

$$\phi(\mathbf{x}, z, t) = \sum_j a_j(\mathbf{k})\mathbf{\Phi}_j(\mathbf{k}, z)\mathrm{e}^{\mathrm{i}(\mathbf{k}\cdot\mathbf{x}+\lambda_j t)}, \tag{1.1.15}$$

where $\lambda_j(\mathbf{k})$ are eigenvalues, $\mathbf{\Phi}_j(\mathbf{k}, z)$ are corresponding normalized eigenvectors (or eigenfunctions when the vertical structure is present), and a_j are scalar *amplitudes*. The dependence $\lambda_j(\mathbf{k})$ is called a *dispersion relation*. In the absence of a vertical structure, the dispersion relation is obtained by direct substitution $\partial_t \to \lambda$, $\nabla \to \mathrm{i}\mathbf{k}$ and finding the eigenvalues of the matrix $\mathbf{L}(\lambda, \mathbf{k})$ grouped in n spectral bands. When z-dependence is present, the dispersion relation is obtained by solving a boundary problem for an ODE in z with appropriate boundary conditions.

In an isotropic system, the eigenvalues depend only on the absolute value $k = |\mathbf{k}|$ rather than on the direction of the wave vector $\mathbf{k}$. In a spatially infinite system, each eigenvalue is blown up into a continuous spectral band. If the region is finite, each spectral band includes only a discrete (infinite countable) set of eigenvalues. These eigenvalues come closer when the characteristic spatial scale L increases, and the band becomes continuous in the limit $L \to \infty$. The stationary state is *stable* when all spectral bands lie entirely in the left complex half-plane, i.e., for any j, k, $\mathrm{Re}\lambda_j(k) < 0$.

Let the matrix $\mathbf{L}$ depend on some parameter μ, and assume that this stability condition is satisfied when μ is sufficiently small. Increasing μ sets the spectral bands $\lambda_j(k;\, \mu)$ into motion. We will refer to a band containing the eigenvalue with the largest real part as the *leading* band and label the eigenvalues from this band as λ_0; this band will be responsible for the onset of instability when it first occurs at some $\mu = \mu_0$. Provided the function $\lambda_0(k)$ is continuous, the spectrum of the linearized system (1.1.14) satisfies at this value of the parameter for some k_0 the condition of *marginal stability*

$$\mathrm{Re}\lambda_0(k_0) = 0, \qquad \mathrm{d}(\mathrm{Re}\lambda_0)/\mathrm{d}k = 0 \ \text{ at } \ k = k_0. \tag{1.1.16}$$

We can distinguish among four possibilities, depending on whether the eigenvalues from the leading band are real or complex and whether the rightmost edge of this band corresponds to $k_0 = 0$ or $k_0 \neq 0$. The *leading* mode, which becomes marginally stable at the critical point $\mu = \mu_0$, is

- stationary homogeneous when $k_0 = 0$, $\mathrm{Im}\lambda_0(0) = 0$,
- oscillatory homogeneous when $k_0 = 0$, $\mathrm{Im}\lambda_0(0) \neq 0$,
- stationary inhomogeneous when $k_0 \neq 0$, $\mathrm{Im}\lambda_0(k_0) = 0$,
- oscillatory inhomogeneous when $k_0 \neq 0$, $\mathrm{Im}\lambda_0(k_0) \neq 0$.

A typical dispersion relation $\lambda_0(k)$ in a spatially extended system just beyond a bifurcation point may have one of the forms shown in Fig. 1.2. In all cases, a narrow band of wavenumbers adjacent to the maximum becomes unstable. Since the dispersion relation should be, generically, parabolic near the maximum, and the leading eigenvalue can be assumed to depend linearly on a chosen bifurcation parameter μ, the width of the excited band scales as the square root of the deviation from the bifurcation point $\mu - \mu_0$.

A spectral band of a finite width can be modeled by allowing the amplitude to change on an extended spatial scale $\mathbf{X}$ large compared to both k_0^{-1} and any "natural" length scale characteristic to the underlying system. If the maximum of the dispersion relation is at $k = 0$ (curve 1 in Fig. 1.2a), the bifurcation does not break spatial symmetry, and the dependence on the extended scale is symmetric. The equation of the lowest order in $\nabla_{\mathbf{X}}$ obtained near a bifurcation at zero eigenvalue should be therefore just a diffusion equation with a polynomial nonlinearity, e.g., (1.1.2). In a similar way, the complex Ginzburg–Landau equation (1.1.6) with a complex order parameter and complex coefficients is obtained at a supercritical Hopf bifurcation when the dependence of the real part of the leading pair of complex conjugate eigenvalues on the wavenumber has the same form.

If the dispersion relation is shaped as curve 2 in Fig. 1.2a, with the maximum attained at $k \ll 1$, the representative equation is the Swift–Hohenberg equation (1.1.3), which has been originally introduced *ad hoc* as a simplest pattern-generating model. When the bifurcation occurs at a finite $k = k_0$, and the dispersion relation looks as in Fig. 1.2b, the amplitude, or "order parameter" characterizing the pattern, should be defined as the envelope of a weakly distorted basic structure with the wavenumber $k_0 = O(1)$. Its dependence on the extended scale should be in this case

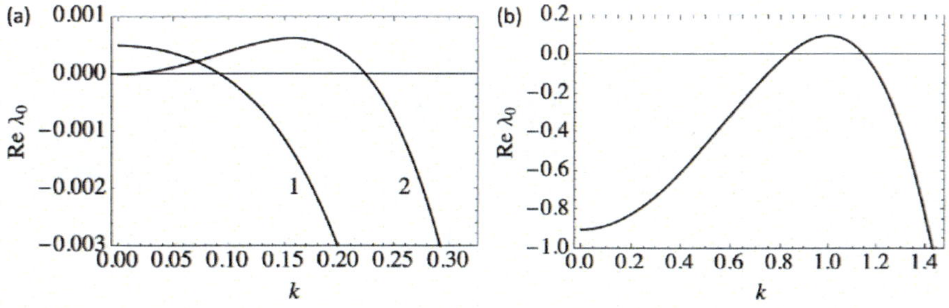

Fig. 1.2. (a) Dispersion relations with an unstable band near $k = 0$ with the maximum either at $k = 0$ (curve 1) or at a small k (curve 2). (b) Dispersion relation with an unstable band at a finite $k = k_0$

anisotropic due to a preferred direction set by the wave vector characterizing the emerging structure (more on this in Sect. 1.3.4).

When the parameter μ further increases beyond the point of marginal stability, one can expect that the growth of perturbations in the leading mode would lead (provided it saturates at a certain amplitude) to a new steady state of *the same* character, i.e., stationary at Im $\lambda_0(k_0) = 0$, oscillatory at Im $\lambda_0(k_0) \neq 0$, homogeneous at $k_0 = 0$, and inhomogeneous at $k_0 \neq 0$. The linear analysis, of course, gives but a hint at the existence of this "nontrivial" state in the supercritical region, and further nonlinear analysis is required to assure that this state does actually exist.

1.1.4 Evolution of Perturbations

The overall state of a weakly perturbed system is described by the Fourier integral

$$\phi(\mathbf{x}, z, t) = \sum_j \int a_j(\mathbf{k})\, e^{\lambda_j(\mathbf{k})t + i\mathbf{k}\cdot\mathbf{x}} \mathbf{\Phi}_j(z; \mathbf{k})\, d\mathbf{k}. \tag{1.1.17}$$

This cumbersome formula is of little use as it stands, but becomes practical if one is able to identify physically relevant modes. Near a bifurcation point, those are slow modes belonging to the leading band. There would be several slow modes when the bifurcation is *degenerate*. One can further assume that the amplitudes of slow modes, rather than being constant, are weakly time-dependent and are modulated on an extended spatial scale, i.e., depend on *extended* temporal and spatial coordinates $T, \mathbf{X}$:

$$\phi(\mathbf{x}, z, t; \mathbf{X}, T) = \sum_j a_j(\mathbf{X}, T; \mathbf{k})\, e^{\lambda_j(\mathbf{k}_j)t + i\mathbf{k}_j\cdot\mathbf{x}} \mathbf{\Phi}_j(z; \mathbf{k}_j). \tag{1.1.18}$$

The summation is carried our here over the slow modes only, and integration over a spectral band is replaced by spatial dependence of the amplitudes, which is equivalent to accounting for a segment of the band close to the value $k = k_0$ at the bifurcation point.

Since eigenvalues λ_j depend on $k_j = |\mathbf{k}_j|$ only in an isotropic system, an entire ring of slow modes may appear near a symmetry-breaking bifurcation in two dimensions. This seems to indicate that integration over angular directions of $\mathbf{k}_j$ should be reintroduced in (1.1.18). This is, however, not necessary, since, as a result of competition among differently directed modes caused by nonlinear interactions, only a small number of modes would survive (more on this in Sect. 1.5.1).

While linear analysis is the necessary tool for detecting instabilities and identifying slow modes, one can question the utility of linear theories for the study of actual dynamic behavior. Indeed, the dynamics is not interesting in stable situations when the system just relaxes back to the basic state, while in unstable situations exponential growth of perturbations promptly makes nonlinear terms essential. Nevertheless, linear theories can sometimes be used beyond the instability limit. One such case is the retreat of an unstable state under the onslaught of an alternative stable solution.

It turns out that the speed of the retreat and even the character of the eventually established stable state is determined by the dynamics at the leading edge of the propagating front where linear theory is still useful (see Sect. 2.1.4).

Another case is a *convective* instability in systems of finite spatial extent. The instability is called convective when the band of unstable eigenvalues does not encompass $k = 0$. If the instability is observed at a finite wavelength and frequency, the relevant eigenfunctions close to criticality are waveforms $a(T)\,e^{i(\mathbf{k}\cdot\mathbf{x}-\omega t)}$, where $\omega = \text{Im}\,\lambda$. Under these conditions, all perturbations are washed out with a characteristic speed equal to the group velocity $d\omega/dk$. Therefore, while the real part of λ remains small, perturbations may not succeed to grow to a finite amplitude, provided the spatial extent of the system is sufficiently small. Such an instability is called *convective*. Under these conditions, the unstable state may persist, provided the intensity of external perturbations is not too large, and dynamics of deviations from the unstable state presented by (1.1.17) or (1.1.18) can be reasonably well described in the linear approximation. This will not happen when a mode with $k = 0$ is unstable, and the instability is *absolute*.

1.1.5 Multiscale Expansion

In the vicinity of a bifurcation manifold, the original system of equations can be reduced to a simple universal form that is characteristic to the particular type of bifurcation and retains the qualitative features of dynamic behavior of the underlying system in an adjacent parametric domain. The reduced equations are called *normal forms* in the mathematical literature, while the term preferred by physicists is *amplitude equations*. Normal forms are ODEs describing dynamics of the amplitudes of slow modes, while amplitude equations may include differential terms operating on an extended spatial scale. The approach to their derivation also differs: mathematicians base it on "near-identity transformation", while the favorite tools of physicists is *multiscale expansion*. The advantage of the latter approach lies in a possibility to connect directly the parameters of amplitude equations and the characteristic time of their evolution to parametric regions of the underlying system. A more abstract mathematical approach may be preferable when one wishes to obliterate these details and concentrate on qualitative features of underlying dynamics.

The formal procedure of multiscale expansion starts with expanding both variables $\mathbf{u}$ and parameters $\mathbf{p}$ in powers of a dummy small parameter ϵ near some known HSS $\mathbf{u} = \mathbf{u}_0(\mathbf{p}_0)$:

$$\mathbf{u} = \mathbf{u}_0 + \epsilon\mathbf{u}_1 + \epsilon^2\mathbf{u}_2 + \cdots, \qquad \mathbf{p} = \mathbf{p}_0 + \epsilon\mathbf{p}_1 + \epsilon^2\mathbf{p}_2 + \cdots \tag{1.1.19}$$

All nonlinear terms are expanded in powers of deviations $\mathbf{u} - \mathbf{u}_0$; e.g., the general nonlinear operator (1.1.1) is expanded as

$$\mathcal{H}(\mathbf{u}, \partial_t\mathbf{u}, \partial_z\mathbf{u}, \nabla\mathbf{u}) = \epsilon\mathcal{L}(\partial_t, \partial_z, \nabla)(\mathbf{u} - \mathbf{u}_0) + \epsilon^2\mathcal{L}_2(\partial_t, \partial_z, \nabla)(\mathbf{u} - \mathbf{u}_0)^2 + \cdots \tag{1.1.20}$$

No zero-order term appears here, since $\mathbf{u}_0(\mathbf{p}_0)$ must verify (1.1.1). The operator $\mathcal{L}$ is the linearization of $\mathcal{H}$. Slow modes appear in the spectrum of this operator in the parametric domain close to the point $\mathbf{p}_0$ lying on a bifurcation manifold of the original system where $\mathcal{L}$ has an eigenfunction with an eigenvalue with vanishing real part, $\mathcal{L}\mathbf{U} = \mathrm{i}\omega\mathbf{U}$. It is always assumed that all fast modes are decaying, so that at $\mathbf{p} = \mathbf{p}_0$ there are no eigenvalues with a positive real part.

Next, we introduce a hierarchy of time scales t_j rescaled by the factor ϵ^j and expand the time derivative in a series of partial derivatives, $\partial/\partial t_j \equiv \partial_j$. In the same way, a hierarchy of spatial scales is introduced to describe weak spatial modulations of slow modes, and derivatives in the horizontal plane are expanded as $\nabla = \sum \epsilon^j \nabla_j$.

$$\partial/\partial t = \partial_0 + \epsilon\partial_1 + \epsilon^2\partial/\partial t_2 + \cdots, \qquad \nabla = \nabla_0 + \epsilon\nabla_1 + \epsilon^2\nabla_2 + \cdots, \tag{1.1.21}$$

where we denote for brevity $\partial_j = \partial/\partial t_j$. On consecutive stages of the expansion labeled by the index k, we shall arrive at inhomogeneous equations of the type

$$\mathcal{L}(\partial_0, \partial_z, \nabla_0, \mathbf{p}_0)\mathbf{u}_k = \boldsymbol{\Psi}, \tag{1.1.22}$$

$$\boldsymbol{\Psi} = \sum_l \mathcal{L}_2(\partial_0, \partial_z, \nabla_0, \mathbf{p}_0)\mathbf{u}_l, \mathbf{u}_{k-l} + \sum_{l,m} \mathcal{L}_3(\partial_0, \partial_z, \nabla_0, \mathbf{p}_0)\mathbf{u}_l, \mathbf{u}_m, \mathbf{u}_{k-l-m}$$
$$+ \left[\mathcal{L}(\partial_1, \partial_z, \nabla_0, \mathbf{p}_0) + \mathcal{L}(\partial_0, \partial_z, \nabla_1, \mathbf{p}_0) + \mathcal{L}(\partial_0, \partial_z, \nabla_0, \mathbf{p}_1)\right]\mathbf{u}_{k-1} + \ldots,$$

where the operator $\mathcal{L}$ has an eigenvalue with the zero real part. These equations have to be solved in turn, and an amplitude equation appears as the lowest-order nontrivial solvability condition.

By Fredholm alternative, an inhomogeneous equation is solvable if the inhomogeneity $\boldsymbol{\Psi}$ does not project on the corresponding eigenfunction $\mathbf{U}$. This solvability condition is expressed as $\langle \boldsymbol{\Phi}^\dagger, \boldsymbol{\Psi} \rangle = 0$, where $\langle \ldots \rangle$ is an appropriate scalar product and $\boldsymbol{\Phi}^\dagger$ is the eigenfunction with zero eigenvalue of the *adjoint* operator $\mathcal{L}^\dagger$. At $\mathbf{k} \neq 0$ and/or $\omega \neq 0$, the operator $\mathcal{L}$ acts in the space of periodic complex-valued vector-functions with the scalar product defined as

$$\langle \boldsymbol{\Phi}^\dagger, \boldsymbol{\Psi} \rangle \sim \int\int \boldsymbol{\Phi}^\dagger(x,t) \cdot \boldsymbol{\Psi}(x,t)\, \mathrm{d}t\, \mathrm{d}x, \tag{1.1.23}$$

Additional restrictions on parametric deviations may be imposed when needed to make the orders of magnitude of different terms compatible. Practical computation of the coefficients of the amplitude equations is a tedious task, but their general structure is predictable, and relatively simple cases, lacking a vertical structure, can be worked out explicitly following the procedure outlined in the next section for a general RDS. Since the problem is fully formalized, it can be handled by symbolic computation programs (Pismen and Rubinstein, 1999a; Wang et al, 2014).

1.2 Reaction–Diffusion Systems

1.2.1 General Structure of RDS

A general space-dependent model of a simple structure is the *reaction–diffusion* system (RDS) involving transport and nonlinear interactions but no advection. It arises in different contexts from conservation laws of a general form

$$\partial_t u_i = -\nabla \cdot \mathbf{j}_i + f_i(\mathbf{u}). \tag{1.2.1}$$

Commonly, the field variables u_i arranged in the array $\mathbf{u}$ are concentrations of chemical species or population densities; then (1.2.1) represents the balance of respective species where $\mathbf{j}_i$ are fluxes and f_i are production rates arranged as components of an array $\mathbf{f}(\mathbf{u};\mathbf{p})$, also dependent on an array of parameters $\mathbf{p}$. Both arrays, of vectors in the phase space, such as $\mathbf{u}$ or $\mathbf{f}$, and vectors in the real space, such as $\mathbf{j}$, are set in boldface, but there should be no way to mix them up. RDS of this kind are commonly studied in the 2D setting of Fig. 1.1a, which is realized on an extended catalytic surface or in a thin layer where reactant gradients in the vertical direction can be neglected. Equations of similar structure can be written for other thermodynamic variables, such as temperature and electric potential, stemming, respectively, from conservation of energy and electric charge.

The fluxes $\mathbf{j}_i$ may, generally, depend on gradients of all field variables[1]:

$$\mathbf{j}_i = -D_{ij}\nabla u_j. \tag{1.2.2}$$

The *diffusivity* matrix D_{ij} is, generally, nondiagonal, and may depend on the local values of state variables. This dependence can be expressed in a less formal way by separating diffusion proper (which originates in random motion of noninteracting particles) from *migration* under the action of combined molecular interaction forces. If the latter can be expressed through the gradient of a thermodynamic potential $\mathcal{H}$, we obtain:

$$\mathbf{j}_i = -D_{ij}\nabla u_j - M_i\nabla\mathcal{H}. \tag{1.2.3}$$

If the dependence $\mathcal{H}(\mathbf{u})$ is *local*, one can express $\nabla\mathcal{H}$ through ∇u_i explicitly, recovering the form (1.2.2) with the effective diffusivity $D_{ij} + M_i\, \partial\mathcal{H}/\partial u_j$. Explicit migration fluxes can be written also for nonequilibrium transport processes, such as chemotaxis, which are not derivable from a potential. Equations (1.2.2), (1.2.3) exclude any externally imposed flow, although transport processes may involve a net mass flux, and we don't need here to be concerned with momentum balance.

Presuming the diffusivities arranged in the diffusivity matrix $\mathbf{D}$ to be constant, a general RDS can be written in the form

$$\mathbf{\Theta} \cdot \partial_t \mathbf{u} = \mathbf{D} \cdot \nabla^2 \mathbf{u} + \mathbf{f}(\mathbf{u};\mathbf{p}), \tag{1.2.4}$$

The capacitance matrix $\mathbf{\Theta}$ (the inverse of the mobility matrix $\mathbf{\Gamma}$) can be eliminated by rescaling $\mathbf{D} \to \mathbf{\Gamma} \cdot \mathbf{D}$ and $\mathbf{f} \to \mathbf{\Gamma} \cdot \mathbf{f}$, so that (1.2.4) becomes

$$\partial_t \mathbf{u} = \mathbf{D} \cdot \nabla^2 \mathbf{u} + \mathbf{f}(\mathbf{u};\mathbf{p}). \tag{1.2.5}$$

However, it might be advantageous to retain capacitances or mobilities, in order to distinguish among three types of parameters:

- elements of the array $\mathbf{p}$ influencing stationary states as well as dynamics;
- mobilities or capacitances influencing dynamic behavior but not equilibria;

[1] Summation over repeated indices is presumed from now on

- diffusivities affecting spatially inhomogeneous solutions only.

The full form (1.2.4) is essential in systems with widely separated time scales where fast components are quasistationary.

An RDS can describe a truly open 2D system, such as the one shown in Fig. 1.1a, either when both diffusion and interactions are taking place on a surface or when the thickness of a bulk layer in the direction of external fluxes is so small that the composition can be assumed uniform in the "vertical" direction. Then the nonlinear functions in (1.2.1) should account for both internal interactions, e.g., chemical reaction rates, and external ("vertical") fluxes due to exchange with the environment. We shall suppose here that these functions do not depend on time and spatial variables explicitly. Then an RDS always has one or more "trivial" homogeneous stationary solutions (HSS) coinciding with fixed points $\mathbf{f}(\mathbf{u};\, \mathbf{p}) = 0$ of a respective dynamical system. This solution may become unstable in a certain parametric domain, giving way to spatially inhomogeneous and/or persistently time-dependent states. The idea of spontaneous pattern formation in RDS has been first put forward by Turing (1952). Chemical wave patterns have been first observed by Zhabotinsky[2] (1964). These phenomena necessarily require some reactants to be *autocatalytic*, i.e., accelerate their own production at certain circumstances.

1.2.2 Examples of RDS

Specific RDS differ by algebraic nonlinearities $\mathbf{f}(\mathbf{u})$, and emerge as spatially dependent extension of dynamical systems $\dot{\mathbf{u}} = \mathbf{f}(\mathbf{u})$. Particular chemical or biological models are often constructed on the basis of the mass action law, which stipulates that reaction rates are proportional to concentrations of reactants. The functions $f_i(\mathbf{u})$ defined in this way are necessarily polynomials, which, of course, may seem unduly restrictive. On the other hand, dynamic behavior of a polynomial system may not differ qualitatively from that of a more algebraically complicated system. The mass action principle has been used to construct a number of artificial easily solvable models specifically designed to illustrate certain features of dynamics. The simplest form of this kind is (1.1.2), where the cubic polynomial can be attributed to a chemical scheme including a monomolecular and a trimolecular reaction, although, since triple collisions are highly improbable, a cubic term would not reflect the actual reaction mechanism.

This disguise would be dismissed by a physicist, who would call u an "order parameter" and use (1.1.2) to describe a transition between a "disordered" phase with $u = 0$ and an "ordered" phase with $u \neq 0$. In particular, the gas–liquid equilibrium in the van der Waals fluid (van der Waals, 1894) is described in the mean field approximation by a single RDS equation with u standing for density and a function $f(u)$ with two stable zeros corresponding to equilibrium densities of two stable phases. For a mathematician, the polynomial in (1.1.2) is a rescaled "normal form"

[2] In the original experiments, there were no external fluxes and chemical waves decayed with time. Persistent wave patterns in a gel (eliminating advection) have been observed much later by Ouyang and Swinney (1991)

satisfying the condition of inversion symmetry. This symmetry can be broken by adding a constant or quadratic term. Similar equations may emerge as a reduction of a more complex system, as in dynamics of planforms in Sect. 1.5.

A simple extension of (1.1.2) to two-dimensional (2D) phase space is the FitzHugh–Nagumo (FN) dynamical system (FitzHugh, 1961; Nagumo et al, 1962), which has been first suggested, in a somewhat different form, as a model of nerve pulse propagation:

$$\dot{u} = u(1 - u^2) - v, \qquad \gamma^{-1}\dot{v} = -\nu - v + \mu u. \tag{1.2.6}$$

This model, supplemented by the diffusional term, may generate both stationary and wave patterns, as well as oscillations and multiple stationary states, depending on the choice of the ratios of the mobilities and diffusivities, as well as of the coupling parameter μ and the bias parameter ν. Its forms rescaled in two characteristic ways will be studied in detail in Ch. 3.

A more artificial "chemical" system is the *Brusselator* model (Lefever et al, 1967). Its algebraic part can be constructed by combining two fictitious chemical reactions $2A + B \to 3A$, $B \to A$. It is further assumed that the species A, but not B, is supplied from and removed to the environment[3]. This leads, after appropriate rescaling, to the system of equations

$$\dot{u} = \nu - (\mu + 1)u + u^2 v, \qquad \gamma\dot{v} = \mu u - u^2 v. \tag{1.2.7}$$

The real purpose of this construction is to have an autocatalytic system which always has a unique and easily computable stationary state $u = \nu,\ v = \mu/\nu$; this state can be made by additional rescaling independent of the parameters of the system. This is, of course, rather unrealistic, but very convenient, as it allows one to avoid the cumbersome task of finding equilibria, and to concentrate upon more interesting questions of symmetry breaking and dynamics. This are just the earliest and most thoroughly studied example of various artificial chemical schemes designed for this aim. Continuing this trend, Rössler (1976) designed several three-variable dynamical systems with simple algebraic functions to generate chaotic dynamics.

Various models of population dynamics are also based on the mass action law and contain polynomials of the general form $\nu_i u_i + u_i \sum_j \mu_{ij} u_j$. The common feature of these models is the presence of trivial "extinction" states with vanishing u_i. The interactions are competitive when both μ_{ij} and μ_{ji} are negative, cooperative when both are positive, and prey–predator when μ_{ij} and μ_{ji} have opposite signs. All these dynamical systems can be converted to standard RDS by adding diffusional terms.

Many realistic chemical models have a more complicated form. The source of complexity is in multistage chemical mechanisms, which may have characteristic rates differing by many orders of magnitude. Some chemical stages rapidly reach equilibrium, which enforces certain algebraic relations among the concentrations of reacting species. Using these equilibrium relations in kinetic equations of slower stages leads to dynamic equations with non-polynomial functions $f_i(\mathbf{u})$ even when all

[3] In the original derivation, two more reactions were added in lieu of the exchange

original equations are based on the mass action law. This is characteristic to catalytic or enzyme reactions, where the fast stage is adsorption of reactants on the catalyst surface or attachment to an enzyme molecule, and the slow stage is further chemical transformation. For example, a bimolecular reaction may proceed according to the following mechanism:

$$\mathrm{A}_i + \mathrm{E} \rightleftharpoons \mathrm{A}_i^*, \qquad \mathrm{A}_1^* + \mathrm{A}_2^* \to \mathrm{P} + 2\mathrm{E},$$

where E is a catalytic site, P is a non-adsorbing product, and asterisks denote adsorbed forms of the reactants A_i. Assuming adsorption equilibrium, the kinetic equations for the concentrations u_i of the reactants A_i in an open system reduce to the form

$$\partial_t u_i = D_i \nabla^2 u_i - \frac{k u_1 u_2}{(1 + \mu_1 u_1 + \mu_2 u_2)^2} + \nu_i (1 - u_i). \tag{1.2.8}$$

An interesting feature of this system is its autocatalytic character made apparent under conditions when strong adsorption of one of the reactants *reduces* the reaction rate by suppressing adsorption of the other reactant; thus, if $\mu_1 u_1$ far exceeds both $\mu_2 u_2$ and unity, the nonlinear term in (1.2.8) is approximately proportional to u_2/u_1, so that $\partial f_1/\partial u_1 > 0$. Both multiple stationary states and oscillations are possible under these conditions.

In non-isothermal reactions, one of the dependent variables is temperature; a typical temperature dependence of reaction rates is exponential, stemming from the Arrhenius law. If the reaction is exothermic, this dependence has an autocatalytic character, as the reaction rate grows with temperature causing its further increase. This system generates multiple stationary states and strongly nonlinear oscillations, but is not as convenient for analysis as artificial models.

1.2.3 Bifurcation of Stationary States

If spatial dependence is suspended, an RDS reduces to a dynamical system. A nonlinear dynamical system arises also when a spatially-dependent problem is reduced to dynamics of a limited number of planforms (see Sect. 1.5). Homogeneous stationary states (HSS) of a dynamical system, also called *fixed points*, form branches of solutions $\mathbf{u}(\mathbf{p})$ in the parametric space. A smooth branch of solution can be constructed starting from a known solution $\mathbf{u} = \mathbf{u}_0$ satisfying $\mathbf{f}(\mathbf{u}_0; \mathbf{p}_0) = 0$. Near this point, $\mathbf{f}(\mathbf{u}; \mathbf{p})$ can be expanded in the Taylor series in both variables and parameters as in (1.1.19) to obtain

$$\begin{aligned} \mathbf{f}(\mathbf{u}; \mathbf{p}) &= \mathbf{f}(\mathbf{u}_0; \mathbf{p}_0) + \epsilon\,(\mathbf{f}_\mathbf{u}\mathbf{u}_1 + \mathbf{f}_\mathbf{p}\mathbf{p}_1) \\ &+ \epsilon^2 \left(\mathbf{f}_\mathbf{u}\mathbf{u}_2 + \mathbf{f}_\mathbf{p}\mathbf{p}_2 + \tfrac{1}{2}\mathbf{f}_{\mathbf{uu}}[\mathbf{u}_1, \mathbf{u}_1] + \mathbf{f}_{\mathbf{up}}[\mathbf{u}_1, \mathbf{p}_1] + \tfrac{1}{2}\mathbf{f}_{\mathbf{pp}}[\mathbf{p}_1, \mathbf{p}_1]\right) + \cdots, \end{aligned} \tag{1.2.9}$$

where the arrays of first ($\mathbf{f}_\mathbf{u},\ \mathbf{f}_\mathbf{p}$), second ($\mathbf{f}_{\mathbf{uu}},\ \mathbf{f}_{\mathbf{up}},\ \mathbf{f}_{\mathbf{pp}}$), etc. (e.g., $\mathbf{f}_{\mathbf{uuu}}$) derivatives with respect to variables and parameters are evaluated at the reference point $\mathbf{u} = \mathbf{u}_0$, $\mathbf{p} = \mathbf{p}_0$. Higher derivatives are defined as arrays of a higher tensor rank: thus, mth derivative acts at m vectors (state arrays) written to the right of it in brackets. We shall also need to expand the time derivative in a series of partial derivatives $\partial_k \equiv \partial/\partial t_k$.

The hierarchy of equations approximating the dynamics near the reference point in successive orders is obtained in the consecutive orders of the expansion (1.2.9). The first-order equation is

$$\partial_0 \mathbf{u}_1 = \mathbf{f_u} \mathbf{u}_1 + \mathbf{f_p} \mathbf{p}_1. \tag{1.2.10}$$

The square array $\mathbf{f_u}$ is called a *Jacobi matrix*. If its inverse $\mathbf{f_u}^{-1}$ exists, we can find the shift of the stationary solution $\mathbf{u}_1 = -\mathbf{f_u}^{-1}\mathbf{f_p}\, \mathbf{p}_1$ corresponding to a parametric shift $\mathbf{p}_1$. The homogeneous linear equation

$$\mathcal{L}\mathbf{u}_1 \equiv (\partial_0 - \mathbf{f_u})\mathbf{u}_1 = 0 \tag{1.2.11}$$

obtained by setting $\mathbf{p}_1 = 0$ governs stability of the stationary state $\mathbf{u} = \mathbf{u}_0$ to infinitesimal perturbations.

Stability of a fixed point is determined by the location of the *leading* eigenvalue in the complex plane, i.e., that with the largest real part. It is usually possible to detect a parametric domain sufficiently close to thermodynamic equilibrium where there is a single stable stationary state, and follow changes of dynamics along a certain path in the parametric space. At a parametric point where the leading eigenvalue of $\mathbf{f_u}$ vanishes, (1.2.11) is satisfied by $\mathbf{u}_1 = a\mathbf{U}$, where $\mathbf{U}$ is the respective eigenvector and a is an arbitrary *amplitude*. The Jacobi matrix becomes *singular* at this point; therefore it marks not only the loss of stability but impossibility to continue the solution branch. Midway along a branch of stationary solutions, stability may be lost as a result of the Hopf bifurcation[4] when a pair of complex eigenvalues crosses the imaginary axis in the complex plane. A locus of either bifurcation is a *codimension one* hypersurface in the parametric space. Additional conditions may define bifurcation manifolds of a higher codimension.

On consecutive stages of the expansion, we shall arrive at inhomogeneous equations of the type $\mathbf{L}\mathbf{u} = \boldsymbol{\Psi}$, where the operator $\mathbf{L} = \partial_0 - \mathbf{f_u}$ is singular. By Fredholm alternative, an inhomogeneous equation is solvable if the inhomogeneity $\boldsymbol{\Psi}$ does not project on the corresponding eigenfunction $\boldsymbol{\Phi}$. This solvability condition is expressed as $\langle \boldsymbol{\Phi}^\dagger, \boldsymbol{\Psi} \rangle$, where $\langle \ldots \rangle$ is an appropriate scalar product and $\boldsymbol{\Phi}^\dagger$ is the eigenfunction with zero eigenvalue of an appropriate adjoint operator.

We will extend this procedure to symmetry-breaking bifurcations in RDS in Sect. 1.3.3, but restrict here to bifurcations of stationary solution branches. In this case, $\boldsymbol{\Phi}^\dagger$ is just the eigenvector $\mathbf{U}^\dagger$ of the transposed matrix $\mathbf{f_u}^\dagger$ normalized to $\mathbf{U}^\dagger \cdot \mathbf{U} = 1$. The second-order equation derived from (1.2.9) is

$$\mathbf{L}\mathbf{u}_2 + \partial_1 a \mathbf{U} = \mathbf{f_p}\mathbf{p}_1 + \tfrac{1}{2} a^2 \mathbf{f_{uu}}[\mathbf{U}, \mathbf{U}] = 0. \tag{1.2.12}$$

Its solvability condition is

$$\partial_1 a = \kappa_0 + \kappa_2 a^2, \quad \kappa_0 = \mathbf{U}^\dagger \cdot \mathbf{f_p}\mathbf{p}_1, \quad \kappa_2 = \tfrac{1}{2}\mathbf{U}^\dagger \cdot \mathbf{f_{uu}}[\mathbf{U}, \mathbf{U}]. \tag{1.2.13}$$

This defines dynamics near the locus of a *fold* bifurcation (Fig. 1.3a). Two solutions exist when κ_0 and κ_2 have opposite signs, and none otherwise. Thus, a pair of

[4] It is also called, more fairly, Andronov–Hopf or Poincaré–Andronov–Hopf

solutions emerges or disappears when κ_0 switches its sign with a shift of some parameter of the system. Only one of these solutions may be stable. Stationary states with all eigenvalues real and of the same sign are called *nodes*, and those with real eigenvalues of different signs, *saddles*; therefore this is also called a saddle-node bifurcation.

Nongeneric situations may arise either because of intrinsic symmetries of the system or "accidentally" on a manifold of a higher codimension. It may happen that the matrix $\mathbf{f_p}$ vanishes identically. This would be the case when $\mathbf{u}_0$ is a "trivial" solution that remains constant at all values of parameters. This happens, for example, in equations of population dynamics where trivial solutions with at least one population density vanishing exist at any values of parameters. Then the solvability condition of (1.2.12) is satisfied identically, and first-order parametric deviations can be retained. The second-order equation derived from (1.2.9) with $\mathbf{f_p}$, $\mathbf{f_{pp}}$ set to zero is then

$$\mathbf{L}\mathbf{u}_2 + \partial_1 \mathbf{u}_1 = a\mathbf{f_{up}}[\mathbf{U}, \mathbf{p}_1] + \tfrac{1}{2}a^2\mathbf{f_{uu}}[\mathbf{U}, \mathbf{U}] = 0, \tag{1.2.14}$$

The solvability condition of this equation is

$$\partial_1 a = \kappa_1 a + \kappa_2 a^2, \quad \kappa_1 = \mathbf{U}^\dagger \cdot \mathbf{f_{up}}[\mathbf{U}, \mathbf{p}_1], \quad \kappa_2 = \tfrac{1}{2}\mathbf{U}^\dagger \cdot \mathbf{f_{uu}}[\mathbf{U}, \mathbf{U}]. \tag{1.2.15}$$

This equation, describing a *transcritical* bifurcation, has two stationary solutions on both sides of the bifurcation manifold (Fig. 1.3b); one of them is the trivial solution

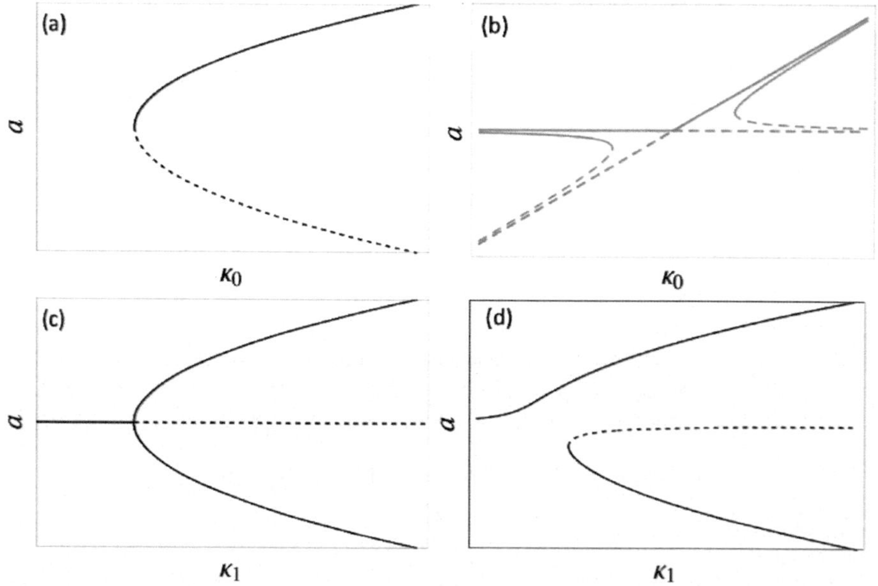

Fig. 1.3. Dependence of the amplitude on a bifurcation parameter. Solid lines denote stable and dashed lines, unstable states. (a) Fold bifurcation. (b) Transcritical bifurcation, showing also its splitting due to a weak perturbation. (c) Supercritical pitchfork bifurcation and (d) its asymmetric unfolding

$a = 0$. In a generic case, a weak perturbation, adding to (1.2.15) a small constant term $\epsilon\kappa_0$, splits the transcritical bifurcation into two folds.

Solutions of either (1.2.13) or (1.2.15) escape to "infinity", i.e., most likely, to an attractor lying far away from the vicinity of $\mathbf{u}_0$. This can be corrected by locating a manifold of codimension two, called a *cusp* singularity where κ_2 vanishes, and constructing a *cubic* amplitude equation in its vicinity in the next order of the expansion. This coefficient may also vanish identically because of the symmetry of the original problem to inversion of $\mathbf{u}$. In order to continue the expansion, parametric deviations should be restricted in such a way that the dependence on t_1 be suppressed. First-order parametric deviations $\mathbf{p}_1$ *parallel* to the bifurcation manifold should be restricted by the condition $\kappa_1 = 0$, but they are superfluous, since they correspond just to gliding into a closer vicinity of another point on the codimension two bifurcation manifold in a higher-dimensional parametric space. Further on, we shall therefore set $\mathbf{p}_1$ to zero identically.

The dynamics unfolding on a still slower time scale t_2 should be determined from the third-order equation

$$\mathbf{f_u u}_3 = \partial_2 a\,\mathbf{U} - \mathbf{f_p p}_3 - a\mathbf{f_{up}}[\mathbf{U}, \mathbf{p}_2] - a\mathbf{f_{uu}}[\mathbf{U}, \mathbf{u}_2] - \tfrac{1}{6}a^3\mathbf{f_{uuu}}[\mathbf{U}, \mathbf{U}, \mathbf{U}]. \quad (1.2.16)$$

The second-order function $\mathbf{u}_2$ has to be found by solving Eq. (1.2.12), now reduced to the form

$$\mathbf{f_u u}_2 = -\mathbf{f_p p}_2 - \tfrac{1}{2}a^2\mathbf{f_{uu}}[\mathbf{U}, \mathbf{U}]. \quad (1.2.17)$$

Only the solution of the inhomogeneous equation, which does not project on the eigenvector $\mathbf{U}$, is relevant. It can be expressed as a quadratic form in a containing two suitable vectors, $\mathbf{u}_2 = \mathbf{U}_{22} + a^2\mathbf{U}_{20}$. The solvability condition of (1.2.16) is obtained then in the form

$$\partial_2 a = \kappa_0 + \kappa_1 a - \kappa_3 a^3, \quad (1.2.18)$$

$$\kappa_0 = \mathbf{U}^\dagger \cdot \mathbf{f_p p}_3, \qquad \kappa_1 = \mathbf{U}^\dagger \cdot \mathbf{f_{up}}[\mathbf{U}, \mathbf{p}_2] + \mathbf{U}^\dagger \cdot \mathbf{f_{uu}}[\mathbf{U}, \mathbf{U}_{22}],$$
$$\kappa_3 = \tfrac{1}{6}\mathbf{U}^\dagger \cdot \mathbf{f_{uuu}}[\mathbf{U}, \mathbf{U}, \mathbf{U}] + \mathbf{U}^\dagger \cdot \mathbf{f_{uu}}[\mathbf{U}, \mathbf{U}_{20}]. \quad (1.2.19)$$

This equation presents parametric unfolding of dynamics in the vicinity of a cusp singularity. The dynamics is confined to the vicinity of $\mathbf{u}_0$, provided $\kappa_3 > 0$. The symmetric form at κ_0 is a *pitchfork* bifurcation (Fig. 1.3c). A weak asymmetry splits the solution branches as in Fig. 1.3d.

The procedure can be continued further if a sufficient number of free parameters are available, arriving at equations with a larger number of fixed points, which present unfoldings of higher-order singularities attained on manifolds of higher codimension. This restricts the dynamics to an ever more diminutive slice of the parametric space where more and more stationary states and transitions among them crowd in. Singularities of the fourth and fifth order have flamboyant names: "butterfly" and "swallowtail". Analysis of parametric dependence of multiple stationary states has been a subject of a huge body of mathematical literature (Golubitsky and Schaeffer, 1985). Numerous studies of bifurcation of equilibria have classified in minute detail a variety of cuts that can be made through bifurcation manifolds (containing differently inclined folds, isolas, etc.), as well as their modifications due to innate or accidental symmetries of the underlying system. A catchy term *catastrophe theory* had been coined (Thom, 1975) to describe the sequence of bifurcations of equilibria. The philosophical balloon growing from this theory was ridiculed by Arnold (1984).

1.2.4 Global Bifurcations

There are only two kinds of generic (codimension one) bifurcations in dynamical systems that can be detected by local linear analysis: a fold, or saddle-node, bifurcation at zero eigenvalue and a Hopf bifurcation at imaginary eigenvalue creating a periodic orbit. However, other kinds of transitions, which cannot be detected by local analysis alone, are necessary to reconstruct a *global* generic picture of qualitative changes of behavior, including bifurcation of stationary states and periodic orbits, and to advance to chaotic dynamcs emerging in three phase space dimensions. These phenomena are outside the scope of this book, and are discussed in a similar style elsewhere (Pismen, 2021a), but we briefly review here global bifurcations, which may be relevant when a spatially distributed system is reduced to a dynamical system for amplitudes of dominating modes.

Global analysis of dynamical systems aims at defining all kinds of bifurcations affecting either the number of attractors or their basin boundaries. While the local analysis in the preceding subsection started from a *stable* stationary state (fixed point) and concentrated on finding other attractors in its vicinity, both stable and *unstable* states are relevant for global analysis. A fixed point in the n-dimensional phase space has, generically, m eigenvalues with positive real parts and $n - m$ eigenvalues with negative real parts. The subspaces spanned by the respective eigenvectors form an m-dimensional *unstable manifold* and an $(n - m)$-dimensional *stable manifold*. Trajectories of the dynamical system approach the fixed point along its stable manifold and escape from its vicinity along its unstable manifold. A periodic orbit in the n-dimensional phase space has, generically, an m-dimensional unstable manifold ($m = 0$ for a stable orbit) and an $(n - m - 1)$-dimensional stable manifold. The remaining direction is tangential to the trajectory. The system is neutrally stable to tangential displacements, i.e., phase shifts of the periodic motion.

Complete global analysis is possible for two-variable dynamical systems, which have two kinds of attractors only: fixed points or periodic orbits. The fixed points may be of three kinds: *nodes* (either stable or unstable), which have two real eigenvalues of the same sign, *saddles*, which have two real eigenvalues with opposite signs, and *foci* (either stable or unstable), which have two complex conjugate eigenvalues. The three types of global codimension one bifurcations in a two-variable dynamical system are *saddle-loop* bifurcation; saddle-node infinite period (*sniper*), or Andronov, bifurcation, and saddle-node bifurcation of periodic orbits.

A saddle-loop bifurcation takes place when a trajectory going out of a saddle point along its unstable manifold returns to the same point along its stable manifold. Such a trajectory, called *homoclinic*, does not exist generically but may be found by adjusting a single parameter of the system. A homoclinic trajectory can be viewed as a periodic orbit with the infinite period. Starting from a *supercritical* Hopf bifurcation

(a)

(b)

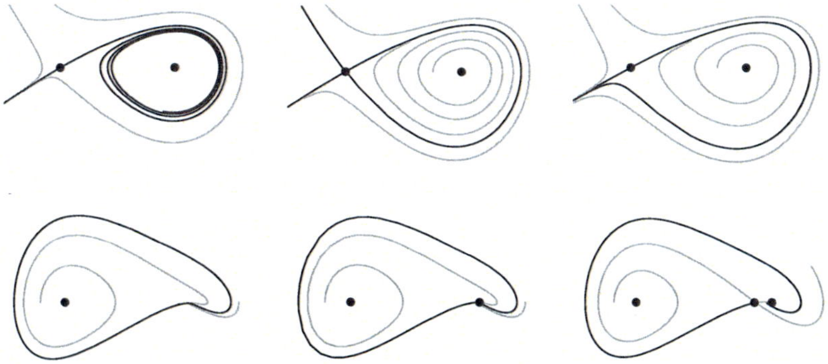

Fig. 1.4. Parametric evolution of a periodic orbit through a saddle-loop (a) and sniper (b) bifurcation

in a system with at least two fixed points, one of which is a saddle, and changing a suitable adjustable parameter, one can follow the parametric evolution of the emerging periodic orbit and observe that both the amplitude and period increase, until the orbit hits the saddle point and turns into a homoclinic (Fig. 1.4a). Beyond the bifurcation, all trajectories are deflected to a faraway attractor.

An *unstable* periodic orbit emerging at a *subcritical* Hopf bifurcation may also undergo a saddle-loop bifurcation, which now affects, however, only basin boundaries, rather than the attractors themselves. An unstable orbit forms the boundary of the attraction basin of the fixed point it surrounds. Following the saddle-loop bifurcation, this basin opens up, and the fixed point may now attract trajectories starting from far removed points in the phase plane. This picture can be read from the same Fig. 1.4a with inverted direction of flow. A sequence going through the formation of either stable or unstable homoclinic orbit can be traced by changing a dynamic parameter that does not affect the fixed point.

A sniper bifurcation occurs when a pair of stationary states – a node and a saddle – emerges on a periodic orbit. The orbits in the phase plane at the bifurcation point are shown in Fig. 1.4b. Like in the case of a saddle-loop bifurcation, the period diverges at the bifurcation point; the sequence going through this bifurcation should be traced by changing one of the parameters affecting the stationary states, since an emerging pair, rather than a "preexisting" saddle, is involved here. The third kind of a global bifurcation is a birth or annihilation of a pair of periodic orbits, one stable and one unstable. In 3D, a periodic orbit may undergo an additional Hopf bifurcation leading to quasiperiodic motion around a torus (commonly further splitting into resonances with rational ratios of frequencies) or undergo period doubling, eventually tending to chaotic motion.

An analytical insight into global dynamics is possible near a manifold in the parametric space where the underlying linearized system has more than one eigenvalue with vanishing real part. Two real modes with unbroken symmetry ($k = 0$, $\omega = 0$)

interact in the vicinity of a double zero singularity, also known as the Takens–Bogdanov bifurcation. In the absence of special symmetries, the bifurcation manifold is of codimension two, and the eigenvalue is degenerate geometrically as well as algebraically. Geometric degeneracy means that there is only one eigenvector $\mathbf{U}$ satisfying $\mathbf{L}_0\mathbf{U} = 0$. The orthogonal basis spanned by this vector and the rest of eigenvectors of $\mathbf{L}_0$ should be then complemented by an additional vector $\mathbf{V}$ satisfying $\mathbf{LV} = \mathbf{U}$. The matrix representation of $\mathbf{L}$ is reduced in this basis to the normal Jordan form where the cell $\begin{pmatrix} 0 & 1 \\ 0 & 0 \end{pmatrix}$ corresponds to the double zero eigenvalue. It is particularly interesting since in its vicinity one can study analytically global bifurcations, which otherwise can be detected only numerically (Pismen, 2021a).

If a_1, a_2 are, respectively, the amplitudes of the modes with the eigenvectors $\mathbf{U}$, $\mathbf{V}$, a straightforward expression for a normal form with the specified linear part is

$$\dot{a}_1 = a_2 + f_1(a_1, a_2), \qquad \dot{a}_2 = f_2(a_1, a_2), \tag{1.2.20}$$

where the polynomials $f_i(a_1, a_2)$ account for nonlinear interactions as well as small parametric deviations from the codimension two bifurcation manifold. This expression can be obtained by projecting the original equations on the eigenvectors $\mathbf{U}$, $\mathbf{V}$, expanding in Taylor series and truncating at a certain order; however, the formal orders of magnitudes of various terms would not match.

Equation (1.2.20) can be further simplified by choosing new variables $x = a_1$, $y = a_2 + f_1(a_1, a_2)$. Taking note that the time variable in (1.2.20) should be slow, we observe that the two amplitudes x, y should be of different orders of magnitude, so that $y \ll x$, and therefore one can neglect all terms nonlinear in y. Then (1.2.20) is rewritten as

$$\dot{x} = y, \qquad \dot{y} = f(x) + y\,g(x), \tag{1.2.21}$$

where $f(x)$, $g(x)$ are, respectively, nth and mth order polynomials. All terms in these equations except the last one balance at $O(\epsilon^n)$ if the amplitude x is scaled as $O(\epsilon)$ and the slow time variable as $O(\epsilon^{-(n-1)/2})$. Then the last term in the second equation is scaled as $O(\epsilon^{m-(n-1)/2})$. If $m > \frac{1}{2}(n-1)$, (1.2.21) is a perturbed Hamiltonian system with weakly dissipative corrections, which is particularly convenient for analysis (Pismen, 2021a).

Other codimension two degenerate bifurcations may involve "accidental" coincidence of monotonic and Hopf or of two Hopf bifurcations with different frequencies. The amplitude equations are constructed in a straightforward way as combinations of the amplitude equations for the respective elementary bifurcations with added interaction terms allowed by symmetry. For example, the pair of amplitude equations for the Hopf-monotonic degeneracy truncated at the third order is

$$\begin{aligned} \dot{a}_1 &= \kappa_0 + \kappa_1 a_1 - a_1^3 + \mu_2 |a_2|^2 + \mu_3 a_1 |a_2|^2, \\ \dot{a}_2 &= a_2 \left(\lambda_0 + \lambda_1 a_1 + \lambda_2 a_1^2 - \lambda_3 |a_2|^2\right), \end{aligned} \tag{1.2.22}$$

where the amplitude a_1 and parameters μ_i are real, while the amplitude a_2 and parameters λ_i are complex.

This equation system, as well as a similar one for a nonresonant double Hopf degeneracy, is not interesting by itself, since, although the phase space dimension is formally, respectively, three and four, the phase of the complex amplitude falls out, and dynamics is essentially two-dimensional. However, chaotic behavior may be generated in this case under the influence of weak phase-dependent perturbations. Near the Hopf-monotonic degeneracy, chaos is likely to arise when a homoclinic orbit exists. Near the double Hopf degeneracy, phase dependence can be caused by coupling the two oscillatory modes.

1.3 Space-Dependent Amplitude Equations

1.3.1 Turing Instability

All four kinds of instabilities listed in Sect. 1.1.3 are realized in RDS, but we are mostly interested here in the last two kinds, where spatial patterns arise. Consider the general RDS (1.2.5) in the infinite plane and suppose that it has an HSS $\mathbf{u} = \mathbf{u}_0$ satisfying $\mathbf{f}(\mathbf{u}_0) = 0$. Without loss of generality, we can assume $\mathbf{u}_0 = 0$, since the stationary values can be taken as the origins of the respective variables. Thus, we assume that the vector-function $\mathbf{f}(\mathbf{u})$ vanishes at the origin. We shall now suppose the deviations from the trivial state to be arbitrarily small, and linearize (1.2.4) to obtain

$$\mathbf{\Gamma}^{-1}\partial_t \mathbf{u} = \left(\mathbf{D}\nabla^2 + \mathbf{f}_\mathbf{u}\right) \cdot \mathbf{u}, \tag{1.3.1}$$

where $\mathbf{f}_\mathbf{u}$ is the Jacobi matrix of the vector-function $\mathbf{f}(\mathbf{u})$ with the elements $F_{ij} = \partial f_i/\partial u_j$ evaluated at $\mathbf{u} = 0$.

Solutions of (1.3.1) are sought for in the harmonic form $\mathbf{u} = \sum_j \mathbf{U}_j\, \mathrm{e}^{\lambda_j t + \mathrm{i}\mathbf{k}\cdot\mathbf{x}}$. Using this in (1.3.1), we see that nontrivial solutions are obtained when $\lambda(k^2)$ are eigenvalues of the matrix

$$\mathbf{L} = \mathbf{\Gamma} \cdot (\mathbf{f}_\mathbf{u} - k^2\mathbf{D}). \tag{1.3.2}$$

In the trivial case of a scalar (one-component) RDS, the leading mode is always stationary and homogeneous, and the point of marginal stability is attained at $\lambda_0 = k_0 = 0$ when the determinant of the Jacobi matrix $\mathbf{f}_\mathbf{u}$ vanishes. But already for two coupled equations, a leading mode with $k_0 \neq 0$ is possible, and a symmetry-breaking instabilityat $\lambda_0(k_0) = 0$ would lead to bifurcation of a stationary pattern.

Consider a general two-component RDS with a constant diagonal matrix of diffusivities in the rescaled form

$$\partial_t u = \nabla^2 u + f(u,v), \qquad \partial_t v = \gamma[\kappa\nabla^2 v + g(u,v)], \tag{1.3.3}$$

where $\kappa = D_v/D_u$, $\gamma = \gamma_v/\gamma_u$ and $f(u,v)$, $g(u,v)$ are source functions depending on the state variables u and v. Then the explicit form of the matrix $\mathbf{L}$ in (1.3.2) is

$$\mathbf{L} = \begin{pmatrix} f_u - k^2 & f_v \\ \gamma g_u & \gamma(g_v - \kappa k^2) \end{pmatrix}, \tag{1.3.4}$$

where κ, γ are the ratios of the diffusivities and mobilities of the two species, and all derivatives $f_u \equiv \partial_u f$, etc. are computed at the applicable HSS. Two conditions of marginal stability are provided by vanishing trace and determinant of (1.3.4):

$$\begin{aligned} \mathrm{Tr}(\mathbf{L}) &= f_u + \gamma g_v - (1+\gamma\kappa)k^2 = 0, \\ \gamma^{-1}\det(\mathbf{L}) &= \kappa k^4 - (\kappa f_u + g_v)k^2 + \Delta = 0, \end{aligned} \tag{1.3.5}$$

where $\Delta = (f_u g_v - f_v g_u)$. The trace condition corresponds to the onset of an oscillatory instability (Hopf bifurcation). It is possible only in the presence of positive feedback, or, in chemical terms, when at least one of the species is *autocatalytic* ($f_u > 0$ or $g_v > 0$). Clearly, since the trace decreases with k, this instability can occur only in the homogeneous mode. However, the determinant, may have a minimum at

$$k_0^2 = \frac{\kappa f_u + g_v}{2\kappa}. \tag{1.3.6}$$

This minimum occurs at a real k_0 under the same conditions ($f_u > 0$ or $g_v > 0$) when the Hopf bifurcation is possible. A stationary inhomogeneous mode, leading to spatial symmetry breaking, is excited only if the condition of the Hopf bifurcation has not yet been satisfied. We have to presume therefore $f_u + \gamma g_v < 0$, which means that spatial symmetry breaking would occur when *only one* of the species is autocatalytic. To be definite, let $f_u > 0$ and $g_v < 0$. Then spatial symmetry breaking (called *Turing* bifurcation when applied to RDS) preempts the Hopf bifurcation at $\gamma\kappa > 1$, i.e., when the autocatalytic species is *less diffusive*. Thus, for spatial symmetry breaking in a two-component system, one needs a combination of a slowly diffusing *activator* and a rapidly diffusing *inhibitor* (Turing, 1952).

The determinant at the locus of Turing bifurcation must vanish:

$$\gamma^{-1}\det(\mathbf{L})_{k=k_0} = -f_v g_u - \frac{(\kappa f_u - g_v)^2}{4\kappa} = 0. \tag{1.3.7}$$

This has to occur while the system is stable to homogeneous perturbations, i.e., when $\det(\mathbf{L}) > 0$ at $k = 0$. At the homogeneous instability threshold $\Delta = 0$, (1.3.7) is transformed as

$$-f_u g_v - \frac{(\kappa f_u - g_v)^2}{4\kappa} = -\frac{(\kappa f_u + g_v)^2}{4\kappa}, \tag{1.3.8}$$

which is negative definite, so that the homogeneous instability is always precluded by the Turing bifurcation whenever the latter takes place.

As a simple representative example, consider the symmetric version of the FitzHugh–Nagumo (FN) system (1.2.6) supplemented by diffusional terms

$$\partial_t u = \nabla^2 u + u(1-u^2) - v, \qquad \gamma^{-1}\partial_t v = \kappa\nabla^2 v - v + \mu u. \tag{1.3.9}$$

For the trivial solution $u = v = 0$, we have $f_u = 1$, $f_v = -1$, $g_v = -1$, $g_u = \mu$ with

$$\begin{aligned} \mathrm{Tr}(\mathbf{L}) &= 1 - \gamma - k^2(1+\gamma\kappa), \\ \gamma^{-1}\det(\mathbf{L}) &= \mu - 1 + (1-\kappa)k^2 + \kappa k^4. \end{aligned} \tag{1.3.10}$$

The Turing bifurcation takes place at

$$k^2 = \frac{\kappa - 1}{2\kappa}, \quad \mu_c = \frac{(1+\kappa)^2}{4\kappa}. \tag{1.3.11}$$

Under these conditions, $\mu > 1$, so that $\det(\mathbf{L}) > 0$ at $k = 0$, and the homogeneous instability is suppressed.

The competition between the Hopf and Turing instability depends on the ratio of mobilities γ, and both coincide at a codimension-two point in the parametric space. A general study of spatiotemporal dynamics near the Turing-Hopf bifurcation was carried out by Meixner et al (1997) and extended to secondary bifurcations at this singular point by Ledesma-Durán and Aragón (2019).

1.3.2 Wave Instability

The number of state variables should be at least three to obtain an inhomogeneous oscillatory instability leading to a wave pattern emerging directly from the trivial state although, as we shall see in Ch. 3, stable wave patterns also exist in two-component systems. A general analysis for a three-component RDS extending (1.3.12) is too complicated to extract transparent results, and we restrict to a relatively simple example (Or-Gil et al, 1998) adding a third component to the FN system.

$$\partial_t u = \nabla^2 u + u(1-u^2) - v - w, \quad \partial_t v = \gamma_1 \left(\kappa_1 \nabla^2 v + \mu u - v\right),$$
$$\partial_t w = \gamma_2 \left(\kappa_2 \nabla^2 w + \nu u - w\right). \tag{1.3.12}$$

Even in this particular case, analysis and presentation of results is thwarted by proliferation of parameters, and we further specify it to retain only parameters important for detecting a wave bifurcation. The essential parameters for oscillatory instabilities are mobilities, but the diffusivities of v and w can be assumed equal. Then the Jacobi matrix obtained after linearization near the trivial solution and Fourier transform becomes

$$\mathbf{L} = \begin{pmatrix} 1-k^2 & -1 & -1 \\ \mu\gamma_1 & -\gamma_1(1+\kappa k^2) & 0 \\ \nu\gamma_2 & 0 & -\gamma_2(1+\kappa k^2) \end{pmatrix}. \tag{1.3.13}$$

Vanishing determinant of $\mathbf{L}$ detects the homogeneous instability at $\mu + \nu = 1$. The Turing bifurcation takes place at

$$k^2 = \frac{\kappa - 1}{2\kappa}, \qquad \mu + \nu = \frac{(\kappa+1)^2}{4\kappa} > 1, \tag{1.3.14}$$

so that, as in Sect. 1.3.1, it prevails over the homogeneous instability.

In order to locate oscillatory bifurcations at either zero (Hopf) or a finite (wave) wavenumber, we apply the Routh–Hurwitz stability criterion based on the characteristic polynomial

$$\det(\lambda \mathbf{I} - \mathbf{L}) = \lambda^3 + a_1\lambda^2 + a_2\lambda + a_3, \tag{1.3.15}$$

where $\mathbf{I}$ is the identity matrix. For a three-variable system with a negative real eigenvalue $-r$ and a pair of complex conjugate eigenvalues $-p \pm iq$, we have

$$a_1 = 2p + r, \quad a_2 = 2pr + p^2 + q^2, \quad a_3 = r(p^2 + q^2). \tag{1.3.16}$$

Clearly, all these coefficients are positive for a stable stationary state. The last one is the determinant of $\mathbf{L}$, which vanishes at a pitchfork or Turing bifurcation. The locus of a Hopf or wave bifurcation is determined by the combination

$$H(k) = a_1 a_2 - a_3 = 2p\left[(p + r)^2 + q^2)\right], \tag{1.3.17}$$

which has the same sign as the real part of both complex eigenvalues.

For the Jacobi matrix (1.3.13), the Hopf bifurcation locus is defined by

$$H(0) = \mu\gamma_1(\gamma_1 - 1) + \nu\gamma_2(\gamma_2 - 1) + (\gamma_1 - 1)(\gamma_2 - 1)(\gamma_1 + \gamma_2) = 0 \tag{1.3.18}$$

It is advantageous to identify the locations of pitchfork and Hopf bifurcations, i.e., to restrict computations to the vicinity of the double zero singularity. This allows us to eliminate one algebraic parameter by setting

$$\nu = \frac{\gamma_2(\gamma_1 - 1)}{\gamma_2 - \gamma_1}. \tag{1.3.19}$$

With these simplifications, we obtain

$$\begin{aligned} H(k) &= \gamma_1(\gamma_1 - \gamma_2)^{-1}(\gamma_1 - 1)[\mu + \gamma_2 - 1) - \mu\gamma_2] \\ &+k^2(\gamma_1 - \gamma_2)^{-1}\left\{\gamma_2^2(\kappa - 1)(\gamma_2 - 1) + \gamma_1^2[\mu + \kappa + \gamma_2 - 2 - \kappa\gamma_2(\mu + 2)]\right. \\ &\quad \left. + \gamma_1\gamma_2[2\kappa\gamma_2(\gamma_2 - 1) - \mu] + \gamma_1^3[1 + \kappa(\mu + 3\gamma_2 - 2)\right\} \\ &+k^4\left\{\gamma_2[(1 + 2\kappa(\gamma_2 - 1) - \kappa^2\gamma_2] + \gamma_1[1 + \kappa^2\gamma_2(3\gamma_2 - 2)] + 2\kappa(2\gamma_2 - 1)]\right. \\ &\quad \left. + \kappa\gamma_1^2[2 + \kappa(3\gamma_2 - 1)]\right\} + k^6\kappa(\gamma_1 + \gamma_2)(1 + \kappa\gamma_1)(1 + \kappa\gamma_2). \end{aligned} \tag{1.3.20}$$

The wave bifurcation locus is located by the conditions $H(k^2) = H'(k^2) = 0$, $H''(k^2) > 0$. There are three roots satisfying the first relation. If all of then are real and positive, two of them are minima with $H''(k) > 0$. Otherwise, a pair of roots may be complex or a minimum would escape into the negative domain. As

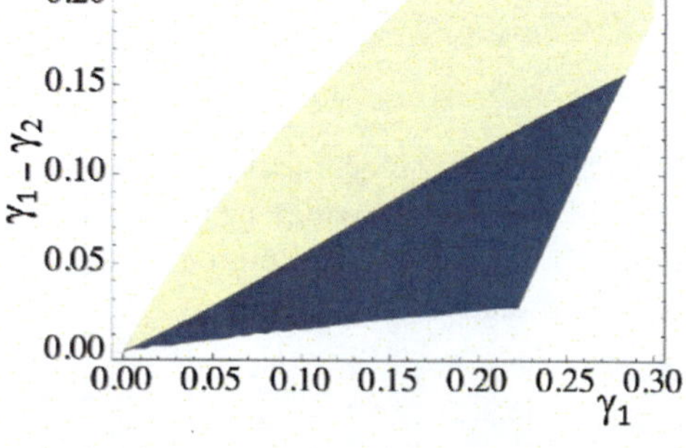

Fig. 1.5. The blue area marks the region in the plane γ_1, γ_2 where the wave instability is dominant at $\kappa = 2$. Its border with the yellow area its the location of the wave–Turing degeneracy. Both regions and the white triangle below constitute the domain where the wavenumber is real, changing between about 0.55 and 0.9.

analytical expressions are extremely cumbersome (and too long to be presented here), the parametric dependence of both the wavenumbers and the values of the parameter μ can be studied only graphically. A test at $\kappa = 2$ detects a parametric region where the wave bifurcation is dominant. It is restricted to small mobilities, as shown in Fig. 1.5. The reader may appreciate how cumbersome is an elementary study of a wave instability, even in a drastically simplified model suitable for this task. Yang et al (2002) studied wave and Turing bifurcations in the Brusselator model (1.2.7) with an added third variable and computed some resulting complex patterns, but this publication does not contained details of the bifurcation analysis.

1.3.3 Derivation of Amplitude Equations

We extend now the approach of Sect. 1.2.3 to symmetry-breaking bifurcations and long-scale spatial and temporal modulation of amplitudes near pitchfork and Hopf bifurcation loci. As we do not need now to discuss shifts of stationary states, which are taken care of in Sect. 1.2.3, we will consider a general RDS in the form (1.2.4), singling out a capacitance matrix $\mathbf{\Theta} = \mathbf{\Gamma}^{-1}$ and suppressing the dependence on parameters $\mathbf{p}$ affecting the algebraic part only,

$$\mathbf{\Theta}\partial_t \mathbf{u} = \mathbf{D}\nabla^2\mathbf{u} + \mathbf{f}(\mathbf{u}), \tag{1.3.21}$$

supposing that the nonlinear vector-function satisfies $\mathbf{f}(\mathbf{u}) = 0$ at $\mathbf{u} = \mathbf{u}_0$. Expanding this function yields in the consecutive orders

$$\begin{aligned}\mathbf{f}(\mathbf{u}) &= \mathbf{f}(\mathbf{u}_0) + \epsilon \mathbf{f}_\mathbf{u}\mathbf{u}_1 + \epsilon^2\left(\mathbf{f}_\mathbf{u}\mathbf{u}_2 + \tfrac{1}{2}\mathbf{f}_{\mathbf{uu}}[\mathbf{u}_1,\mathbf{u}_1]\right)\\ &+ \epsilon^3\left(\mathbf{f}_\mathbf{u}\mathbf{u}_3 + \mathbf{f}_{\mathbf{uu}}[\mathbf{u}_1,\mathbf{u}_1] + \tfrac{1}{6}\mathbf{f}_{\mathbf{uuu}}[\mathbf{u}_1,\mathbf{u}_1,\mathbf{u}_1]\right) + \cdots,\end{aligned} \tag{1.3.22}$$

where the arrays of first ($\mathbf{f}_\mathbf{u}$), second ($\mathbf{f}_{\mathbf{uu}}$), and third ($\mathbf{f}_{\mathbf{uuu}}$) derivatives with respect to variables are evaluated at the reference point $\mathbf{u} = \mathbf{u}_0$. The diffusivity and capacitance matrices are expanded likewise as $\mathbf{D} = \mathbf{D}_0 + \epsilon\mathbf{D}_1 + \ldots$, $\mathbf{\Theta} = \mathbf{\Theta}_0 + \epsilon\mathbf{\Theta}_1 + \ldots$.

Let the linear operator $\mathbf{L}_0 = \mathbf{\Theta}_0\partial_0 - \mathbf{D}_0\nabla_0^2 - \mathbf{f}_\mathbf{u}$ have an eigenvalue with the vanishing real part at a bifurcation point. Then the first-order equation $\mathbf{L}_0\mathbf{u}_1 = 0$ is solved by

$$\mathbf{u}_1 = a\,\mathbf{\Phi}(\mathbf{x}, t_0) = a\,\mathbf{U}(\omega, k)\mathrm{e}^{\mathrm{i}(\omega t_0 + \mathbf{k}\cdot\mathbf{x})}, \tag{1.3.23}$$

where a is a scalar amplitude, as well as by its complex conjugate $\overline{\mathbf{u}}_1 = \overline{a}\overline{\mathbf{\Phi}}$. The eigenfunctions of the adjoint operator $\mathbf{L}_0^\dagger = -\mathbf{\Theta}_0^\dagger\partial_0 + \mathbf{D}_0^\dagger\nabla_0^2 + \mathbf{f}_\mathbf{u}{}^\dagger$ are $\mathbf{\Phi}^\dagger = \mathbf{U}^\dagger\mathrm{e}^{\mathrm{i}(\omega t_0 + \mathbf{k}\cdot\mathbf{x})}$ and its complex conjugate normalized by the condition $\mathbf{\Phi}^\dagger\cdot\mathbf{\Theta}_0\cdot\overline{\mathbf{\Phi}} = 1$.

Applying also the expansion (1.1.21) of the time and space derivatives, we arrive in the second order at the inhomogeneous equation

$$\mathbf{L}_0\mathbf{u}_2 = \mathbf{L}_1\mathbf{u}_1 + \tfrac{1}{2}a^2\mathbf{f}_{\mathbf{uu}}[\mathbf{U},\mathbf{U}]\mathrm{e}^{2\mathrm{i}(\omega t_0 + \mathbf{k}\cdot\mathbf{x})} + \mathrm{c.c} + |a|^2\mathbf{f}_{\mathbf{uu}}[\mathbf{U},\overline{\mathbf{U}}], \tag{1.3.24}$$

where c.c. stands for the complex conjugate of the preceding expression. The inhomogeneity of this equation contains both the principal harmonic contributed by the linear term with the operator[5]

[5] The long-scale dependence is applicable to the pitchfork and Hopf bifurcations and should be corrected to symmetry-breaking bifurcations – see Sect. 1.3.4

$$\mathbf{L}_1 = -\boldsymbol{\Theta}_0\partial_1 - \mathrm{i}\omega\boldsymbol{\Theta}_1 + \mathbf{D}_1\nabla_0^2 \tag{1.3.25}$$

and double-harmonic and constant components stemming from the quadratic term. The scalar products of the latter with the eigenfunction $\boldsymbol{\Phi}^\dagger(t_0)$ and its complex conjugate vanish. The solvability condition of (1.3.24) requires therefore just vanishing of the linear harmonic term, and is trivially satisfied when first-order parametric deviations $\boldsymbol{\Theta}_1, \mathbf{D}_1$ vanish and $\partial_1 a = 0$, so that the amplitude evolves only on a still slower scale t_2.

The second-order function $\mathbf{u}_2$ has to be found by solving the remaining part of (1.3.24). The solution is

$$\begin{aligned}\mathbf{u}_2 &= |a|^2\mathbf{V}_0 + \tfrac{1}{2}a^2\,\mathrm{e}^{2\mathrm{i}\omega t_0}\mathbf{V}_2 + \text{c.c.}, \qquad (1.3.26)\\ \mathbf{V}_0 &= -\mathbf{f_u}^{-1}\cdot\mathbf{f_{uu}}[\mathbf{U},\overline{\mathbf{U}}], \quad \mathbf{V}_2 = \left(\mathrm{i}\omega\boldsymbol{\Theta}_0 + D_0k^2 - \mathbf{f_u}\right)^{-1}\cdot\mathbf{f_{uu}}[\mathbf{U},\mathbf{U}].\end{aligned}$$

The third-order equation is

$$\mathbf{L}_0\mathbf{u}_3 = (\mathbf{D}_0\nabla_1^2 - \boldsymbol{\Theta}_2\partial_0)\mathbf{u}_1 + \mathbf{f_{uu}}[\mathbf{u}_1,\mathbf{u}_2] + \tfrac{1}{6}\mathbf{f_{uuu}}[\mathbf{u}_1,\mathbf{u}_1,\mathbf{u}_1]. \tag{1.3.27}$$

The amplitude equation is obtained as the solvability condition of this equation. Only the part of the inhomogeneity containing the principal harmonic contributes to the solvability condition, which takes the form

$$\dot{a} = \eta\nabla^2 a + \mu a - \nu|a|^2 a. \tag{1.3.28}$$

Here the dot stands for the time derivative on the extended scale, and the scaling indices are suspended. The coefficients, following from (1.3.27), are

$$\begin{aligned}\eta &= \overline{\mathbf{U}}^\dagger\cdot\mathbf{D}_0, \quad \mu = -\overline{\mathbf{U}}^\dagger\cdot(\mathbf{D}_2k^2 + \mathrm{i}\omega\boldsymbol{\Theta}_2),\\ \nu &= -\overline{\mathbf{U}}^\dagger\cdot\left(\tfrac{1}{2}\mathbf{f_{uuu}}[\mathbf{U},\mathbf{U},\overline{\mathbf{U}}] + \mathbf{f_{uu}}[\mathbf{U},\mathbf{V}_0] + \mathbf{f_{uu}}[\overline{\mathbf{U}},\mathbf{V}_2]\right). \qquad (1.3.29)\end{aligned}$$

All coefficients are real when the emerging pattern is stationary (at $\omega = 0$) and complex in the case of a Hopf or wave bifurcation at $\omega \neq 0$. More convoluted expressions are obtained when appropriately scaled deviations of parameters of the algebraic functions are included; in particular, the parameter μ is supplemented by the term $\overline{\mathbf{U}}^\dagger\cdot\mathbf{f_{up}}[\mathbf{U},\mathbf{p}_2]$ dependent on their second-order deviations from the bifurcation point.

The general form of the amplitude equation resulting from this procedure could have been predicted by scaling and symmetry considerations: it is, indeed, identical, upon rescaling, to (1.1.6), and in the case of a stationary pattern further reduces to (1.1.7), with all coefficients eliminated by rescaling. Therefore we expect it to be applicable to bifurcations in systems differing from RDS, including those possessing vertical structure, like convective systems of Sect. 1.4, even though computing its coefficients would be then far more difficult. For RDS, locating bifurcation points involves only elementary algebra, but it also becomes quite a cumbersome task as the number of variables increases, even for simplest models. Explicit amplitude equations for several simple reaction schemes are listed by Ipsen et al (2000).

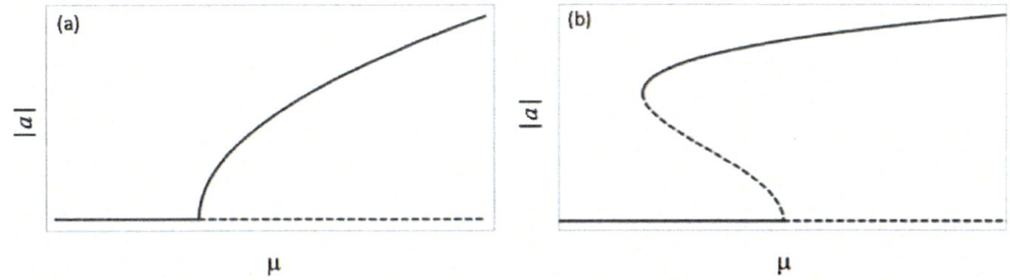

Fig. 1.6. Dependence of the real amplitude on a bifurcation parameter near a supercritical (a) and subcritical (b) bifurcations. Solid and dashed lines denote stable and unstable states, respectively

The amplitude equation (1.3.28) has smooth bounded solutions only at $\mathrm{Re}\,\nu > 0,\ \mathrm{Re}\,\eta > 0$. Then the bifurcation is *supercritical*, and the amplitude of an undistorted pattern at $\eta = 0$ depends on the parameter μ as in Fig. 1.6a. At $\mathrm{Re}\,\nu < 0$, the expansion can be continued to higher orders, after readjusting the scaling of slow time, extended length, and parametric deviations, near a parametric point where the real part of ν vanishes, which can be found by adjusting parameters of the matrices $\mathbf{D}_0$, $\mathbf{\Theta}_0$ or algebraic parameters of $\mathbf{f}(\mathbf{u})$. This leads to a quintic equation, on a slower $O(\epsilon^{-4})$ time scale and longer $O(\epsilon^{-2})$ spatial scale, describing dynamics in the vicinity of a subcritical bifurcation:

$$\dot{a} = \eta\nabla^2 a + \mu a - \nu|a|^2 a - \varrho|a|^4 a. \tag{1.3.30}$$

The bifurcation diagram at $\mathrm{Re}\,\varrho > 0,\ \mathrm{Re}\,\nu < 0$, has the form shown in Fig. 1.6b. Since both cubic and quintic amplitude equations retain the symmetry to phase shifts, all transitions depend only on real parts of their coefficients.

At $\mathrm{Re}\,\eta < 0$, (1.3.28) has no smooth solutions, and an amplitude equation quartic in ∇ on a *shorter* spatial scale extended by $\epsilon^{-1/2}$ can be derived near a parametric point with $\mathrm{Re}\,\eta = 0$:

$$\dot{a} = \left(\eta_0\nabla^4 + \eta_1\nabla^2 + \mu\right) a - \nu|a|^2 a, \tag{1.3.31}$$

where the coefficient η_1 is of $O(\epsilon)$. One can also combine (1.3.30), (1.3.31) to a quintic fourth-order equation.

Although the above derivations apply to bifurcations at both zero and imaginary eigenvalues, the differences are substantial. In the former case, when the coefficients are *real*, the amplitude equations possess a gradient structure. Thus, (1.3.28) is derivable then from the energy integral:

$$\dot{a} = -\frac{\partial \mathcal{F}}{\partial \bar{a}}, \qquad \mathcal{F} = \int \left(\eta\nabla a \cdot \nabla\bar{a} - \mu|a|^2 + \frac{\nu}{2}|a|^4\right) \mathrm{d}\mathbf{x}. \tag{1.3.32}$$

One can use (1.3.32) to compare the energy of different solutions. It is zero for the trivial one ($a = 0$), and should be negative for solutions corresponding to stable patterns. The gradient structure is retained in higher-order equations, with the energy integral supplemented by the term $\frac{1}{3}\varrho|a|^6$ for the quintic (1.3.30) and by the term $-\frac{1}{2}\eta_0\nabla^2 a\nabla^2\overline{a}$ for (1.3.31).

Although amplitude equations with real coefficients have been derived starting from a system far from equilibrium, the emergence of a gradient structure, absent in an original RDS, means that they behave near a bifurcation point similar to systems relaxing to equilibrium. The amplitude equations at $\omega \neq 0$, lacking a gradient structure, generate far more interesting and variegated dynamics (see Ch. 4).

1.3.4 Anisotropy of Long-Scale Spatial Dependence

The dependence on long-scale coordinates is substantially different in the case of symmetry-breaking bifurcations. It must have an *anisotropic* form, the source of anisotropy being the direction of the wave vector itself. The form of the spatial differential operator replacing the Laplacian is determined by the requirement that all directions of the wave vector $\mathbf{k}$ at a symmetry-breaking bifurcation in an isotropic system should be *equivalent*. Adding to a wave vector $\mathbf{k}_0$ a small longitudinal component, say, $\epsilon q_\parallel$ perturbs the wavenumber k by $O(\epsilon)$, while the same effect is caused by the transverse component $\epsilon^{1/2}q_\perp$, so that the overall correction is $q = \epsilon(q_\parallel + q_\perp^2)$.

This suggests that long-scale derivatives should be scaled *anisotropically*. Indeed, if the derivatives in the directions parallel and normal to k are scaled as $\partial_\parallel \sim \epsilon$, $\partial_\perp \sim \epsilon^{1/2}$, the equivalence of all directions is ensured by the operator $\heartsuit = \partial_\parallel + \mathrm{i}\partial_\perp$ acting on the scale extended by $O(\epsilon)$:

$$\heartsuit \exp\{\mathrm{i}[(k + \epsilon q_\parallel x_\parallel + \epsilon^{1/2}q_\perp x_\perp]\} = \mathrm{e}^{\mathrm{i}k}[1 + \mathrm{i}\epsilon(q_\parallel + q_\perp^2)]. \tag{1.3.33}$$

Accordingly, we expect that the applicable long-scale amplitude equation modifying (1.3.28) should read

$$\dot{a} = \eta\heartsuit^2 a + \mu a - \nu|a|^2 a. \tag{1.3.34}$$

with the coefficients defined as before by (1.3.29).

This amplitude equation was originally derived in the context of Rayleigh–Bénard convection (see Sect. 1.4.2), and is known as the Newell–Whitehead–Segel (NWS) equation (Newell and Whitehead, 1969; Segel, 1969). The coefficients are real near a Turing or Rayleigh bifurcation at zero eigenvalue, and can be rescaled to unity. However, it is applicable as well in the vicinity of a wave bifurcation, in which case the coefficients are complex. Similar to RGL, the NWS equations obeys the general equation of gradient dynamics (1.1.9) with the Lagrangians

$$\mathcal{L}_{\mathrm{NWS}} = \overline{\heartsuit u} \cdot \heartsuit u + \tfrac{1}{2}|u|^2(1 - \tfrac{1}{2}|u|^2). \tag{1.3.35}$$

The an anisotropic mixed-order differential operator $\heartsuit$ precisely accounts for the equivalence of all structures with identical wavenumbers, independently of the direction of the wave vector. However, the NWS equation is not covariant and is quite difficult to analyze or compute. Indeed, even in a weakly distorted pattern, the orientation of the coordinate axes in the operator $\heartsuit$ depends on the local phase gradient, so that this operator is in fact strongly nonlinear. The covariant formulation

(Pismen, 1989) in a coordinate frame aligned with the pattern is also ill-suited for computations.

Following the derivations in Sect. 1.3.3 and adding a parametric dependence as in Sect. 1.2.3, the coefficients of (1.3.34) can be computed rather easily for the FN system (1.3.9) with $\mathbf{\Gamma}$ rescaled to the unity matrix and $\mathbf{D}_0$ diagonal with the components $1, \gamma\kappa$. Applying the bifurcation conditions (1.3.11) leads to the Jacobi matrix at the bifurcation point

$$\mathbf{L}_0 = \begin{pmatrix} (1+\kappa)/\kappa & -(1+\kappa)^2/\kappa \\ \beta & -\frac{1}{2}(1+\kappa) \end{pmatrix}. \tag{1.3.36}$$

The eigenvectors with the zero eigenvalue of $\mathbf{L}_0$ and its transpose satisfying $\mathbf{U} \cdot \mathbf{U}^\dagger = 1$ are

$$\mathbf{U} = \left(\frac{1+\kappa}{2\beta}, 1 \right), \qquad \mathbf{U}^\dagger = \left(\frac{2\kappa\beta}{\kappa^2 - 1}, -\frac{1}{\kappa - 1} \right). \tag{1.3.37}$$

The only nonlinear term is $\mathbf{f}_{111} = -3$, so that the second-order function vanishes identically, and the coefficients of (1.3.34) are computed, keeping $\beta = \beta_0 + \epsilon^2\beta_2$ as the only perturbed parameter, as

$$\eta = \mathbf{U}^\dagger \cdot \mathbf{D}_0 \cdot \mathbf{U} = \frac{\kappa(\gamma - 1)}{\kappa - 1}, \quad \mu = \mathbf{U}^\dagger \cdot \begin{pmatrix} 0, & 0 \\ \beta_2, & 0 \end{pmatrix} \cdot \mathbf{U} = -\frac{\beta_2}{\beta_0} \frac{\kappa + 1}{2(\kappa - 1)},$$
$$\nu = -\frac{3}{2} \mathrm{U}_1^\dagger \mathrm{U}_1^3 = -24 \left(\frac{\beta_0 \kappa}{\kappa^2 - 1} \right)^4. \tag{1.3.38}$$

1.4 Convective Systems

1.4.1 Convection Driven by Heat or Mass Transfer

Fluid flows are commonly driven externally and are extremely sensitive to instabilities, eventually developing into hydrodynamic turbulence, the most fascinating phenomenon that likely will be never understood in all details necessary for such practical tasks as, say, weather forecast. Studies of hydrodynamic instabilities have a long history, and their detailed analysis is outside the framework of this book, but we will briefly review some simple cases when emerging flow patterns (*convection cells*) are supported by a uniform energy input from the "vertical" direction, along the coordinate z as in Fig. 1.1. Convective problems, unlike RDS, necessarily possess vertical structure, as in Fig. 1.1b, which makes their analysis far more difficult even in simplest settings, as it involves solving z-dependent ODEs.

The first experiment demonstrating spontaneous formation of convective patterns has been carried out by Michael Faraday (1831) who observed standing surface waves in a vertically oscillating fluid or sand layer. This is an example of *parametric waves* driven by a spatially uniform oscillatory input but generating patterns of convective cells and oscillatory motion with frequencies different from the driving frequency.

Leaving aside purely hydrodynamic instabilities, we concentrate here on convective patterns driven by heat or mass transfer. The most famous experiment in the formation of patterns of this kind has been carried out by Bénard (1900), who observed hexagonal convection cells in a thin layer of whale oil heated from below. The mechanism was explained by Lord Rayleigh: the light warm fluid raises upward, cools there, and descends. Rayleigh (1916) carried out linear stability analysis and computed the critical temperature difference across the layer, above which convection starts. The theory turned out, however, to be wrong in respect to Bénard's actual experiments, where convection patterns were caused by surface tension gradients driving the fluid along its free surface from locations warmed up by ascending currents to cooler patches. The convection patterns are similar, and the term Bénard–Marangoni convection distinguishes it from the gravity-driven Bénard–Rayleigh convection, which is more common, and is the only possible one when a fluid is confined between two solid plates.

Rayleigh's theory (Chandrasekhar, 1961, Gershuni and Zhukhovitsky, 1976, Normand et al, 1977) is based on the forced Navier–Stokes equation of a viscous incompressible fluid,

$$\partial_t \mathbf{v} + (\mathbf{v} \cdot \nabla)\mathbf{v} = -\rho^{-1}\nabla p + \nu\nabla^2\mathbf{v} - G\,\widehat{\mathbf{e}}_z, \quad \nabla \cdot \mathbf{v} = 0. \tag{1.4.1}$$

Here p is pressure, ρ is density, ν is kinematic viscosity, $\widehat{\mathbf{e}}_z$ is the unit vector in the vertical direction, and $G = g\alpha(T_0 - T)$ is the forcing term proportional to a temperature deviation from a reference level T_0; g is the acceleration of gravity and α is the heat expansion coefficient. The 3D velocity vector $\mathbf{v}$ will be further separated into the horizontal ($\mathbf{v}_\perp$) and vertical (w) velocity components. We will also distinguish between the horizontal and vertical coordinates, $\mathbf{x}$ and z, and the respective derivatives, $\nabla_\perp$ and ∂_z. The flow equation is coupled to the heat transport equation

$$\partial_t T + (\mathbf{v} \cdot \nabla)T = \kappa\nabla^2 T, \tag{1.4.2}$$

where κ is thermal diffusivity.

Common boundary conditions are of the two kinds. On a solid wall, a *no-slip* condition $\mathbf{v} = 0$ is applicable. Making use of the continuity equation $\nabla \cdot \mathbf{v} = \nabla_\perp \cdot \mathbf{v} + \partial_z w = 0$, it can be converted into boundary conditions involving only the vertical velocity: $w = \partial_z w = 0$. Rayleigh (1916) used instead the non-physical but computationally easier free-slip condition. On a free surface, as long as its deformation is neglected, only the vertical velocity vanishes, $w = 0$, but the horizontal velocity can be driven by a force applied along the interface, absent in Rayleigh's model. Bénard–Marangoni convection is caused by a linear dependence of surface tension σ on temperature, $\mathrm{d}\sigma/\mathrm{d}T = -\gamma$. In both cases, instability may emerge when the fluid is heated from below. Marangoni convection can be also caused by a concentration gradient along the free interface when a species with a lower surface tension is supplied from the bulk of the fluid.

Linear analysis provides a sufficient condition for *instability*. The alternative is the method of energy that provides a sufficient condition for *stability*. Unlike the former, the latter is not universally applicable, but works only when the problem under study is *self-adjoint* in an applicable functional space and can be defined by varying some energy functional. When the method of energy works, coincidence of both results testifies absence of subcritical instabilities. This is the case for the problem of Bénard–Rayleigh convection in its common formulation (Joseph, 1965). Contrariwise, for the Bénard–Marangoni convection the method of energy yields a lower limit (Davis, 1969), pointing out that a subcritical instability exists below the limit established by linear theory. Weakly nonlinear analysis of convective problems can be carried out, in principle, along the same lines as in Sect. 1.3.3, but is too cumbersome, since it necessarily involves integration over the vertical structure. The structure of the amplitude equations is identical to that for RDS, and the only practical consequence of these derivations, distinguishing between subcritical and supercritical instabilities, is proven by the method of energy when it is applicable.

1.4.2 Convection Driven by Buoyancy

When the boundary conditions at confining plates are identical, the problem is symmetric, and it is convenient to place them at $z = \pm d/2$. In a quiescent fluid, the temperature profile is linear, $\widehat{T} = T_0 - \Delta z/d$, where Δ is the temperature difference between the lower and upper plates. It is convenient to use as the dynamic variable, instead of temperature, the scaled temperature deviation $\theta = (T - \widehat{T})/T^*$, where the temperature scale is taken as $T^* = \kappa\nu/\alpha g d^3$. The physical meaning of this scaling can be understood by viewing T^*/d as the temperature gradient effectively overcoming the dissipative action of viscosity and diffusivity. Near the instability threshold, the advective term in (1.4.2) reduces to $w\Delta$ and is counteracted by the diffusive flux of the temperature perturbation. The commonly used thermal boundary conditions are $\theta(\pm d/2) = 0$, which presumes perfect control of temperature of the confining walls.

In the Boussinesq approximation, only the crucial buoyancy effect of thermal expansion is taken into account, so that the fluid is still viewed as incompressible. The dimensionless form of (1.4.1), (1.4.2) with length scaled by the layer thickness d and time, by d^2/ν reads

$$\partial_t \mathbf{v} + (\mathbf{v} \cdot \nabla)\mathbf{v} = -\nabla p + \nabla^2 \mathbf{v} + \theta\, \widehat{\mathbf{e}}_z, \tag{1.4.3}$$

$$\mathrm{Pr}\, \partial_t \theta + (\mathbf{v} \cdot \nabla)\theta = \mathrm{Ra}\, w + \nabla^2 \theta. \tag{1.4.4}$$

It contains two dimensionless parameters, Rayleigh and Prandtl numbers:

$$\mathrm{Ra} = \frac{\alpha g \Delta d^3}{\kappa \nu}, \qquad \mathrm{Pr} = \frac{\nu}{\kappa}. \tag{1.4.5}$$

Near the onset of convection, both velocity and temperature deviation are small, and nonlinear terms in (1.4.3), (1.4.4) should be omitted. Pressure can be eliminated by taking the curl of (1.4.3), thereby replacing velocity by vorticity $\boldsymbol{\omega} = \nabla \times \mathbf{v}$. Then the linearized hydrodynamic equation takes the form

$$\partial_t \boldsymbol{\omega} = \nabla^2 \boldsymbol{\omega} + \nabla \times (\theta\, \widehat{\mathbf{e}}_z). \tag{1.4.6}$$

Taking the curl again and applying the incompressibility condition to the vertical component of the resulting equation yields

$$\partial_t \nabla^2 w = \nabla^4 w + \nabla_\perp^2 \theta, \tag{1.4.7}$$

where $\nabla_\perp$ denotes the gradient in the horizontal plane. The thermal equation (1.4.4) becomes

$$\Pr \partial_t \theta = \mathrm{Ra}\, w + \nabla^2 \theta. \tag{1.4.8}$$

As in Sect. 1.1.3, we are looking for a solution in the spectral form (1.1.15) with $\mathbf{k}, \mathbf{x}$ defined in the horizontal plane and the vertical coordinate singled out. Denoting the deviations by the array $\phi = \{w, \theta\}$ and their spectral amplitudes as $\boldsymbol{\Phi} = \{W(z), \Theta(z)\}$ leads to the linearized system in the explicit form

$$\lambda(\partial_z^2 - k^2)W = (\partial_z^2 - k^2)^2 W - k^2 \Theta, \tag{1.4.9}$$

$$\lambda \Pr \Theta = \mathrm{Ra}\, W + (\partial_z^2 - k^2)\Theta. \tag{1.4.10}$$

The linearized equation system with no-slip boundary conditions is symmetric and self-adjoint in an appropriate function space (Chandrasekhar, 1961). Therefore eigenvalues λ must be real, and the bifurcation takes place at $\lambda = 0$. Then the Prandtl number falls out, and the critical value of the Rayleigh number is defined by the condition of existence of nontrivial solutions of the reduced linear system

$$(\mathrm{d}_z^2 - k^2)^2 W = k^2 \Theta, \qquad (k^2 - \mathrm{d}_z^2)\Theta = \mathrm{Ra}\, W. \tag{1.4.11}$$

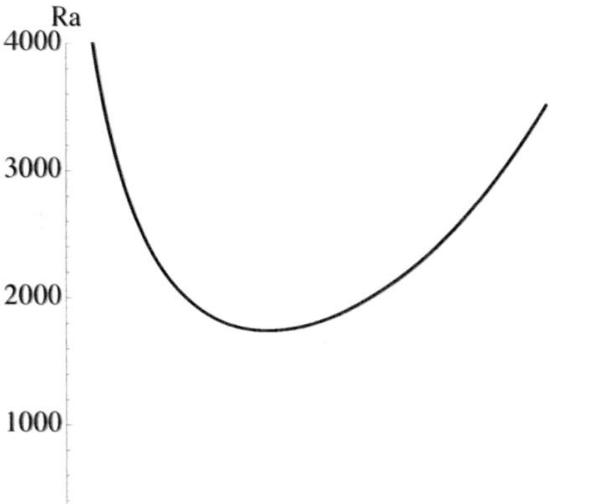

Fig. 1.7. The approximate dependence of the critical Rayleigh number on the wavenumber k

This is a six-order problem, which has to satisfy three boundary conditions at each of the confining walls. In a layer bounded by solid walls with perfectly controlled temperature those are (see Sect. 1.4.1) $W = \Theta = \partial_z W = 0$ at $z = \pm 1/2$. The analytical solution is cumbersome, as it involves a combination of exponents $\mathrm{e}^{\pm r_i z}$, where r_i are roots of the equation $(r^2 + k^2)^3 - k^2 \mathrm{Ra} = 0$, solved by $r^2 = -k^2 + (k^2 \mathrm{Ra})^{1/3} \iota$, where ι is one of the three cubic roots of 1. The resulting dependence $\mathrm{Ra}(k)$ has a minimum at $k_0 \approx 3.12$, $\mathrm{Ra}_0 \approx 1708$. The characteristic length of incipient convection rolls π/k is close to unity, i.e., to the thickness of the convection cell. The approximate analytical expression (Manneville, 1990) is shown in Fig. 1.7:

$$\mathrm{Ra} = \frac{18}{17k^2}(k^2 + 10)(k^4 + 24k^2 + 504). \tag{1.4.12}$$

The amplitude equation at the bifurcation point is the NWS equation (1.3.34). The coefficients were computed numerically by Cross (1980). Oscillatory convection may emerge in binary fluid mixtures, where the temperature and concentration fields interact through the Soret effect, the solute flux $\mathbf{j}_s = -k\nabla\theta$ driven by the temperature gradient and the coupled Dufour effect, which, however, is commonly omitted, as it is negligible in liquid mixtures. The amplitude equations at the oscillatory symmetry breaking instability were derived for the coupled flow, temperature and concentration equation in the Boussinesq approximation by Schöpf and Zimmermann (1993).

1.4.3 Convection Driven by Marangoni Flow

Alternatively, convection may be caused by the Marangoni effect: the temperature dependence of the surface tension σ on the free upper surface that induces surface flow proportional to the surface tension gradient. The same effect can be caused by the concentration gradient of a species diffusing from the bulk and evaporating from the free surface. Flow in the bulk of the layer obeys the same equation (1.4.3) but the gravitational term $\theta\,\widehat{\mathrm{e}}_z$ falls out when buoyancy is negligible, as it is common when convection is driven by a concentration rather than temperature gradient. To be definite, we will still talk of temperature, but the derivations equally apply to a concentration field.

In the absence of symmetry, it is more convenient to place the lower boundary at $z = 0$ and the upper boundary, viewed for the moment as flat, at $z = 1$. A free boundary is apt to deform. The depressed area warms up, as it comes closer to the heated lower boundary, and the depression strengthens due to the tangential Marangoni flow, so that runaway instability may lead to the formation of dry spots, unless counteracted by surface tension and gravity. At this stage we avoid these strong effects. Weak surface deformations near the instability threshold bring about only a small correction to the critical Marangoni number. However, they may induce an additional *long-scale* instability (Takashima, 1981; Golovin et al, 1994).

It is difficult in practice to maintain a fixed temperature at the free surface. Rather than fixing temperature, the thermal boundary condition has to balance the thermal flux $-\kappa\partial_z T$ and the heat loss to the gas phase $\beta T(d)$, where β is the heat transfer coefficient and the gas temperature is set to zero. The ratio of both fluxes is characterized by the Biot number $\mathrm{Bi} = \beta d/\kappa$. In a quiescent fluid with $T(0) = \Delta$, the temperature profile is linear, and it follows from the identity $T(d) = T(0)+d\partial_z T = -\kappa\partial_z T/\beta$ that the unperturbed temperature gradient is $\partial_z T = -B\Delta$ where $B = \mathrm{Bi}/(1+\mathrm{Bi})$. The dimensionless heat exchange condition on the free surface for the temperature perturbation is

$$\partial_z\theta = -\mathrm{Bi}\,\theta \equiv -\frac{B\theta}{1-B}. \tag{1.4.13}$$

The hydrodynamic boundary condition on the interface has to satisfy the tangential stress condition, which balances the Marangoni stress due to temperature dependence of surface tension, $-\gamma\nabla_\perp\theta$ and the tangential component of viscous stress tensor $\boldsymbol{\sigma}$. If viscosity of the gas phase is neglected, its elements are $\sigma_{ij} = \rho\nu(\partial_i v_j + \partial_j v_i)$, projecting on a flat interface as $\nabla_\perp w + \partial_z \mathbf{v}_\perp$. Applying the 2D divergence and making use of the continuity equation converts the tangential balance to the form involving

w only:

$$\rho\nu(\nabla_\perp^2 - \partial_z^2)w = \gamma\nabla_\perp^2\theta. \tag{1.4.14}$$

The dimensionless form of (1.4.14) with temperature scaled by Δ includes the Marangoni number Ma:

$$(\nabla_\perp^2 - \partial_z^2)w = \mathrm{Ma}\nabla_\perp^2\theta \quad \text{at} \quad z = 1, \qquad \mathrm{Ma} = \frac{\gamma\beta\Delta d}{\rho\nu^2}. \tag{1.4.15}$$

When buoyancy is neglected, stability analysis is simplified, since the thermal field affects the flow only through the boundary conditions on the free surface. After transforming to the spectral amplitudes W, Θ, (1.4.9) and (1.4.10) are replaced by

$$\lambda(\partial_z^2 - k^2)W = (\partial_z^2 - k^2)^2W, \tag{1.4.16}$$
$$(\lambda\,\mathrm{Pr} + k^2 - \partial_z^2)\Theta = BW. \tag{1.4.17}$$

The boundary conditions are

$$\partial_z\Theta + \frac{B\Theta}{1-B} = 0 \quad \text{at } z = 1, \tag{1.4.18}$$
$$\partial_z^2W + \mathrm{Ma}\nabla_\perp^2\Theta = 0 \quad \text{at } z = 1, \tag{1.4.19}$$
$$W(1) = 0, \quad W = \Theta = \partial_zW = 0 \quad \text{at } z = 0. \tag{1.4.20}$$

For the bifurcation to a stationary pattern, the solution of the bulk flow equation $(\partial_z^2 - k^2)^2W = 0$ is readily expressed as

$$W(z) = a[(1 + qz)\sinh kz - kz\cosh kz], \quad q = k\coth k - 1, \tag{1.4.21}$$

containing an indefinite amplitude a. The thermal field defined by (1.4.17) and satisfying $\Theta(0) = 0$ is expressed as

$$\Theta(z) = \left[b + \frac{aB}{8k^2}(3 + 2qz + 2k^2z^2)\right]\sinh kz - \frac{aBz}{4k}(3 + qz)\cosh kz. \tag{1.4.22}$$

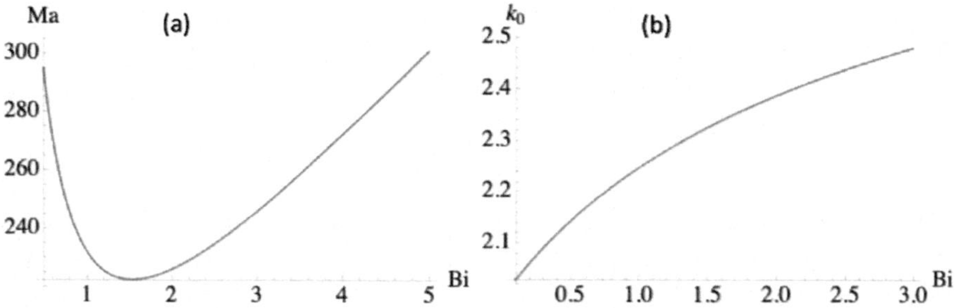

Fig. 1.8. The dependence of the critical Marangoni number Ma (a) and the critical wavelength k_0 (b) on the Biot number Bi

Eliminating the integration constant b with the help of the thermal boundary condition (1.4.15) leads to an expression proportional to the arbitrary amplitude a:

$$\begin{aligned}\Theta(z) &= \frac{aB}{4k^3}\left[\frac{F(z)}{(B-1)k\cosh k - B\sinh k} - qz(3+qz)\cosh kz\right],\\ F(z) &= q[(B-1)z-1] + (B-1)k^2(z^2-1) - 3\}k\cosh k\\ &\quad + [q - Bqz + k^2(B(2+q-z^2) - 1 - q)]\sinh k \sinh kz. \end{aligned} \tag{1.4.23}$$

Then (1.4.19) is resolved to obtain the value of the Marangoni number at the instability threshold as a function of the wavenumber k:

$$\mathrm{Ma} = \frac{4k(1+\mathrm{Bi})(k\,\mathrm{Bi}\cosh k + \sinh k)(\sinh 2k - 2k)}{\mathrm{Bi}(\sinh^3 k - k^3\cosh k)}. \tag{1.4.24}$$

The function $\mathrm{Ma}(k)$ passes a minimum at some $k = k_0$, which defines the critical point and the wavelength of the incipient pattern. The dependences of k_0 and of the critical Marangoni number $\mathrm{Ma}(k_0)$ on the Biot number Bi are shown in Fig. 1.8.

1.5 Dynamics of Planforms

1.5.1 Interaction of Degenerate Modes

Symmetry-breaking bifurcations are degenerate due to spatial symmetries. In an isotropic system, an arbitrary number of differently directed modes with k = idem and identical vectors $\mathbf{U}$ can be excited beyond the bifurcation point. A combination of these modes can give rise to a variety of distinct *planforms*. Competition among the modes that determines the pattern selection is governed by nonlinear interactions.

Interaction coefficients between different modes for a general RDS (1.2.5) can be computed following the same procedure as in Sect. 1.3.3. The second-order equation (1.3.24) (with the first-order operator $\mathbf{L}_1$ now omitted) acquires then additional terms, extending to

$$\begin{aligned}\mathbf{L}\mathbf{u}_{i2} &= |a_i|^2\mathbf{f_{uu}}[\mathbf{U},\overline{\mathbf{U}}] + \tfrac{1}{2}a_i^2\mathbf{f_{uu}}[\mathbf{U},\mathbf{U}]\mathrm{e}^{2\mathrm{i}(\mathbf{k_i}\cdot\mathbf{x}+\omega t_0)} \\ &\quad + a_i\bar{a}_j\mathbf{f_{uu}}[\mathbf{U},\overline{\mathbf{U}}]\mathrm{e}^{\mathrm{i}(\mathbf{k}_i-\mathbf{k}_j)} + a_i a_j\mathbf{f_{uu}}[\mathbf{U},\mathbf{U}]\mathrm{e}^{\mathrm{i}(\mathbf{k}_i-\mathbf{k}_j)+2\omega t_0]} + \text{c.c.}\end{aligned} \tag{1.5.1}$$

This adds to the second-order function mixed terms $a_i\bar{a}_j\mathbf{V}_{0,ij}\mathrm{e}^{\mathrm{i}(\mathbf{k}_i-\mathbf{k}_j)\cdot\mathbf{x})}$ + c.c., $a_i a_j\mathbf{V}_{2,ij}\mathrm{e}^{\mathrm{i}[2\omega t+(\mathbf{k}_i+\mathbf{k}_j)\cdot\mathbf{x}]}$ + c.c., where

$$\begin{aligned}\mathbf{V}_{0,ij} &= \left(\mathbf{D}_0|\mathbf{k_i}-\mathbf{k_j}|^2 - \mathbf{f_u}\right)^{-1}\cdot\mathbf{f_{uu}}[\mathbf{U},\overline{\mathbf{U}}],\\ \mathbf{V}_{2,ij} &= \left(2\mathrm{i}\omega\mathbf{\Gamma}_0 + \mathbf{D}_0|\mathbf{k}_i+\mathbf{k}_j|^2 - \mathbf{f_u}\right)^{-1}\cdot\mathbf{f_{uu}}[\mathbf{U},\mathbf{U}].\end{aligned} \tag{1.5.2}$$

The coefficients η, μ, as well as the self-interaction coefficient ν in the amplitude equations obtained in the third order are the same as in (1.3.29), and are identical for all modes, but they are supplemented now by cross-interaction terms. The evolution equations for the amplitudes a_i have the general form

$$\dot{a}_i = \eta \heartsuit^2 a_i + \mu a_i - \nu |a_i|^2 a_i - a_i \sum_{i \neq j} \nu_{ij} |a_j|^2, \tag{1.5.3}$$

$$\nu_{ij} = -\overline{\mathbf{U}}^\dagger \left(\mathbf{f_{uuu}}[\mathbf{U}, \mathbf{U}, \overline{\mathbf{U}}] + \mathbf{f_{uu}}[\mathbf{U}, \mathbf{V}_0 + \mathbf{V}_{0,ij}] + \mathbf{f_{uu}}[\overline{\mathbf{U}}, \mathbf{V}_2 + \mathbf{V}_{0,ij}] \right).$$

All coefficients ν_{ij} should have a positive real part to prevent runaway to infinity.

In the case of a bifurcation at zero (but not imaginary) eigenvalue, the system (1.5.3) has a gradient structure, and is derivable from the energy integral

$$\mathcal{F} = \int \sum_i \left(\eta \heartsuit a_i \cdot \heartsuit \overline{a}_i - \mu |a_i|^2 \mathrm{d}\mathbf{x} + \frac{\nu}{2} |a_i|^4 \right) \mathrm{d}\mathbf{x} + \frac{1}{2} \int \sum_{i \neq j} \nu_{ij} |a_i|^2 |a_j|^2 \mathrm{d}\mathbf{x}. \tag{1.5.4}$$

One can use (1.5.4) to compare the energy of different solutions. It is zero for the trivial one (all $a_j = 0$), and should be negative for solutions corresponding to stable patterns.

The amplitude equations applicable to regular patterns (with the term $\eta \heartsuit^2 a_i$ suspended) effectively turn underlying PDEs into dynamical systems for the amplitudes of constituent modes. The preferred number of modes depends on the relative strength of nonlinear suppression by self- and cross-interactions. Suppose we start with a single-mode solution. Its stationary amplitude defined by (1.3.28) is $\rho_1 = \mu/\nu$. The equation for the amplitude of another mode, viewed as an infinitesimal perturbation of the single-mode solution, is

$$\dot{\rho}_2 = \mu \rho_2 \left(1 - \nu_{12}/\nu\right). \tag{1.5.5}$$

Clearly, the one-mode state is stable when the cross-interaction coefficient is larger by its absolute value than the self-interaction coefficient. For a stationary pattern, this is confirmed by comparing the energy of one- and two-mode solutions. In the case cross-mode suppression is weaker, the preferred number of modes will be affected by the angular dependence of $\mathbf{V}_{0,ij}$ and $\mathbf{V}_{1,ij}$.

Unfortunately, analytical expression for the interaction coefficients are too involved and lack transparency even in simplest cases, but they can be checked numerically when all parameters of a problem are given. They depend on the angle between interacting modes and are typically more suppressive than self-interactions. Take, for example, two modes forming a small angle. Then the array $\mathbf{V}_{0,ij}$ will be close to $\mathbf{V}_0$ when the modes are almost parallel or to $\mathbf{V}_2$ when they are almost antiparallel; likewise, $\mathbf{V}_{1,ij}$ will be close to either $\mathbf{V}_2$ or $\mathbf{V}_0$. Comparing (1.5.3) with (1.3.29) we see that the coefficient ν_{ij} should be close to 2ν. Clearly, such a mode will never be excited. There are more chances that an additional mode will be directed at a larger angle, e.g., $\pi/2$, forming a regular quadratic pattern, but in any case combinatorics works to enhance mutual suppression. Newell and Pomeau (1993) imagined a "turbulent crystal" formed when all possible degenerate modes are excited, having the same amplitudes but arbitrary fluctuating phases. Sadly, this nice idea is unrealistic, since alternative modes tend to be mutually damping, unless they are *in resonance* (see the next subsection).

Interaction of degenerate modes is modified in the case of bifurcation of wave patterns with $\omega, \mathbf{k} \neq 0$. The difference is in the presence of a non-vanishing *group velocity* $\mathbf{c} = \partial_{\mathbf{k}}\omega$. It is irrelevant in the case when a single wave is excited, since the amplitude equation remains then applicable in the moving frame, but this is impossible when, due to degeneracy, waves propagating in the different directions (e.g., opposite directions in 1D) are excited. When two waves are present, each wave, viewed in its own moving frame, samples the *average* amplitude of its counterpart propagating in this frame with a fast speed, and wave interactions are expected to differ from the standard situation described above.

Generically, $|\mathbf{c}| = O(1)$, and the advective term $(\mathbf{c}_j - \mathbf{c}_i) \cdot \heartsuit a_i$ would dominate the amplitude equations (1.5.3), unless on a manifold of a higher codimension in the parametric space. Stability of standing waves in this special case was analyzed in detail by Riecke and Kramer (2000). Amplitude equations with $\mathbf{c} = 0$ arise in other applications where $u_\pm$ are identified as intensities of two components of a light beam with opposite circular polarizations or densities of two mixed superfluids or Bose–Einstein condensates (Pismen, 1999; Amengual et al, 1997).

Knobloch and de Luca (1990) overcame the scale mismatch problem for counter-propagating waves in 1D (marked by the indices $\pm$) by introducing an intermediate time scale t_1 extended by ϵ, and $O(\epsilon^2)$ amplitude corrections $b_\pm$ to the amplitudes $a_\pm$. Setting in the amplitude equations (1.5.3) $a_\pm \to a_\pm + \epsilon b_\pm$ and transforming the equations of both modes to the frames $z_\pm = t_1 \mp cx$ moving with the respective group velocities brings them to the form

$$2c\,\partial_{z_\pm} b_\pm = -\dot{a}_\pm + \eta\partial_x^2 a_\pm + \mu a_\pm - \nu|a_\pm|^2 a_\pm - \nu_\pm|a_\mp|^2 a_\pm. \tag{1.5.6}$$

The orders of magnitude are balanced here, and the amplitude correction is eliminated by integrating over the propagation direction, leading to the amplitude equations in the standard form (1.3.28) but with the renormalized linear growth coefficient, $\mu \to \mu + \nu_\pm\langle|a_\mp|^2\rangle$, which retain only *global* coupling carried by the spatial average of the alternative mode. This derivation can be extended to noncollinear modes in 2D. The result is trivial if the two modes are identical, but is relevant for interactions of modulated waves, as will be detailed in Sect. 5.2.5.

1.5.2 Resonant Interactions

The multiscale formalism chooses in each order of the expansion combinations of modes satisfying the *resonance* conditions. The product of amplitudes $\overline{a}_j$, $\overline{a}_k$, etc. may appear in the equation for the amplitude a_i if the respective wave vectors $\mathbf{k}_j, \mathbf{k}_l, \ldots$ add up to $\mathbf{k}_i$. In other words, the vectors $-\mathbf{k}_i, \mathbf{k}_j, \mathbf{k}_l, \ldots$ should form a polygon. This condition ensures that the modes in question are in resonance. It is satisfied in the third order of the bifurcation expansion for any two amplitudes a_i, a_j and their complex conjugates, since the respective wave vectors form a parallelogram. Otherwise, the product of modes rapidly oscillates and is averaged out to zero.

A special case is the resonance encountered in the *second* order of the expansion between *stationary* modes with wave vectors forming a regular triangle. The quadratic nonlinearity contains a harmonic with the basic wavenumber k if the excited modes include a triplet with the wave vectors directed at the mutual angle $2\pi/3$. Their amplitudes a_j, $j = \{0, \pm 1\}$ will be further numbered modulo three. Then retaining first-order deviations in (1.5.1) and collecting resonant terms leads to the solvability condition[6] of the second-order equation on the $O(\epsilon^{-1})$ time scale:

$$\dot{a}_j = \mu a_j + \kappa \bar{a}_{j-1} \bar{a}_{j+1}, \tag{1.5.7}$$

where, according to (1.5.1), $\mu = k^2 \mathbf{U}^\dagger \cdot \mathbf{DU}$ and $\kappa = \mathbf{U}^\dagger \cdot \mathbf{f_{uu}}[\mathbf{U}, \mathbf{U}]$. These expressions may be modified when parametric deviations are included, as in (1.2.14). The quadratic term could vanish either under special conditions (at a bifurcation manifold of a higher codimension) or due to inversion symmetry, like, in particular, in the classical problem of Rayleigh–Bénard convection (Sect. 1.4.2) with symmetric boundary conditions. The stationary space-independent solution of (1.5.7) with all three modes excited with equal real amplitudes ρ_j is a regular *hexagonal* pattern, alternative to the singe-mode *striped* pattern.

Presenting the complex amplitudes in the polar form $a_j = \rho_j \exp(\mathrm{i}\theta_j)$ and separating the real and imaginary parts reduces three complex equations (1.5.7) to *four* real equations, since the phases appear only in the combination $\Theta = \sum \theta_j$:

$$\dot{\rho}_j = \rho_j \left[\mu - \rho_j^2 - \nu\left(\rho_{j-1}^2 + \rho_{j+1}^2\right)\right] + \kappa \rho_{j-1}\rho_{j+1} \cos\Theta, \tag{1.5.8}$$

$$\dot{\Theta} = -\sin\Theta \sum_j \frac{\rho_{j-1}\rho_{j+1}}{\rho_j}. \tag{1.5.9}$$

The remaining two phase degrees of freedom are translational Goldstone modes, since shifting phases in an arbitrary way while keeping their sum constant is equivalent to shifting the entire pattern in the plane. However, the total phase Θ is dynamically important, since it determines the action of the quadratic term in (1.5.8). According to (1.5.9), Θ relaxes to zero when $\kappa > 0$, thereby rendering the quadratic term destabilizing. If, on the opposite, $\kappa < 0$, the stable equilibrium is $\Theta = \pi$, and the term is destabilizing again. This is why the hexagonal pattern comprising three resonant modes always bifurcates subcritically, and why hexagonal planforms are ubiquitous in 2D pattern-forming systems (Palm, 1960).

Since the quadratic term is always destabilizing, cubic terms are necessary to arrive at a stable finite-amplitude solution. In order to obtain a cubic amplitude equation with balanced orders of magnitudes, the coefficient κ should be assumed small, $\kappa \sim O(\epsilon)$. Deviations of diffusivities should be again reduced to $O(\epsilon^2)$, and the evolution on the $O(\epsilon^{-1})$ time scale frozen, as in the preceding subsection. In order to compute cross-interaction coefficients, we need again to solve the second-order equation (1.5.1), where the two-mode combinations with $a_j \bar{a}_{j\pm 1}$ contribute harmonics with the wavenumber $\sqrt{3}k$. The terms containing $a_{j+1}a_{j-1}$ and their complex conjugates produce resonant harmonics, and their resonant contribution must be small to suppress the quadratic term in this order. They may, however, contribute an $O(\epsilon^2)$ term containing arrays $\mathbf{V}_{1,j}$ satisfying $\mathbf{U}^\dagger \cdot \mathbf{V}_{1,j} = 0$. Thus, the solution of the second-order equation is

[6] Long-scale spatial modulation is suspended here, to be discussed in Sect. 4.5.1

$$\mathbf{u}_2 = \sum_j \Big(|a_j|^2 \mathbf{V}_{0,j} + a_j^2 \mathbf{V}_{2,j} e^{2i\mathbf{k}_j\cdot\mathbf{x}} + a_{j+1}a_{j-1}\mathbf{V}_{1j} e^{-i\mathbf{k}_j\cdot\mathbf{x}}$$
$$+ a_j \bar{a}_{j\pm1} \mathbf{V}_{3,j} e^{-i(\mathbf{k}_j - \mathbf{k}_{j\pm1})\cdot\mathbf{x}} + \text{c.c.} \Big), \tag{1.5.10}$$

where $\mathbf{V}_{0,j}$, $\mathbf{V}_{2,j}$ are defined as in (1.3.26) and $\mathbf{V}_{3,j} = (3k^2\mathbf{D}_0 - \mathbf{f_u})^{-1} \cdot \mathbf{f_{uu}}[\mathbf{U},\mathbf{U}]$. The arrays $\mathbf{V}_{k,j}$ are real and, by symmetry, identical for all modes, but the indices are retained to trace their contributions to the resulting third-order amplitude equations:

$$\dot{a}_j = a_j \left[\mu - \nu_j |a_j|^2 - \nu_{j\pm} \left(|a_{j-1}|^2 + |a_{j+1}|^2 \right) \right] + \kappa \bar{a}_{j-1} \bar{a}_{j+1}, \tag{1.5.11}$$
$$\nu_j = -\mathbf{U}^\dagger \left(\tfrac{1}{2} \mathbf{f_{uuu}}[\mathbf{U},\mathbf{U},\mathbf{U}] + \mathbf{f_{uu}}[\mathbf{U},\mathbf{W}_j] \right), \tag{1.5.12}$$
$$\mathbf{W}_j = \mathbf{V}_{0,j} + \mathbf{V}_{2,j} + \mathbf{V}_{1,j+1} + \mathbf{V}_{1,j-1} + \mathbf{V}_{3,j+1} + \mathbf{V}_{3,j-1},$$
$$\nu_{j\pm} = -\mathbf{U}^\dagger \left(\mathbf{f_{uuu}}[\mathbf{U},\mathbf{U},\mathbf{U}] + \mathbf{f_{uu}}[\mathbf{U},\mathbf{W}_{j\pm}] \right), \tag{1.5.13}$$
$$\mathbf{W}_{j\pm} = \mathbf{V}_{0,j\pm} + \mathbf{V}_{2,j\pm} + +\mathbf{V}_{1,j} + \mathbf{V}_{1,j\mp1} + \mathbf{V}_{3,j} + \mathbf{V}_{3,j\mp1}.$$

Take note of the contributions to the self-interaction coefficient ν_j additional to (1.3.29). All interaction coefficients should be positive.

The system (1.5.11), like (1.5.3), has a gradient structure, and is derivable from the energy integral $\mathcal{F} = \int \mathcal{L} d\mathbf{x}$, with the potential

$$\mathcal{L} = \frac{1}{2} \sum_j \left(-\mu |a_j|^2 + \frac{\nu}{2} |a_j|^4 \right) + \frac{\nu_{j\pm}}{2} \sum_{j\neq k} |a_j|^2 |a_k|^2 - \kappa (a_0 a_1 a_2 + \text{c.c.}). \tag{1.5.14}$$

The general interaction potential, including any number of modes and valid as well in higher dimensions is

$$\mathcal{L} = -\frac{\mu}{2} \sum |a_j|^2 - \kappa \sum a_i a_j a_k + \sum \nu_{ijkl} a_i a_j a_k a_l + \cdots \tag{1.5.15}$$
$$= -\frac{\mu}{2} \sum \rho_j^2 - \kappa \sum \rho_i \rho_j \rho_k \cos\Theta_{ijk} + \sum \nu_{ijkl} \rho_i \rho_j \rho_k \rho_l \cos\Theta_{ijkl} + \cdots,$$

where the summation is carried out over all closed polygons formed by the wave vectors of extant modes, and $\Theta_{ijk} = \theta_i + \theta_j + \theta_k$ and Θ_{ijkl} are sums of phases around the respective triangles or rhombi. This representation is most transparent, since every term can be associated with a triangle or rhombus formed by the respective wave vectors.

All cubic terms in (1.5.15) are based on identical equilateral triangles, which may differ only by orientation; therefore they are universal and all come with the same coefficient. The cubic terms are phase-dependent, and, as observed above, at $\kappa > 0$ the energy is minimal when the sum of the phases of modes forming an equilateral triangle is zero or (modulo 2π). When phases adjust in this way, the triplet interactions are destabilizing; this is why symmetry-breaking transitions leading to the formation of stationary patterns are always subcritical (of the first order). The emerging pattern is stabilized by quartic terms, which depend on angles between interacting modes and are specific for particular systems. The quartic terms are phase-dependent only when the rhombus formed by the respective wave vectors is nonplanar.

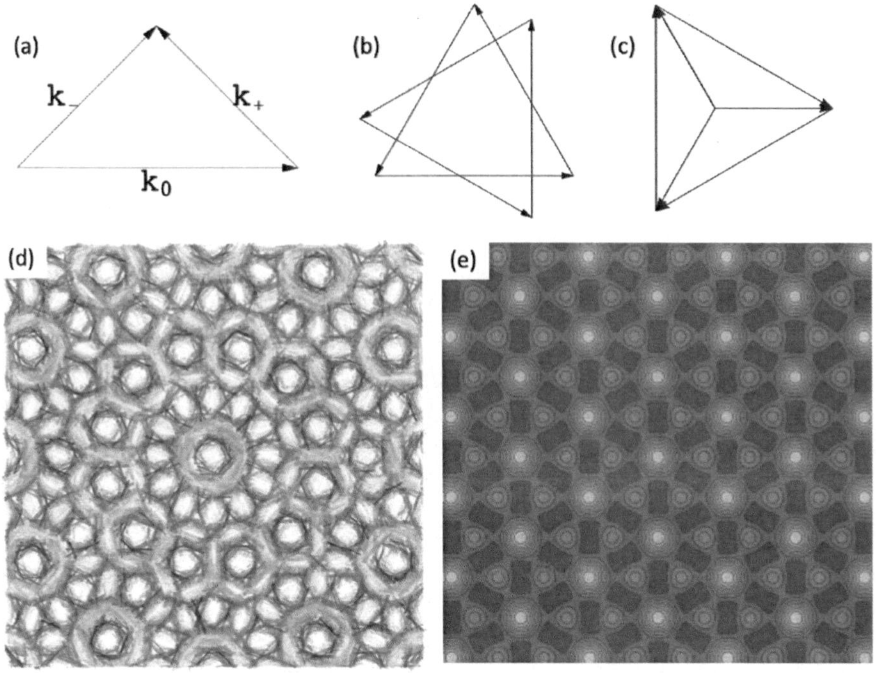

Fig. 1.9. (a) An isosceles resonance. (b) Two resonant triangles. (c) Resonant triangles formed by two triplets of modes with the scale ratio $\sqrt{3}$. (d) Dodecagonal quasicrystalline pattern. (e) Hexagonal superlattice.

As noted in the end of Sect. 1.3.3, stationary patterns behave in the same way as crystalline solids, and the same triple resonance determines the preferred structure of crystals (Alexander and McTague, 1978). In three dimensions, triplet interactions favor a combination of modes forming one of regular polyhedra with triangular faces – tetrahedron, octahedron, or icosahedron. The former two correspond to the same body-centered cubic crystalline lattice, and the latter, to a quasicrystalline structure. One could expect that these structures are likely to have the lowest energy close to the bifurcation point, i.e., crystallization, especially in the subcritical region, in the same way as the hexagonal structure is preferable in 2D. Since the pattern boosted by the resonant terms can be stabilized only by non-universal higher-order interactions, other structures can be actually chosen by the evolving system, and both experiment and numerical simulations show that the face-centered cubic structure is usually favored (Groh and Mulder, 1999). The special character of the icosahedral combination of modes may be responsible for the formation of quasicrystals with pentagonal symmetry (Shechtman et al, 1984), which was once seen as a sensational discovery contradicting the dogma of classical crystallography.

In two dimensions, a most likely (but still exotic) possibility is a "dodecagonal" quasicrystalline pattern formed by a combination of two resonant triangles turned by

$\pi/6$ relative to each other, as shown in Fig. 1.9b (Malomed et al, 1989). A piece of such a six-mode pattern, comprising almost symmetric dodecagonal and pentagonal "flowers" but never repeating itself, is shown in Fig. 1.9d. Fixing two phase relations to satisfy resonance conditions in the two resonant triangles still leaves four phase degrees of freedom. Thus, in addition to the two translational modes, there are two Goldstone "phasons" modifying the pattern in a nontrivial way. The quasicrystalline structure may be lost when some constituent modes are suppressed by higher-order interactions. Various transitions among possible patterns (one-mode striped, two-mode squared, three-mode hexagonal, and six-mode dodecagonal) in the six-mode system of amplitude equations, dependent on the angular dependence of stabilizing quartic interactions, were studied by Echebarria and Riecke (2001).

Other resonant structures may appear "accidentally" at degenerate bifurcations of a higher codimension when planforms with different wavenumbers k are excited simultaneously. This can be achieved in a most natural way in two-layer systems where the wavelength of the excited pattern depends on the thickness of each layer, as in Bénard–Rayleigh convection (Proctor and Jones, 1988), or on different diffusivities, as in a pattern-forming chemical system (Yang et al, 2002). More possibilities arise in nonlinear optics where spatial symmetry breaking may occur on different wavelengths at rather close values of a control parameter (Pampaloni et al, 1997). The resulting coupled amplitude equations can generate a variety of composite planforms. Proctor and Jones (1988) carried out extensive analysis of the 2:1 resonance, including semi-analytical construction of small-amplitude periodic orbits.

A resonant triangle may include either three Turing modes or one Turing and two wave modes with $\omega \neq 0$. In one dimension, the strongest resonance, appearing already in the second order of the bifurcation expansion, exists when the two wavelengths are in the 2:1 ratio, with the wave modes, if they are present, being the longer ones. Taking into account the necessity to fit the wavenumbers at two degenerate bifurcations to ensure resonance, the codimension of the bifurcation manifold rises to three. In two dimensions, no rigid fitting of wavenumbers is required, since the resonant modes can form an *isosceles* triangle (Fig. 1.9a).

A special case, also requiring exact fit of wavenumbers, is $\sqrt{3}$:1 resonance. Two resonant triplets with the wavenumber ratio $\sqrt{3}$ form three additional resonant triangles when the wave vectors of the two triplets are turned at $\pi/6$ relative to each other (Fig. 1.9c). Two relations connecting the phases of each triplet are fixed by resonance conditions around the two equilateral triangles. The three isosceles triangles give three more relations, only two of which are independent, leaving two translational degrees of freedom. Depending on the sign of the interaction coefficients, the phases around each resonant triangle should sum up to either 0 or π (modulo 2π). For each case, there are several combinations of phases satisfying these conditions, and there is a possibility that, due to mutual frustration, runaway to infinity may be prevented even without suppression by higher-order interactions. Superposition of all resonant modes yields a *superlattice* pattern, such as shown in Fig. 1.9e.

1.5.3 Stripes–Hexagons Competition

Possible stationary solutions of (1.5.11) in 2D are single-mode (striped) and triple-mode (hexagonal) patterns coming un two varieties, "up" and "down" hexagons differing by whether the phase of the quadratic term is 0 or π. All these patterns and their combinations have been observed experimentally (Fig. 1.10).

Rescaling time $t \to t/\kappa t$ and the amplitudes $a_j \to (\kappa\nu)^{-1/2} a_j$ brings (1.5.11) to the dimensionless form containing, besides μ, the rescaled cross-interaction coefficient $\sigma = \nu_3/\nu$:

$$\dot{a}_j = a_j \left[\mu - |a_j|^2 - \sigma\left(|a_{j-1}|^2 + |a_{j+1}|^2\right)\right] + \bar{a}_{j-1}\bar{a}_{j+1}, \tag{1.5.16}$$

The three amplitudes a_j are numbered, as before, modulo three. Self-interaction is assumed to be suppressing; otherwise, expansion to higher orders would be needed. The stationary amplitude of the hexagonal pattern can be computed after setting $\Theta = 0$ and moduli $\rho_j = \text{idem} = \rho$ by solving the equation

$$\mu + \rho - (1 + 2\sigma)\rho^2 = 0. \tag{1.5.17}$$

The solution is

$$\rho = \frac{1 \pm \sqrt{1 + 4\mu(1 + 2\sigma)}}{2(1 + 2\sigma)} \equiv \rho_s. \tag{1.5.18}$$

It exists at $\mu \geqslant -\frac{1}{4}(1 + 2\sigma)^{-1}$, encroaching on the subcritical region $\mu < 0$. The lower limit is the fold (saddle-node) bifurcation point of (1.5.18), which is also the marginal stability point, so that the lower solution, intermediate between the stable trivial solution and the stable upper branch, must be unstable. The lower branch of solutions terminates at the subcritical bifurcation point $\mu = 0$, while the upper branch continues into the supercritical region $\mu > 0$ where it coexists with the single-mode solution (striped pattern) $\rho_0 = \sqrt{\mu}$, $\rho_{\pm 1} = 0$. The dependence of both solutions on μ is shown in Fig. 1.11a.

Stability of the supercritical resonant solution to decay of a pair of modes can be decided by returning to (1.5.16) rewritten in the polar form

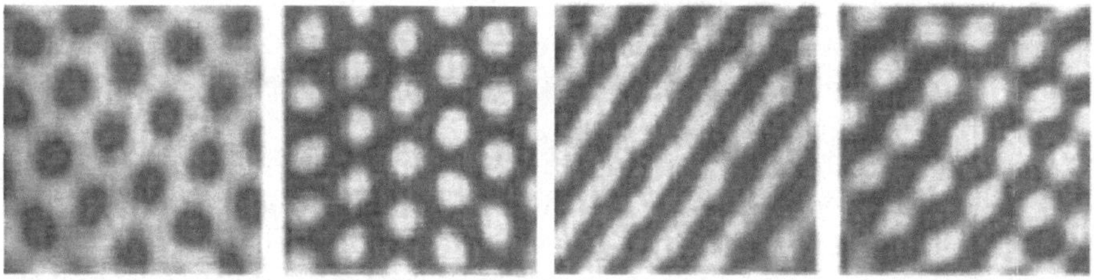

Fig. 1.10. Hexagonal and striped patterns in the Belousov–Zhabotinsky reaction (Ouyang and Swinney, 1991, ©Springer Nature). In the two pictures on the left, the total phase Θ differs by π. The distorted pattern in the rightmost panel represents a state intermediate between hexagons and stripes

$$\dot{\rho}_i = \mu\rho_i + \rho_{i+1}\rho_{i-1} - \rho_i^3 - \sigma\rho_i(\rho_{i+1}^2 + \rho_{i-1}^2). \tag{1.5.19}$$

and linearizing it in the vicinity of the stationary solution $\rho_j = \rho_s =$ idem; the phase variable Θ can be kept out, as it always relaxes to zero. Two modes out or three should decay or be excited together, so their amplitudes can be assumed equal and denoted as $\varrho(t)$, while keeping $\rho(t)$ as the amplitude of the remaining mode. Then (1.5.19) is rewritten as

$$\dot{\rho} = \mu\rho - \rho^3 + \varrho^2 - 2\sigma\rho\varrho^2, \qquad \dot{\varrho} = \varrho(\mu + \rho - \varrho^2 - \sigma(\rho^2 + \varrho^2)). \tag{1.5.20}$$

It is convenient to parametrize the computation by the stationary amplitude ρ_s instead of the bifurcation parameter, expressing the latter as $\mu = -\rho_s + \rho_s^2(1 + 2\sigma)$. Then, linearizing (1.5.20) near the stationary solution $\rho = \varrho = \rho_s$, we arrive at the Jacobi matrix

$$\mathbf{J} = \begin{pmatrix} -\rho_s(1 + 2\rho_s) & 2\rho_s(1 - 2\sigma\rho_s) \\ \rho_s(1 - 2\sigma\rho_s) & -2\rho_s^2(1 + \sigma) \end{pmatrix}. \tag{1.5.21}$$

The trace of this matrix is always negative, whereas the determinant is positive, indicating stability, at

$$\frac{1}{2(1 + 2\sigma)} < \rho_s < \frac{1}{\sigma - 1}, \quad \text{or} \quad -\frac{1}{4(1 + 2\sigma)} < \mu < \frac{2 + \sigma}{(\sigma - 1)^2}. \tag{1.5.22}$$

The lower limit is trivial, as it coincides with the saddle-node bifurcation point in the subcritical region, while the upper limit indicates the instability of the resonant solution to decay of two modes. This limit is singular at $\sigma \to 1$, but we expect mutual suppression to be stronger than self-suppression, as argued in Sect. 1.5.1. The breakdown of the resonance and transition to a single-mode striped pattern shifts to lower μ as σ increases, and the hexagonal pattern is always stable at $\sigma < 1$.

The resonant term facilitates excitation of additional modes, so that stability of the single-mode solution is no longer defined by (1.5.5). Linearizing (1.5.20) near the single-mode solution $\rho = \sqrt{\mu}$, $\varrho = 0$ leads to the stability condition $\mu > \mu_0 = (1+\sigma)^{-2}$. This is a Hopf bifurcation point, below which two more modes are excited. It lies below the upper limit in (1.5.22). This indicates that the Hopf bifurcation is subcritical; at $\mu > \mu_0$, the basins of the two stable solutions are separated by an unstable periodic orbit. The two limits are drawn as functions of σ in Fig. 1.11b, showing the shadowed area where the single-mode and triplet solutions coexist.

We can use (1.5.14) to compare the energies of the striped and hexagonal patterns. The respective potentials are

$$V = -\frac{\mu^2}{4}, \quad V = -\frac{(3\sqrt{1 + 4\mu(1 + 2\sigma)} - 1)(\sqrt{1 + 4\mu(1 + 2\sigma)} + 1)^3}{64(1 + 2\sigma)^3}. \tag{1.5.23}$$

As the bifurcation parameter μ increases, the energy of the striped pattern, which is metastable at μ just above the excitation limit, becomes at some stage lower than that of the hexagonal pattern, which turns from absolutely stable to metastable at this point before reaching its instability limit. Both energies equalize at

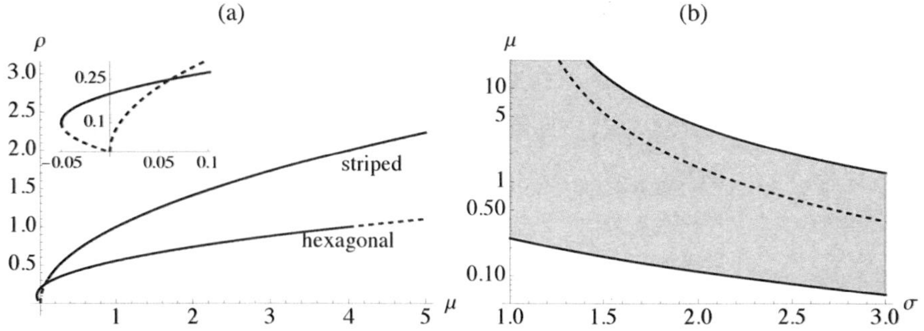

Fig. 1.11. (a) Dependence of the real amplitude of the striped and hexagonal patterns on the bifurcation parameter μ at $\sigma = 2$. The solutions are shown by solid lines when stable and dashed lines when unstable. Inset: the blow-up at $|\mu| \ll 1$. (b) Coexistence region of striped and hexagonal patterns in the plane σ, μ, bounded from above by the instability of the resonant triplet to decay, and from below, by the instability of a single mode to excitation of a resonant pair. The hexagonal pattern has a lower energy below, and the striped, above the dashed line

$$\mu = \frac{1 + 3\sigma + \sqrt{2}(1 + \sigma)^{3/2}}{2(\sigma - 1)^2(1 + 2\sigma)}. \tag{1.5.24}$$

The locus of equal energies is also shown in Fig. 1.11b.

Resonant interactions enhancing a triplet of modes forming an equilateral triangle may become possible also in the presence of inversion symmetry, if a a homogeneous mode is included (Dewel et al, 1995). The specific example treated there is the SH model (1.1.3). The advantage of this system is that Θ may now tend to either 0 or π, depending on whether the hexagonal and global nodes have opposite or identical signs, which may lead to a dynamic competition between "up" and "down" hexagons. However, the two kinds of modes bifurcate in this model, as they do generically, at different values of parameters, and presence of higher harmonics generated by the cubic term complicates their interaction. The existence region of oscillatory solutions does not encompass the entire domain where stationary solutions are unstable, since, in the absence of cubic and higher-order dumping that is apt to stabilize oscillations in an underlying system, and could be detected by a higher-order bifurcation analysis, some trajectories escape to infinity. Chaotic behavior is also possible in this system (de Wit et al, 1996).

1.5.4 Wave–Turing Resonance

Interaction between the Turing and Hopf modes near a degenerate bifurcation was studied by de Wit et al (1996) who also referred to studies in the fluid-mechanical context. Resonant interactions are brought forth in this case by a combination of a Turing mode and two wave modes with the wave vectors satisfying the condition $\mathbf{k}_+ - \mathbf{k}_- + \mathbf{k}_0 = 0$, i.e., forming an isosceles triangle (Fig. 1.9a). This is a special example of resonant interactions allowing for finite solutions in the absence of stabilizing cubic terms (Pismen and Rubinstein, 1999b, Frohoff-Hülsmann et al, 2023).

The general form of the lowest-order resonant amplitude equations is

$$\dot{a}_0 = \mu_0 a_0 + \kappa_0 a_- \overline{a}_+, \quad \dot{a}_+ = \mu_1 a_+ + \kappa_1 a_- \overline{a}_0, \quad \dot{a}_- = \mu_1 a_- + \kappa_1 a_+ a_0, \tag{1.5.25}$$

where $\kappa_0 = \mathbf{U}_0^\dagger \cdot \mathbf{f_{uu}}[\mathbf{U}_1, \overline{\mathbf{U}}_1]$ is real, while $\kappa_1 = \mathbf{U}_1^\dagger \cdot \mathbf{f_{uu}}[\mathbf{U}_1, \mathbf{U}_0]$ is complex. Since the imaginary part of μ_1 can be absorbed in frequency, this parameter can be viewed as real, alongside μ_0, also for wave modes. This system lacks gradient structure, allowing, in principle, for persistent non-stationary behavior. Unlike Sect. 1.5.3, including cubic terms is unnecessary, since the amplitudes may remain finite in this system even without higher-order damping interactions.

Using the polar representation of the complex amplitudes, $a_0 = \rho_0\, \mathrm{e}^{\mathrm{i}\theta_0}$, $a_\pm = \rho_\pm \mathrm{e}^{\mathrm{i}\theta_\pm}$, (1.5.25) can be reduced to a system of four real equations including a single phase combination $\Theta = \theta_0 + \theta_+ - \theta_-$. The system (1.5.25) has two kinds of stationary solutions: symmetric, when $\rho_\pm$ are equal and asymmetric when they are not. Three out of five real parameters of this system can be eliminated by rescaling the amplitudes and time. We choose μ and κ of opposite signs, so that the phase Θ for the symmetric state relaxes to zero stationary value, and to be definite, assume $\kappa_0 < 0$, $\mu_0 > 0$. The parameter κ_1 will be presented in a polar form $\kappa_1 = \kappa \mathrm{e}^{\mathrm{i}\alpha}$, $0 \leqslant \alpha \leqslant \pi$. Using μ_0^{-1} as the time scale, $|\mu_0/\kappa|$ as the scale of ρ_0, and $\mu_0/\sqrt{|\kappa\kappa_0|}$ as a scale of $\rho_\pm$ reduces (1.5.25) to the polar form containing just two parameters $\mu = \mu_1/\mu_0$ and α:

$$\dot{\rho}_0 = -\rho_0 + \rho_+\rho_- \cos\Theta, \qquad \dot{\rho}_\pm = \mu\rho_\pm + \rho_\mp\rho_0 \cos(\Theta \mp \alpha), \tag{1.5.26}$$

$$\dot{\Theta} = -\frac{\rho_+\rho_-}{\rho_0} \sin\Theta - \rho_0 \left[\frac{\rho_+}{\rho_-} \sin(\Theta + \alpha) + \frac{\rho_-}{\rho_+} \sin(\Theta - \alpha)\right]. \tag{1.5.27}$$

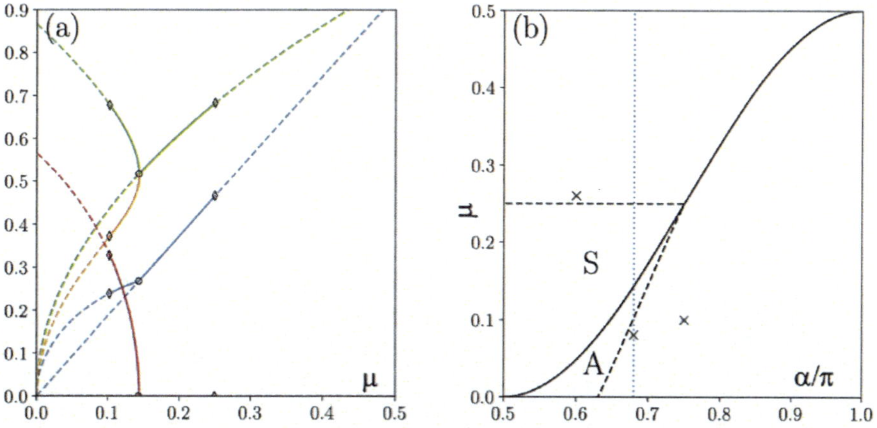

Fig. 1.12. (a) The branches of stationary solutions at $\alpha = 0.68\pi$ as functions of μ. The curves for $\rho_0, \rho_+, \rho_-, \Theta$ are shown in blue, orange, green, and red, respectively. Solid lines show stable, and dashed line unstable solutions. Diamond and circle symbols indicate Hopf and pitchfork bifurcations. (b) Phase diagram with solid and dashed lines indicating pitchfork and Hopf bifurcations, respectively. The regions of stable symmetric and asymmetric stationary states are indicated by the letters S, A. The blue dotted line marks the section in (a). The crosses mark the location of time plots in Fig. 1.13 (Frohoff-Hülsmann et al, 2023)

The stationary values of the amplitudes $\rho_0, \rho_\pm$ obtained by resolving (1.5.26) are

$$\rho_0 = \frac{\mu}{[\cos(\Theta-\alpha)\cos(\Theta+\alpha)]^{1/2}}, \quad \rho_\pm = \left[-\frac{\mu}{\cos(\Theta\pm\alpha)\cos\Theta}\right]^{1/2}. \tag{1.5.28}$$

The last relation implies that asymmetric stationary solutions are restricted to the range $\pi/2 \leqslant \alpha \leqslant \pi$. Plugging (1.5.28) in (1.5.27) brings the equation defining stationary values of Θ to the form

$$-\tan\Theta + \mu[\tan(\Theta-\alpha) + \tan(\Theta+\alpha)] = 0. \tag{1.5.29}$$

It is converted through a chain of trigonometric transformation to a transparent implicit relation

$$(1-2\mu)\cos^2\Theta = \sin^2\alpha. \tag{1.5.30}$$

This implies that acceptable values of μ are confined to the interval $0 < \mu < \frac{1}{2}\cos^2\alpha$. Typical solution branches are drawn in Fig. 1.12a.

Linearizing the fourth-order system (1.5.26), (1.5.27) near the symmetric stationary solution $\rho_0 = \mu$, $\rho_\pm = \sqrt{-\mu/\cos\alpha} \equiv \varrho$ yields the Jacobi matrix

$$\mathbf{L} = \begin{pmatrix} -1 & \varrho & \varrho & 0 \\ \varrho & \mu & -\mu & \varrho^3\sin\alpha \\ \varrho & -\mu & \mu & -\varrho^3\sin\alpha \\ 0 & -2\varrho\sin\alpha & 2\varrho\sin\alpha & 2\mu-1 \end{pmatrix}, \tag{1.5.31}$$

We apply the Routh–Hurwitz criterion to determine linear stability, based on the characteristic polynomial

$$\det(\lambda\mathbf{I} - \mathbf{L}) = \lambda^4 + a_1\lambda^3 + a_2\lambda^2 + a_3\lambda + a_4, \tag{1.5.32}$$

This stability criterion requires the following combinations of a_i to be positive:

$$H_4 = a_4, \quad H_3 = a_1a_2 - a_0a_3, \quad H_2 = (a_1a_2 - a_0a_3)a_3 - a_1^2a_4, \quad H_1 = a_1. \tag{1.5.33}$$

Based on (1.5.31), these expressions are computed as

$$\begin{aligned} H_4 &= -4\mu^2[(-1+4\mu)(1-3\mu+4\mu^2+(1-2\mu)\cos(2\alpha)+\mu\cos(4\alpha))]\sec^4\alpha, \\ H_3 &= 2-12\mu+28\mu^2-16\mu^3-4\mu^2(-1+4\mu)\tan^2\alpha, \\ H_2 &= 4\mu^2[2\mu(1+\tan^2\alpha)-1], \qquad H_1 = 2(1-2\mu). \end{aligned} \tag{1.5.34}$$

The first condition vanishes at the Hopf bifurcation locus. Although looking cumbersome, it has a single real zero $\mu = 1/4$, bounding the stability domain from above. The second condition vanishes at $\mu < \frac{1}{2}\cos^2\alpha$, which is the pitchfork bifurcation locus. The remaining two expressions are positive in the entire domain $0 \leqslant \mu \leqslant 1/4$.

The Jacobi matrix of the asymmetric solution is more involved, and trigonometric relations should be used to replace all Θ dependencies, but otherwise stability is tested as before. The left boundary of the stability domain is the pitchfork bifurcation

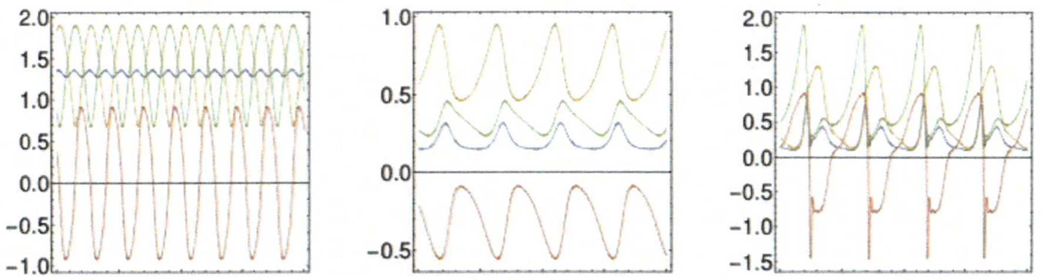

Fig. 1.13. Time plots of oscillations at the points marked by crosses in the parametric plane Fig. 1.12b: $\mu = 0.26$, $\alpha = 0.6\pi$ (left), $\mu = 0.08$, $\alpha = 0.68\pi$ (center), $\mu = 0.1$, $\alpha = 0.75\pi$ (right). The curves for $\rho_0, \rho_+, \rho_-, \Theta$ are shown in blue, orange, green, and red, respectively (Frohoff-Hülsmann et al, 2023)

exchanging stability between the asymmetric and symmetric solution. The right boundary is the Hopf bifurcation locus expressed by the implicit relation

$$(4\mu - 1)(1 - 5\mu + 8\mu^2)\sin^4\alpha + 3(4 - 7\mu + 4\mu^2)(\cos^2\alpha - 2\mu)^2 \\ + [3 + 8\mu(\mu - 2)(1 - 2\mu)](\cos^2\alpha - 2\mu)\sin^2\alpha = 0. \tag{1.5.35}$$

The instability limits of both symmetric and asymmetric stationary solutions, shown in Fig. 1.12b, merge at the double zero singularity located at $\alpha = \frac{3}{4}\pi$, $\mu = \frac{1}{2}$. Take note that the Hopf bifurcations in this system are *secondary* bifurcations on the top of the bifurcation creating this planform. The bifurcations are supercritical; examples of periodic orbits outside the stability domain of both kinds of stationary solutions are shown in Fig. 1.13. The oscillations become more irregular as the values of parameters deviate further from the Hopf bifurcation locus, as seen in the rightmost panel. Further away, trajectories are apt to escape to infinity, which is quite natural in the absence of cubic dumping that could be detected by a higher-order bifurcation analysis. Chaotic solutions are also possible in this system (De Wit et al, 1996).

1.5.5 Resonant Standing Waves

In wave patterns emerging when modes with both $\mathbf{k}$ and ω differing from zero are excited beyond the bifurcation point, degeneracy is possible already in 1D: the two symmetric waveforms, $a^{\pm}e^{i(\pm kx+\omega t)}$ correspond to waves propagating in the opposite directions. Triplet resonance of degenerate waveforms is impossible, since it would require, besides vanishing sum of $\mathbf{k}_j$, that the sum of the three equal frequencies also vanish. However, another type of a resonance leading to nontrivial amplitude dynamics involves interaction of *pairs* of waves propagating in the opposite directions.

Let $a_1^{\pm}$, $a_2^{\pm}$ be the amplitudes of the four waves with identical ω and k propagating in the opposite directions along, respectively, the x_1 and x_2 axes. The amplitude equations are written as

$$\dot{a}_i^{\pm} = a_i^{\pm}\left[\mu - \nu_+|a_i^{\pm}|^2 - \nu_-|a_i^{\mp}|^2 - \nu_{ij}\left(|a_j^{+}|^2 + |a_j^{-}|^2\right)\right] + \kappa\overline{a}_i^{\mp}a_j^{+}a_j^{-}. \tag{1.5.36}$$

The coefficients are computed along the same lines as in Sect. 1.5.1. The mixed quadratic terms for counter-propagating waves in (1.5.2) are now specified as

$$\begin{aligned} &a_i^+ \overline{a}_i^- \mathbf{V}_{0,\pm} \mathrm{e}^{2ikx_i}, \qquad && a_i^+ \overline{a}_j^\pm \mathbf{V}_{0,ij} \mathrm{e}^{\mathrm{i}k(x_i \mp x_j)}, \\ &a_i^+ a_i^- \mathbf{V}_{2,\pm} \mathrm{e}^{2\mathrm{i}\omega t}, && a_i^+ a_j^\pm \mathbf{V}_{2,ij} \mathrm{e}^{\mathrm{i}(2\omega t + k(x_i \pm x_j))}. \end{aligned} \tag{1.5.37}$$

The expressions on the left involve waves propagating in the opposite, and those at the right, in cross directions. The respective arrays $\mathbf{V}$ are computed as

$$\mathbf{V}_{0,\pm} = \left(4k^2\mathbf{D}_0 - \mathbf{f_u}\right)^{-1} \mathbf{f_{uu}}[\mathbf{U}, \overline{\mathbf{U}}], \;\; \mathbf{V}_{2,\pm} = \left(2\mathrm{i}\omega - \mathbf{f_u}\mathbf{I}\right)^{-1} \mathbf{f_{uu}}[\mathbf{U}, \mathbf{U}] = 0, \tag{1.5.38}$$

$$\mathbf{V}_{0,ij} = \left(2k^2\mathbf{D}_0 - \mathbf{f_u}\right) \mathbf{f_{uu}}[\mathbf{U}, \overline{\mathbf{U}}], \;\; \mathbf{V}_{2,ij} = \left(2k^2\mathbf{D}_0 + 2\mathrm{i}\omega\mathbf{I} - \mathbf{f_u}\right)^{-1} \mathbf{f_{uu}}[\mathbf{U}, \mathbf{U}].$$

The coefficients μ, $\nu_+ = \nu$ are the same as in (1.3.29), and ν_{ij} as in (1.5.3) with appropriate arrays (1.5.38). The coefficient ν_- differs from ν_+ only by doubling the first term, and, since all terms tend to be suppressive, it is likely that counter-propagating waves are mutually exclusive. The essential new coefficient at the resonant term is given by

$$\kappa = \overline{\mathbf{U}}^\dagger \left(\mathbf{f_{uuu}}[\mathbf{U}, \mathbf{U}, \overline{\mathbf{U}}] + \mathbf{f_{uu}}[\mathbf{U}, \mathbf{V}_{0,ij} + \mathbf{V}_{2,ij}] + \mathbf{f_{uu}}[\overline{\mathbf{U}}, \mathbf{V}_{2,ii}] \right). \tag{1.5.39}$$

It is easy to see that both wave vectors and frequencies add up to zero along the rectangle formed by the four modes that correspond to the last term. While all other terms are phase-independent, the resonant term depends on the combination of phases $\Theta = \theta_1^+ + \theta_1^- - \theta_2^+ - \theta_2^-$ (Pismen, 1986; Hoyle, 1994). Similar to the triplet resonance in Sect. 1.5.3, they are capable to impede decay of standing wave patterns into single propagating waves and thereby stabilize multimode patterns. All coefficients in (1.5.36) are complex, so that this equation system does not have a gradient structure.

Further on, we concentrate on standing wave solutions with equal real amplitudes of counter-propagating waves $\rho_j = |a_j^\pm|$. Since the imaginary part of μ_1 can be absorbed in frequency, this parameter can be viewed as real, and it suffices to consider only one of the equivalent equations for $a^\pm$. It is convenient to present the coefficient at the resonant term in the polar form $\kappa \mathrm{e}^{\mathrm{i}\alpha}$. Other parameters will be defined as $\sigma_\pm = \nu_{ij} \pm \frac{1}{2}(\nu_+ + \nu_-) = \sigma_r^\pm + \mathrm{i}\sigma_i^\pm$. After transforming the amplitudes to the polar form, the real part of (1.5.36) becomes

$$\begin{aligned} \dot\rho_1 &= \rho_1 \left[\mu - (\sigma_r^+ - \sigma_r^-)\rho_1^2 - (\sigma_r^+ + \sigma_r^-)\rho_2^2 + \kappa\rho_2^2 \cos(\Theta - \alpha)\right], \\ \dot\rho_2 &= \rho_2 \left[\mu - (\sigma_r^+ - \sigma_r^-)\rho_2^2 - (\sigma_r^+ + \sigma_r^-)\rho_1^2 + \kappa\rho_1^2 \cos(\Theta + \alpha)\right]. \end{aligned} \tag{1.5.40}$$

These equations can be rearranged by transforming to the new variables $p = \rho_1^2 + \rho_2^2$, $q = \rho_1^2 - \rho_2^2$:

$$\begin{aligned} \dot p &= 2\mu p - 2\sigma_r^+ p^2 + 2\sigma_r^- q^2 + \kappa(p^2 - q^2)\cos\Theta\cos\alpha, \\ \dot q &= 2\mu q - 2(\sigma_r^+ - \sigma_r^-)pq + \kappa(p^2 - q^2)\sin\Theta\sin\alpha. \end{aligned} \tag{1.5.41}$$

Adding up the imaginary parts of the polar form of (1.5.36) yields the equation of Θ:

$$\dot{\Theta} = -\kappa p \cos\alpha \sin\Theta - 2q(\kappa \sin\alpha \cos\Theta - \sigma_i^-). \tag{1.5.42}$$

Two easily computable stationary solution of (1.5.41), (1.5.42) are synphase ($\Theta = 0$) and antiphase ($\Theta = \pi$) locked states, which exist in the supercritical region $\mu > 0$:

$$q_s = 0, \qquad p_s = \frac{2\mu}{\sigma_r^+ \mp \kappa \cos\alpha}. \tag{1.5.43}$$

Here and in the following, the upper and lower signs correspond to the solutions at $\Theta = 0$ and $\Theta = \pi$, respectively. All characteristics of these states are mapped one onto the other when α is shifted by π.

The analysis is hampered by proliferation of parameters, which is reduced by denoting $\sigma^- = s e^{i\beta}$, $\kappa/s = \psi$, $\sigma_r^+/s = \chi$ and scaling p, q by $\gamma = \mu/s$ and time by μ^{-1}. This transforms (1.5.41), (1.5.42) to

$$\begin{aligned}
\dot{p} &= 2p - 2\chi p^2 + 2q^2 \cos\beta + \psi(p^2 - q^2)\cos\Theta \cos\alpha, \\
\dot{q} &= 2q - 2(\chi - \cos\beta)pq + \psi(p^2 - q^2)\sin\Theta \sin\alpha, \\
\gamma\dot{\Theta} &= -\psi p \cos\alpha \sin\Theta - 2q(\psi \sin\alpha \cos\Theta - \sin\beta).
\end{aligned} \tag{1.5.44}$$

The Jacobi matrix obtained by linearizing this system is ed in the vicinity of the stationary solution, which is now rescaled to $p_s = 2/(\chi \mp \psi \cos\alpha)$, is

$$\mathbf{f_u} = \begin{pmatrix} -2 & 0 & 0 \\ 0 & \dfrac{4\cos\beta \mp 2\psi\cos\alpha}{2\chi \mp \cos\alpha} & \pm\dfrac{4\psi \sin\alpha}{(2\chi \mp \cos\alpha)^2} \\ 0 & \dfrac{2\sin\beta \mp \psi \sin\alpha}{\gamma} & \mp\dfrac{2\psi\cos\alpha}{\gamma(2\chi \mp \cos\alpha)} \end{pmatrix}. \tag{1.5.45}$$

Stability depends on the signs of the determinant and trace of the 2×2 matrix $\mathbf{L}$ in the lower right:

$$\det(\mathbf{L}) = \frac{4\psi[\psi \mp 2\cos(\alpha - \beta)]}{\gamma(2\chi \mp \cos\alpha)^2}, \quad \mathrm{tr}(\mathbf{L}) = \frac{4\gamma\cos\beta \mp 2(\gamma + 1)\psi\cos\alpha}{\gamma(2\chi \mp \cos\alpha)}. \tag{1.5.46}$$

The denominators are always positive; hence, the static and oscillatory instability domains are, respectively,

$$\psi < \pm 2\cos(\alpha - \beta), \qquad \psi < \frac{2\gamma\cos\beta}{(\gamma + 1)|\cos\alpha|}. \tag{1.5.47}$$

The condition $\psi < 2\cos\beta/|\cos\alpha|$ is necessary to keep the growth rate of p at $p = 0$ positive at any values of Θ, so that unphysical negative values be avoided. The same condition insures that the growth rate of p is negative when q is close to the single-wave stationary state $q = \pm p$. Therefore the oscillatory region exists only at $\gamma > 1$.

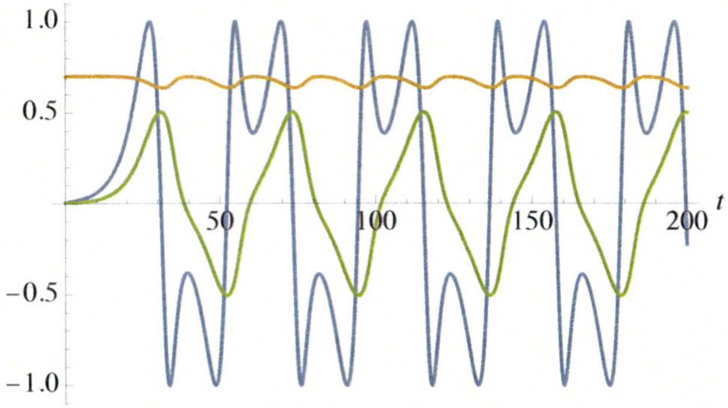

Fig. 1.14. Oscillations at $\gamma = 2$, $\chi = 1.5$, $\kappa = 0.3$, $\alpha = \pi/3$, $\beta = 0.52\pi$. The blue, orange, and green curves show the dependence of Θ/π, p, q, on time, respectively

The Hopf bifurcation does not necessarily leads to steady oscillations. Of course, it must be subcritical, but even then the system may evolve to an alternative locked state or decay to a single standing wave solution $q = \pm p$, which, in turn, is likely to further desintegate into a single propagating wave. An example of oscillations is shown in Fig. 1.14. In this case, oscillations persist in a rather narrow area. If, say, the parameter β is shifted down by about 0.2π, the system is attracted to the single wave $q = p$, while a similar increase brings it to the synphase stationary state.

1.6 Conserved Systems

1.6.1 Nonlocal and Local Description

Nonlinear models possessing two or more stable HSS appear also in closed systems relaxing toward thermodynamic equilibrium. Classical models of phase separation are characterized by gradient dynamics (1.1.9). However, the free energy functional derived from underlying molecular interactions have a more complicated form than those highlighted in Sect. 1.1.1. In a simple case of a single-component fluid equilibrating between the states of low (gas) and high (liquid) densities $u(\mathbf{x})$, this functional is expressed in the density functional approximation (Evans, 1979) as[7]

$$\mathcal{F} = \int u(\mathbf{x})\, d\mathbf{x} \left[f(u) + \tfrac{1}{2} \int_{r>d} U(r)[u(\mathbf{x}+\mathbf{r}) - u(\mathbf{x})]\, d\mathbf{r} \right]. \tag{1.6.1}$$

The first term of this expression gives the free energy of a homogeneous state. The second *nonlocal* term is the distortion energy due to changes of density in space; $U(r) < 0$ is an isotropic attractive pair interaction kernel with a molecular-scale cutoff d, e.g., the modified Lennard-Jones potential with hard-core repulsion:

$$U = -Ar^{-6} \ \text{at}\ r > d, \quad U = \infty \ \text{at}\ r < d. \tag{1.6.2}$$

where d is the nominal hard-core molecular diameter. This interaction kernel gives the free energy density of a homogeneous van der Waals fluid

[7] The notation is chosen here in accordance to a general role of density as an "order parameter"

$$f(u) = T \ln \frac{u}{1 - bu} - au, \tag{1.6.3}$$

where T is temperature (expressed in energy units), $b = \frac{2}{3}\pi d^3$ is the excluded volume, and

$$a = -2\pi \int_d^\infty U(r) r^2 \, \mathrm{d}r = \frac{2\pi A}{3d^3}. \tag{1.6.4}$$

If density changes only slightly over distances comparable with the characteristic interaction length, one can expand

$$u(\mathbf{x} + \mathbf{r}) = u(\mathbf{x}) + \mathbf{r} \cdot \nabla u(\mathbf{x}) + \tfrac{1}{2}\mathbf{rr} : \nabla\nabla u(\mathbf{x}) + \ldots \tag{1.6.5}$$

The terms of this expansion are plugged in (1.6.1) and integrated in spherical coordinates centered on an arbitrary point $\mathbf{x}$. The contribution of the linear term vanishes upon angular integration when the system is isotropic and, as a consequence, the interaction term is spherically symmetrical. The lowest order nonzero contribution is due to the quadratic term:

$$-\tfrac{1}{2}K \int u(\mathbf{x}) \nabla^2 u(\mathbf{x}) \, \mathrm{d}\mathbf{x} = \tfrac{1}{2}K \int |\nabla u(\mathbf{x})|^2 \, \mathrm{d}\mathbf{x}, \tag{1.6.6}$$

where K is the rigidity coefficient:

$$K = -\tfrac{2}{3}\pi \int_d^\infty U(r)\, r^4 \, \mathrm{d}r > 0. \tag{1.6.7}$$

Thus, (1.6.1) simplifies to

$$\mathcal{F} = \int \left[\tfrac{1}{2}K|\nabla u|^2 + V(u) \right] \mathrm{d}\mathbf{x}. \tag{1.6.8}$$

If $V(u)$ is the symmetric quartic potential, the respective Lagrangian reduces to (1.1.10) after rescaling $\mathbf{x} \to \sqrt{K}\mathbf{x}$. This potential is applicable close to a critical point, i.e., a saddle-node bifurcation in terms of Sect. 1.2.3, with the value of u at the bifurcation point set to zero, but at the critical point itself is incapacitated by strong fluctuations.

The differential equation obtained by varying (1.6.8) is

$$\Gamma^{-1}\partial_t u = K\nabla^2 u + f(u), \quad f(u) = -V'(u), \tag{1.6.9}$$

The mobility coefficient $\Gamma > 0$ is, generally, dependent on the order parameter u, but if it is constant, its inverse can be taken as the time scale. The form with $f(u) = u(1 - u^2)$ derived from the quartic potential is commonly called in this context the Allen–Cahn equation, although it has been first derived by van der Waals (1894). It coincides with (1.1.2) after rescaling.

The overall interfacial energy per unit area of a stationary flat interface separating the dense and dilute phases, or *surface tension*, σ is defined by integrating (1.6.8) over the infinite normal axis z. Denoting $P = \frac{1}{2}Kp^2$, $p = \mathrm{d}u/\mathrm{d}z$, the right-hand side of

(1.6.9) can be expressed as $P'(u) - V'(u)$. The values of $V(u_s^{\pm})$ should be equal for the two HSS $u = u_s^{\pm}$ in equilibrium, and P, evidently, vanishes for any HSS; hence, integrating the latter relation proves that the overall distortion energy equals the potential energy at any point. This is formally equivalent to the virial theorem of classical mechanics (where the distortion energy is replaced by the kinetic energy). Thus, the surface tension can be computed as

$$\sigma = K \int_{-\infty}^{\infty} u'(z)^2 \mathrm{d}z = K \int_{u_s^-}^{u_s^+} p(u) \mathrm{d}u = \int_{u_s^-}^{u_s^+} \sqrt{2KV(u)} \mathrm{d}u. \tag{1.6.10}$$

However, the above reduction from the integral to the differential form of spatial dependence, going back to van der Waals (1894), has a disturbing flaw (Pismen, 2001). The interaction has to fall off sufficiently fast to ensure convergence of the integral (1.6.6), as well as of higher order integrals in the expansion. This, unfortunately, is not true for interaction kernels with power decay. Thus, for the commonly used Lennard-Jones potential $U(r) = -Ar^{-6}$, (1.6.7) integrates to $K = \frac{2}{3}\pi/d$, but the integral diverges in the next nonvanishing (fourth) order. All orders of the expansion converge only when the interaction falls of exponentially. Nevertheless, local Landau functionals are widely used because of a great gain in simplicity of computations, compared to the nonlocal expression (1.6.1).

1.6.2 Phase Separation

Equation (1.6.9) is inapplicable when dynamics is constrained by a conservation law. In the case of a one-component fluid discussed here as the simplest example, conservation of mass should be taken care of when dynamic equations are formulated. Evidently, the variable evolving according to (1.6.9) is not conserved. To ensure conserved dynamics, the gradient principle should be formulated in a different way, based on a conservation law for the order parameter u:

$$\partial_t u = -\nabla \cdot \mathbf{j}. \tag{1.6.11}$$

The flux $\mathbf{j}$ should be proportional to the *gradient* of the variation of the free energy functional:

$$\mathbf{j} = -\Gamma(u)\nabla \frac{\delta \mathcal{F}}{\delta u} \equiv -\Gamma(u)\nabla\mu, \tag{1.6.12}$$

where $\mu = \delta\mathcal{F}/\delta u$ is *chemical potential*. As before, the free energy always decreases until a minimum is reached. This is proven by using (1.6.11), (1.6.12) in a general free energy functional and integrating by parts:

$$\frac{\mathrm{d}\mathcal{F}}{\mathrm{d}t} = \int \frac{\delta\mathcal{F}}{\delta u}\frac{\partial u}{\partial t} \mathrm{d}\mathbf{x} = \int \frac{\delta\mathcal{F}}{\delta u} \nabla \cdot \left[\Gamma(u)\nabla\frac{\delta\mathcal{F}}{\delta u}\right] \mathrm{d}\mathbf{x} = -\int \Gamma(u)\left|\nabla\frac{\delta\mathcal{F}}{\delta u}\right|^2 \mathrm{d}\mathbf{x} \leq 0. \tag{1.6.13}$$

Equations (1.6.11), (1.6.12) with $\mathcal{F}$ defined by (1.6.8) combine to the generalized Cahn–Hilliard equation (Cahn and Hilliard, 1959):

$$\partial_t u = -\nabla \cdot (\Gamma(u)\nabla\mu), \tag{1.6.14}$$

$$K\nabla^2 u + f(u) + \mu = 0. \tag{1.6.15}$$

A more commonly used version with Γ = const rescaled to unity is

$$\partial_t u = -\nabla^2 \left[K\nabla^2 u + f(u)\right]. \tag{1.6.16}$$

When $V(u) = -\int f(u)\mathrm{d}u$ is a double-well potential and μ is fixed in such a way that $f(u) + \mu = 0$ has two stable roots (HSS), *spinodal decomposition* takes place, and the system separates into two phases corresponding to the alternative HSS. Stationary solutions of the Cahn–Hilliard equation describing equilibrium interphase boundaries coincide with the stationary solutions of its nonconserved counterpart (2.1.1), but there is a substantial difference: when the order parameter is conserved, a single planar interface cannot propagate.

A flat interface separating the two phases is at equilibrium when both uniform equilibria are at the Maxwell construction, i.e. the minima of are of equal depth. This can be achieved at a certain value of chemical potential, which can be viewed therefore as an adjustable constant ensuring the existence of a stationary front solution. The value of μ required to maintain the equilibrium is obtained by multiplying (1.6.15) by ∇u and integrating across the front. The differential term vanishes upon integration, while the integral of the algebraic part yields the Maxwell condition

$$\mu_s = \frac{V(u_s^+) - V(u_s^-)}{u_s^+ - u_s^-}. \tag{1.6.17}$$

This, together with $\mu_s = -f(u_s^{\pm})$, defines the three unknowns $\mu_s,\ u_s^{\pm}$.

Chemical potential enters the nonlocal description after (1.6.1) is enlarged to the Helmholtz free energy $\mathcal{H} = \mathcal{F} - \mu N$ where N is the number of particles, leading to $\mu = \delta\mathcal{H}/\delta N$. The stationary equation obtained by varying this functional is

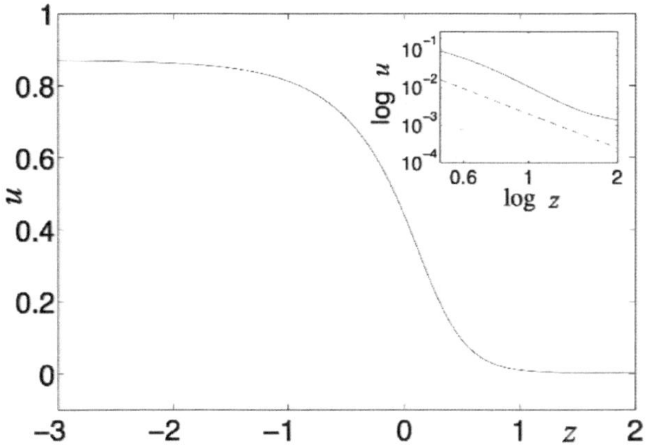

Fig. 1.15. The density profile of the liquid-vapor interface at $\beta = 9$ obtained by numerical solution of (1.6.19). The inset shows the vapor-phase tail of the numerical solution (solid line) compared to the exponential tail of the solution of (1.6.15) depicted by the dashed line (Yochelis and Pismen, 2006)

$$f(u) + \mu + \int_{r>d} U(r)[u(\mathbf{x}+\mathbf{r}) - u(\mathbf{x})]\,\mathrm{d}\mathbf{r} = 0. \tag{1.6.18}$$

The dimensionless form of this equation for the van der Waals fluid with $U(r)$ defined by (1.6.2) and $f(u)$ by (1.6.3) is

$$\frac{1}{1-\rho} - \ln\left(\frac{1}{\rho} - 1\right) - 2\beta\rho - \mu + \frac{3}{4}\beta \int_{-\infty}^{\infty} U(\zeta)[u(z+\zeta) - \rho(z)]\,\mathrm{d}\zeta = 0, \tag{1.6.19}$$

where length scaled by the nominal molecular diameter d, density by b^{-1}, and chemical potential by T. The only remaining dimensionless parameter is the rescaled inverse temperature $\beta = a/(bT)$. A numerical solution of (1.6.19) is plotted in Fig. 1.15. Surface tension is computed in the same way as in the preceding subsection, and, due to a sufficiently fast asymptotic decay of the solution, the numerical result is only slightly different.

1.6.3 Symmetry Breaking in Conserved Systems

The generalization of (1.1.9) to a multicomponent system is

$$\partial_t u_i = -\delta\mathcal{F}(\mathbf{u})/\delta u_i, \quad \mathcal{F} = \int \mathcal{L}(\mathbf{u}, \nabla\mathbf{u}, \ldots)\,\mathrm{d}\mathbf{x}. \tag{1.6.20}$$

Similar to (1.6.16), a system containing several conserved variables with respective mobility and rigidity matrices $\mathbf{\Gamma}, \mathbf{K}$ (further assumed to be constant) differs from a general RDS (1.2.5) by added outside Laplacian:

$$\partial_t u_i = -\mathbf{\Gamma}\nabla^2[\mathbf{K}\nabla^2 u_i + f_i(\mathbf{u}; \mathbf{p})]. \tag{1.6.21}$$

Its linearization, analogous to (1.3.1), is

$$\partial_t \mathbf{u} = -\mathbf{\Gamma}\nabla^2 \left[(\mathbf{K}\nabla^2 + \mathbf{f_u}) \cdot \mathbf{u}_1\right], \tag{1.6.22}$$

where $\mathbf{f_u}$ is the Jacobi matrix computed at a stationary state $\mathbf{u} = \mathbf{u}_0$ and $\mathbf{u}_1$ denotes the array of deviations from the stationary state. Presenting them, as usual, in a harmonic form $\mathbf{u}_1 = \sum_j \mathbf{U}_j\, \mathrm{e}^{\lambda_j t + \mathrm{i}\mathbf{k}\cdot\mathbf{x}}$, we arrive at the eigenvalue problem with the matrix

$$\mathbf{L}_c = -k^2\mathbf{\Gamma} \cdot (\mathbf{f_u} - k^2\mathbf{K}) = -k^2\mathbf{L}, \tag{1.6.23}$$

where $\mathbf{L}$ is defined by (1.3.2) with the replacement $\mathbf{D} \to \mathbf{K}$. Clearly, zero eigenvalues $\lambda(|k|)$ of (1.3.2) are retained in (1.6.23), in addition to the persistent zero eigenvalue at $k = 0$. Moreover, extrema of the determinant $J_c = \det(\mathbf{L}_c)$ and its nonconserved counterpart $J = \det \mathbf{L}$ coincide as well at the Turing or wave bifurcation locus, since $\mathrm{d}(k^2 J)/\mathrm{d}(k^2) = J + k^2 \mathrm{d}J/\mathrm{d}(k^2) = 0$. However, this identity is of little value, since it applies only at a fixed point satisfying $\mathbf{f}(\mathbf{u}_0) = 0$, which is necessary, say, in the context of chemical interactions, but does not apply to conserved systems where arbitrary constant concentrations verify (1.6.21).

By the Onsager reciprocity relations, a system monotonically relaxes to equilibrium, which in this case corresponds to a complete phase separation, when the matrices $\mathbf{f_u}, \mathbf{K}, \mathbf{\Gamma}$ are symmetric. The symmetry of $\mathbf{f_u}$ is always present when the functions f_i are derived from a potential, $f_i = \partial_i V$, implying $\partial_j f_i = \partial_i f_j$. Otherwise, chemical potentials $\mu_i = \delta\mathcal{F}/\delta c_i$ can be formally defined as before, but, in the absence of a potential, $\partial_j \mu_i \neq \partial_i \mu_j$, so that the interactions are *nonreciprocal*.

In this case, the energy is not well defined, and complete phase separation may give way to a pattern with a finite characteristic scale defined near the bifurcation point by the maximum of the dispersion relation $\lambda(|k|)$. Once periodic spatial structures arise at the Turing bifurcation, we expect that hexagonal patterns enhanced by resonant interactions will be prevalent in 2D (Sect. 1.5.2), but also mixed planforms, as in Fig. 1.10, are possible, as well as natural patterns with various defects.

In the case of a two-component system, the coupled Cahn-Hilliard equations reduce, after rescaling time and length, to

$$\partial_t u = -\nabla^2 \left[\nabla^2 u + f(u,v)\right], \qquad \partial_t v = -\gamma\nabla^2 \left[\kappa\nabla^2 v + g(u,v)\right], \tag{1.6.24}$$

where γ is the ratio of mobilities and κ is the ratio of rigidities of the two species. The linearized system is

$$\mathbf{L}_c = k^2 \begin{pmatrix} f_u - k^2 & f_v \\ \gamma g_u & \gamma(g_v - \kappa k^2) \end{pmatrix} = k^2 \mathbf{L}, \tag{1.6.25}$$

where the derivatives $f_u = \partial f/\partial u$, etc. are defined as in Sect. 1.3.1.

Oscillations with zero wavelength are incompatible with conserved dynamics, and are possible only at some finite k, so that only *wave* patterns can be observed. The trace of the Jacobi matrix,

$$\mathrm{Tr}(\mathbf{L}_c) = k^2(f_u + \gamma g_v) - k^4(1+\gamma\kappa), \tag{1.6.26}$$

is positive at small k if $f_u + \gamma g_v > 0$. This condition is the same as in Sect. 1.3.1, but when the system is conserved the trace *increases* with k and, after passing a maximum, vanishes at

$$k_{\mathrm{w}}^2 = \frac{f_u + \gamma g_v}{1+\gamma\kappa}. \tag{1.6.27}$$

before decreasing at larger k, as shown by the dashed line in Fig. 1.16d. Therefore, unlike a nonconserved system, the wave bifurcation becomes possible already in two phase space dimensions, and there is a preferred wavelength of emerging waves (Frohoff-Hülsmann et al, 2021, 2023).

Formally, the trace condition for the onset of oscillations is insensitive to the symmetry of interactions and therefore may be misleading. One could infer that the Hopf bifurcation could take place when at least one variable is autocatalytic, but this is impossible if the matrix $\mathbf{L}$ is symmetric and, hence, derivable from a potential, which would forbid oscillatory relaxation to equilibrium. Since the Jacobi matrix $\mathbf{L}_c$ is sufficiently simple, this can be checked by expressing its eigenvalues explicitly. After some algebra, we obtain

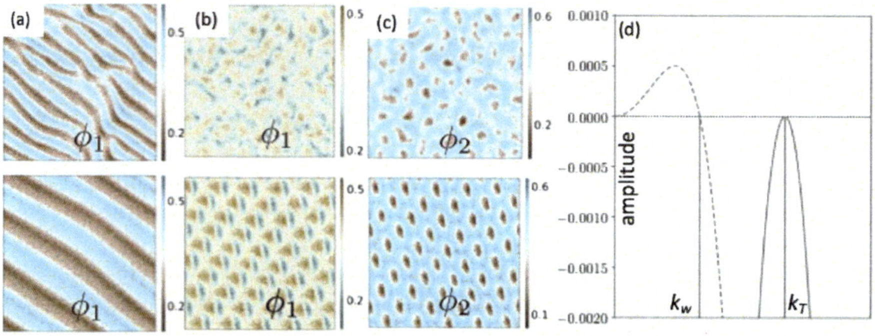

Fig. 1.16. Wave patterns in a nonreciprocal system with the conserved components ϕ_1, ϕ_2 (Saha et al, 2020 CC 4.0). (a) Lamellar pattern with self-propelled density bands. (b), (c) Moving 2D spotted patterns. The levels of the coexisting phases ϕ_1 (a,b) and ϕ_2 (c) are color-coded according to the scales on the right (d) A schematic dependence of the amplitude on the wavenumber k near the Turing (solid line) and wave (dashed line) bifurcations. The vertical lines denote the wavenumbers at the Turing–wave degeneracy (Frohoff-Hülsmann et al, 2023)

$$\lambda = \frac{1}{2}\left[1 \pm \frac{\sqrt{4\gamma f_v g_u + [(f_u - \gamma g_v - k^2(1-\gamma\kappa)]^2}}{f_u - \gamma g_v - k^2(1+\gamma\kappa)}\right]. \tag{1.6.28}$$

Oscillations take place when the expression under the radical is negative, which can happen only when f_v and g_u have different signs, since the second term under the radical is positive definite. Take note that although nonreciprocity is essential for both stationary and oscillatory symmetry-breaking bifurcations, f_v and g_u enter neither (1.6.27) nor (1.3.6). Beyond the wave bifurcation locus, a variety of irregular nonstationary patterns arise. Some lamellar patterns with self-propelled density bands and moving spotlike 2D patterns observed in 2D simulations of a two-component system by Saha et al (2020) are shown in Fig. 1.16a–c.

The wave and Turing bifurcations are incompatible in a two-variable nonconserved system, and remain so in a conserved system in sufficiently large domains, but if the allowed wavenumbers are restricted by the domain size, it might be possible to observe a codimension two bifurcation of stationary and wave patterns leading to their interaction at small amplitudes as in Sect. 1.5.4. The marginal wave bifurcation takes place in a domain of the size $2\pi/k_{\mathrm{w}}$, provided the nonreciprocity parameter $f_v g_u$, while being negative, is bounded from below by

$$|f_v g_u| < \gamma \left(\frac{\kappa f_u - g_v}{1+\gamma\kappa}\right)^2, \tag{1.6.29}$$

which, combined with (1.3.7), translates to

$$\left(\frac{\kappa f_u + g_v}{\kappa f_u - g_v}\right)^2 < \frac{4\gamma\kappa}{1+\gamma\kappa}. \tag{1.6.30}$$

The degeneracy takes place when the threshold of the Turing bifurcation coincides with the threshold of the wave bifurcation at the largest available wavelength, An example of wave-Turing degeneracy in a modified FitzHugh–Nagumo system was treated in detail by Frohoff-Hülsmann et al (2023).

1.6.4 Partially Conserved Systems

A conservation condition may serve as a global constraint imposed on dynamics of a nonconserved species. A general formulation for a two-component system with v but not u conserved, mirroring that of Sects. 1.3.1, 1.6.3 is

$$\partial_t u = \nabla^2 u + f(u,v), \qquad \partial_t v = -\gamma\nabla^2\left[\kappa\nabla^2 v + g(u,v)\right]. \tag{1.6.31}$$

The Jacobi matrix of the linearized system is

$$\mathbf{L} = \begin{pmatrix} f_u - k^2 & f_v \\ \gamma k^2 g_u & \gamma k^2(g_v - \kappa k^2) \end{pmatrix}. \tag{1.6.32}$$

The determinant $J = \det(\mathbf{L})$ can be expressed as

$$J = \gamma k^2(\Delta - Qk^2 + \kappa k^4, \quad \Delta = f_u g_v - f_v g_u, \quad Q = \kappa f_u + g_v. \tag{1.6.33}$$

If the homogeneous instability is suppressed, so that $\Delta > 0$, the determinant is positive at both small and large k but may become negative at intermediate wavelengths if $Q > 0$, i.e., under the same conditions when the Turing bifurcation takes place in a nonconserved system. The extrema of J are located at

$$k_0^2 = \frac{1}{3\kappa}\left(1 \pm \sqrt{1 - 3\kappa\Delta Q^{-2}}\right). \tag{1.6.34}$$

The positive sign corresponds to the minimum. The determinant at the bifurcation locus should vanish:

$$J_{k=k_0} = (3Q-2)\left[1 + \sqrt{1 - 3\kappa\Delta Q^{-2}}\right] - 3\kappa\Delta\left[3 + Q^{-2}\right] = 0. \tag{1.6.35}$$

Even a single conservation law precludes homogeneous oscillations but bifurcation to wave patterns is possible. The trace of $\mathbf{L}$,

$$\mathrm{Tr}(\mathbf{L}_c) = f_u + k^2(\gamma g_v - 1) - k^4\gamma\kappa, \tag{1.6.36}$$

is positive at small k, so that waves prevail over stationary structures in a sufficiently large domain when the nonconserved component is an activator. As in a fully conserved systems, wave-Turing resonance is possible when the largest allowed wavelength is restricted by the wavenumber k_{w} corresponding to the zero of (1.6.36).

For example, for the FN model (1.3.9) with only the inhibitor being conserved, we have $\Delta = \mu - 1$, $Q = \kappa - 1$, and the conditions of J vanishing at its minimum with respect to k defining the locus of the Turing bifurcation are

$$k_0^2 = \frac{1}{3\kappa}\left[1 + \sqrt{1 - \frac{3\kappa(\mu-1)}{(\kappa-1)^2}}\right], \quad \mu + (k^2-1)(\kappa k^2 + 1) = 0. \tag{1.6.37}$$

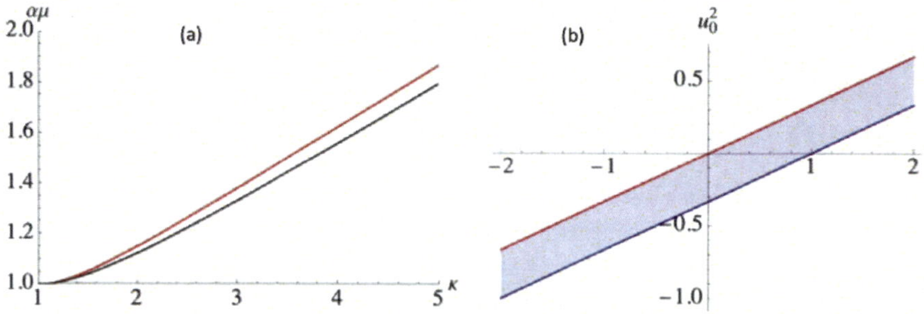

Fig. 1.17. (a) The dependence of μ on κ at the Turing bifurcation locus for the non- or fully conserved (black curve) and partially conserved (red curve) FN system. (b) Stability of a nontrivial HSS $u = u_0$ in the phase field crystal model. Instability is homogeneous below the blue line; the shaded area corresponds to the Turing bifurcation, and the state is stable above the red line

The dependence of μ on κ within the allowed parametric domain of positive Δ and Q is drawn as a parametric plot in Fig. 1.17a; it differs only slightly from that for either nonconserved or fully conserved system.

An extension of a conservation law is global control maintaining a constant level of an integral of some function of dynamic variables, say, $\int Q(u, v)\mathrm{d}\mathbf{x}$ over the entire area (Pismen, 1994; Meerson and Sasorov, 1996). Rather than considering stability of a uniform solution controlled in this way, both studies concentrated on slow motion of sharp boundaries between domains of alternative stationary states controlled by an area integrals involving a single slow variable – the technique similar to that to be employed in Ch. 3.

The most thoroughly studied partially conserved model is the conserved version of the Swift–Hohenberg (SH) equation (1.1.3). It was first brought forward by Mathews and Cox (2000) in the form

$$\partial_t u = -\nabla^2[\mu u - (1 + \nabla^2)^2 u - \nu u^2 - u^3]. \tag{1.6.38}$$

with q^{-1} taken as a length scale and added quadratic term. It qualifies as only *partially* conserved, since a single conservation condition is imposed on a fourth-order equation. The original SH equation (1.1.3), designed with a particular purpose of demonstrating the formation of stationary patterns, became popular due to easy handling, and has spawned various modifications. The conserved version acquired a catchy name "phase field crystal model" after being applied to modeling crystal growth (Elder et al, 2002).

The dispersion relations near the trivial state for the nonconserved (λ) and conserved (λ_c) versions are

$$\lambda = \mu - (1 - k^2)^2, \qquad \lambda_c = k^2[\mu - (1 - k^2)^2]. \tag{1.6.39}$$

In the standard SH equation, the trivial state undergoes the Turing bifurcation at $\mu = 1$, $k_0 = 1$, and in (1.6.38), it takes place at $\mu = 0$, $k_0 = 1$. Oscillations are excluded in both cases, but patterns may coexist with homogeneous stationary states (HSS). In the nonconserved version, two symmetric states $u_0 = \pm\sqrt{\mu - 1}$ exists at $\mu > 1$. Linearization near either solution leads to the dispersion relation

$$\lambda(u_0) = 2(1 + k^2 - \mu) - k^2, \tag{1.6.40}$$

The solutions are stable at $\mu > 3/2$, at which limit they lose stability to a pattern with $|k_0| = 1$.

In the conserved version, HSS are arbitrary, being fixed by the level of chemical potential. Linearization near a nontrivial HSS leads to the dispersion relation

$$\lambda_c(u_0) = k^2[\mu - (1 - k^2)^2 - 3u_0^2]. \tag{1.6.41}$$

An HSS is unstable at small k if $|u_0| < \sqrt{\frac{1}{3}(\mu - 1)}$. Above this line, it loses stability to a pattern with $k_0^2 = \frac{1}{3}(2 + \sqrt{1 + 3\mu - 9u_0^2}$, equal to $\frac{4}{3}$ at this point. As $|u_0|$ further increases, λ_c vanishes at $|u_0| = \sqrt{\mu/3}$, as the optimal wavelength goes down to unity. The HSS at stable above this level (see Fig. 1.17b).

Matthews and Cox (2000) derived the amplitude equation near the symmetry-breaking instability taking place within the shaded region in Fig. 1.17b at the wavelength $1 \leqslant k_0^2 \leqslant \frac{4}{3}$. The general procedure described in Sect. 1.3.3 can be implemented here without cumbersome algebra. The single variable is expanded as $u = u_0 + \sum \epsilon^n u_n$, and the parameter μ as $\mu = \mu_0 + \epsilon^2 \mu_2$. In the case of bifurcation of a single mode, say, along the x direction, the first-order solution (1.3.23) is simply $u_1 = a(\mathbf{X}, T)\exp(ikx)$ with $|k| = k_0$ and the $O(\epsilon)$ amplitude a dependent on long-scale coordinates $\mathbf{X} = \epsilon\mathbf{x}$ and extended time $T = \epsilon^2 t$. The linear operator is $\mathcal{L} = \mathcal{L}_0 + \epsilon^2\mathcal{L}_2$ with

$$\mathcal{L}_0 = \partial_t + \nabla^2(1 + \nabla^2)^2 - \mu_0 - 2\nu u_0 - 3u_0^2, \quad \mathcal{L}_2 = \partial_T + \heartsuit_{\mathbf{X}}^2 - \mu_2. \tag{1.6.42}$$

where long-scale modulations are accounted for by the anisotropic mixed-order differential operator $\heartsuit$ defined in Sect. 1.3.4.

The second-order equation (1.3.24)reduces to

$$\mathcal{L}_0 u_2 = -(\nu + 3u_0)\left(a^2 e^{2ikx} + c.c + 2|a|^2\right), \tag{1.6.43}$$

It is solved by

$$u_2 = |a|^2 V_0 + a^2 V_2 e^{2ikx} + c.c, \qquad V_0 = \frac{2(\nu + 3u_0^2)}{\mu_0 + 2\nu u_0 - 3u_0^2}$$
$$V_2 = \frac{\nu + 3u_0}{k^2(1 - 4k^2)^2 + \mu_0 + 2\nu u_0 - 3u_0^2}. \tag{1.6.44}$$

The third-order equation (1.3.27) becomes, retaining only resonant terms,

$$\mathcal{L}_0 u_3 = \mathcal{L}_2 u_1 + [2(\nu - 3u_0)(V_0 + V_2) - 3]a^2\overline{a}. \tag{1.6.45}$$

Its solvability condition is simply vanishing right-hand side, leading to the amplitude equation

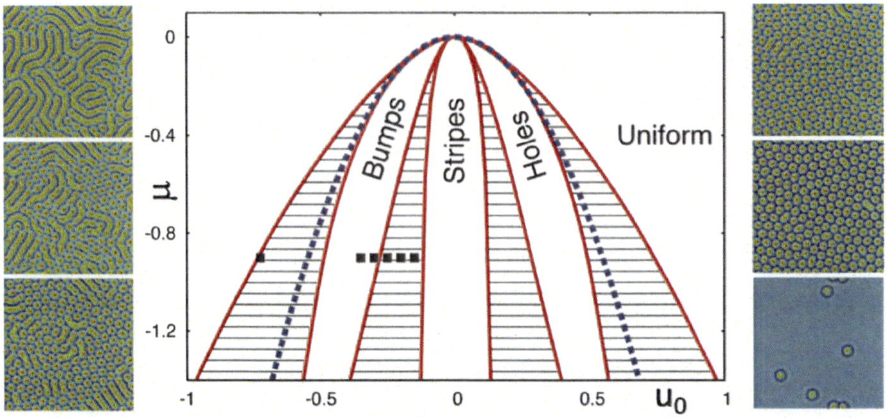

Fig. 1.18. Phase diagram for the conserved SH equation in 2D. The red solid lines show the boundaries of prevailing patterns, , and the gray striped areas show their coexistence regions.The blue dotted line shows the limit of linear stability. The small panels show steady solutions at $\mu = 0.9$ and selected mean order parameter values u_0 indicated by the black squares: from top to bottom, on the left: $u_0 = 0.15$, $u_0 = 0.2, u_0 = 0.25$, and on the right $u_0 = 0.3$, $u_0 = 0.35$, $u_0 = 0.725$ (Thiele et al, 2013, reproduced by permission, ©The American Physical Society)

$$\partial_T a = \heartsuit^2 a + \mu_2 a - [2(\nu - 3u_0)(V_0 + V_2) - 3]|a|^2 a, \tag{1.6.46}$$

Symmetry-breaking instabilities develop into elaborate patterns. In 1D, those are "snaking" steady-state profiles, extending in 2D to irregular lamellar or labyrinthine patterns, such as shown un the upper left panel in Fig. 1.18 (Thiele et al, 2013). The resonance between modes directed at the angle $\pi/3$ to each other favoring the formation of hexagonal patterns in 2D, as discussed in Sect. 1.5.2, takes place also in this model, developing into spotted patterns shown in the right panels of Fig. 1.18.

2

Fronts between Uniform States

2.1 Planar Fronts

2.1.1 Advance into a Metastable State

A single (scalar) reaction-diffusion equation with a function $f(u)$ having two stable homogeneous stationary states (HSS) with $f'(u) < 0$ separated by an unstable state:

$$\partial_t u = D\nabla^2 u + f(u), \tag{2.1.1}$$

The function $f(u)$ may depend on a number of parameters modifying its shape. In particular, the cubic equation (1.2.18) obtained at the cusp singularity with added diffusional term can be seen as the simplest model describing coexistence and competition of alternative stationary states. The diffusivity D can be removed by rescaling the coordinates $\mathbf{x} \to l\mathbf{x}$, where $l = \sqrt{D\tau}$ is the characteristic scale of spatial inhomogeneities in the system and τ is a characteristic time scale, which can be chosen as $\tau^{-1} = f'(\widehat{u})$ for some nominal $u = \widehat{u}$. We shall further recognize l as the characteristic width of a front separating alternative states.

Equation (2.1.1) may model an actual physical system when a single variable or "order parameter" is sufficient for its description. It belongs to a wide class of equations possessing gradient structure (Sect. 1.1.1), and is derivable from the energy functional similar to (1.6.8) with "costs" of the the gradients of the order parameter playing the role of the distortion energy:

$$u_t = -\delta\mathcal{F}/\delta u, \qquad \mathcal{F} = \int \mathcal{L}\,\mathrm{d}\mathbf{x}, \qquad \mathcal{L} = V(u) + \tfrac{1}{2}D|\nabla u|^2. \tag{2.1.2}$$

In this section, we shall consider fronts between two HSS, which are both stable to infinitesimal perturbations and correspond to two alternative minima of the potential $V(u)$. A suitable function $f(u)$ has three zeros, two of which, $u = u_s^{\pm}$ with $f'(u) < 0$), correspond to stable HSS and the middle one, $u = u_s^0$ with $f'(u)u > 0$, to an unstable state. The respective double-well potential is shown in Fig. 2.1. The position and depth of the minima can be regulated by shifting parameters of the model.

L. Pismen, *Patterns and Interfaces in Dissipative Dynamics*, Springer Series in Synergetics,
https://doi.org/10.1007/978-3-031-29579-9_3

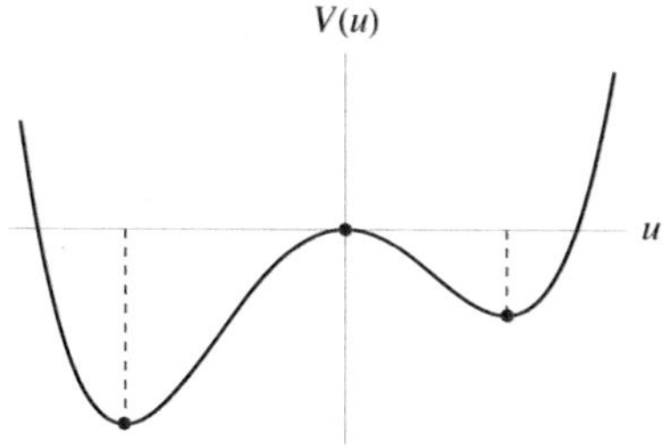

Fig. 2.1. A double-well potential

Suppose that two stable HSS with different energy are approached in spatially removed parts of an extended system. The boundary between them, also called a *front* or a *kink*, is a narrow region of a characteristic width l where the dependent variable changes between the two alternative values $u = u_s^{\pm}$. The state with a lower energy is *globally* stable, while that with a higher energy is *metastable*, and the overall energy of the system will decrease when the boundary between the alternative states propagates in such a way that the area occupied by the globally stable state increases. Since both HSS are attractors, small perturbations of $u_s^{\pm}$ tend to decay. This serves as a localization mechanism, which ensures that a straight-line front in an unrestricted area should move with a constant propagation speed and retain its stationary form as it propagates. The situation is different when the retreating state is *unstable*; this case will be considered in Sect. 2.1.4.

Since propagation is steady, (2.1.1) can be reduces to an ordinary differential equation depending on a single coordinate $z = x - ct$ in the frame *comoving* with the front velocity c, as yet unknown:

$$cu'(z) + Du''(z) + f(u) = 0. \tag{2.1.3}$$

The solution on the infinite line is subject to the boundary conditions $u = u_s^{\pm}$, $u'(z) = 0$ at $z \to \pm\infty$. This problem is overdetermined, so that the solution exists only at a single value of c; thus, the propagation speed is obtained as a solution of a nonlinear eigenvalue problem. This is made transparent when (2.1.3) is viewed as a damped mechanical system, where the coordinate z is playing the role of time. In this interpretation, diffusivity is replaced by mass; $-f(u)$ is a force with the potential $-V(u)$, and c is a *friction* coefficient. Since the sign of the potential is interchanged, *stable* equilibria of the reaction-diffusion equation are mapped onto *unstable* mechanical equilibria and vice versa; the energy integral of a one-dimensional reaction-diffusion system corresponds to the action integral of a mechanical system. The front solution corresponds to a heteroclinic trajectory connecting the equilibria $u = u_s^{\pm}$. In mechanical terms, this would be a trajectory of a Newtonian particle starting from the higher maximum of the potential $-V(u)$, dissipating a part of its energy through friction, and stopping precisely at the lower maximum. This can be made possible only by adjusting the friction coefficient in a unique way.

A more precise proof is based on a "counting" argument. Linearizing (2.1.3) in the vicinity of either stable solution leads to the equation with constant coefficients

$$cu'(z) + Du''(z) - \alpha u = 0. \tag{2.1.4}$$

The coefficients $\alpha = -f'(u_s^{\pm})$ are negative, in view of the presumption that both equilibria are stable. The solutions of (2.1.4) are combinations of exponents $u \propto \exp(\lambda_{\pm} z)$, with the eigenvalues

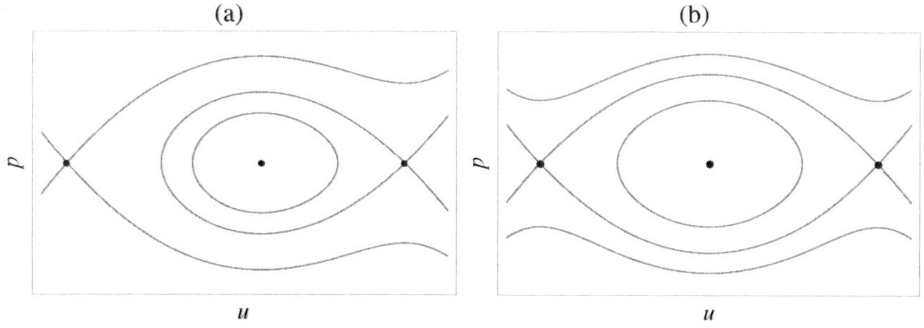

Fig. 2.2. Generic trajectories in the phase plane $u, p = u'(z)$ (a) and a nongeneric set of trajectories containing a heteroclinic orbit (b)

$$\lambda_{\pm} = \frac{c}{2D}\left(-1 \pm \sqrt{1 + 4\alpha D/c^2}\right). \tag{2.1.5}$$

For $\alpha > 0$, both pairs of eigenvalues $\lambda_{\pm}$ are real and of opposite signs, so that both HSS are *saddles* when viewed as equilibria of (2.1.3). There is a single trajectory leaving or entering the saddle point. The front solution is a heteroclinic trajectory connecting both equilibria. Such a trajectory does not exist generically (Fig. 2.2), but can be only obtained by adjusting a single parameter of the equation. The adjustable parameter is the propagation speed c, which is unknown a priori. Since the motion slows down near the equilibrium, the heteroclinic trajectory takes infinite time to traverse; accordingly, the kink solution exists, strictly speaking, only on an infinite line. However, the equilibrium is approached *exponentially*, and therefore, for all practical purposes, the deviation from a HSS becomes negligible at distances of several basic length units $l = \sqrt{D\tau}$.

The sign of c can be determined by multiplying (2.1.3) by $u'(z)$ and integrating over the infinite line, The integral of the diffusional term vanishes, leading to the relation

$$c\int_{-\infty}^{\infty}[u'(z)]^2 \mathrm{d}z = -\int_{-\infty}^{\infty} f(u)u'(z)\mathrm{d}z = V(u_s^+) - V(u_s^-). \tag{2.1.6}$$

Since the integral on the left-hand side is positive, the sign of c coincides with the sign of the difference between the energies of the two competing states. The state with the lower energy advances, and the energy of the entire system is lowered due to increasing area occupied by the more favorable state. The potential difference equals to the derivative of the energy with respect to the front position $\zeta(t) = \zeta_0 + ct$, which can be interpreted as a "thermodynamic force" acting on the front. Therefore the speed can be presented as the ratio of the thermodynamic force and the *dissipation integral*:

$$\frac{\mathrm{d}\zeta}{\mathrm{d}t} = \mathcal{I}^{-1}\frac{\partial \mathcal{F}}{\partial \zeta}, \qquad \mathcal{I} = \int_{-\infty}^{\infty}[u'(z)]^2 \mathrm{d}z = \int_{u_s^-}^{u_s^+} p(u)\mathrm{d}u, \tag{2.1.7}$$

where $p(u) = u'(z)$. The dissipation integral is proportional to the distortion energy due to a single front on the infinite line, obtained by integrating the differential term in (2.1.2), which is equal to $\frac{1}{2}D\mathcal{I}$.

2.1.2 Computation of the Propagation Speed

As a rule, the exact value of the propagation speed can be computed only numerically. Equation (2.1.3) is solved most easily in the phase plane variables, using u as an independent, and $p = u'(z)$ as a dependent variable. Then it is rewritten as

$$cp + Dpp'(u) + f(u) = 0. \tag{2.1.8}$$

Since p vanishes when a HSS is approached, this equation has to be integrated with the boundary conditions $p(u_s^+) = p(u_s^-) = 0$. In this formulation, it is obvious that the first-order ODE with two boundary conditions is overdetermined. One could argue that a simple transformation to the form (2.1.8) might have replaced lengthy arguments from the preceding subsection; this transformation is, however, possible only when the unknown function $u(x)$ is monotonic.

The propagation speed can be found by integrating (2.1.8) starting from either u_s^+ or u_s^- and finding the value of c by shooting. However, this non-autonomous equation is singular at $u = u_s^{\pm}$, since the coefficient at the derivative vanishes there. A numerical solution cannot start precisely at the equilibrium point. Instead, the initial point should be slightly shifted along the unstable manifold of the saddle point of (2.1.8). Taking as a starting point $u = u_s^- + \delta$ with $\delta \ll 1$ and choosing the sign of p to be positive, we set the initial condition

$$p(u_s^- + \delta) = \lambda_+(u_s^-)\delta, \tag{2.1.9}$$

where the eigenvalue $\lambda_+(u_s^-)$ is defined by (2.1.5) with $\alpha = -f'(u_s^-)$. Using some chosen starting value of c, we continue the integration until either u reaches the maximum value u_s^+ or p vanishes. In the former case, the guessed value of c is too small and in the latter, too large (see Fig. 2.3). An efficient numerical procedure may start with the upper and lower a priori bounds of c and converge to the true value by iteratively subdividing the interval, until $p(u_s^+)$ drops below a set small target value. Once the function $p(u)$ is found, the profile $u(x)$ is computed (in an implicit form) by quadrature: $x = \int p^{-1}\mathrm{d}u$.

The propagation speed can be computed analytically when $f(u)$ is cubic:

$$f(u) = -(u - u_s^+)(u - u_s^-)(u - u_s^0), \tag{2.1.10}$$

where u_s^0 is the unstable middle root. The solution satisfying both boundary conditions is sought for in the form

$$p(u) = b(u_s^+ - u)(u - u_s^-). \tag{2.1.11}$$

When this *ansatz* is substituted in (2.1.8), the equation factors out as

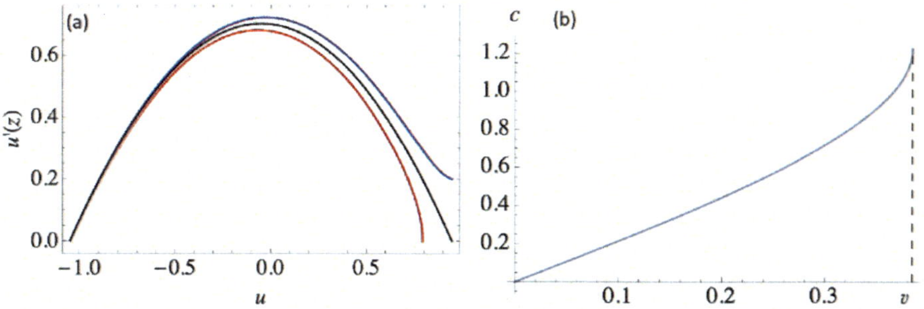

Fig. 2.3. (a) Trial solutions with an underestimated (blue) overestimated (red) and exact (black) values of the propagation speed. (b) The dependence of the propagation speed on the parameter v for the cubic function $f(u) = u - u^3 - v$

$$(u - u_s^+)(u - u_s^-)[(2b^2D - 1)u + bc + u_s^0 - b^2D(u_s^- + u_s^+)] = 0. \qquad (2.1.12)$$

The bracketed linear form has just enough adjustable coefficients to make it vanish. This yields

$$b = \pm 1/\sqrt{2D}, \qquad c = \sqrt{2D}\left[u_s^0 - \tfrac{1}{2}(u_s^- + u_s^+)\right]. \qquad (2.1.13)$$

The speed is assumed to be positive when the lower state advances. The two signs of b correspond to the two alternative orientations of the front.

For the cubic in the form $f(u) = u - u^3 - v$, the speed can be written explicitly using a convenient trigonometric expressions for the roots:

$$u_s^\pm = \pm(2/\sqrt{3})\sin(\psi \mp \pi/3), \quad u_s^0 = (2/\sqrt{3})\sin\psi, \quad \psi = \tfrac{1}{3}\arcsin\left(\tfrac{3}{2}\sqrt{3}\,v\right). \qquad (2.1.14)$$

Using (2.1.14) in (2.1.13) yields

$$c = \sqrt{6D}\sin\left[\tfrac{1}{3}\arcsin\left(\tfrac{3}{2}\sqrt{3}\,v\right)\right]. \qquad (2.1.15)$$

The dependence of the propagation speed on the parameter v is shown in Fig. 2.3b. Note that the curve becomes vertical at the bifurcation point $v = 2/3^{3/2}$ where the roots u_s^0 and u_s^+ merge, terminating there at the maximum value $c_m = \sqrt{3D/2}$. The curve continues antisymmetrically into the region $v < 0$.

The good luck does not extend to higher-order polynomials. Take $f(u)$ of the order $2n - 1$; itr should be odd to prevent runaway to infinity in the dynamical system $u'(t) = f(u)$. Then $p(u)$ must be of th3 order n to match the orders of $f(u)$ and $p(u)p'(u)$. After the boundary conditions are satisfied, $n - 1$ free parameters are left in the polynomial $p(u)$. Those should be used, together with c, to cancel $2(n-1)$ terms of the polynomial remaining after factoring. The numbers match only for $n = 2$, i.e. for a cubic $f(u)$. In more complicated cases, only numerical solutions are available, but Goriely (2006) derived some analytical estimates. Propagating front solutions for various equation systems were recently reviewed by Sánchez-Garduño and Castellanos (2022).

In the special case when both energies $V(u_s^{\pm})$ are equal, called *Maxwell construction*, the propagation speed is zero, so that both states coexist at equilibrium. This may be achieved by adjusting some parameter v at a certain level $v = v_0$, e.g. $v = 0$ for the cubic form. Equation (2.1.3) with vanishing propagation speed is rewritten in phase plane variables as

$$Dpp'(u) + f(u) = 0, \tag{2.1.16}$$

which can be expressed, as in Sect. 1.6.1, in the form $P'(u) - V'(u) = 0$ with $P(u) = \frac{1}{2}Dp^2(u)$. This defines the interfacial energy per unit area of a stationary front, which can be computed as

$$\sigma = D\int_{-\infty}^{\infty} u'(x)^2 \mathrm{d}x = D\int_{u_s^-}^{u_s^+} p(u)\mathrm{d}u = \int_{u_s^-}^{u_s^+} \sqrt{2DV(u)}\mathrm{d}u. \tag{2.1.17}$$

This expression is proportional to the dissipation integral (2.1.7). $\sigma = D\mathcal{I}$. The energy of the front can be also called *surface tension*, stressing the analogy to phase transitions in equilibrium systems.

The stationary front profile can be computed in quadratures by integrating (2.1.16):

$$p(u) = \pm\sqrt{\frac{2}{D}}\sqrt{V(u) - V(u_s^-)}, \tag{2.1.18}$$

whereby the front profile is expressed in an implicit form

$$x(u) = \sqrt{\frac{D}{2}}\int_{u_s^0}^{u} [V(u) - V(u_s^-)]^{-1/2}\mathrm{d}u. \tag{2.1.19}$$

The origin $x = 0$ may be arbitrary, but, to be definite, we have placed it at the location of the unstable intermediate HSS u_s^0. A useful analytical form is the kink solution of the symmetric cubic (KPP) equation (1.1.2) derived from the symmetric quartic potential

$$V(u) = \tfrac{1}{4}\left(1 - u^2\right)^2. \tag{2.1.20}$$

The inhomogeneous stationary solution of this equation dependent on a single coordinate x normal to the front is elementary:

$$p(u) = \left(1 - u^2\right)/\sqrt{2}, \qquad u(x) = \tanh\left(x/\sqrt{2}\right). \tag{2.1.21}$$

2.1.3 Nonmonotonic Solutions

Besides the heteroclinic (front) solution, (2.1.3) has a continuous family of periodic solutions, which correspond to wave trains of varying wavelength. The respective trajectories in the phase plane (u, p) shown in Fig. 2.4a lie within the heteroclinic loop and surround the unstable equilibrium $u = u_s^0$. Another family of solutions are

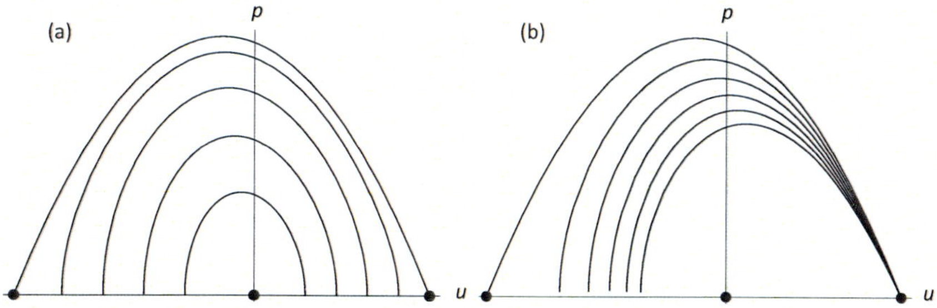

Fig. 2.4. (a) Phase plane trajectories $p(u)$ corresponding to unstable wave trains. (b) A family of homoclinic trajectories corresponding to unstable pulses propagating with different speeds. Only half-trajectories are shown. The limiting heteroclinic trajectory is shown in both figures

homoclinic trajectories approaching at $z \to \pm\infty$ the metastable HSS, interpreted as a propagating pulses, shown in Fig. 2.4b. All these solutions are dynamically unstable when viewed as solutions of the evolution equation (2.1.1).

Stationary periodic patterns can be computed in quadratures as in the preceding subsection, though expressions are somewhat more complicated. The solution of (2.1.16) should be now written as

$$p(u) = \pm\sqrt{2/D}\sqrt{V(u) - V(u_{\min})}, \tag{2.1.22}$$

$$x(u) = \sqrt{\frac{D}{2}} \int_{u_{\min}}^{u} \frac{\mathrm{d}u}{\sqrt{V(u) - V(u_{\min})}}, \tag{2.1.23}$$

where $u_{\min}$ and $u_{\max}$ are the minimum and maximum values of u satisfying the condition $V(u_{\min}) = V(u_{\max})$. The last formula gives in an implicit form the profile $u(x)$ for a half-period of the solution. The admissible values of V lie between the potential of the unstable state $V(u_s^0)$ and the potential of the metastable state, i.e. the larger of $V(u_+^0)$ and $V(u_-^0)$. The lower bound of this interval corresponds to a homoclinic orbit (a critical nucleus) when the two minima are unequal and to a heteroclinic orbit (a static front) when $V(u_+^0) = V(u_-^0)$. The total wavelength is

$$L = \sqrt{2D} \int_{u_{\min}}^{u_{\max}} \frac{\mathrm{d}u}{\sqrt{V(u) - V(u_{\min})}}. \tag{2.1.24}$$

One can see that it diverges when $V(u)$ is extremal, i.e. for $u_{\min} = u_\pm^s$.

Stability of a stationary space-dependent solution, say, $\mathbf{u} = \mathbf{u}_0(x)$ can be tested by considering the system's reaction to a weak perturbation $u_1(x,t)$, which can be presented, as usually, in the spectral form $u_1(x,t) = \sum \mathrm{e}^{\lambda_i t}\phi_i(x)$, where λ_i are eigenvalues and $\phi_i(x)$ are the corresponding eigenvectors. Then (2.1.1) is linearized on the line to

$$\lambda u_1 = D u_1''(x) + f'(u_0)u_1. \tag{2.1.25}$$

This is a simple example of the classical Sturm–Liouville problem. The operator in the left-hand side is *self-adjoint*, and its eigenvalues λ_i are real, and can be ordered in such a way that $\lambda_0 > \lambda_1 > \lambda_2 > \ldots > -\infty$ and each corresponding eigenfunction $\phi_i(x)$ has exactly i zeroes. The left-hand side is annulled by the derivative $u_1 = u_0'(x)$, as proven by differentiating the stationary solution with respect to the coordinate. Thus, $u_0'(x)$ is the eigenfunction of (2.1.1) with zero eigenvalue. If $u_0(x)$ is a monotonic stationary front, this eigenfunction, corresponding to translations of the front, is monotonic, and therefore zero is the largest eigenvalue, and the solution is stable to all perturbations modifying its shape. If, on the other side, $u_0(x)$ is non-monotonic, $u_0'(x)$ has at least one zero, and therefore there must exist another, monotonic, eigenfunction with a positive eigenvalue, which signifies that a stationary state having an extremum must be unstable. Stationary nonmonotonic solutions are possible only in systems with at least two variables, where they arise as a result of the Turing instability (Sect. 1.3.1).

However, relaxation of an unstable may be quite slow. On an interval far exceeding the characteristic front thickness, alternative HSS separated by fronts of opposite polarity (kinks and "antikinks") may arise at different locations following a phase transition. Consider a kink and an antikink centered at $x = \pm L/2$. Each front is only weakly perturbed by the exponential tail of its counterpart when they are removed at a distance of several characteristic diffusional lengths $l = \sqrt{D\tau}$. In the absence of interactions, each front would be described by the respective stationary solutions, $u_0(x \pm L/2)$. As the separation L decreases, the energy of the kink/antikink pair, equal at infinite separation to $2\sigma = 2D\mathcal{I}$, as defined by (2.1.17), decreases as well. The derivative $\partial\mathcal{F}/\partial L$ can be interpreted as a "force" bringing the two fronts together. In non-equilibrium thermodynamics this is known as the *Peach–Köhler* force.

The drift velocity can be computed in the most transparent way starting from the expression for the dissipation rate (1.1.12). The computation is possible when the drift is slow, so that the profile of u always remains quasistationary and is independent of the history of motion, so that it can be expressed by a certain function depending on the current value of L as a parameter. Under these conditions, (1.1.12) can be rewritten as

$$\frac{\partial\mathcal{F}}{\partial L}\frac{\mathrm{d}L}{\mathrm{d}t} = -\int\left(\frac{\delta\mathcal{F}}{\delta u}\right)^2 \mathrm{d}x. \tag{2.1.26}$$

An approximate solution describing a quasistationary kink/antikink pair can be obtained by matching the *inner* solution near the front and the *outer* solution in the region between the two fronts where u is close to u_s^+ (see Fig. 2.5a). The inner solution is the standard front profile $u^{(i)} = u_0(L/2 \pm x)$, while the outer solution is $u^{(o)} = u_s^+ - \widetilde{u}$, where the deviation $\widetilde{u}$ verifies the linearized equation $\widetilde{u}''(x) - \alpha\widetilde{u} = 0$ with $\alpha = -f'(u_s^{\pm}) > 0$. The symmetric outer solution in the overlap region is $\widetilde{u} \propto \cosh(\sqrt{\alpha}x)$ with a proportionality constant that has to match the outer asymptotics of the inner solution. For the kink centered at $x = -L/2$, the latter has the form

$$\lim_{x\to\infty} u^{(i)} = u_s^+ - 2a\mathrm{e}^{-\sqrt{\alpha}L/2}\cosh(\sqrt{\alpha}x). \tag{2.1.27}$$

For example, for the cubic $f(u) = u(1-u^2)$, $\alpha = 2$ and $a = 2$, as can be seen by computing the asymptotics of (2.1.21) at $x \to \infty$. The matching solution in the

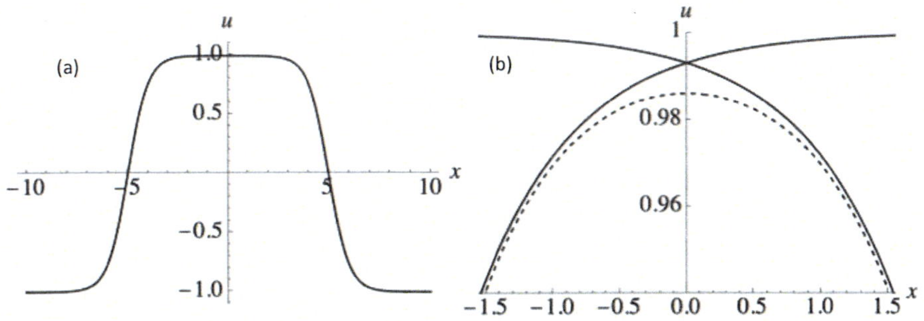

Fig. 2.5. (a) The profile of interacting fronts. (b) The blow-up of the matching solution (2.1.28) in the overlap region (dashed line). The solid lines are tails of well separated stationary fronts.

overlap region is

$$\widetilde{u} = -2a\mathrm{e}^{-\sqrt{\alpha}L/2}\cosh(\sqrt{\alpha}x). \tag{2.1.28}$$

By symmetry, this also matches the antikink centered at $x = L/2$. For example, for the cubic $f(u) = u(1-u^2)$, $\alpha = 2$ and $a = 2$, as can be seen by computing the asymptotics of (2.1.21) at $x \to \infty$. This solution is drawn in Fig. 2.5b.

Combining the inner and outer solutions matched at some position $x = \pm X$, such that $\exp[\sqrt{\alpha}(X - L/2)] \ll 1$, we compute the energy

$$\begin{aligned}
\mathcal{F} &= 2\int_{-\infty}^{L/2-X}[u_0'(x)]^2\mathrm{d}x + \int_0^X \{[\widetilde{u}'(x)]^2 + 2V[u^{(o)}(x)]\}\mathrm{d}x \\
&= 2\mathcal{I} - \int_{L/2-X}^{\infty} a^2\alpha\mathrm{e}^{-\sqrt{\alpha}(2x+L)}\mathrm{d}x + 2\int_0^X a^2\alpha\mathrm{e}^{-\sqrt{\alpha}L}\cosh(2\sqrt{\alpha}X)\mathrm{d}x \\
&= 2\mathcal{I} - a^2\sqrt{\alpha}\mathrm{e}^{-\sqrt{\alpha}L}.
\end{aligned} \tag{2.1.29}$$

The transformation of the outer integral takes into account that the potential can be presented near $u = u_s^+$ as $V = \frac{1}{2}\alpha\widetilde{u}^2$. The result is independent of the matching position X, provided the above condition is satisfied, so that $\cosh(2\sqrt{\alpha}X) \approx \frac{1}{2}\exp(2\sqrt{\alpha}X)$. As expected, $\partial\mathcal{F}/\partial L > 0$, and the dependence on separation is exponential as long as separation is large compared with the effective front thickness.

Differentiating the final expression in (2.1.29) with respect to L gives the thermodynamic force on the left-hand side of (2.1.26). This force can be compensated by shifting the parameters to lower the energy of the state in the segment between the two fronts. Since the total energy gain is proportional to the length of the region occupied by the preferred state, the stationarity condition is $\partial\mathcal{F}/\partial L = \Delta V = V(u_s^-) - V(u_s^+)$. For example, for a simple parametric shift $f(u) \to f(u) + v$ with $v \ll 1$, the energy gain for the upper state is $\Delta V = v(u_s^+ - u_s^-)$. Using (2.1.29), we see that the domain with the width L is stationary at

$$v = \widetilde{v} = \frac{a^2\sqrt{\alpha}}{u_s^+ - u_s^-}\mathrm{e}^{-\sqrt{\alpha}L}. \tag{2.1.30}$$

The equilibrium is unstable, but can be maintained if the parameter v is adjusted dynamically, for example by maintaining the ratio of domain lengths of the alternative states (see Sect. 2.2.4).

Returning to the dynamical problem, it remains to evaluate the right-hand side of (2.1.26). According to the gradient dynamics principle (2.1.2), the variation $\delta\mathcal{F}/\delta u$ equals in the evolving system to the time derivative $u_t \approx u_0'(x)\mathrm{d}L/\mathrm{d}t$ with an inverted sign. Thus, (2.1.26) can be rewritten as

$$\frac{\mathrm{d}L}{\mathrm{d}t} = -\frac{\partial\mathcal{F}}{\partial L}\Big/\int_{-\infty}^{\infty}[u_0'(x)]^2\mathrm{d}x = -\tfrac{1}{2}\mathcal{I}^{-1}a^2\alpha \mathrm{e}^{-\sqrt{\alpha}L}. \tag{2.1.31}$$

The dissipation integral is replaced here by twice the standard value for two noninteracting fronts, since the correction due to interactions is exponentially small.

This formula can be extended to an arbitrary array of interacting fronts centered at $x = \zeta_j$. Since interactions decrease exponentially with separation, only nearest-neighbor interactions are relevant. The generalization of (2.1.31) for a symmetric case $\alpha = f'(u_s^{\pm}) = f'(u_s^{-})$ is

$$\mathrm{d}_t\zeta_j = \tfrac{1}{2}\mathcal{I}^{-1}a^2\alpha\left[\mathrm{e}^{-\sqrt{\alpha}(\zeta_{j+1}-\zeta_j)} - \mathrm{e}^{-\sqrt{\alpha}(\zeta_j-\zeta_{j-1})}\right]. \tag{2.1.32}$$

In an asymmetric case when $\alpha_{\pm} = f'(u_s^{\pm})$ differ, one should replace in this formula α by α_+ in the first term and by α_- in the second term if the front at the jth position is a kink, and the other way around if it is an antikink. Equation (2.1.32) does not describe the last stages of interaction at short distances and the annihilation of the colliding kink and antikink, but it is clear that the duration of these stages is much shorter than the time of slow approach of fronts originally separated at distances far exceeding the characteristic diffusional length. Interaction of fronts of opposite polarity leading to their annihilation provides the mechanism of Ostwald ripening in 1D.

2.1.4 Propagation into an Unstable State

The problem of front propagation into an *unstable* state arises naturally in population dynamics. A stationary state where a competitively advantageous species is absent is formally unstable to infinitesimal perturbations but will be nevertheless preserved at any location until this species is introduced there. This will typically happen as the habitat boundary propagates outward, i.e., via a propagating front connecting the stable and unstable states. The difficulty of this problem, which has motivated the pioneering work of Kolmogorov et al (1937), lies in the nonuniqueness of a solution describing a steadily propagating front. Therefore the propagation speed cannot be computed in a straightforward way by solving an ODE. It is selected dynamically, and the full nonstationary problem should be considered to understand the selection mechanism.

If the receding state is unstable, the coefficient α in the linearized equation (2.1.4) is positive. Then both eigenvalues (2.1.5) are real negative at $c > c_{\min} = \frac{1}{2}\sqrt{\alpha D}$;

below this value they are complex with a negative real part. This means that at $c > c_{\min}$ the unstable state, viewed as an equilibrium point of (2.1.3), is a stable node, and the unstable manifold of the advancing stable HSS connects generically to this point (see Fig. 2.6). The generic connection persists also at smaller positive velocities, but the approach to the equilibrium is oscillatory. Thus, there is a continuum of one-dimensional front solutions, and restricting to monotonic solutions gives only a lower bound $c = c_{\min}$. The question is, which speed is actually chosen by the dynamically evolving system. Moreover, it is not evident a priori that transition from an unstable to a stable state should be carried by a propagating front retaining its quasistationary shape.

Any perturbation of an unstable state is destined to grow; therefore, dynamics of transition from an unstable to a stable state should strongly depend on initial conditions. The kind of an initial condition that is likely to generate a propagating front is a perturbation localized in a narrow region. This kind of a perturbation is typical for many physical problems. When a perturbation of this kind is introduced into a system that had been rapidly quenched into an unstable state, the transition into a stable state is effected locally, and a sharp front propagating into the unstable state is formed. In problems of population dynamics (which historically were the first to prompt the interest in front propagation dynamics), a localized perturbation corresponds to a locally introduced successful mutant population that displaces an old genome. In other systems, the stable state may nucleate at the boundary or at a local inhomogeneity. In the absence of other perturbations, the global transition to the stable state is effected through front propagation.

The method developed for the reaction-diffusion equation by Kolmogorov et al. (1937), and later generalized to pattern-forming systems (Dee and Langer, 1983, Ben-Jacob et al., 1985; see Sect. 4.4.1), is based on the asymptotic linear analysis of the leading edge of the propagating front. The front is called "pulled" when its propagation speed is determined by growth of perturbations at its leading edge.

Consider growth of a perturbation of an unstable state $u = u_s^0$. While the perturbation remains small, the linear theory applies, and the solution can be presented as

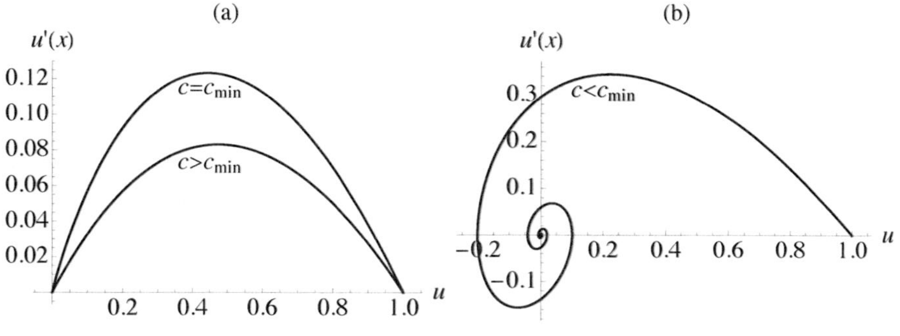

Fig. 2.6. Trajectories in the phase plane $u, p = u'(z)$ connecting a stable and an unstable equilibrium at $c \geqslant c_0$ (a) and $c \leqslant c_0$ (b)

a Fourier integral

$$u(x,t) - u_s^0 = \frac{1}{2\pi}\int_{-\infty}^{\infty} \widehat{u}(k)\mathrm{e}^{\mathrm{i}[kx-\omega(k)]t}\mathrm{d}k. \tag{2.1.33}$$

The dispersion relation $\omega(k)$ follows from a linearized evolution equation. If the latter has a general form

$$\partial_t u = \mathcal{L}(\partial_x^2)u + \alpha(u - u_s^0), \tag{2.1.34}$$

the dispersion relation obtained in the vicinity of $u = u_s^0$ is

$$\omega = \mathrm{i}[\alpha + \mathcal{L}(-k^2)] = \mathrm{i}(\alpha - Dk^2). \tag{2.1.35}$$

In particular, $\mathcal{L} = -Dk^2$ for the linearized reaction-diffusion equation.

To determine the front speed c, we evaluate (2.1.33) asymptotically for large x, t, such that $z = x - ct = O(1)$. This solution should describe the leading edge of the front (where the linear theory still applies), advancing with the speed c, as yet undetermined. The integral (2.1.33) transformed to the frame $z = x - ct$ comoving with the propagating front takes the form

$$u(\xi,t) - u_s^0 = \frac{1}{2\pi}\int_{-\infty}^{\infty} \widehat{u}(k)\mathrm{e}^{\mathrm{i}[kz-\Omega(k)t]}\mathrm{d}k, \tag{2.1.36}$$

where $\Omega(k) = \omega(k) - ck$ is the Doppler-shifted frequency. If the original perturbation is sufficiently localized, all Fourier modes are present, and $\widehat{u}(k)$ should be a smooth $O(1)$ function. The integral (2.1.36) can be evaluated using the method of steepest descent in the complex k-plane, which implies that the integral is dominated by the largest saddle point $k = \kappa$ of $\Omega(k)$. This point should be an extremum of $\Omega(k)$, and the imaginary part of $\Omega(\kappa)$ must vanish if the solution expressed by (2.1.36) is expected to neither grow nor decay in time.

$$\Omega'(k) = \omega'(k) - c = 0, \qquad \mathrm{Im}[\omega(k) - ck] = 0. \tag{2.1.37}$$

Since c is real, the imaginary part of $\omega(\kappa)$ should be constant, so that

$$c = \omega_r'(\kappa) = \omega_i/\kappa_i, \quad \omega_i'(\kappa) = 0. \tag{2.1.38}$$

where the subscripts r, i denote the real and imaginary parts, respectively. It follows from the last relation in (2.1.38) that the real part of κ must vanish; thus, as expected, no oscillations arise here, and both ω and κ are purely imaginary. The approximation implies that the Fourier integral is dominated at late stages of evolution by this particular value of k. The solution at the leading edge of the front can be now presented as

$$u - u_s^0 \propto \mathrm{e}^{\mathrm{i}[\kappa z-\Omega(\kappa)t]} = \mathrm{e}^{-\kappa_i(x-ct)}, \tag{2.1.39}$$

so that the imaginary part of κ determines the rate of decay in the direction of propagation.

According to (2.1.38), the selected propagation speed coincides with the group velocity $\widehat{c} = \omega_r'(\kappa)$ that determines the propagation speed of perturbations. In a frame moving with a speed faster than $\widehat{c}$, the solution would appear to be decaying in time; thus, one can say that propagation plays a stabilizing role, in a sense that one can outrun a growing perturbation by moving sufficiently fast. If $\widehat{c}$ were larger than the speed c with which the front moves, perturbations would run ahead of the front and cause the profile to be unstable. The selected speed is such that the perturbation neither grows nor decays, i.e., it satisfies the condition of *marginal stability* (Dee and Langer, 1983).

2.1.5 Stability of the Leading Edge

In order to gain more physical understanding of the stability of the leading edge of the propagating front and of the way the leading edge of the propagating front evolves to a marginally stable profile, we generalize (2.1.39), following van Saarloos (1988, 2003), to $u = \mathrm{e}^{\mathrm{i}\theta(x,t)}$, where $k = \partial\theta/\partial x$ is allowed now to change slowly in space and time. The evolution equation of θ, rewritten in a frame propagating with the velocity c, has a general form

$$\partial_t\theta = ck - \omega(k) + \mathcal{M}(\partial_x, \partial_x^2, \ldots)k, \tag{2.1.40}$$

where $\mathcal{M}$ is a nonlinear differential operator derivable from $\mathcal{L}$. One can easily see that the imaginary part of this equation is indeed verified by θ_i = const with $k = \kappa$ when the propagation speed is defined by (2.1.38). To check stability of this solution, we linearize (2.1.40) to obtain the equation for a perturbation $\widetilde{\theta}$:

$$\partial_t\widetilde{\theta} = [c(\kappa) - \widehat{c}_r(\kappa)]\partial_x\widetilde{\theta} + \mathcal{M}_{\partial_x}(\kappa)\partial_x^2\widetilde{\theta} + \ldots. \tag{2.1.41}$$

Omitted terms contain higher derivatives with respect to x. The profile is stable if perturbations bounded at $x \to \infty$, that have the form $\widetilde{\theta} \sim \mathrm{e}^{\mathrm{i}\widetilde{k}x}$ with $\widetilde{k}_i > 0$, decay with time. For small $\widetilde{k}$, only the first term in (2.1.41) is relevant, and therefore the *necessary* condition for stability is

$$\mathrm{Re}\{\mathrm{i}[c(\kappa) - \widehat{c}(\kappa)]\widetilde{k}\} < 0 \quad \text{for all } \widetilde{k} \ll 1,\ \widetilde{k}_i > 0. \tag{2.1.42}$$

Separating the real and imaginary parts gives the necessary stability conditions

$$c > \widehat{c}(\kappa), \qquad \widehat{c}_i(\kappa) = 0. \tag{2.1.43}$$

The first condition coincides with the intuitive statement in the preceding subsection that the velocity c of a stable front must exceed the propagation speed of small perturbations that is determined by the real part of the group velocity $\widehat{c}$. Vanishing imaginary part of the group velocity is the condition, that has appeared already in (2.1.38), insuring stability to changes of a local wave number of the emerging pattern.

A nonlinear analysis of (2.1.40) is required to describe the evolution to the marginally stable profile. This analysis has also been carried out by van Saarloos (1988) in the long-scale approximation, assuming a slow spatial variation of k. Suppose that the amplitude of the propagating solution is monotonically decreasing, so that $\vartheta = \theta_i$ is a monotonically increasing function of x. It is advantageous then to write the evolution equation of k taking ϑ, instead of x, as an independent variable changing in the direction of propagation. The transformation to the new variable is carried out using the relations

$$(\partial_x)_t = k_i(\partial_\vartheta)_t, \qquad (\partial_t)_x = (\partial_t)_\vartheta + (\vartheta_t)_x(\partial_\vartheta)_t,$$

where $(\partial_x)_t$ denotes the differentiation with respect to x at $t = \text{const}$, etc., and the imaginary part of the right-hand side of (2.1.40) should be substituted for $(\vartheta_t)_x$. Then differentiating (2.1.40) with respect to ϑ yields

$$\partial_t k = [c(k) - \widehat{c}(k)]\, k_i \partial_\vartheta k + \text{ (higher order terms)}. \tag{2.1.44}$$

The bracketed combination, that we denote now as $C(k)$, is the same as in (2.1.41), except that it is now computed at a variable rather than constant value of k. When small terms, containing higher order derivatives with respect to ϑ and powers of $\partial_\vartheta k$, are omitted, we are left with what appears to be a standard first-order quasilinear PDE, but is generally a system of equations for k_i, k_r. The standard case is, however, recovered when the stability condition $\widehat{c}_i = 0$ holds. Using this condition to fix the relation between k_r and k_i, C, which is now real, can be written as a function of k_i only. Then it follows from (2.1.44) that k is conserved along a characteristic $\partial_t\vartheta = -k_i C(k_i)$. Since $k_i > 0$, and C must be positive for a stable profile, the characteristics are directed toward *decreasing* ϑ. This supports our notion that the leading edge plays the premier role in shaping the emerging pattern: the causal propagation along the characteristics is directed toward the *stable* state. When $C'(k_i) < 0$, steeper parts of the profile propagate back slower, and therefore dominate the leading edge of the front. Under these conditions, the front steepens in the course of evolution, and approaches the slope k_i corresponding to either a minimum of $C(k_i)$ or the marginal value $C(k_i) = 0$.

2.1.6 Pulled and Pushed Fronts

The leading edge analysis allows us to predict with a high precision both the propagation velocity and the wavelength of the emerging pattern. Some caveats are, however, in order. The above analysis has all common drawbacks of both linear and long-scale theories. The front stable according to the leading edge criteria can be destabilized by some "invasion" mode involving rapid changes of the slope, or modified by nonlinear interactions to give way to a more rapidly propagating front, as shown below.

The dispersion relation for the reaction–diffusion equation is

$$\omega(k) = \mathrm{i}(\alpha - Dk^2),$$

so that $\widehat{c} = -2\mathrm{i}Dk$. The second condition (2.1.38) yields simply $k_r = 0$; thus, as expected, no oscillations arise here, and both ω and $k = \mathrm{i}q$ are purely imaginary. Using $c = \omega/k = \alpha/q + Dq$ and $\widehat{c} = 2Dq$, we see that $C(q) = \alpha/q - Dq$ vanishes at $q = \sqrt{\alpha/D}$, which coincides with the minimum on the dependence $c(q)$. Thus, the marginally stable propagation speed $c = 2\sqrt{\alpha D}$ turns to be in this case also the *slowest* speed satisfying the stability condition (2.1.43).

This simple result of the leading edge theory becomes, under certain circumstances, rather unlikely. Consider equation (2.1.3) with the standard cubic nonlinearity $f(u) = u - u^3 - v$. There is an exact solution determining the propagation speed or the front separating two stable states (Sect. 2.1.2). As one of the competing states approaches the bifurcation point $v = 2/3^{3/2}$ where it becomes marginally stable, the propagation speed grows (see Fig. 2.2b), reaching the maximum value $c = \sqrt{3D/2}$ at the bifurcation point itself. On the other hand, the propagation speed into the same marginally stable state can be computed, as above, by approaching it along the branch of *unstable* states. Along this branch, the propagation speed predicted by the leading edge theory, $c = \frac{1}{2}\sqrt{\alpha D}$ *decreases* with the decreasing derivative $\alpha = f'(u_s^0)$ and *vanishes* at the marginally stable point.

The clue to this apparent contradiction is that the special propagating solution obtained in Sect. 2.1.2 is also applicable to a front separating a stable and an unstable state. Such a front, determined by the full nonlinear solution, is called "pushed", as it is "pushed ahead" by an advancing stable state rather than "pulled along" by perturbations growing at the leading edge. If, say, the state $u = u_s^+$ is approached at $x \to -\infty$, this solution is expressed as

$$u'(x) = -(u - u_s^0)(u_s^+ - u)/\sqrt{2D}, \qquad c = \sqrt{D/2}(u_s^+ + u_s^0 - 2u_s^-), \quad (2.1.45)$$

where $u_s^\pm$, u_s^0 are defined by (2.1.14). The asymptotic value of the slope at the leading edge following from (2.1.45) is

$$\widehat{q} = \left| \frac{\mathrm{d}\ln(u - u_s^0)}{\mathrm{d}\xi} \right|_{\xi\to\infty} = \frac{p(u_s^0)}{u - u_s^0} = \frac{u_s^+ - u_s^0}{\sqrt{2}}, \qquad (2.1.46)$$

which *increases* when u_s^0 is shifted toward the marginally stable point $v = -2/3^{3/2} \approx -0.3849$, contrary to a *decrease* of $q = \sqrt{f'(u_s^0)}$ predicted by the leading edge theory. At $v > \widehat{v} \approx -0.2078$, the slope $\widehat{q}$ exceeds q (Fig. 2.7a).

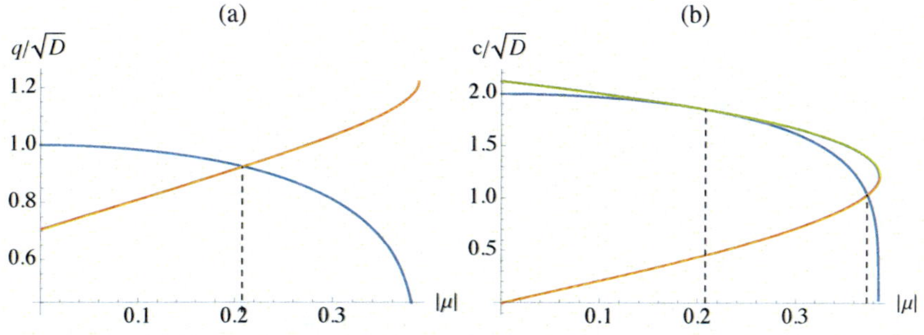

Fig. 2.7. The asymptotic slope (a) and velocity (b) of pulled fronts (blue curve) and pushed fronts, with the green and orange curves corresponding to the advance of the stable state to the unstable and metastable states, respectively

The system chooses the propagating solution with the steepest slope, which now, unlike the generic situation in the linear leading edge theory, is also moving faster (Fig. 2.7b). The plot of the propagation speed defined by (2.1.45) against v is tangential at $v = \widehat{v}$ to the plot of the generic propagation speed $c = 2Dq$. The transition between the generic pulled front and the pushed front determined by the special nonlinear solution (2.1.45) occurs at $v = \widehat{v}$ (Ben Jacob et al., 1985). Beyond this value, the pushed front, which is still faster but has a lower slope than the pulled one, has no physical significance.

The reason why competition with nongeneric pushed fronts has not surfaced in the mathematically rigorous theories of Kolmogorov et al. (1937) and Aronson and Weinberger (1978) is that the special nonlinear solution (2.1.45), apparently, cannot be attained as a result of evolution starting from localized initial condition with $u \geq u_s^0$. In problems of population dynamics, which gave motivation to the early theories, states with $u < u_s^0 = 0$ are nonphysical, and, in any case, solutions of (2.1.4) with $\alpha > 0$ starting from strictly nonnegative u initial conditions remain nonnegative at all times.

On the other hand, if deviations from $u = u_s^0$ of any sign are allowed, evolution to the pushed front can be understood in the following way (van Saarloos, 1989). Consider a propagating front of the state $u = u_s^-$, which is metastable at v slightly above the bifurcation value $v = -2/3^{3/2}$, into the unstable state $u = u_s^0$, being trailed by a transition front from u_s^- to the globally stable state u_s^+, as shown in Fig. 2.8. Sufficiently close to the bifurcation point, (actually, at $|v| > 0.3730$ – see Fig. 2.7b), the trailing front moves faster, and catches on. As a result, the intermediate metastable state is exterminated, and u becomes a monotonically decreasing function of ξ, while the front evolves to the nongeneric profile (2.1.45). A more complex mechanism of relaxation to a pushed front is required at $0.2078 < |v| < 0.3730$ when a pulled front propagates into the unstable state faster than the globally stable state displaces the metastable one. This indicates an important role of perturbations bringing the system locally below the level $u = u_s^0$. While in the final state $u = u_s^0$ is nonnegative, transient states oscillating around $u = u_s^0$ are necessary for evolution to the rapidly propagating pushed front (Powell et al., 1991).

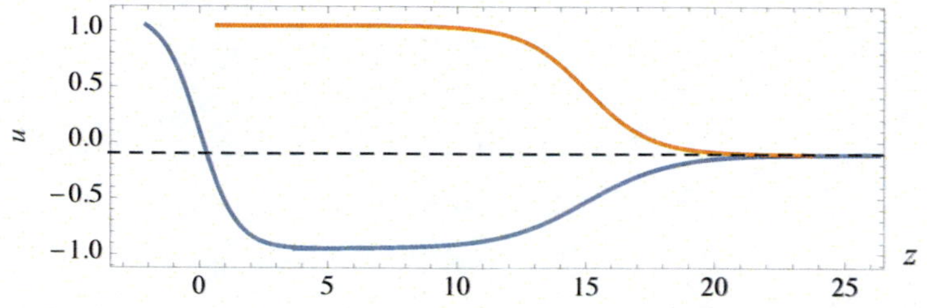

Fig. 2.8. A nonlinear front trailing a pulled front (blue curve) and a pushed front formed after the nonlinear front has caught up (orange curve)

2.2 Weakly Curved Fronts

2.2.1 Aligned Coordinate Frame

The front solution is neutrally stable to translations along the x-axis. This *Goldstone mode* is weakly perturbed when the translation is weakly nonuniform, so that the front becomes curvilinear but the curvature radius still far exceeds the characteristic front thickness.

Propagation of a weakly curved front should be studied using a coordinate frame aligned with its deformed shape. First, we have to define the nominal front position, replacing a diffuse transitional region by a planar curve $\mathcal{C}$. The most natural choice is to fix it as the locus of the unstable zero of the function $f(u)$. The curve chosen in this way can be approximated locally by a circular segment, and (2.1.3) rewritten in polar coordinates r, ϕ. The local coordinates transverse and parallel to the front can be expressed as $x = r - R$ and $y = R\phi$, and since the curvature radius R far exceeds the characteristic diffusional scale, $x \ll R$ in the transitional region where u differs significantly from both alternative HSS. Then the 2D Laplacian can be approximated as

$$\nabla^2 = r^{-1}\partial_r(r\partial_r) + r^{-2}\partial_\phi^2 = \partial_x^2 + \kappa\partial_x + \partial_y^2 + O(\kappa^2), \tag{2.2.1}$$

where $\kappa = R^{-1} \ll 1$ is the local curvature.

Generally, any planar curve $\mathcal{C}$ can be defined at any moment of time t in a parametric form $\mathbf{X}(s;t)$, where s is the arc length. The *tangent* to $\mathcal{C}$ at any point is the derivative $\mathbf{X}_s = \mathbf{l}$. Since $\mathbf{l}$, by construction, is a unit vector, the vector $\mathbf{l}_s$ must be orthogonal to $\mathbf{l}$. By definition, the direction of $\mathbf{l}_s$ is the *normal* $\mathbf{n}$, and its absolute value is the *curvature* κ. Alternatively, the curve can be parametrized by any coordinate y; the infinitesimal arc length $\mathrm{d}s$ is defined as $\mathrm{d}s = g\,\mathrm{d}y$, where $g = |\mathbf{X}'(y)|$ is the metric factor. Then the unit tangent vector is computed as $\mathbf{l} = \mathbf{X}'(s) = g^{-1}\mathbf{X}'(y)$.

The coordinate frame aligned at any point of the curve with the tangent and normal vectors is set up in the following way. By convention, the x-axis has its origin on the curve and is directed from the region occupied by the down state to the region occupied by the up state, so that the boundary conditions $u = u_s^{\pm}$ are approached asymptotically far from the front location. The vector $\mathbf{n}$ is directed, by definition, toward the local center of curvature removed at the distance $|\kappa|^{-1}$. In view of these

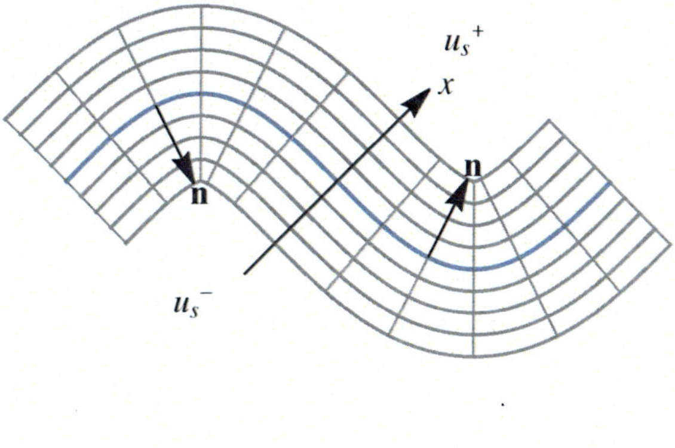

Fig. 2.9. Construction of the aligned coordinate frame. The coordinate lines are shown in gray. *Arrows* show the local directions of the normal $\mathbf{n}$ and the x-axis. Note of a singularity developing on the concave side

direction conventions, the curvature κ should be considered positive when the normal is directed toward the down state, and negative otherwise. The coordinate lines $x =$ const are obtained by shifting the curve along the normal by a constant increment, as shown in Fig. 2.9. Evidently, this shift causes the length to increase on convex, and to decrease on concave side of the curve. The length element is computed as

$$\mathrm{d}s^2 = \mathrm{d}x^2 + (1+\kappa x)^2 \mathrm{d}y^2. \tag{2.2.2}$$

Taking the metric tensor from this expression, the Laplacian is expressed as

$$\nabla^2 = (1+\kappa x)^{-1} \left[\partial_x (1+\kappa x)\partial_x + \partial_y (1+\kappa x)^{-1} \partial_y\right]. \tag{2.2.3}$$

The aligned system is well defined only sufficiently close to the curve, due to a singularity developing on the concave side at a distance of $O(\kappa^{-1})$. If the curvature radius far exceeds the diffusional range l, the aligned system remains regular in the region where u changes between the alternative equilibria $u = u_s^{\pm}$. If the thickness of this transitional region is $l = O(1)$, the curvature correction can be treated perturbatively when $\kappa \ll 1$. We set $\kappa = O(\epsilon)$, expressing the order of magnitude through a book-keeping small parameter ϵ.

The state variable(s) should change slowly in the tangential direction. The appropriate scaling compatible with the scaling of the curvature can be estimated by computing the curvature of a slightly bended curve defined in the Cartesian coordinates X, Y by the equation $X - \zeta(Y) = 0$. Assuming $\zeta'(Y) \ll 1$, the components of the normal vector are, in the leading order, $n_X = 1,\ n_Y = -\zeta'(Y)$, and the curvature is computed, again in the leading order, as $\kappa = \zeta''(Y)$. Thus, one should assume $\partial_Y = O(\sqrt{\epsilon})$ to obtain $\kappa = O(\epsilon)$. Adopting the same scaling for the derivatives with respect to the transverse coordinate in the aligned frame, we rewrite the Laplacian (2.2.3) as

$$\nabla^2 = \partial_x^2 + \epsilon(\kappa\partial_x + \partial_y^2) - \epsilon^2 x\left(\kappa_y \partial_y + 2\kappa\partial_y^2 + \kappa^2\partial_x\right) + O(\epsilon^3). \tag{2.2.4}$$

Up to the first order, the general derivation is equivalent to the simplified version using the local polar coordinates, but in the second order there is an additional term due to transverse variation of curvature.

In 3D, one can define in a similar way a coordinate system aligned with a surface $\mathbf{X}(y^i;t)$ parametrized by coordinates $y^i,\ i = 1,2$. The surface metric tensor is defined as $\gamma_{ij} = (\partial_i \mathbf{X})(\partial_j \mathbf{X})$, where $\partial_i = \partial/\partial y^i$. As in 2D, the surface frame is extended into a neighboring layer by translating the surface along the normal. The length element in the aligned frame is expressed as

$$\mathrm{d}s^2 = \mathrm{d}x^2 + \left(\gamma_{ij} + x\kappa_{ij} + x^2 \kappa_i^k \kappa_{kj}\right) \mathrm{d}y^i \mathrm{d}y^j, \tag{2.2.5}$$

where κ_{ij} are elements of the curvature tensor; the indices are raised and lowered with the help of the surface metric tensor γ_{ij} and summation over repeated lower and upper indices is presumed.

An orthogonal frame can be constructed on a surface by choosing coordinate axes directed locally along the principal directions of the curvature tensor. Then the Laplacian is expressed as

$$\nabla^2 = \frac{1}{\sqrt{g}}\left[\partial_x \sqrt{g}\,\partial_x + \partial_1 \frac{1+\kappa_2 x}{1+\kappa_1 x}\partial_1 + \partial_2 \frac{1+\kappa_1 x}{1+\kappa_2 x}\partial_2\right], \tag{2.2.6}$$

where κ_i are the two principal curvatures and $\sqrt{g} = (1+\kappa_1 x)(1+\kappa_2 x)$. The approximate expression employing the scaling $\kappa_i = O(\epsilon)$, $\partial_i = O(\sqrt{\epsilon})$ is

$$\nabla^2 = \partial_x^2 + \epsilon(\kappa\partial_x + \nabla_\perp^2) + O(\epsilon^2), \tag{2.2.7}$$

where $\kappa = \kappa_1 + \kappa_2$ is twice the mean curvature and $\nabla_\perp^2 = \partial_1^2 + \partial_2^2$ is the transverse Laplacian. The second-order term in this expansion is computed as

$$-\epsilon^2 x\left[(\kappa_1^2+\kappa_2^2)\partial_x + \partial_1(\kappa_1-\kappa_2)\partial_1 + \partial_2(\kappa_2-\kappa_1)\partial_2 + 2(\kappa_1\partial_1^2 + \kappa_2\partial_2^2)\right].$$

The general covariant form of the transverse part of the bracketed expression is $2\partial_i\kappa^{ij}\partial_j - \partial^i\kappa\partial_i$.

2.2.2 Expansion Near the Maxwell Construction

Near the Maxwell construction $v = v_0$, a curvature correction may become comparable to a slow speed of a straight-line front. Both the deviation $v - v_0$ and the propagation speed must be of $O(\epsilon)$ to match the orders of magnitude. The difference between the energies of both stable uniform states should be small, $V(u_s^+) - V(u_s^-) = \epsilon V_1$. Thus, we set $f(u) = f_0(u) + \epsilon f_1(u)$. Transforming the time derivative in the comoving frame as $\partial_t \to \partial_t - \epsilon c\partial_x$ and expanding the state variable as $u = u_0(x) + \epsilon u_1(x,y) + \ldots$ yields, in the zero order

$$Du_0''(x) + f_0(u_0(x)) = 0. \tag{2.2.8}$$

This equation is verified by a stationary front solution $u_0(x)$ satisfying the asymptotic condition $u = u_s^\pm$ at $x \to \pm\infty$.

The first-order equation is

$$Du_1''(x) + f_0'(u_0(x))u_1 + \Psi(x,\, y) = 0, \tag{2.2.9}$$

which contains the inhomogeneity

$$\Psi(x,\, y) = (c + D\kappa)u_0'(x) + f_1(u_0(x)). \tag{2.2.10}$$

The term containing the curvature κ of a planar curve or the mean curvature of an interface in 3D comes from the expansion of the Laplacian (2.2.4) or (2.2.7). The propagation speed is determined by the solvability condition of (2.2.9). The inhomogeneity $\Psi(x,\, y)$ should be orthogonal to the Goldstone eigenmode $u_0'(x)$ that corresponds to the translational symmetry of a stationary front:

$$\int_{-\infty}^{\infty} u_0'(x)\Psi(x,y)\mathrm{d}x = 0. \tag{2.2.11}$$

This yields the *eikonal equation*

$$c = c_0 - D\kappa, \tag{2.2.12}$$

where $c_0 = V_1/\mathcal{I}$ is the speed of the straight-line front (cf. (2.1.6) applied in the vicinity of Maxwell construction). The dissipation integral $\mathcal{I}$ defined by (2.1.7) is proportional to the interfacial energy per unit length (surface tension) $\sigma = D\mathcal{I}$, as shown in Sect. 2.1.2.

At the Maxwell construction ($f_1 = V_1 = 0$), the motion is driven exclusively by curvature: $c = -D\kappa$. The front propagates along the curvature vector in such a way that a convex region occupied by one of the alternative equilibria shrinks. The stationary condition $c_0 = D\kappa$ defines the radius of a critical nucleus of a preferred state; this stationary state is, evidently, unstable, as the droplet with the curvature larger than critical shrinks, while beyond the critical radius it grows indefinitely. A more explicit form of the stationary condition, $V(u_s^+) - V(u_s^-) = \sigma\kappa$, is linked to the Gibbs–Thomson law relating the equilibrium pressure or temperature with the radius of a droplet (see also Sect. 2.3.1).

The speed of the curvature-driven motion can be derived directly from the variational principle (2.1.2). We observe that both the standard profile of the order parameter across the front $u_0(x)$ and the distortion energy per unit length of a weakly curved front arc, in the leading order, independent of curvature. Therefore, the energy functional is proportional to the length (in 2D) or area (in 3D) of the front. In 3D it is computed as

$$\mathcal{F} = D\mathcal{I} \int \mathrm{d}S = D\mathcal{I} \int \sqrt{g}\, \mathrm{d}y_1 \mathrm{d}y_2. \tag{2.2.13}$$

Since energy is totally determined in this approximation by geometry of the interface, its variation with respect to the order parameter can be expressed in purely geometrical terms. Consider a Cartesian frame $(X, \mathbf{Y})$ where the X-axis is aligned with the normal at a chosen point and $\mathbf{Y}$ spans the tangent plane at the same point. An adjacent segment of the interface can be defined by the equation

$$\Phi(X, \mathbf{Y}) \equiv X - \zeta(\mathbf{Y}) = 0. \tag{2.2.14}$$

In this parametrization, the metric factor is $g = |\nabla\Phi|^2 = 1 + |\nabla_\perp\zeta|^2$ and the normal vector is $\mathbf{n} = g^{-1/2}\nabla\Phi$. We take note that only displacements normal to the front are significant, while parallel displacements lead only to reparametrization. The variation can be therefore computed as

$$\delta\mathcal{F} = \int g^{-1/2}\nabla_\perp\zeta \cdot \delta\nabla_\perp\zeta\, \mathrm{d}^2\mathbf{Y} = \int \nabla_\perp \mathbf{n}\, \delta\zeta\, \mathrm{d}^2\mathbf{Y} = \int \kappa\, \delta X\, \mathrm{d}^2\mathbf{Y}. \tag{2.2.15}$$

One can add here a shape-independent gain $\delta\mathcal{F} = -V_1\, \delta X$ due to advancement of a more favorable state, which is proportional to a weak deviation from the Maxwell construction. The relaxation equation (2.1.2) can be rewritten therefore, introducing also a mobility coefficient Γ, as

$$c = \frac{\mathrm{d}X}{\mathrm{d}t} = -\Gamma\frac{\delta\mathcal{F}}{\delta X} = \Gamma(V_1 - D\mathcal{I}\kappa). \tag{2.2.16}$$

This coincides with (2.2.12) if one sets $\Gamma = \mathcal{I}^{-1}$.

2.2.3 Burgers Equation

For a weakly distorted planar front defined by (2.2.14), the eikonal equation (2.2.12) can be rewritten explicitly in the Cartesian frame $(X, \mathbf{Y})$ propagating with a constant speed c_0. In accordance to the scaling of the transverse coordinates in the preceding subsection, we presume that the transverse gradients denoted by the 2D operator $\nabla_\perp$ are small compared to those in the propagation direction, and scale as $\nabla_\perp = O(\sqrt{\epsilon})$. The extended time variable is scaled accordingly as $\partial_t = O(\epsilon)$.

The projection of the normal displacement on the X-axis is expressed as

$$c_0 \left(1 + |\nabla_\perp \zeta|^2\right)^{-1/2} = c_0 \left(1 - \tfrac{1}{2}|\nabla_\perp \zeta|^2\right) + O(\epsilon^2),$$

and the curvature is computed as $\kappa = -\nabla_\perp^2 \zeta + O(\epsilon^2)$. Using this in (2.2.12) transformed to the comoving frame yields in the leading order the Burgers equation

$$\partial_t \zeta = D\nabla_\perp^2 \zeta - \tfrac{1}{2} c_0 |\nabla_\perp \zeta|^2. \tag{2.2.17}$$

The diffusional term acts in such a way that convex segments retreat and concave ones advance, so that the front tends to relax to a straight-line or planar shape. The nonlinear term is purely advective and reflects the tendency of the interface to propagate normal to itself. This term is responsible for asymmetry due to the propagation direction. Since segments with a larger slope lag behind, the front tends to be smoother on the onward and steeper on the back side, except at inevitable "shock" locations, as seen in Fig. 2.10.

At the Maxwell construction ($c_0 = 0$), (2.2.17) degenerates into a linear diffusion equation. Otherwise, it can be linearized with the help of the Hopf–Cole transformation $\zeta = -(2D/c_0) \ln \chi$ transforming it to a linear diffusion equation

$$\partial_t \chi = D\nabla_\perp^2 \chi. \tag{2.2.18}$$

The "hydrodynamic" form of the Burgers equation obtained by applying the gradient operator is

$$\partial_t \mathbf{v} + \mathbf{v} \cdot \nabla \mathbf{v} = D\nabla^2 \mathbf{v}, \tag{2.2.19}$$

where $\mathbf{v} = c_0^{-1} \nabla \zeta$ is interpreted as "velocity", and the $\perp$ sign is omitted. This form bears a superficial resemblance to the Navier–Stokes equation, but lacks pressure and is not coupled to the continuity equation. One can say that the Burgers equation describes a "supercompressible" fluid with pressure independent of density (and therefore irrelevant). The Burgers equation has been originally introduced (Burgers,

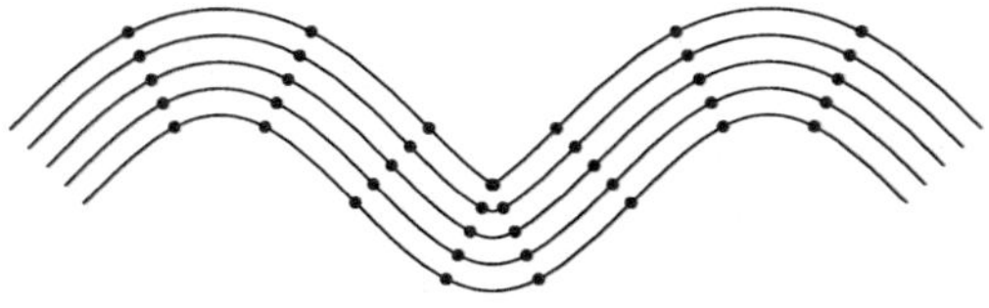

Fig. 2.10. Successive positions of a front propagating upward with a constant normal velocity. Dots mark successive positions of chosen points. Note a singularity developing on the concave segment

1948) in this form as a model replacement of the Navier–Stokes equation. It is, however, ill suited for studies of hydrodynamic phenomena, especially turbulence, as the velocity field defined in this way is *potential*, while the heart of hydrodynamic turbulence is *vorticity*.

In the "inviscid" limit $D = 0$, which is approached far from the Maxwell construction when the front propagates with a constant speed and the effect of curvature is negligible, (2.2.19) degenerates into a first-order PDE that generates moving discontinuities – "shocks" forming boundaries between front segments with different slopes. A small diffusional term with $D = O(\epsilon)$ resolves discontinuities on an $O(\sqrt{\epsilon})$ scale.

The characteristic equation of the first-order PDE is $\mathrm{d}\mathbf{v}/\mathrm{d}t = \mathbf{v}$, which means that segments with a higher slope propagate faster and thus conquer and obliterate segments with a lower slope. To compute the propagation speed of a shock, to be denoted as q, we rewrite the degenerate form of (2.2.19) in a 2D frame comoving with the shock as

$$q\,\mathrm{d}\mathbf{v}/\mathrm{d}z = \tfrac{1}{2}\,\mathrm{d}|\mathbf{v}|^2/\mathrm{d}z, \tag{2.2.20}$$

where the z axis is directed normally to the shock. Integrating this across the shock and taking into account that the component of $\mathbf{v}$ parallel to the shock must be continuous, while the normal component may jump between the values $v_z = v_\pm$, yields $q = \frac{1}{2}(v_+ - v_-)$, i.e., the shock propagates with the speed equal to the arithmetical mean of the slopes on the two sides. The dynamics of shocks is in fact a simple geometrical consequence of normal propagation of the interface with a constant speed masked by a chain of mathematical transformations of the eikonal equation (2.2.12).

The Burgers equation is one of the most extensively studied universal equations. A question of interest is the law of decay of originally imposed inhomogeneities (see e.g. Bec et al., 2000). Another well-studied topic is dynamics in the presence of noise described by the Burgers equation with an added random term, commonly called Kardar–Parisi–Zhang (KPZ) equation (Kardar et al, 1986).

2.2.4 Ripening under Global Control

Dependence of the front propagation speed on curvature plays a primary role in long-scale evolution of patterns. Consider a pattern consisting of two-dimensional "droplets" of a minority phase (say, $u = u_s^-$) immersed in a continuum of an alternative majority phase. Since front interactions decay exponentially with separation, their influence on the dynamics of fronts that bound droplets separated at a distance much larger than the characteristic diffusional length should be negligible compared to the influence of curvature. Thus, well-separated droplets will relax to the circular shape and evolve according to (2.2.12) that gives preference to growth of larger droplets. For any value of $c_0 > 0$ (i.e., under conditions favoring growth of the minority phase), there is a critical radius $R_c = D/c_0$, such that droplets with larger radii $R > R_c$ grow, while droplets with $R < R_c$ shrink and eventually disappear.

If c_0 remains fixed, the globally stable phase, initially a minority, eventually conquers the entire plane. This can be prevented if there is an additional global variable, which is dynamically adjusted to maintain the total area occupied by each phase constant. This variable should act to reduce the value of c_0 in the course of evolution in order to compensate the decrease of average curvature. As a result, the critical radius grows and, as it overtakes the radius of growing droplets, the latter start to shrink and eventually disappear. The final result of evolution minimizing the total energy of the system under the constant area constraint would be a single circular droplet. This *ripening* process (similar to Ostwald ripening in two-phase systems, see Sect. 2.3.4) may, however, take exceedingly long time at its later stages, when the radii of the majority of surviving droplets become very large compared to the thickness of the interfacial layer. It is interesting therefore to compute the distribution of droplet sizes at late stages of ripening when the system has been evolving sufficiently long to obliterate the initial size distribution, but the number of droplets still remains sufficiently large to justify statistical approach.

The asymptotic analysis follows the classical work by Lifshitz and Slyozov (1958) who have considered ripening in 3D with a growth rate different from that used here. We shall return to this case in Sect. 2.3.4; see also a generalized analysis by Giron et al. (1998). Let $f(R)$ be the number density of droplets with the radius R; this function is not normalized, since the total number $\int f(R)\mathrm{d}R$ decreases in time as smaller droplets collapse. The evolution equation for $f(R)$ is

$$\partial_t f + \partial_R(cf) = 0, \tag{2.2.21}$$

where the growth rate $c(R)$ is expressed, using (2.2.12) and denoting $c_0 = D/R_c(t)$, as

$$c(R) = D\left(\frac{1}{R_c} - \frac{1}{R}\right). \tag{2.2.22}$$

An asymptotic distribution, if it exists, should admit some quasistationary representation. Quasistationarity may be preserved relative to the time-dependent critical radius. A suitable variable, replacing R, is therefore the dimensionless radius $\rho = R/R_c$. The growth of the respective dimensionless area is proportional to

$$\frac{1}{2}\frac{\mathrm{d}(\rho^2)}{\mathrm{d}t} = \frac{D}{R_c^2}(\rho - 1) - \frac{\rho^2}{R_c}\frac{\mathrm{d}R_c}{\mathrm{d}t}. \tag{2.2.23}$$

This expression is further simplified by using R_c as a measure of time. A convenient transformed time variable is $\tau = \ln(R_c/R_0)^2$, where R_0 is any suitable length scale. Denoting

$$q(t) = \frac{1}{2D}\frac{\mathrm{d}(R_c^2)}{\mathrm{d}t}, \tag{2.2.24}$$

we rewrite (2.2.23) as

$$\frac{1}{2}\frac{\mathrm{d}(\rho^2)}{\mathrm{d}\tau} = \frac{\rho - 1}{q(\tau)} - \rho^2. \tag{2.2.25}$$

It turns out that there is a unique asymptotic value of q at $\tau \to \infty$ compatible with the area conservation condition. The right-hand side of (2.2.25) is negative at both small and large values of ρ, but becomes positive in an intermediate range when $q < 1/4$

(see Fig. 2.11a). If such a range existed, the area covered by droplets with respective radii would grow with time without limit, which is forbidden, as it contradicts the area conservation condition. Conversely, if q were negative everywhere, the area occupied by all droplets would keep shrinking, which is again incompatible with the area conservation. The only remaining possibility is that the asymptotic value is $q_\infty = 1/4$; then the right-hand side of (2.2.25) reaches the zero maximum value at $\rho = 2$. This defines the maximum dimensionless radius in the asymptotic distribution. Indeed, if droplets with a larger radius are present in the initial distribution, their radius, notwithstanding their continuous growth, draws closer to the critical radius, until they are swept to the vicinity of the point $\rho = 2$.

The function $q(\tau)$ should approach its limiting value $q_\infty = 1/4$ from above in such a way that a small gap decreasing with time remain between the maximum of the right-hand side of (2.2.25) and the ρ axis. The droplets from the tail of the distribution filter slowly through this gap in the course of evolution. The time-dependent correction to the asymptotic value affects the form of the distribution near the point $\rho = 2$, but this does not affect the rest of the distribution, since, as we shall see, the contribution of the vicinity of this point to the total area is negligible.

The asymptotic size distribution can now be obtained by solving (2.2.21), which is rewritten in the transformed variables as

$$\partial_\tau f + \partial_\rho (cf) = 0, \qquad c(\rho) = -\frac{2}{\rho}(2-\rho)^2. \tag{2.2.26}$$

The general solution of (2.2.26) is

$$f(\rho,\tau) = -\frac{\chi(\tau+\psi)}{c(\rho)}, \qquad \psi(\rho) = -\int \frac{\mathrm{d}\rho}{c(\rho)}, \tag{2.2.27}$$

where $\chi(\ldots)$ is an arbitrary function that has to be obtained from the normalization condition

$$\pi \int_0^{R_{\max}} R^2 f(R)\mathrm{d}R = S_0, \tag{2.2.28}$$

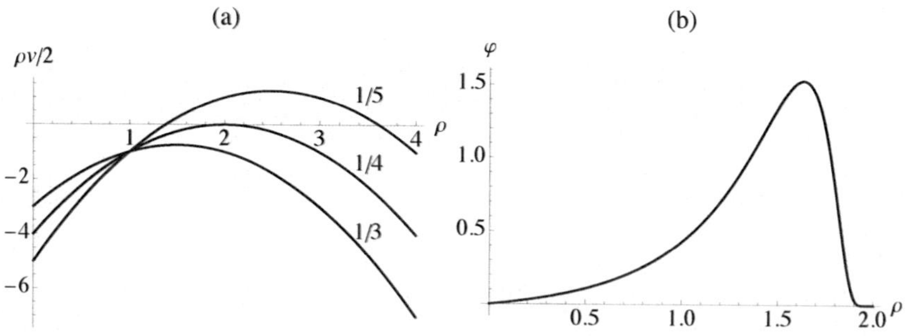

Fig. 2.11. (a) Plot of the right-hand side of (2.2.25); the values of q are indicated at the respective curves. (b) The normalized distribution function $\phi(\rho)$.

or, in dimensionless variables,

$$e^{\tau}\int_0^2 \rho^2 f(\rho)\mathrm{d}\rho = y \equiv \frac{S_0}{\pi R_0^2}. \tag{2.2.29}$$

It is clear from the latter expression that the only choice making the left-hand side independent of τ is $\chi(\tau) = A\mathrm{e}^{-\tau}$ with an appropriate constant A. Computing the integrals yields

$$f(\rho,\tau) = \tfrac{1}{2}A\mathrm{e}^{-\tau-1/(2-\rho)}\rho(2-\rho)^{-5/2}, \tag{2.2.30}$$

$$A = 2y \Big/ \int_0^2 \rho^3(2-\rho)^{-5/2}\mathrm{e}^{-1/(2-\rho)}\mathrm{d}\rho = 1.159y. \tag{2.2.31}$$

The distribution becomes independent of τ when it is normalized to unity, $f(\rho,\tau) = B\mathrm{e}^{-\tau}\varphi(\rho)$, where $\int \varphi(\rho)\mathrm{d}\rho = 1$ and B is a normalization constant. The normalized distribution is shown in Fig. 2.11b. To translate the result back to physical variables, one has to keep in mind that R_c increases asymptotically proportionally to $\sqrt{t}$ and $\mathrm{e}^{-\tau} \propto t^{-1}$. Convergence to the asymptotic distribution may be, however, impaired for initial distributions with a very long tail (Niethammer and Pego, 1999).

2.3 Fronts in Conserved Systems

2.3.1 Inner and Outer Equations

Interfaces separating alternative stable homogeneous states (HSS) in closed systems relaxing toward thermodynamic equilibrium considered in (Sect. 1.6.2) are similar in many respects to fronts separating alternative HSS in non-equilibrium systems to interphase boundaries in equilibrium systems, and such terms as surface tension, critical nucleus and Maxwell construction (defining a stationary front) applied above in the RDS context originated in classical studies of equilibrium systems.

Any interphase boundary is essentially a *mesoscopic* structure. While the material properties vary smoothly at macroscopic distances along the interface, the gradients in the normal direction are steep, approaching a molecular scale. This brings about a contradiction between the need for a macroscopic description and the necessity to take into consideration microscopic factors that come to influence the motion on incommensurately larger scales.

The characteristic width $l \sim \sqrt{K}$ of an interphase boundary defined by (1.6.16) is typically short on a macroscopic scale. Across this thin layer, the order parameter switches between the two alternative values corresponding to the same constant value of chemical potential. The latter is constant everywhere when the two phases are at equilibrium, while under nonequilibrium conditions it varies on a longer scale L that may be determined by the geometry of the system and external fluxes. Problems of this kind, containing a small parameter $\epsilon = l/L$, should be solved by matching expansions in ϵ in regions characterized by two widely separated scales: the *inner* region localized at the interface, and the *outer* region spreading out to the bulk of both alternative phases. This approach is suitable for tracking slow motion of the interphase boundary at times far exceeding the characteristic relaxation time to a stationary profile of the order parameter in the interfacial layer (Pego, 1989; Karma and Rappel, 1998).

In the outer region, we extend the coordinates by the factor ϵ^{-1}. Accordingly, the gradient operator is transformed to $\epsilon\nabla$; the matching time scale should be extended by ϵ^{-2}. Then the distortion term in (1.6.15) becomes of $O(\epsilon^2)$, and can be neglected. Marking the conserved order parameter u and the chemical potential μ in the two bulk regions corresponding to the two alternative HSS by the superscripts $\pm$, we write the outer equations as

$$\partial_t u^{\pm} = \nabla \cdot \left[\Gamma(u^{\pm})\nabla\mu^{\pm}\right], \qquad \mu^{\pm} = -f(u^{\pm}). \tag{2.3.1}$$

Combining both equations, we see that the evolution in the outer regions obeys the linear diffusion equation

$$\partial_t u^{\pm} = \nabla \cdot \left[D(u^{\pm})\nabla u^{\pm}\right]. \tag{2.3.2}$$

The effective diffusivities $D^{\pm} = -\Gamma(u^{\pm})f'(u^{\pm})$ are positive, since $f'(u^{\pm}) < 0$ on a branch of stable equilibria; hence, the diffusion equation is well posed.

In the inner region, we shall use, as in Sect. 2.2.2, the aligned comoving frame. The characteristic scale along the axis x normal to the nominal front position is the short scale l, while the coordinates $\mathbf{y}$ parallel to the interface are scaled by L. If the curvature radius is of $O(L)$, the curvature is written as $\epsilon\kappa$ when measured on the short inner scale. Inasmuch as the interfacial layer is assumed to be locally at equilibrium, the interface is expected to move under the influence of long-scale changes of chemical potential. Therefore, the propagation speed should be measurable on a long scale, so that the Péclet number based on the long length scale L and a typical value of the diffusivity, $\mathrm{Pe} = cL/D$ should be at most of $O(1)$. The chemical potential within the front region should differ from the Maxwell construction by $O(\epsilon)$, and is expressed as $\mu = \mu_s + \epsilon\mu_1$. Using this in (1.6.14), (1.6.15), and expanding also the order parameter, $u = u_0 + \epsilon u_1 + \ldots$, we obtain the inner equations in successive orders of ϵ.

The zero-order equations are

$$\mu_0''(x) = 0, \qquad u_0''(x) + f(u_0(x)) + \mu_0 = 0. \tag{2.3.3}$$

The solution is a stationary front that exists at $\mu_0 = \mu_s = \text{const}$, where μ_s is the equilibrium chemical potential defined by (1.6.17). The profile of the stationary front is computed exactly as in Sect. 2.1.2, and is most conveniently expressed in phase plane variables $u = u_0(x)$, $p(u) = u_0'(x)$ as a function $p(u)$. Formally, one could still add to μ_0 a linear term, but this would be incompatible with matching conditions when propagation is slow.

In the first order, (1.6.14) reduces to

$$\frac{\mathrm{d}}{\mathrm{d}x}\left[\Gamma(u_0)\mu_1'(x) + cu_0(x)\right] = 0. \tag{2.3.4}$$

The solutions of the inner and outer equations should be matched at a distance from the front that is large on the inner but small on the outer scale; the result must be independent of a precise matching position within this range. The matching conditions are

$$\lim_{x\to\pm\infty} \mu_1'(x) = \mathbf{n}\cdot\nabla\mu^{\pm} \equiv -j^{\pm}/\Gamma(u^{\pm}), \tag{2.3.5}$$

where $\mathbf{n}$ is the normal to the interface (directed, by convention, in the same way as the x-axis), and the fluxes $j^{\pm}$ on both sides of the interface are computed as the inner limit of the outer solution.

Integrating (2.3.4) and using (2.3.5) yields

$$c(u^+ - u^-) = -\Gamma(u^+)\mathbf{n}\cdot\nabla\mu^+ + \Gamma(u^-)\mathbf{n}\cdot\nabla\mu^- \equiv j^+ - j^-. \tag{2.3.6}$$

The right-hand side of (2.3.6) is the difference of the fluxes on the two sides; thus, this condition defines the speed of the local interface displacement required to ensure the mass conservation. Since the variable part of μ is restricted in the interfacial layer to $O(\epsilon)$, the order parameter in the matching regions may deviate from the equilibrium values $u_s^{\pm}(\mu_s)$ satisfying $\mu_s = f(u_s^{\pm})$ by no more than $O(\epsilon)$. Using again a linearization of $f(u)$, the mass conservation condition (2.3.6) can be rewritten as

$$c = -\frac{D_s^+\mathbf{n}\cdot\nabla u^+ - D_s^-\mathbf{n}\cdot\nabla u^-}{u_s^+ - u_s^-}. \tag{2.3.7}$$

Since $f(u)$ is, generally, asymmetric and mobilities depend on u, the effective diffusivities $D_s^{\pm} = -\Gamma(u_s^{\pm})f'(u_s^{\pm})$ are distinct on the two sides of the interface.

2.3.2 Solvability Condition and Matching

It remains to determine the relation between the first-order correction to the interfacial chemical potential $\mu^{\pm}$ and the fluxes $j^{\pm}$. The respective values of the order parameter $u^{\pm} = \overline{u}^{\pm}$, which can be used as interfacial boundary conditions for the outer equation (2.3.1), are defined by the near-equilibrium linearized relation

$$\mu^{\pm} - \mu_s = -f'(u_s^{\pm})(u^{\pm} - u_s^{\pm}). \tag{2.3.8}$$

Integrating (2.3.4) twice yields

$$\mu_1(x) = \overline{\mu}_1 - \int_0^x \frac{j_1 + cu_0(x)}{\Gamma(u_0(x))}\,\mathrm{d}x, \tag{2.3.9}$$

where $\overline{\mu}_1 = \mu_1(0)$ and j_1 are integration constants. The latter equals to the flux through the interface, which is determined by the matching conditions (2.3.5) yielding either of the equivalent expressions

$$j_1 = -cu^{\pm} - \Gamma(u^{\pm})\mathbf{n}\cdot\nabla\mu^{\pm} = j^{\pm} - cu^{\pm} = \frac{j^- u_s^+ - j^+ u_s^-}{u_s^+ - u_s^-}. \tag{2.3.10}$$

The first-order solution (2.3.9) is used in the first-order expansion of (1.6.15):

$$K\left[u_1''(x) + \kappa u_0'(x)\right] + f'(u_0)u_1 + \mu_1 = 0. \tag{2.3.11}$$

This equation has the same form as (2.2.9), and is subject to the solvability condition (2.2.11). Multiplying this inhomogeneous equation by the Goldstone eigenmode $u_0'(x)$ and integrating yields

$$\overline{\mu}_1(u_s^+ - u_s^-) = j_1\mathcal{J}_0 + c\mathcal{J}_1 - \kappa K\mathcal{I}. \tag{2.3.12}$$

Here $\mathcal{I}$ is the dissipation integral defined by (2.1.7), and the integrals $\mathcal{J}_k$ are computed as

$$\begin{aligned}\mathcal{J}_k &= \int_{-\infty}^{\infty} u_0'(x)\mathrm{d}x \int_0^x \frac{u_0^k(\widetilde{x})}{\Gamma(u_0(\widetilde{x}))}\mathrm{d}\widetilde{x} = \int_{u_s^-}^{u_s^+}\mathrm{d}u \int_{u_s^0}^{u} \frac{\widetilde{u}^k}{p(\widetilde{u})\Gamma(\widetilde{u})}\mathrm{d}\widetilde{u} \\ &= \int_{u_s^0}^{u_s^+} \frac{\widetilde{u}^k(u_s^+ - \widetilde{u})}{p(\widetilde{u})\Gamma(\widetilde{u})}\mathrm{d}\widetilde{u} - \int_{u_s^-}^{u_s^0} \frac{\widetilde{u}^k(\widetilde{u} - u_s^-)}{p(\widetilde{u})\Gamma(\widetilde{u})}\mathrm{d}\widetilde{u} \equiv \mathcal{J}_k^+ - \mathcal{J}_k^- .\end{aligned} \tag{2.3.13}$$

The lower integration limit for $\widetilde{u}$ corresponding to the nominal origin $x = 0$ may be arbitrary, but, to be definite, we have chosen it to coincide with the unstable intermediate HSS u_s^0. If $\Gamma = \text{const}$ and $u_0'(x)$ is symmetric (which follows from the symmetry of the potential $V(u)$), the integral $\mathcal{J}_0$ vanishes, and the constant j_1 falls out. The dependence of the excess chemical potential on curvature in (2.3.12) is immediately recognized as the Gibbs–Thomson relation, while the other two terms give dynamic corrections due to the flux through the interface.

The limit of the inner solution (2.3.9) at $x \to \pm\infty$ is expressed after using (2.3.10) and rearranging the integral as

$$\lim_{x\to\pm\infty} \mu_1(x) = \overline{\mu}_1 + c\mathcal{J}_0^\pm - j^\pm\mathcal{K}^\pm + x\,\mathbf{n}\cdot\nabla\mu^\pm, \tag{2.3.14}$$

where we have separated the converging integrals $\mathcal{J}_0^\pm$, defined by (2.3.13), and

$$\mathcal{K}^\pm = \int_{x_0}^{\pm\infty}\left[\frac{1}{\Gamma(u_0(x))} - \frac{1}{\Gamma(u_s^\pm)}\right]\mathrm{d}x = \int_{u_s^0}^{u_s^\pm}\frac{1}{p(u)}\left[\frac{1}{\Gamma(u)} - \frac{1}{\Gamma(u_s^\pm)}\right]\mathrm{d}u. \tag{2.3.15}$$

One can check by differentiating the integrals $\mathcal{J}_k^\pm$, $\mathcal{K}^\pm$ with respect to the variable limit that the limiting value of $\mu(x)$ is indeed independent of the value of the order parameter at the origin.

On the other hand, the chemical potential in the outer regions close to the interface, i.e., at some distance along the normal $\mathbf{n}$, which is small when measured on the outer scale, is presented by expanding the outer solution in the Taylor series as

$$\mu^\pm = \overline{\mu}^\pm + x\,\mathbf{n}\cdot\nabla\mu^\pm. \tag{2.3.16}$$

The last terms in (2.3.14) and (2.3.16) match, and the remaining constant terms yield two matching conditions for computing $\overline{\mu}^\pm$:

$$\overline{\mu}^\pm - \mu_s = \overline{\mu}_1 + c\mathcal{J}_0^\pm - j^\pm\mathcal{K}^\pm, \tag{2.3.17}$$

or, using (2.3.6), (2.3.10), and (2.3.12),

$$\overline{\mu}^{\pm} - \mu_s = c\,\frac{\mathcal{J}_1 + u_s^+ \mathcal{J}_0^- - u_s^- \mathcal{J}_0^+}{u_s^+ - u_s^-} + j^{\pm}\left(\frac{\mathcal{J}_0}{u_s^+ - u_s^-} - \mathcal{K}^{\pm}\right) - \frac{\kappa K \mathcal{I}}{u_s^+ - u_s^-}. \tag{2.3.18}$$

This expression yields local relations between the values of chemical potential and fluxes on both sides of the front, which can serve as the boundary conditions for outer equations.

This completes the reduction of the Cahn–Hilliard equation to the classical sharp interface Stefan problem, which is fully determined by two linear diffusion equations (2.3.2) with the interfacial velocity defined by (2.3.7) and the boundary conditions (2.3.18) at the moving interface corresponding to equilibrium values of the order parameter shifted by $O(\epsilon)$. If the problem is defined in a bounded region, it should be complemented by boundary conditions at external boundaries. In the presence of external fluxes, the chemical potential μ and the respective values of u may substantially deviate, respectively, from μ_s and $u_s^{\pm}$ far from the interface. This is still compatible with the adopted scaling, as long as the gradients remain weak.

In the symmetric case, $u_s^+ = -u_s^- = u_s$, $\mathcal{J}_0^+ = \mathcal{J}_0^-$, $\mathcal{J}_0 = 0$, the relation (2.3.18) simplifies to

$$\overline{\mu}^{\pm} - \mu_s = (2u_s)^{-1}\left[c\left(\mathcal{J}_1 + 2u_s \mathcal{J}_0^{\pm}\right) - \kappa K \mathcal{I}\right] - j^{\pm}\mathcal{K}^{\pm}. \tag{2.3.19}$$

If, in addition, Γ = const, the integrals $\mathcal{K}^{\pm}$ vanish, and $\overline{\mu}^{\pm}$ = idem = $\overline{\mu}$. Then (2.3.19) can be recast as a mobility relation, or eikonal equation, similar to (2.2.12):

$$c = c_0 - \frac{K\mathcal{I}}{\mathcal{J}_1 - 2u_s\mathcal{J}_0^{\pm}}\,\kappa, \quad c_0 = \beta\left(\overline{\mu} - \mu_s\right), \quad \beta = \frac{2u_s}{\mathcal{J}_1 - 2u_s\mathcal{J}_0^{\pm}}, \tag{2.3.20}$$

where c_0 is the propagation speed of a flat interface proportional to the deviation from the Maxwell construction $\overline{\mu} - \mu_s$, and β is the effective mobility.

For the standard case of a cubic $f(u) = u(1 - u^2)$, we have $\mu_s = 0$, $u_s^{\pm} = \pm 1$, $f'(u_s^{\pm}) = -2$. With $\Gamma = D/2$ = const, the integrals $\mathcal{K}^{\pm}$ vanish, and the other integrals are evaluated using, in accordance to (2.1.21), $p(u) = (1-u)^2/\sqrt{2K}$. This yields $\mathcal{I} = \frac{2}{3}\sqrt{2/K}$, $\mathcal{J}_0^{\pm} = \sqrt{2K}\Gamma^{-1}\ln 2$, $\mathcal{J}_1 = 2\sqrt{2K}\Gamma^{-1}(1-\ln 2)$, and (2.3.20) reduces to

$$c = c_0 - \frac{D\kappa}{6}, \qquad c_0 = -\frac{D}{2\sqrt{2K}}\left(\overline{\mu}^{\pm} - \mu_s\right) = -\frac{D}{\sqrt{2K}}\left(u^{\pm} \mp 1\right). \tag{2.3.21}$$

where $u^{\pm}$ are the values of the order parameter on the two sides of the interface.

Although (2.2.12) and (2.3.20) or (2.3.21) have the same form, the substantial distinction of conserved systems is the additional restriction of the propagation speed by the conservation condition (2.3.7), which, unlike (2.2.12), involves the *gradient* of the order parameter. The conservation condition is commonly viewed as determining the velocity of the interface, while (2.3.20) or more generally (2.3.18) as the condition for a deviation of the order parameter from its standard equilibrium value at a stationary planar interface. In particular, when extant gradients are weak, the propagation speed may be far less than either term on right-hand side of (2.3.20). Then the kinetic contribution to the shift of the order parameter can be neglected and (2.3.21) reduces to the Gibbs–Thomson relation defining the equilibrium of a curved interface.

2.3.3 Propagation and Coarsening in 1D

Following a rapid quench into an unstable state, nuclei of alternative states are formed, being locally triggered by weak inhomogeneities or fluctuations. This is followed by slow evolution, as phase domains coarsen to reduce the interfacial energy. In the case when formation of nuclei is suppressed, spinodal decomposition may be effected by a propagating front invading an unstable homogeneous state. The speed of the front is selected, as in Sect. 2.1.4, at its leading edge, and can be computed with the help of the dispersion relation obtained by linearization in the vicinity of the unstable state.

We shall work with the rescaled 1D form of the Cahn–Hilliard equation

$$\partial_t u = -\partial_x^2 \left[u_{xx} + f(u) \right]. \tag{2.3.22}$$

The dispersion relation analogous to (2.1.35) is

$$\omega(k) = \mathrm{i} \left(\alpha k^2 - k^4 \right), \tag{2.3.23}$$

where $\alpha = f'(u_0) > 0$ when the homogeneous state $u = u_0$ is unstable, e.g., at $|u_0| < 1/\sqrt{3}$ for the standard function $f(u) = u(1-u^2)$. The group velocity is

$$\widehat{c} = \omega'(k) = 2\mathrm{i}k(\alpha - 2k^2). \tag{2.3.24}$$

Using here $k = k_r + k_i$ and separating the real and imaginary part of $\widehat{c}$, ω, we find that the condition $\widehat{c}_i = 0$ in (2.1.38) has a nontrivial solution

$$2k_r^2 = \alpha + 6k_i^2. \tag{2.3.25}$$

The indices r and i denote here, as in Sect. 2.1.4, the real and imaginary parts. Using this relation in the condition of equality of the phase and group velocities, we find that the only positive solution defining the value of k_i (i.e., the slope of the marginally stable profile) is

$$k_i^2 = \frac{1}{6} \left(\alpha - 3 + \sqrt{9 - 6\alpha + \tfrac{5}{2}\alpha^2} \right). \tag{2.3.26}$$

The propagation speed is now evaluated as

$$\begin{aligned} c = \widehat{c}_r &= -2k_i(\alpha + k_i^2 - 3k_r^2) \\ &= \frac{1}{9} \left(6 + \alpha - 4\sqrt{9 - 6\alpha + \tfrac{5}{2}\alpha^2} \right) \left(\alpha - 3 + \sqrt{9 - 6\alpha + \tfrac{5}{2}\alpha^2} \right)^{1/2} . \end{aligned} \tag{2.3.27}$$

Because of the mass conservation, the profile behind the invading front cannot be monotonic, and the phase separation in the front region can only occur through the formation of two-phase domains. The modulated pattern formed behind the propagating front further coarsens, as seen in Fig. 2.12. While this process can only be

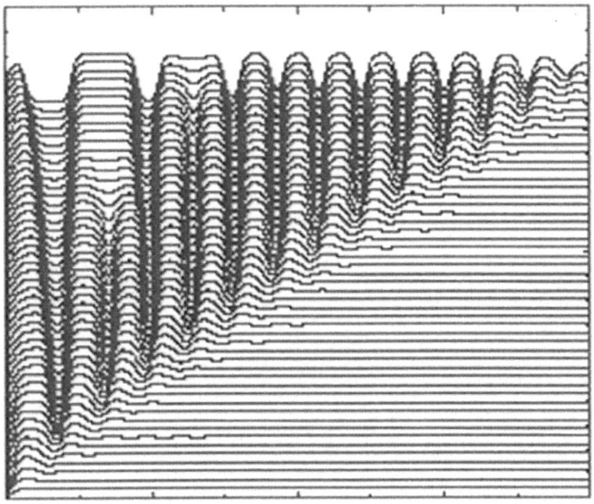

Fig. 2.12. Space–time plot of simulations results for a pulled front propagating into the unstable state $u = 0.4$ (van Saarloos, 2003, reproduced by permission, ©Elsevier). The lines denote the profiles of u at successive time intervals, and are shifted upward relative to each other. Take note of the coarsening of the pattern behind the propagating front

simulated numerically, its later stages, when the domains size far exceeds the characteristic diffusional range which defines the width of the interface between the two alternative phases, can be described through a rational expansion in the scale ratio and presented in terms of front interactions as in Sect. 2.1.3 (Kawasaki and Ohta, 1982).

It is reasonable to assume that, due to slow-down of the fronts and increasing domain sizes, the Peclet number $\mathrm{Pe} = cL/D$ based on the propagation speed c and a typical length L of phase domains becomes small at late stages of coarsening. Then a quasistationary approximation applies also to the "outer" regions and, since mobilities $\Gamma^{\pm}$ can be considered constant in each phase domain, (2.3.1) is replaced by the stationary diffusion equation $\mu_{xx} = 0$. The immediate corollary is that the fluxes $j^{\pm} = -\Gamma^{\pm}\mu_x$ are constant within each phase domain bounded by fronts with opposite polarity (a kink and an antikink).

Consider a sequence of domains with kink positions ζ_n and lengths L_n. We shall restrict to a symmetric case; an asymmetric case is treated in a similar way, but the algebra is more involved because of the necessity to distinguish between $\overline{\mu}^{\pm}$, as well as taking into account different asymptotics of approach to the alternative states. The driving force of the coarsening process stems from the dependence of the energy of the fronts on their separation computed in Sect. 2.1.3. The equilibrium chemical potential μ_s at a stationary front should be replaced therefore by a shifted value ensuring stationarity of the front in the presence of interactions. Taking into account nearest-neighbor interactions only is sufficient in view of the exponential decay of interactions. One can use therefore (2.1.30) applied to interactions with two neighboring fronts. If the front at the nth position is a *kink*, the potential shift is

$$\widetilde{\mu}_n = \frac{a^2\sqrt{\alpha}}{u_s^+ - u_s^-}\left(\mathrm{e}^{-\sqrt{\alpha}L_{n+1}} - \mathrm{e}^{-\sqrt{\alpha}L_{n-1}}\right). \tag{2.3.28}$$

For an *antikink*, this formula should be taken with the opposite sign.

The most straightforward way to the description of coarsening in a system of interacting fronts is to use the material balance and mobility relations derived in the two preceding subsections. The fluxes within each domain are

$$j_n = -\frac{\Gamma^{\pm}}{L_n}\left(\overline{\mu}_{n+1} - \overline{\mu}_n\right). \tag{2.3.29}$$

The chemical potentials at the fronts and propagation velocities are defined by (2.3.6) and (2.3.20):

$$c_n = \frac{j_n^+ - j_{n-1}^-}{2u_s} = \beta\left(\overline{\mu}_n - \widetilde{\mu}_n\right). \tag{2.3.30}$$

The signs in these expressions pertain to kinks; since, following our common sign convention, all formulas in the preceding subsections were derived for kinks with the lower state on the left, the signs of velocities should be inverted for antikinks with the opposite orientation. The system (2.3.29), (2.3.30) comprises $3N$ equations defining fluxes within phase domains, chemical potentials at the fronts and propagation velocities. The system is closed on a ring, and has to be supplemented by boundary conditions for chemical potentials or fluxes in a bounded domain.

As a simple example, consider a single "bump" with the length L (a paired kink and antikink, as in Fig. 2.5)a subject to external fluxes, j_+ from the left and j_- to the right. Denoting the chemical potentials at the kink on the left and the antikink on the right as $\overline{\mu}_\pm$, their displacement rates as $\dot{\zeta}_\pm = \pm c_\pm$, and the flux between them as j, the system (2.3.29), (2.3.30) reduces to five equations

$$j = -\frac{\Gamma}{L}\left(\overline{\mu}_+ - \overline{\mu}_-\right), \qquad \dot{\zeta}_\pm = \frac{j - j_\pm}{2u_s} = \mp\beta\left(\overline{\mu}_\pm - \widetilde{\mu}\right), \tag{2.3.31}$$

where $\widetilde{\mu}(L)$ is given by (2.1.30). The solution leads to the dynamic equations for the domain length L and the position of the domain center ζ:

$$\dot{L} = \dot{\zeta}_- - \dot{\zeta}_+ = \frac{j_+ - j_-}{2u_s}, \qquad \dot{\zeta} = \frac{\dot{\zeta}_- + \dot{\zeta}_+}{2} = -\frac{\beta L(j_+ + j_-)}{2(\beta u_s L + \Gamma)}. \tag{2.3.32}$$

The first relation is trivial, and can be obtained simply by adding up the material balance conditions for the two fronts. The second relation, which describes the migration of the "bump" toward the source of material, requires solving (2.3.31) and depends on the mobilities as well as on the domain size.

The potential shift $\widetilde{\mu}$ falls out of (2.3.31), and may come into the action only when coarsening is possible, which requires at least three fronts. For a sequence of two kinks flanking an antikink on an infinite line with the no-flux conditions at $x \to \pm\infty$, (2.3.6) and (2.3.20) give the system of eight equations for j_n, $n = 0, ...3$, $\overline{\mu}_n$, c_n, with $j_0 = j_3 = 0$ and $\widetilde{\mu}_n(L_n)$ defined by (2.3.28) where L_0 and L_3 should be set to ∞. The result is

$$c_1 = \frac{\beta\Gamma(2\beta u_s L_2\widetilde{\mu}_2 - \Gamma\mu_1)}{4\beta^2 u_s^2 L_1 L_2 - \Gamma^2}, \qquad c_3 = -\frac{\beta\Gamma(2\beta u_s L_1\widetilde{\mu}_1 - \Gamma\mu_2)}{4\beta^2 u_s^2 L_1 L_2 - \Gamma^2},$$
$$c_2 = \frac{\beta\Gamma[\widetilde{\mu}_2(\Gamma + 2\beta u_s L_2) - \widetilde{\mu}_1(\Gamma + 2\beta u_s L_1)]}{4\beta^2 u_s^2 L_1 L_2 - \Gamma^2}. \tag{2.3.33}$$

In a symmetric case, $L_1 = L_2 = L,\ \widetilde{\mu}_1 = \widetilde{\mu}_2 = \widetilde{\mu}$, this reduces to

$$c_1 = -c_3 = \beta \Gamma \widetilde{\mu}/(2\beta u_s L + \Gamma), \qquad c_2 = 0, \tag{2.3.34}$$

so that the two kinks move toward the antikink until they annihilate. When the symmetry is broken, the antikink moves in the direction of the closest kink, so that eventually the domain with a larger value of $\widetilde{\mu}$ disappears. Of course, the theory breaks down just before the annihilation event when the domain size becomes small, but this final stage is very brief.

2.3.4 Ostwald Ripening

In the absence of external fluxes, the system would evolve to a state close to equilibrium after a period of $O(L^2/D)$, where L is a characteristic size of domains corresponding to the alternative coexisting phases. At the same time, the interfaces would evolve locally to minimize the surface energy. If the volume ratio of the two alternative phases strongly deviates from unity, the resulting intermediate state would consist of well-separated spherical droplets of the minority phase immersed in a continuum of the majority phase – a situation we have already encountered in Sect. 2.2.4.

Assuming local equilibrium governed by the Gibbs–Thomson relation, the residual difference in local chemical potentials will only be caused by a scatter of droplet radii R, and the upper estimate of the interfacial propagation velocity defined by (2.3.6) or (2.3.7) will be $c \sim lRD/L$, where $l = \sqrt{K}$ is the characteristic width of the interface and L is a typical distance between droplets. This estimate is by $O(\epsilon)$ lower that the a priori estimate $c \sim D/L$. This, first of all, justifies neglecting the kinetic contribution to (2.3.18) or (2.3.20), thereby validating the above assumption of local equilibrium. Moreover, the characteristic time scale, estimated before as $D/c^2 \sim L^2/D$, now increases by the factor $(R/l)^2 \gg 1$. Therefore, the time derivative in the outer equation (2.3.1) or (2.3.2) can be neglected and the diffusion equation replaced by the Laplace equation

$$\nabla^2 \mu = 0. \tag{2.3.35}$$

This is called the *Hele–Shaw* limit, by analogy with viscous fingering phenomena where the motion of an interface is determined by the pressure field obeying the Laplace equation.

The equilibrium deviation of the order parameter at the interface $\widetilde{u}$ is determined by the Gibbs–Thomson relation for the equilibrium chemical potential at the interface,

$$\overline{\mu} = \mu_s + bl/R, \tag{2.3.36}$$

where b is a numerical constant equal to $\frac{2}{3}\sqrt{2}$ in the standard case of the cubic $f(u)$.

If $\mu_c \equiv \mu_s + bl/R_c$ is a prevailing value of chemical potential far from the interface, one can easily compute the flux toward a spherical droplet at the origin by solving (2.3.35) in symmetric spherical geometry with the boundary conditions $\mu(R) = \overline{\mu}$, $\mu(\infty) = \mu_c$:

$$\mu(r) = \begin{cases} \overline{\mu} + (\mu_c - \overline{\mu})(1 - R/r) & \text{at } r \geq R \\ \overline{\mu} & \text{at } r \leq R. \end{cases} \tag{2.3.37}$$

Using (2.3.36), (2.3.37) in the interfacial balance condition (2.3.6) yields the growth rate of the droplet

$$c = \frac{\mathrm{d}R}{\mathrm{d}t} = \frac{bl\Gamma}{R^2(u_s^+ - u_s^-)}\left(\frac{R}{R_c} - 1\right). \tag{2.3.38}$$

Thus, as in Sect. 2.2.4, larger droplets with radii $R > R_c$ grow, while smaller ones with $R < R_c$ shrink and eventually disappear; the bulk chemical potential μ_c slowly changing in the course of evolution plays here the role of a global control variable. This is the classical process of Ostwald ripening. The asymptotic distribution of droplet sizes at late stages of the ripening process has been computed in the original work by Lifshitz and Slyozov (1959) for this particular case, which differs from that considered in Sect. 2.2.4 by the space dimension and the dependence of the growth rate on the droplet size.

The quasistationary asymptotic size distribution is computed along the same lines as in Sect. 2.2.4. The variable characterizing the asymptotic distribution, similar to (2.2.23), is the dimensionless radius $\rho = R/R_c$. The growth of the respective dimensionless volume is proportional to

$$\frac{1}{3}\frac{\mathrm{d}(\rho^3)}{\mathrm{d}t} = \frac{bl\Gamma}{R_c^3}(\rho - 1) - \frac{\rho^3}{R_c}\frac{\mathrm{d}R_c}{\mathrm{d}t}. \tag{2.3.39}$$

This expression is further transformed by introducing a new time variable $\tau = \ln(R_c/R_0)^3$ (with an arbitrary R_0 setting the origin) and denoting

$$q(t) = \frac{u_s^+ - u_s^-}{3bl\Gamma}\frac{\mathrm{d}(R_c^3)}{\mathrm{d}t}. \tag{2.3.40}$$

Then (2.3.39) is rewritten as

$$\frac{1}{3}\frac{\mathrm{d}(\rho^3)}{\mathrm{d}\tau} = \frac{\rho - 1}{q(\tau)} - \rho^3. \tag{2.3.41}$$

As in Sect. 2.2.4, there is a unique asymptotic value of q at $\tau \to \infty$ compatible with the volume conservation condition. The right-hand side of (2.3.41) is negative at both small and large values of ρ, and reaches the zero maximum at $\rho = 3/2$ when $q = 4/27$. The latter must be the asymptotic value of q at $\tau \to \infty$, while the location of the maximum on the ρ axis defines the maximum dimensionless radius in the asymptotic distribution. The function $q(\tau)$ approaches the limiting value from above in such a way that a small gap, decreasing with time, remain between the maximum of the right-hand side of (2.3.41) and the ρ axis as in Fig. 2.11a. The asymptotic size distribution can be obtained as in Sect. 2.2.4 by solving (2.2.21) with $c(R)$ given by (2.3.38) after transforming this equation to the new variables ρ, τ.

2.3.5 Phase Field Model

Phase field models have been developed primarily for the description of pattern formation phenomena in solidification. They include two variables: a nonconserved order parameter (*phase field*), which is used to distinguish between the solid and fluid phases, and a conserved control field that determines relative thermodynamic stability of the two phases and may be physically identified with enthalpy. Unlike the models of gas–liquid interface, phase field models of solidification are usually viewed as a computational device enabling to avoid discontinuities and to solve PDEs instead of tracking a sharp interphase boundary, and are not expected to describe the actual structure of a diffuse interfacial layer. The only requirement is to choose model parameters in a thermodynamically and kinetically consistent way, so that the sharp interface limit of the theory would faithfully describe the motion of the interphase boundary.

The free energy functional for an isotropic system is written in the form (Langer, 1986; Karma and Rappel, 1998)

$$\mathcal{F} = \int \left[V(u) + \tfrac{1}{2}\lambda v^2 + \tfrac{1}{2}K|\nabla u|^2 \right] \mathrm{d}\mathbf{x}, \tag{2.3.42}$$

where u is the phase field and $v = W + H(u)$ is a control field (which may be identified with temperature or concentration) related to the conserved "enthalpy" W; $V(u)$ and $H(u)$ are potentials of a suitable form, K is the rigidity coefficient defined as in Sect. 1.6.1, and λ is a coupling constant. As in Sect. 2.1.1, $V(u)$ is a double-well potential with two minima corresponding to the two alternative phases; the standard choice is the symmetric quartic (2.1.20). The biasing potential $H(u)$ should be odd; its form is not very important, and may be a matter of convenience. A linear function, $H(u) = bu$, appears to be a natural choice, but using a higher order odd function

$$H = b \int_0^u (1 - u^2)^k \mathrm{d}u \tag{2.3.43}$$

gives an advantage of keeping the potential minima fixed at $u = \pm 1$. With $k = 1$, the control variable, while remaining within the limits $|v| < (b\lambda)^{-1}$, just shifts the unstable equilibrium of the phase field; a model with $k = 2$ has also been used in computations, even though it introduces a spurious fourth root. The coefficient b can be related in the thermal context to the ratio of the solidification heat to heat capacity.

The variational equations for the nonconserved and conserved fields are

$$\frac{\partial u}{\partial t} = -\chi \frac{\delta \mathcal{F}}{\delta u}, \quad \frac{\partial W}{\partial t} = \frac{D}{\lambda} \nabla^2 \frac{\delta \mathcal{F}}{\delta W}, \tag{2.3.44}$$

where we have introduced kinetic coefficients χ and $\Gamma = D/\lambda$, presumed to be constant. These equations coincide, up to coefficients, with (2.1.2) and (1.6.11), (1.6.12). As in Sects. 1.6.2, 2.1.1, the free energy decreases monotonically in the course of evolution.

Taking the variations of (2.3.42) explicitly yields the equations for the phase and control fields:

$$\chi^{-1}\partial_t u = K\nabla^2 u + f(u) + \lambda v g(u), \tag{2.3.45}$$
$$\partial_t v = D\nabla^2 v - \tfrac{1}{2} g(u)\partial_t u, \tag{2.3.46}$$

where $f(u) = -V'(u)$, $g(u) = -H'(u)$. One should note the way the two fields are coupled. The control field skews the effective potential of the phase field $\widehat{V}(u) = V(u) + \lambda v H(u)$ in (2.3.45), thereby making one of the minima lower and setting the interphase boundary into motion. At the same time, a moving boundary layer where u switches between the two alternative values contributes a source in the equation of the control field, imitating the heat release or density change at the interface.

An advantage of the phase field model, compared to the Cahn–Hilliard model, is that it disentangles the phase field, now just a nonconserved tag, from conserved variables (density or enthalpy). As a result, equilibrium is enforced only on the interphase boundary, while the bulk may remain in a nonequilibrium (supersaturated or undercooled) state. At the same time, conservation laws are enforced, with due account of density change or heat release accompanying the phase transition, and the system evolves in a thermodynamically consistent way toward a minimum of free energy.

The sharp interface limit can be obtained much in the same way as in Sect. 2.3.1, using a wide separation between the characteristic width of the interphase boundary $l = \sqrt{K}$ applicable in the *inner* region and the characteristic length scale of the control field $L = l/\epsilon$ applicable in the *outer* region (Karma and Rappel, 1998). If no extrinsic scale is imposed, the latter can be estimated as $L = D/c_0$, where c_0 is a characteristic velocity of the interface. The matching time scale is $T = D/c_0^2$. Since the last term in (2.3.46) vanishes far from the interface, the outer equations are just diffusion equations for the control fields on either side of the interface, denoted as $v^{\pm}$, and are made parameterless by the chosen scaling:

$$\partial_t v^{\pm} = \nabla^2 v^{\pm}. \tag{2.3.47}$$

In the inner region, we shall use again the aligned comoving frame and scale the derivatives along and across the axis x normal to the nominal front position, respectively, by $l = \sqrt{K}$ and $L = l/\epsilon$. The curvature is then of $O(L^{-1})$, and is written as $\kappa = K^{-1/2}\epsilon\kappa_1$ when measured on the short inner scale, and the velocity is expressed as $c = (D/L)c_1 = \epsilon\, DK^{-1/2}c_1$. Retaining the terms up to the first order, we write the inner equations as

$$u''(x) + \epsilon(\tau c_1 + \kappa_1)u'(x) + f(u) + \lambda v g(u) = 0, \tag{2.3.48}$$
$$v''(x) + \epsilon\left[(c_1 + \kappa_1)v'(x) + \tfrac{1}{2}c_1 g(u)u'(x)\right] = 0, \tag{2.3.49}$$

where $\tau = D/\chi K$ is a dimensionless capacitance factor, further assumed to be of $O(1)$.

The solution of the inner equations is sought for as an expansion in ϵ:

$$u = u_0 + \epsilon u_1 + \cdots, \qquad v = v_0 + \epsilon v_1 + \cdots. \tag{2.3.50}$$

The zero-order equations are

$$u_0''(x) + f(u_0) + \lambda v_0 g(u_0) = 0, \qquad v_0''(x) = 0. \tag{2.3.51}$$

The solution is again a stationary front that exists only at $v_0 = 0$. This compels the zero-order control field at the interface to be at the level corresponding to the Maxwell construction. In the case of a liquid–solid interface, this just tells that the interface should be at the melting point.

The first-order equation of v reduces to the form

$$\frac{\mathrm{d}}{\mathrm{d}x}\left[v_1'(x) - \tfrac{1}{2}c_1 H(u_0)\right] = 0, \tag{2.3.52}$$

which can be integrated to

$$v_1(x) = \overline{v}_1 + j_1 x + \tfrac{1}{2}c_1 \int_0^x H(u_0(x_1))\mathrm{d}x_1, \tag{2.3.53}$$

where $\overline{v}_1 = v_1(0)$, j_1 are integration constants. This solution has to be matched asymptotically at $x \to \pm\infty$ with the solution of the outer equation (2.3.47):

$$\lim_{x\to\pm\infty} v_1'(x) = j_1 + \tfrac{1}{2}c_1 H^{\pm} = \epsilon^{-1}\mathbf{n}\cdot\nabla v^{\pm}, \tag{2.3.54}$$

where $H^{\pm} = H(u_s^{\pm})$ and the derivative $\mathbf{n}\cdot\nabla$ along the normal $\mathbf{n}$ (directed, by convention, from the solid to the liquid) is computed at the matching interface. Combining the matching conditions on either side and using (2.3.53) yields the condition of interfacial balance, which is more transparent when rewritten in the dimensional form

$$\tfrac{1}{2}c\left(H^{+} - H^{-}\right) = D(\mathbf{n}\cdot\nabla v^{+} - \mathbf{n}\cdot\nabla v^{-}). \tag{2.3.55}$$

The first-order equation of u is again of the form (2.2.9) with the inhomogeneity

$$\Psi(x) = (\tau c_1 + \kappa_1)u_0'(x) + \lambda v_1 g(u_0(x)). \tag{2.3.56}$$

The solvability condition (2.2.11) requires the inhomogeneity to be orthogonal to the translational Goldstone eigenmode $u_0'(x)$. The contribution due to the linear term in (2.3.53) vanishes, as both $u_0'(x)$ and $g(u_0(x))$ are even in x. Collecting the contributions due to other terms, the solvability condition is computed as

$$\lambda\left(H^{+} - H^{-}\right)\overline{v}_1 = \mathcal{I}(\tau c_1 + \kappa_1) + \tfrac{1}{2}c_1\mathcal{J}, \tag{2.3.57}$$

where $\mathcal{I}$ is defined by (2.1.7) and

$$\mathcal{J} = \int_{-\infty}^{\infty} u_0'(x)\,\mathrm{d}x \int_0^x H(u_0(x_1))\,\mathrm{d}x_1. \tag{2.3.58}$$

The values of the control field to be used as interfacial boundary conditions for the outer equation should be of $O(\epsilon)$ to match its value in the inner region. They are denoted as $v_s^{\pm} = \epsilon\,\overline{v}^{\pm}$ and obtained by matching the inner limit of the outer solution

$$v^{\pm} = \epsilon\,\overline{v}^{\pm} + x\,\mathbf{n}\cdot\nabla v^{\pm} \tag{2.3.59}$$

with the outer limit of the inner solution (2.3.53) presented using (2.3.54) as

$$\lim_{x\to\pm\infty} v_1(x) = \overline{v}_1 + \tfrac{1}{2}c_1\mathcal{K}^{\pm} + x\,\mathbf{n}\cdot\nabla v^{\pm}, \tag{2.3.60}$$

where

$$\mathcal{K}^{\pm} = \int_0^{\pm\infty}\left[H^{\pm} - H(u_0(x))\right]\mathrm{d}x. \tag{2.3.61}$$

This yields

$$\overline{v}^{\pm} = \overline{v}_1 + \tfrac{1}{2}c_1\mathcal{K}^{\pm}. \tag{2.3.62}$$

The limits $v^{\pm}$ are identical when both $H(u)$ and $u_0(x)$ are antisymmetric, and can be written explicitly using (2.3.57) and omitting $\pm$ sign as

$$\overline{v} = c_1\left[\tfrac{1}{2}\mathcal{K} + \frac{\mathcal{I}_T + \frac{1}{2}\mathcal{J}}{\lambda\,(H^+ - H^-)}\right] + \frac{\kappa_1\mathcal{I}}{\lambda\,(H^+ - H^-)}, \tag{2.3.63}$$

or, reverting to a dimensional form

$$v_s = \frac{c\sqrt{K}}{D}\left[\tfrac{1}{2}\mathcal{K} + \frac{\mathcal{I}_T + \frac{1}{2}\mathcal{J}}{\lambda\,(H^+ - H^-)}\right] + \frac{\kappa\sqrt{K}\mathcal{I}}{\lambda\,(H^+ - H^-)}. \tag{2.3.64}$$

The integrals appearing in the above expressions are computed most easily by using u_0 as the integration variable instead of x; for the standard $u_0(x)$ given by (2.1.21), the transformation uses the expression for the derivative $p(u) = u'(x) = (1-u^2)/\sqrt{2}$. For $H(u)$ defined by (2.3.43) with $k = 1$, this yields $\mathcal{K}^{\pm} = \mathcal{K} = b/\sqrt{2}$. Other integrals computed with the same functions are $\mathcal{I} = \frac{2}{3}\sqrt{2}$, $\mathcal{J} = \frac{1}{3}b\sqrt{2}$. This completes the reduction to the Stefan problem determined by two linear diffusion equations (2.3.47) with the common value $v^{\pm} = v_s$ on the interface defined by (2.3.64) and the interfacial velocity fixed by (2.3.55).

The interfacial condition (2.3.64) can be again rearranged into a mobility relation similar to (2.2.12) or (2.3.21), which, in turn, reduces to a Gibbs–Thomson relation when the kinetic contribution to the shift of the value of the conserved field at the interface can be neglected. This happens, for example, at $b \gg 1$ when the coefficient at c in (2.3.64) becomes very large, while the coefficient at κ remains of $O(1)$.

2.4 Instabilities of Interphase Boundaries

2.4.1 Instabilities due to Coupling to Control Field

It is quite common for nonequilibrium chemically reacting systems that characteristic spatial and temporal scales associated with different reactants and other dynamic variables, like temperature or potential, are widely separated due to great differences in respective reaction and diffusion rates. This scale separation has been profitably exploited by the analytical theory in Sects. 2.3.2, 2.3.5 where a sharp interface limit has been derived using rational approximation techniques based on the method of matched asymptotic expansions. From the point of view of numerics, the role of scale separation is ambivalent. On the one hand, it makes computations very difficult, requiring high resolution in narrow mobile front regions. On the other hand, the need in a fine grid can be eliminated if the short-scale field is eliminated and replaced by equations of motion of sharp fronts dependent on the local front geometry and on the long-range fields.

The equation of motion of a weakly curved front – the eikonal equation (2.2.12) or (2.2.16) – can be coupled to an additional long-scale "control" field that determines the speed of a straight-line front c_0 or the energy gain V_1. Assuming a linear dependence on the control variable v, the equation of front motion can be then written in the form

$$c = c_0(v - l\kappa), \tag{2.4.1}$$

where l is the capillary length and the scale of v is chosen in such a way that a planar interface is stationary at $v = 0$. Equations of motion of this kind remain valid as long as the characteristic length of the control variable far exceeds the front thickness and, consequently, the control field remains almost constant across the front.

Under certain conditions, coupling to the control field causes the front to lose stability and become corrugated (Mullins and Sekerka, 1963; Langer, 1980). This instability may eventually lead to the formation of stationary, moving, or dynamic structures composed of domains with prevailing alternative states of the short-scale field separated by sharp fronts and immersed in the field of the control variable changing on a longer scale commensurable with the domain size.

Consider a weakly distorted planar front in 3D; the front position is defined as a function $\zeta(\mathbf{y}, t)$ of transverse coordinates $\mathbf{y}$ and time. The normal velocity c is approximated in the leading order by the derivative $\partial\zeta/\partial t$ if the displacement ζ is small, or, more precisely, $|\nabla_\perp\zeta(\mathbf{y})| \ll 1$, where $\nabla_\perp$ is the gradient operator in the transverse plane. Under the same conditions, the interfacial curvature is expressed as $\kappa = -\nabla_\perp^2\zeta$. Suppose that the control field v obeys a linear equation and is fast as well as long-scale, so that it adjusts instantaneously to front displacements. Then the value of the control variable at the front can be expressed using an appropriate Green's function $G(r)$ as a functional of the front position:

$$v(\mathbf{y}) = \int_0^\infty G(|\mathbf{r}|)\zeta(\mathbf{y} - \mathbf{r})d\mathbf{r}. \tag{2.4.2}$$

The eikonal equation (2.4.1) can be written now using the approximate expression for the interfacial curvature in a closed form

$$c_0^{-1}\frac{\partial\zeta}{\partial t} = l\nabla_\perp^2\zeta + \int_0^\infty G(|\mathbf{r}|)\zeta(\mathbf{y} + \mathbf{r})d\mathbf{r}. \tag{2.4.3}$$

This integro-differential equation can be further reduced to a local form if the wavelength of interfacial deformations far exceeds the characteristic scale of the control field. Then the deformation can be expanded as

$$\zeta(\mathbf{y}+\mathbf{r}) = \zeta(\mathbf{y}) + \mathbf{r}\cdot\nabla_\perp\zeta(\mathbf{y}) + \mathbf{rr} : \nabla_\perp\nabla_\perp\zeta(\mathbf{y}) + \dots \tag{2.4.4}$$

Due to the symmetry of the Green's function, the only terms in the expansion contributing to the integral in (2.4.3) are those proportional to $r = |\mathbf{r}|$. Retaining the terms up to the fourth order, we have

$$c_0^{-1}\partial_t\zeta = \beta_0\zeta + (l+\beta_2)\nabla_\perp^2\zeta + \beta_4(\nabla_\perp^2)^2\zeta, \tag{2.4.5}$$

$$\beta_j = \int_0^\infty r^{j+1}G(r)\mathrm{d}r. \tag{2.4.6}$$

The expansion is valid only if $G(r)$ decays sufficiently fast, so that the integrals (2.4.6) converge. The dispersion relation for perturbations with the transverse wavenumber k generated by (2.4.5) is

$$c_0^{-1}\lambda = \beta_0 - (l+\beta_2)k^2 + \beta_4 k^4. \tag{2.4.7}$$

Since (2.4.5) has been derived by long-scale expansion, this dispersion relation is valid for $k \ll 1$ and is duly balanced only when β_0 and $l+\beta_2$ are small, $|\beta_0| \ll |l+\beta_2| \ll |\beta_4| = O(1)$.

The first term on the right-hand side of (2.4.5) vanishes when the front is neutrally stable to parallel displacements. This is a usual situation in a homogeneous infinitely extended system, but in a finite system the front can be either stabilized or destabilized by the gradient of a control field or due to the influence of boundary conditions. A practically important example is instability of diffusionally limited deposition due to an increase of the deposition flux on surface segments drawn closer to the source of deposited particles. This instability may be balanced on shorter scales by surface tension, but is usually not saturated and leads to fingering, formation of dendrites, etc.

A more important *symmetry-breaking* instability independent of extrinsic factors is caused by the reversal of the sign of the effective surface tension at $\beta_2 < -l$. If $\beta_0 = 0$, this instability first arises at long wavelengths and is stabilized at shorter scales by the last term, provided $\beta_4 < 0$. The growth rate reaches a maximum at some $k \neq 0$, which remains small only close to the point of marginal instability, $\beta_2 = -l$. This instability, leading to the structure formation on a moving interface, will be studied in more detail further in this section. Dynamic instabilities, leading to interfacial oscillations and waves, are possible when the response of the control field to surface displacements is delayed.

2.4.2 Mullins–Sekerka Instability

When the interface propagation involves a conserved field, the velocity is primarily governed by a conservation condition at the interface, while (2.4.1) determines the deviation of the order parameter from its standard equilibrium value at a stationary planar interface. We take as a starting point the sharp interface limit of the phase field model, and consider first a planar interface propagating along the x-axis with a constant speed c. The conserved variable v obeys the 1D diffusion equation written in the comoving frame as

$$cv'(x) + Dv''(x) = 0. \tag{2.4.8}$$

The boundary conditions are

$$c = -D[v'(+0) - \beta v'(-0)], \qquad v(0) = c/c_0, \quad v(\infty) = v_\infty. \tag{2.4.9}$$

The conditions at the interface placed at $x = 0$ are given by (2.3.55) and (2.4.1), but the former is generalized by allowing the diffusivities to be different on the two sides of the interface; β is the ratio of diffusivities. This problem can be viewed as a model of solidification where $v = T_m - T$ is interpreted as the deviation[1] of temperature T from its equilibrium value at the interface (the melting temperature) T_m. In this context, the asymptotic value of v in the liquid bulk far from the interface v_∞ is called *undercooling*. An alternative is a "chemical" interpretation, where v is the concentration of a contaminant influencing the melting temperature. In this case, one can expect a concentration discontinuity across the interface, which can be formally eliminated by choosing different scales of v in the two phases (more on this in Sect. 2.4.4).

The solution satisfying (2.4.8), (2.4.9) is

$$v(x) = \begin{cases} c/c_0 & (x \leqslant 0) \\ v_\infty + (c/c_0 - v_\infty)\mathrm{e}^{-(c/D)x} & (x \geqslant 0), \end{cases} \tag{2.4.10}$$

where the propagation speed and undercooling are constrained by the relation $c = c_0(v_\infty - 1)$. In a practically important case $c \ll c_0$, i.e., negligible *kinetic undercooling*, the interfacial temperature remains at the equilibrium melting point, $v(0) = 0$. Then the solution exists at an *arbitrary* speed, but at the *fixed* value of undercooling $v_\infty = 1$. Of course, this remains true only as long as propagation is sufficiently slow.

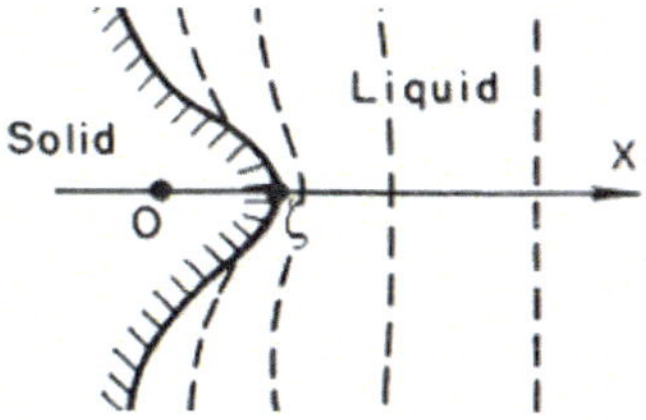

Fig. 2.13. Schematic illustration of the Mullins–Sekerka instability, showing levels v = const for a perturbed interface shifted to the point ζ. Adapted from Langer (1980), reproduced by permission, ©The American Physical Society

The property of Galilean invariance of a propagating planar interface is justified physically, since faster propagation generates larger temperature gradients sufficient to dissipate the increased heat of solidification at a higher rate, so that the interface might remain at melting temperature. It can be seen, however, as an indication of a likely transverse instability. Indeed, the interface would become highly corrugated if its different segments were allowed to propagate with different velocities; this tendency is counteracted by surface tension but enhanced by heat transport, since advanced convex segments dissipate solidification heat more efficiently (see a schematic illustration in Fig. 2.13).

We shall study the stability of a planar front to symmetry-breaking perturbations assuming that the motion is slow, so that the interface remains at equilibrium and (2.4.1) can be replaced by the Gibbs–Thomson relation $v = l\kappa$. Both this relation

[1] The choice of the sign follows our common convention placing a higher HSS at $x > 0$

and the interfacial balance condition (2.3.55), in spite of their innocuous form, are nonlinear, since the direction of the normal and curvature κ are dependent on the position of the interface, and have to be linearized assuming the perturbations to be small. As in the preceding subsection, the perturbed front position is defined as a function $\zeta(\mathbf{y}, t)$ of the transverse coordinate $\mathbf{y}$ and time. The perturbations of the normal velocity $\widetilde{c} = \partial_t \zeta$ and curvature $\kappa = -\nabla_\perp^2 \zeta$ are approximated in the same way as before; the approximation holds as long as the amplitude of the perturbation is much smaller than its wavelength. The control variable is expanded in the vicinity of the unperturbed position as $v = v_s + \widetilde{v}$, where $v_s(x)$ is the basic solution (2.4.10) with some fixed c, written now as

$$v_s(x) = \begin{cases} 0 & (x \leqslant 0) \\ 1 - \mathrm{e}^{-(c/D)x} & (x \geqslant 0), \end{cases} \tag{2.4.11}$$

and $\widetilde{v}(x, \mathbf{y}, t)$ is a perturbation.

Then the Gibbs–Thomson relation is reduced to the form

$$l \nabla_\perp^2 \zeta + [\widetilde{v}(0, \mathbf{y}, t) + \mathrm{d}_x v_s(0) \zeta] = 0. \tag{2.4.12}$$

Under the same assumption of slow propagation, the perturbation field can be assumed to be quasistationary in the comoving frame, and described by the stationary equation

$$D^{\pm}(\nabla_\perp^2 \widetilde{v} + \partial_x^2 \widetilde{v}) + c \partial_x \widetilde{v} = 0, \tag{2.4.13}$$

where $D^+ = D$, $D^- = \beta D$. This equation has to be solved both in the liquid ($x < 0$) and solid ($x > 0$) domains, subject to the boundary condition (2.4.12) at $x = 0$, as well as the asymptotic conditions $\widetilde{v} \to 0$ at $x \to \pm\infty$. The resulting fluxes $D^{\pm} \partial_x \widetilde{v}(\pm 0, \mathbf{y})$ on the liquid and solid sides are used in the perturbed interfacial balance condition

$$\partial_t \zeta = D \left[\partial_x \widetilde{v}(+0) - \beta \partial_x \widetilde{v}(-0) + \mathrm{d}_x^2 v_s''(+0) \zeta \right] \tag{2.4.14}$$

to obtain the variable part of the normal velocity $\partial_t \zeta$.

We are looking for a solution in the spectral form

$$\zeta(y, t) = \chi(k) \mathrm{e}^{\lambda t + \mathrm{i}\mathbf{k} \cdot \mathbf{y}}, \qquad \widetilde{v}(x, y, t) = \psi^{\pm}(x; k) \mathrm{e}^{\lambda t + \mathrm{i}\mathbf{k} \cdot \mathbf{y}}, \tag{2.4.15}$$

where $\psi^{\pm}$ denote the solutions in the liquid and solid domains. Using (2.4.15) in (2.4.12), (2.4.13) we compute

$$\psi^+(x) = (c/D + l k^2) \chi \mathrm{e}^{-q^+ x}, \qquad \psi^-(x) = l k^2 \chi \mathrm{e}^{q^- x}, \tag{2.4.16}$$

where $q^{\pm}$ are the only positive roots of $D^{\pm}(q^2 - k^2) \mp cq = 0$:

$$q^{\pm} = \frac{c}{2D^{\pm}} \left[\sqrt{1 + \left(\frac{2kD^{\pm}}{c} \right)^2} \pm 1 \right]. \tag{2.4.17}$$

Plugging this in (2.4.14) yields the dispersion relation $\lambda(k^2)$ that determines stability of the basic solution:

$$\lambda = -lk^2(D^+q^+ + D^-q^-) + cq^+ - c^2/D. \tag{2.4.18}$$

As expected, λ vanishes at $k = 0$, which reflects the translational symmetry of the front. Stability to long-scale perturbations is determined therefore by the sign of the derivative $\mathrm{d}\lambda/\mathrm{d}(k^2)$ at $k = 0$, which is positive at $D/c > l$, i.e., when the diffusion length $L = D/c$ exceeds the capillary length. This is always true in practical situations satisfying the above assumption of slow propagation; moreover, this is a necessary condition for the validity of the sharp interface limit. The dispersion relation (2.4.18) can be approximated then as

$$\lambda \approx kc\left[1 - (1+\beta)lLk^2\right]. \tag{2.4.19}$$

The "optimal" wavelength of the fastest growing perturbation is of the same order of magnitude as the geometrical mean of the diffusional and capillary lengths, and gradually decreases with growing propagation speed. This trend is, however, reversed at higher velocities when the approximation (2.4.18) breaks down, so that the "optimal" wavelength diverges as the stability limit $L = l$ is approached.

If the quasistationarity assumption is abandoned, $q^{\pm}$ become dependent on λ and the dispersion relation becomes implicit. Additional nonstationary effects appear when kinetic undercooling cannot be neglected. Generally, oscillatory instability becomes possible under these conditions. We shall not consider this case here, as rather nontypical for solidification fronts, but similar nonstationary effects will be further discussed in Sect. 3.1.3.

2.4.3 Instability of a Curved Interface

The above analysis can be generalized to the case of a curved interface representing, for example, a growing spherical nucleus of solid phase. Under these conditions, the problem is, strictly speaking, always nonstationary, since the radius keeps growing as the interface propagates. Nevertheless, the perturbation field $\widetilde{v}$ remains quasistationary as long as propagation is slow, and can be obtained by solving the diffusion equation in spherical coordinates x, θ, ϕ. The solution is expressed as a combination of spherical harmonics:

$$\widetilde{v}(x,\theta,\phi) = \begin{cases} \psi^-_{nm}(x/R)^n P^m_n(\cos\theta)\mathrm{e}^{\mathrm{i}m\phi} & \text{at } r \leqslant R \\ \psi^+_{nm}(R/x)^{n+1} P^m_n(\cos\theta)\mathrm{e}^{\mathrm{i}m\phi} & \text{at } r \geqslant R, \end{cases} \tag{2.4.20}$$

where R is the instantaneous radius of the solid sphere and P^m_n are Legendre polynomials. The components of the radial perturbation of the interface are

$$\zeta(\theta,\phi) = x(\theta,\phi) - R = \sum_{n,m} \chi_{nm} P^m_n(\cos\theta)\mathrm{e}^{\mathrm{i}m\phi}, \tag{2.4.21}$$

and the approximate expression for the curvature of a weakly perturbed sphere is

$$\kappa = 2(R+\zeta)^{-1} - \nabla_\perp^2 \zeta = 2R^{-1} + R^{-2} \sum_{n,m} [n(n+1) - 2] \chi_{nm} P_n^m(\cos\theta) e^{im\phi}. \tag{2.4.22}$$

Take note that the curvature correction vanishes at $n = 1$, since this mode corresponds to a translation of the sphere without change of form. The mean curvature is independent of the azimuthal index m, and therefore the azimuthal dependence can be further omitted.

The spherically symmetric basic solution is expressed in the quasistationary approximation as

$$v_s(x) = \begin{cases} 2l/R & (x \leqslant R) \\ v_\infty + x^{-1}(2l - Rv_\infty) & (x \geqslant R). \end{cases} \tag{2.4.23}$$

The asymptotic undercooling is no more forced to be equal to unity, and the propagation velocity is computed as

$$c_s = \frac{D}{R}\left(v_\infty - \frac{2l}{R}\right) = \frac{2lD}{R^2}\left(\frac{R}{R_c} - 1\right). \tag{2.4.24}$$

The velocity vanishes at $R = R_c = 2l/v_\infty$, which defines the radius of a critical nucleus; further on, it is convenient to eliminate v_∞ in favor of R_c.

Using (2.4.22), (2.4.23) in the Gibbs–Thomson relation $v(R+\zeta) = l\kappa$ expanded to the first order yields

$$\psi_n^+ = l R^{-2}\left[n(n+1) - 2R/R_c)\right]\chi_n, \qquad \psi_n^- = l R^{-2}(n+2)(n-1)\chi_n. \tag{2.4.25}$$

The dispersion relation is now obtained from the interfacial balance condition (2.4.14):

$$\begin{aligned} \lambda &= DlR^{-3}\left[-n(n+1)^2 + 2(n+1)R/R_c - 2\beta n - 4(R/R_c - 1)\right] \\ &= (n-1)DlR^{-3}\left\{-[4 + (3+2\beta)n + (1+\beta)n^2] + 2R/R_c\right\}. \end{aligned} \tag{2.4.26}$$

As expected, λ vanishes at $n = 1$. Instability in the symmetric mode $n = 0$, leading to the collapse of a small nucleus, is observed at $R/R_c < 2$. The most dangerous symmetry breaking instability occurs in the dipole mode $n = 2$ at $R/R_c > 7 + \beta$; thus, the nucleus loses its spherical shape when it is still not much larger than the critical nucleus.

Extending the analysis to more irregular shapes, we can conjecture that the propagating interface tends to destabilize whenever the local curvature radius exceeds R_c. This causes the growing solid to acquire peculiar dendrite forms[2], as seen in Fig. 2.14. The dendrite tip propagates, apparently retaining its shape, while new protuberances, eventually developing into similar propagating tips, develop further along the stem. The entire structure may keep growing and developing tertiary, etc., branches, as long as it is not foiled by collisions of branches and retarded transport.

[2] Actual dendrite forms observed experimentally cannot be reproduced without taking into account anisotropy of crystalline solid interfaces.

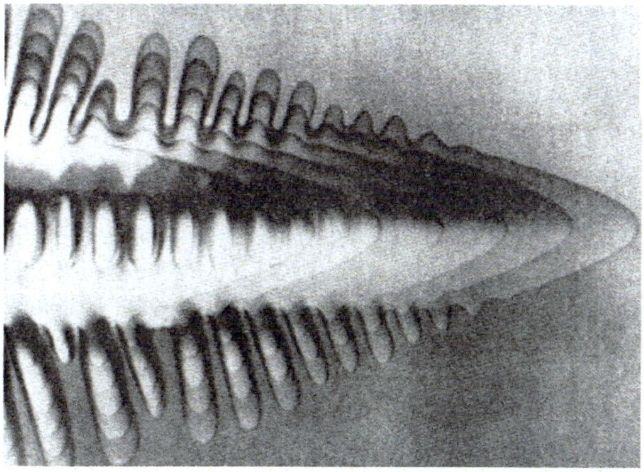

Fig. 2.14. Multiexposure photograph of a growing dendrite (Langer, 1980, reproduced by permission, ©The American Physical Society)

An interesting question is the existence of a self-preserving form that might model the growing tip. This problem is relevant also for other applications where the interface propagation is controlled by a field obeying the Laplace equation, and has been first studied in the context of fingering of the interface of a less viscous fluid propagating into a more viscous one. An infinite family of solutions parametrized by the propagation speed has been found by Ivantsov (1947) for the case of vanishing surface tension; see also an elegant solution method given by Kurtze (1987). The radius of the tip is determined by the velocity, but the selection mechanism for both as yet remains unclear. A nonvanishing surface tension destroys the self-similarity of the problem and renders the analytical theory hopeless. It is commonly believed that surface tension selects a unique solution out of Ivantsov's infinite family, but it is possible that the selection mechanism is dynamic, as in propagation into an unstable state Sect. 2.1.4, and may operate also in the absence of surface tension, as was suggested in a controversial paper by Mineev-Weinstein (1998).

2.4.4 Directional Solidification

The experimental setup used for purification of solids by remelting includes a planar front moving in an imposed temperature gradient (Fig. 2.15a). Two control fields are involved here: temperature and the contaminant concentration. In the simplest formulation of this problem, both the melt and the solid are assumed to be good heat conductors, so that the heat of solidification can be neglected and temperature can be assumed to be a linear function of the coordinate, $T = \Theta x$. This explicit coordinate dependence breaks the translational invariance of the system.

The emergence of spontaneous surface corrugations in this system can be understood in the following way (Wollkind and Segel, 1970). Suppose a protuberance has formed on a solid surface, bringing it into the region of higher temperatures. At a fixed contaminant concentration, this protuberance would melt, but, due to enhanced contaminant transport, the local concentration would decrease and the melting temperature rise, so that the protuberance may keep growing, counteracting the surface

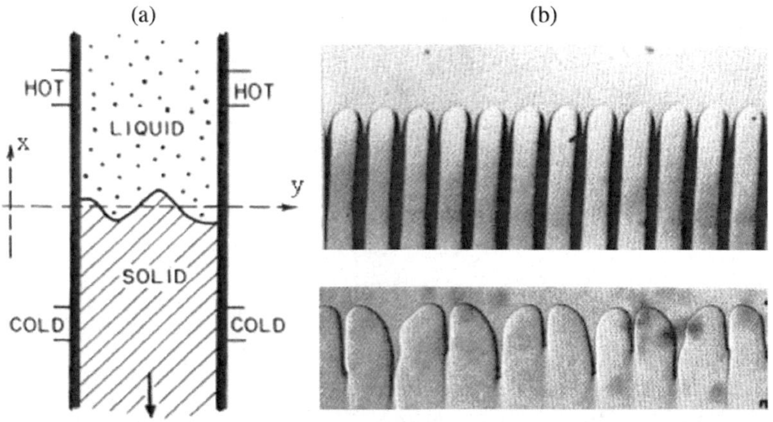

Fig. 2.15. (a) Directional solidification. (a) Schematic of experimental setup (adapted from Langer, 1980). (b) Experimental cellular patterns (Jamgotchian et al., 1993; reproduced by permission, ©The American Physical Society)

tension. Some typical cellular patterns – a regular one close to the point of marginal instability and a more complicated "doublet" structure – are shown in Fig. 2.15b.

The contaminant concentration in the liquid phase is described by the diffusion equation

$$\partial_t v = D\nabla^2 v. \tag{2.4.27}$$

Diffusion in the solid is neglected here. The boundary conditions depend on the distribution coefficient $K < 1$ equal to the ratio of the equilibrium contaminant concentrations in the solid and in the liquid. The flux at the interface determined by the material balance is

$$-D\mathbf{n} \cdot \nabla v(\zeta) = c(1 - K)v, \tag{2.4.28}$$

where $\mathbf{n}$ is the normal to the interface and c is the normal velocity. The rejected material diffuses back into the melt. The asymptotic contaminant concentration far from the interface $v(\infty)$ will be taken as unity. The problem is closed by the equation for the equilibrium melting temperature, which, taking into account the postulated linear dependence of temperature on the coordinate as well as the Gibbs–Thomson relation between the melting temperature and interfacial curvature $(T_m^0 - T_m)/\Theta = \sigma\kappa$, just relates the front position to the local curvature κ and the contaminant concentration in the solid at the interface:

$$\zeta = \alpha[1 - Kv(\zeta)] - \sigma\kappa, \tag{2.4.29}$$

where the constant $\alpha = \Theta^{-1}\mathrm{d}T_m/\mathrm{d}v > 0$ characterizes a linear dependence of melting temperature on concentration.

The stationary concentration field $v_0(x)$ ahead of a planar interface steadily propagating with the speed c_0 verifies the equation in the comoving frame where the interface position is fixed at $\zeta = 0$:

$$Dv_0''(x) + c_0 v_0'(x) = 0, \quad v_0(\infty) = K, \quad -Dv_0'(0) = c_0(1-K)v_0(0). \quad (2.4.30)$$

The contaminant concentration at the interface is taken as the scale of v. The solution is

$$v_0(x) = K + (1-K)\mathrm{e}^{-c_0 x/D}. \quad (2.4.31)$$

We shall look for a perturbed solution in a spectral form

$$\zeta(\mathbf{y},t) = \chi(\mathbf{k},\lambda)\mathrm{e}^{\lambda t + \mathrm{i}\mathbf{k}\cdot\mathbf{y}}, \qquad v(x,\mathbf{y},t) = v_0(x) + \psi(x;\mathbf{k},\lambda)\mathrm{e}^{\lambda t + \mathrm{i}\mathbf{k}\cdot\mathbf{y}}. \quad (2.4.32)$$

The dispersion relation $\lambda(k^2)$ determines stability to infinitesimal perturbations. Using (2.4.32) in (2.4.27) yields

$$D\psi''(x) + c_0\psi'(x) - (\lambda + Dk^2)\psi = 0. \quad (2.4.33)$$

The boundary conditions for (2.4.33) are set at $x = \zeta$; therefore when (2.4.28) and (2.4.29) are expanded and linearized, one should use the value of the zero-order function and its derivative at this location, $v_0(\zeta) = v_0'(0)\zeta$, $v_0'(\zeta) = v_0''(0)\zeta$. Since the zero-order solution is independent of the transverse coordinates, deviations of the normal from the x-axis do not affect the linearized boundary conditions, so that one can replace $\mathbf{n}\cdot\nabla \approx \partial_x$ and express the normal velocity as $c \approx c_0 + \partial\zeta/\partial t$. Then (2.4.28) yields

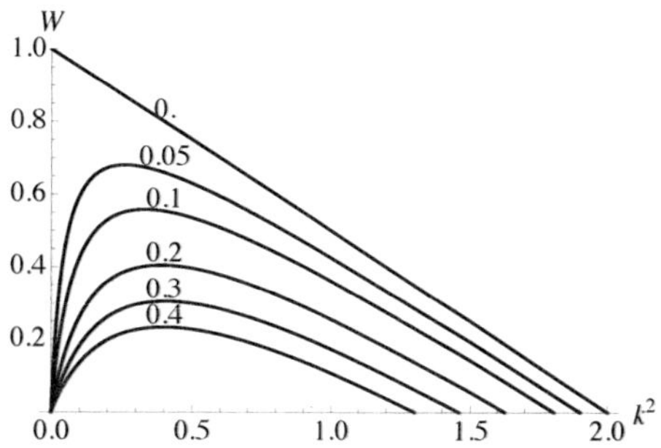

Fig. 2.16. The marginal instability locus at $\beta = 1/2$. The numbers indicate the values of K

$$-D[\psi'(0) + v_0''(0)\chi] = (1-K)[c_0\psi(0) + \lambda v_0(0)\chi + c_0 v_0'(0)\chi]. \quad (2.4.34)$$

The three terms on the right-hand side are due to the concentration perturbation, perturbation of the normal speed, and displacement of the interface. The condition (2.4.29) reduces to

$$\chi = -\alpha K[\psi(0) + v_0'(0)\chi] - \sigma k^2\chi. \quad (2.4.35)$$

The dispersion relation is obtained in an implicit form by solving (2.4.33) with the boundary conditions (2.4.34), (2.4.35) and the asymptotic condition $\psi(\infty) = 0$. A convenient dimensionless expression is obtained using D/c_0 as the length scale and D/c_0^2 as the time scale and introducing the dimensionless parameters

$$W = \frac{D}{\alpha c_0(1-K)}, \qquad \beta = \frac{Wc_0^2\sigma}{D^2} = \frac{c_0\sigma}{(1-K)\alpha D}. \quad (2.4.36)$$

The result is

$$\lambda = (1 - W - \beta k^2)\left(K - \tfrac{1}{2} + \sqrt{\tfrac{1}{4} + \lambda + k^2}\right) - K. \tag{2.4.37}$$

The dependence of the critical value of W at the point of marginal instability ($\lambda = 0$) on the wavenumber at different values of K is shown in Fig. 2.16. The instability sets on at a finite K at sufficiently small values of W below $W_c = 1$.

2.4.5 Long-Scale Instability

At $K \ll 1$, the instability occurs at long wavelengths and W close to unity, which implies $\alpha c_0/D = O(\epsilon^2)$. The marginal instability locus can be approximated in this parametric range as

$$W = 1 - \beta k^2 - (1 + k^2/K)^{-1}. \tag{2.4.38}$$

The maximum is reached at $k^2 = O(\epsilon) \ll 1$ and $1 - W = O(\epsilon)$ when $K = O(\epsilon^2)$. The dispersion relation (2.4.37) can be approximated near the maximum by a quartic form similar to (2.4.7). Setting

$$W = 1 - \epsilon w, \qquad K = \epsilon^2 q, \tag{2.4.39}$$

and rescaling $k \to \sqrt{\epsilon}k$, $\lambda \to \epsilon^2\lambda$, balances the dispersion relation on the $O(\epsilon^2)$ scale as

$$\lambda = -q + k^2(w - \beta k^2). \tag{2.4.40}$$

The dispersion relation reaches a maximum at $k^2_{\max} = w/2\beta$, and the instability threshold is $w = 2\sqrt{q\beta}$. The linear growth of corrugations beyond the instability threshold is described by (2.4.5), suggesting the formation of a transverse pattern with the wavelength $2\pi/k_{\max}$.

A nonlinear equation describing the evolution of long-wave transverse instabilities on a slow time scale can be constructed by expanding the original equations and boundary conditions in powers of the scale ratio (Sivashinsky, 1983). It is advantageous to work in the aligned coordinate frame. By its construction (Sect. 2.2.1), the x-axis is normal to the interface at each location, and the moving interface is always located at $x = 0$. Due to the lack of translational invariance in the directional solidification problem, the local displacement in the laboratory frame ζ cannot be excluded. It is not presumed anymore to be infinitesimally small, but has to be of $O(\epsilon)$ to match the orders of magnitude in further expansion, and is rescaled accordingly: $\zeta \to \epsilon\zeta$. As a consequence, the x-axis remains in the leading order directed along the temperature gradient as before, but a slight disalignment between the two axes has to be taken into account in higher orders of the expansion.

Based on the results of linear analysis, we restrict to the parametric range (2.4.39) and scale the derivatives with respect to transverse coordinates and time as $\nabla_\perp = O(\sqrt{\epsilon})$ $\partial_t = O(\epsilon^2)$. Using the velocity of the flat interface c_0 to fix the basic (short) length scale D/c_0 and the short time scale D/c_0^2, we express the normal front propagation velocity in the aligned frame as

$$c = 1 + \epsilon^3 \left(\zeta_t + \tfrac{1}{2}|\nabla_\perp \zeta|^2\right). \tag{2.4.41}$$

Successive approximations are constructed by expanding $v = v^{(0)} + \epsilon v^{(1)} + \cdots$. Retaining the terms up to $O(\epsilon^3)$, the diffusion equation (2.4.27) and the boundary conditions (2.4.28), (2.4.29) are rewritten in the rescaled dimensionless variables as

$$\epsilon^2 v_t = v_{xx} + \left(c - \epsilon^2 \nabla_\perp^2 \zeta\right) v_x + \epsilon \nabla_\perp^2 v, \tag{2.4.42}$$

$$-v_x(0) = (c - \epsilon^2 q)v(0), \qquad v(\infty) = \epsilon^2 q, \tag{2.4.43}$$

$$v(0) = -\left[1 - \epsilon w - \epsilon^2\left(q + \tfrac{1}{2}|\nabla_\perp \zeta|^2\right)\right]\epsilon\zeta + \epsilon^2 \beta \nabla_\perp^2 \zeta. \tag{2.4.44}$$

The equations and boundary conditions obtained in successive orders of the expansion have a general form

$$\partial_x^2 v^{(j)} + \partial_x v^{(j)} = \Psi_j(x, \mathbf{y}, t), \tag{2.4.45}$$

$$\partial_x v^{(j)}(0) + v^{(j)}(0) = A_j(\mathbf{y}, t), \quad v^{(j)}(0) = V_j^0(\mathbf{y}, t), \quad v^{(j)}(\infty) = V_j^\infty. \tag{2.4.46}$$

The problem is overdetermined, and a solution exists only when a certain solvability condition is satisfied. The general solution satisfying the boundary conditions at $x = 0$ is

$$v^{(j)} = V_j^0 \mathrm{e}^{-x} + A_j\left(1 - \mathrm{e}^{-x}\right) + \int_0^x \left(1 - \mathrm{e}^{\xi - x}\right)\Psi_j(\xi)\mathrm{d}\xi. \tag{2.4.47}$$

Assuming $\Psi_j(\infty) \to 0$, the solvability condition is therefore

$$V_j^\infty = A_j + \int_0^\infty \Psi_j(\xi)\mathrm{d}\xi. \tag{2.4.48}$$

The zero-order solution, in accordance with (2.4.30), (2.4.39), is $v^{(0)} = \mathrm{e}^{-x}$. In the first order, (2.4.42) – (2.4.44) reduce, in view of (2.4.41), to

$$v_{xx}^{(1)} + v_x^{(1)} = 0, \quad -v_x^{(1)}(0) = v^{(1)}(0), \quad v^{(1)}(0) = -\zeta, \quad v^{(1)}(\infty) = 0. \tag{2.4.49}$$

As $A_1 = \Psi_1 = 0$, (2.4.48) is satisfied automatically. The solution $v^{(1)} = -\zeta \mathrm{e}^{-x}$ exists for an arbitrary ζ, so that the displacement remains indefinite in this order.

The second-order equation and boundary conditions are

$$\partial_x^2 v_{xx}^{(2)} + \partial_x v^{(2)} - \partial_x v^{(0)} \nabla_\perp^2 \zeta + \nabla_\perp^2 v^{(1)} = 0, \tag{2.4.50}$$

$$-\partial_x v^{(2)}(0) = v^{(2)}(0) - q v^{(0)}(0), \quad v^{(2)}(\infty) = q, \quad v^{(2)}(0) = w\zeta + \beta \nabla_\perp^2 \zeta. \tag{2.4.51}$$

The inhomogeneous terms in (2.4.50) cancel, so that $\Psi_2 = 0$, and, as $A_2 = q$, (2.4.48) is satisfied again. This is, of course, a consequence of the particular choice of the values of K and W in (2.4.39). The solution is

$$v^{(2)} = q + \mathrm{e}^{-x}(w\zeta + \beta \nabla_\perp^2 \zeta - q). \tag{2.4.52}$$

The evolution equation of ζ is finally obtained as the solvability condition of the third-order problem. The relevant inhomogeneities are

$$\begin{aligned}
\Psi_3 &= \partial_t v^{(1)} + \partial_x v^{(1)} \nabla_\perp^2 \zeta - \partial_x v^{(0)} \left(\partial_t \zeta + \tfrac{1}{2}|\nabla_\perp \zeta|^2\right) - \nabla_\perp^2 v^{(2)} \\
&= \mathrm{e}^{-x} \left(\zeta \nabla_\perp^2 \zeta + \tfrac{1}{2}|\nabla_\perp \zeta|^2 - w \nabla_\perp^2 \zeta - \beta \nabla_\perp^4 \zeta\right), \qquad (2.4.53) \\
A_3 &= q v^{(1)}(0) - v^{(0)}(0) \left(\zeta_t + \tfrac{1}{2}|\nabla_\perp \zeta|^2\right) = -\left(q\zeta + \partial_t \zeta + \tfrac{1}{2}|\nabla_\perp \zeta|^2\right).
\end{aligned}$$

Using this in (2.4.48) yields the desired equation for ζ, which can be written in the form

$$\partial_t \zeta + q\zeta + \nabla_\perp \cdot [(w - \zeta)\nabla_\perp \zeta] + \beta \nabla_\perp^4 \zeta = 0. \qquad (2.4.54)$$

The linear part of (2.4.54) yields the dispersion relation (2.4.40), while the nonlinear terms make the effective diffusivity dependent on the "order parameter" ζ. The equation is reminiscent of the Kuramoto–Sivashinsky (KS) equation (1.1.5), but the substantial difference is the presence of the "order parameter" ζ itself rather than only its derivatives. The dynamics is very much different compared to the damped KS equation (with an added linear algebraic term), which was studied as a model of transition to spatio-temporal chaos (Chaté and Manneville, 1987). Integrating (2.4.54) across the front yields simply $\mathrm{d}\langle\zeta\rangle/\mathrm{d}_t = -q\langle\zeta\rangle$; thus, the average displacement $\langle\zeta\rangle$ decays to zero with time, in contrast to both the translationally invariant form (1.1.5) and its damped version.

The effective diffusivity $\zeta - w$ increases with growing ζ, suggesting that the interface should be smoother on the segments convex in the direction of higher temperatures and sharper on concave segments left behind, as commonly seen in solidification profiles observed experimentally (Fig. 2.15b). When the diffusivity becomes negative, the interface is stabilized by the fourth-order term. In actual patterns, which are not faithfully modeled by long-scale equations, the instability on the back side may never saturate, leaving long grooves where the contaminant may be trapped.

The standard KS equation (1.1.5) can be obtained, under rather restrictive conditions, for an unstable solidification front in a translationally invariant system. A derivation in a more general context of reaction-diffusion systems with separated scales will be given in Sect. 3.2.6.

3 Structures with Sharp Interfaces

3.1 Rectilinear Structures

3.1.1 Basic Model and Scaling

A reaction-diffusion model with separated scales suitable for generation of a variety of patterns is a two-component system including a short-range activator u and a long-range inhibitor v, written in a general form

$$\gamma_u^{-1}\partial_t u = D_u \nabla^2 u + f(u,v), \quad \gamma_v^{-1}\partial_t v = D_v \nabla^2 v + g(u,v). \tag{3.1.1}$$

Using the characteristic time and length scales of the long-range component (respectively, γ_v^{-1} and $\sqrt{D_v}$) as basic units, and denoting the ratios of the characteristic time and length scales of the two variables $\gamma = \gamma_v/\gamma_u$, $\epsilon = \sqrt{\gamma D_u/D_v}$, we have

$$\gamma \partial_t u = \epsilon^2 \nabla^2 u + f(u,v), \quad \partial_t v = \nabla^2 v + g(u,v). \tag{3.1.2}$$

If the function $f(u,v)$ has at a given value of v two stable zeros, say, $u = u_\pm(v)$, the short-scale variable may switch between these two values across a front of the characteristic thickness ϵ. The Turing instability leading to spontaneous separation of domains with alternative HSS requires the activator to be short-range and the inhibitor to be long-range (Sect. 1.3.1). The boundary between alternative domains, or a *front* of the characteristic thickness ϵ is relatively sharp at $\epsilon \ll 1$. The long-scale variable remains almost constant across a narrow front, and therefore the derivation of the normal front velocity in Sects. 2.1.2 and 2.2.2 remains in force. The front, generally, will propagate with a characteristic speed $c^* = \epsilon/\gamma$. This speed can be further reduced by keeping the function $f(u,v)$ close to the Maxwell construction.

The time scale ratio γ can be chosen in a particular way to make possible the formation of nontrivial static and dynamic structures involving immobilized or propagating fronts. The essential parameter is the ratio of the characteristic time that the front takes to advance at a distance of $O(1)$ and the characteristic time of the inhibitor, $\tau^* = \gamma/\epsilon$. If $\tau^* \gg 1$, the inhibitor is fast and adjusts to the front motion quasistationarily. This case is conducive to the formation of stationary structures. In

L. Pismen, *Patterns and Interfaces in Dissipative Dynamics*, Springer Series in Synergetics,
https://doi.org/10.1007/978-3-031-29579-9_4

the opposite case $\tau^* \ll 1$, when the inhibitor is very inertial, immobilized fronts are unstable and mobile structures are typical. The transition between both types of behavior occurs at $\tau^* = O(1)$, i.e., under conditions when the activator is still fast as well as short-range.

Another feature influencing the propagation speed is the strength of coupling between the activator and the inhibitor. The most interesting case for the study of stationary or slowly evolving structures in more than one dimension is that of *weak* coupling, $|\partial f/\partial v| = O(\epsilon)$. Then the nonlinear function in (3.1.2) can be written as $f(u,v) = f_a(u) + \epsilon v f_b(u)$, where $f_a(u)$ is antisymmetric and $f_b(u)$ is symmetric. In this case, the system remains close to the Maxwell construction in a wide range of v, and the characteristic front propagation speed, slowing down by a factor of $O(\epsilon)$, becomes comparable to a correction due to the interfacial curvature with a radius measurable on the long inhibitor scale. The substantial advantage of keeping the system nearly symmetric is making the algebra easier without affecting the results in a qualitative way. The essential time scale ratio is modified in this case to $\tau = \gamma/\epsilon^2$. This is the basic scaling to be used in a larger part of this chapter. A different scaling, suitable for rapidly propagating fronts, will be introduced in Sect. 3.4.1.

A particular form of (3.1.2) obeying this scaling, which we shall explore in detail, is the FitzHugh–Nagumo dynamical system (1.2.6) extended to the RDS with a cubic $f(u,v)$ linearly biased by v (i.e., $f_b(u) = -1$) and a linear $g(u,v)$:

$$\epsilon^2 \partial_t u = \epsilon^2 \nabla^2 u + u - u^3 - \epsilon v, \tag{3.1.3}$$

$$\tau^{-1} \partial_t v = \nabla^2 v - v - \nu + \mu u. \tag{3.1.4}$$

The choice of the cubic function in (3.1.3) ensures the existence of two stable homogeneous stationary states (HSS) at $v \sim O(1)$. Other symmetric functions $f(u)$ with three zeros would be suitable as well. Besides ϵ, the three parameters that cannot be eliminated by rescaling are the coupling parameter μ, the bias parameter ν, and the time scale ratio τ. The scaling of the bias and differential terms in (3.1.3) is chosen to suit the eikonal equation (2.2.12), which is rewritten in the adopted scaling as

$$c = bv - \kappa, \tag{3.1.5}$$

where b is a numerical parameter. For the cubic function, $b = 3/\sqrt{2}$, which follows from the expansion of (2.1.15) near the Maxwell construction at $v = 0$.

The relevant range of parameters can be suggested by qualitative analysis. Two stable roots of (3.1.3) always exist in the limit $\epsilon \to 0$ at $|\mu|, |\nu| \ll 1/\epsilon$. The system is excitable when μ is positive, so that the null isoclines of the two variables are situated as in Fig. 3.1. Under these conditions, both stationary states would retreat when placed in contact with an alternative HSS of the short-scale equation at the same value of v. With this choice of signs, v is depleted in the lower state $u_s^- < 0$ and produced in the upper state $u_s^+ > 0$, while, in contrast, the lower state advances at $v > 0$ and the upper at $v < 0$. Similar considerations can be extended to other functions $f(u)$.

Such conditions are conducive to the formation of stable solitary structures[1]. Consider, for example, a finite region with $u < 0$ immersed within a continuum of positive u. If it starts to shrink, the level of v would raise and, as a result, the front may be immobilized again. If the dimension is higher than one, the question is whether this stabilizing action wins over curvature-driven shrinking, which is accelerated inversely proportional to its radius. At $\mu < 0$, neither global stationary state is excitable. Stationary solitary structures, which can be formally constructed also in this case, correspond to unstable critical nuclei of one of alternative states. The parameter τ plays, as we shall further see, a special role, influencing the transition from stationary to mobile structures.

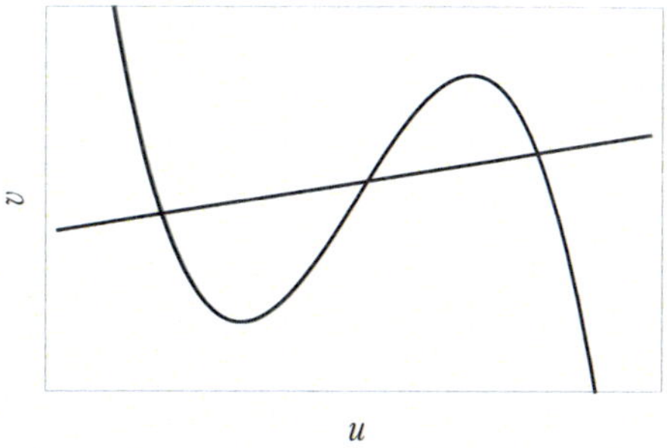

Fig. 3.1. Null isoclines of the FN system

Weak coupling in (3.1.2) is convenient for the analysis of stationary or slowly involving structures, but it is unsuitable when propagation is fast. A convenient form of a strongly coupled system contains a quadratic symmetric part $f_b(u) = -(1-u^2)$, where v is constrained to $|v| \leq 1$. This leads to

$$f(u,v) = -(u+1)(u-1)(u-v). \tag{3.1.6}$$

With this choice, $u_s^{\pm} = \pm 1$, and only the unstable intermediate state $u_s^0 = v$ is v-dependent. The disadvantage is, however, a discontinuous zigzag form of the null isocline replacing a smooth S-shaped curve in Fig. 3.1. This may bring about non-physical effects, such as an unlimited front propagation speed.

3.1.2 Stationary Fronts

Various stationary structures can be obtained in the limit $\epsilon \to 0$ following a common strategy (Ohta et al, 1989). We solve the long-scale equation (3.1.4) replacing u by its quasistationary values $u_s^{\pm} = \pm 1 + O(\epsilon)$ in the regions separated by a front, which is assumed to be of negligible thickness on the characteristic length scale of the long-range variable. The position of the front is further defined by requiring the values of v to annul the propagation speed given by (3.1.5). Since only solving the linear equation for the slow variable is involved, the results are not restricted to the cubic form, but are applicable to any symmetric function $f(u)$ with two stable roots.

A 1D solution corresponding to a single stationary front on the infinite line verifies the equation

$$v''(x) - v - \nu \mp \mu u = 0, \tag{3.1.7}$$

[1] Localized solutions of nonlinear equations are sometimes called *dissipative solitons* (Purwins et al, 2010), but, unlike classical solitons in integrable conservative systems, such structures do not retain their identity upon collisions

with the sign of the coupling term coincides with the sign of x. The matching conditions at the short-scale discontinuity $x = 0$ are given by requiring v to be continuous and smooth. The solution is

$$v = \begin{cases} \mu(e^x - 1) - \nu & \text{at } x \leqslant 0, \\ \mu(1 - e^{-x}) - \nu & \text{at } x \geqslant 0. \end{cases} \tag{3.1.8}$$

The stationarity condition is just $v(0) = 0$, which implies $\nu = 0$. Naturally, a single stationary front is possible only when the system is symmetric.

Stability of a stationary front to infinitesimal displacements is checked in a standard way by linearizing the eikonal equation (3.1.5). We consider a straight-line front parametrized by the coordinate y and denote the instantaneous displacement of the front from its unperturbed position toward the region occupied by the up state as $\zeta(y, t)$. As long as the amplitude of the perturbation is much smaller than its wavelength, the normal vector defining the direction of the front propagation is almost parallel to the x-axis, and the propagation speed is expressed as $\zeta_t = \epsilon c$. In the same approximation, the curvature is given by $\kappa = -\zeta_{yy}$. Expanding also the long-range variable in the vicinity of the unperturbed position, we reduce (3.1.5) to the form

$$\zeta_t = \zeta_{yy} + b\left[\widetilde{v}(0, y, t) + v_s'(0)\zeta\right]. \tag{3.1.9}$$

Here $\widetilde{v}$ is the perturbation of the long-range variable, and $v_s'(0) = -j$ is determined by the stationary flux at the front following from the stationary solution (3.1.8).

The correction to v due to a weak perturbation of the front can be computed in a simple way by observing that any shift of the front position by an infinitesimal increment ζ is equivalent to switching the sign of the source term in (3.1.4) in a narrow region near $x = 0$. This introduces a source localized at the unperturbed front position and proportional to its displacement in the equation for the perturbation $\widetilde{v}$, which is written as

$$\tau^{-1}\partial_t \widetilde{v} = \nabla^2 \widetilde{v} - \widetilde{v} - 2\mu\zeta(y, t)\delta(x). \tag{3.1.10}$$

Although the perturbation is habitually presumed to be arbitrarily small for the purpose of linear stability analysis, the method is actually also applicable to finite perturbations of the front position, provided they are small on the extended scale, i.e., $\zeta \leqslant O(\epsilon)$.

As usual, we are looking for a solution in the spectral form

$$\zeta(y, t) = \chi(k, \lambda)e^{\lambda t + iky}, \quad \widetilde{v}(x, y, t) = \psi(x; k, \lambda)e^{\lambda t + iky}. \tag{3.1.11}$$

The dispersion relation $\lambda(k^2)$ determines stability to infinitesimal perturbations. Using (3.1.11) in (3.1.10) yields

$$\psi''(x) - q^2\psi = 2\mu\chi(k, \lambda)\delta(x), \tag{3.1.12}$$

where $q^2 = 1 + \lambda/\tau + k^2$. The solution of this equation is

$$\psi(x) = -\mu\chi(k, \lambda)q^{-1}e^{-|qx|}. \tag{3.1.13}$$

Plugging (3.1.8), (3.1.11), (3.1.13) in (3.1.9) yields the dispersion relation

$$\lambda = -k^2 + b\mu(1 - q^{-1}) \,. \tag{3.1.14}$$

As expected, λ vanishes at $k = 0$, which reflects the translational symmetry of the front, but, since q is dependent on λ, this dispersion relation is still unresolved. At $k = 0$, it has, besides $\lambda = 0$, a finite root

$$\lambda = b\mu + \tfrac{1}{2}\left[\sqrt{\tau(\tau + 4b\mu)} + \tau\right] , \tag{3.1.15}$$

which is positive, indicating instability, at $b\mu > 2\tau$. This relation defines the locus of *traveling* instability. This and further results are made transparent by defining $\tau_c = b\mu/2$. Stability at $\tau_c < \tau$ can be tested by expanding (3.1.14) assuming $\lambda \sim k^2 \ll 1$. This yields in the leading order

$$\lambda = -k^2\tau\frac{b\mu - 2}{b\mu - 2\tau} \equiv -k^2\tau\frac{\tau_c - 1}{\tau_c - \tau} \,. \tag{3.1.16}$$

The denominator must be negative in the case of a stable front; therefore, the remaining stability condition is $\tau_c < 1$. An easy way to arrive at the same result is to observe that the location of a bifurcation at zero eigenvalue should not depend on τ and test it at $\tau \to \infty$. Then the stability condition $b\mu < 2$, or $\tau_c < 1$, is obtained directly by expanding (3.1.16). Above this limit, the zigzag instability takes place.

It remains to investigate whether an oscillatory instability may take place here. Setting $\lambda = \mathrm{i}\omega\tau$ and separating the real and imaginary parts of (3.1.14) yields

$$k^2 = b\mu\left(1 - \chi^{-1}\cos\alpha\right) + \mathrm{i}\left(b\mu\chi^{-1}\sin\alpha - \omega\tau\right) , \tag{3.1.17}$$

where $\chi = [(1 + k^2)^2 + \omega^2]^{1/4}$, and α is the half of the phase angle of the complex number $1 - q^{-1}$:

$$\alpha = \frac{1}{2}\arg(1 + k^2 + \mathrm{i}\omega/\tau) = \frac{1}{2}\arctan\frac{\omega}{(1 + k^2)\tau}, \quad |\alpha| \leqslant \frac{\pi}{4}. \tag{3.1.18}$$

The instability is possible when the imaginary part of the above expression vanishes, while its real part is positive. The imaginary part comes to zero at

$$\frac{b\mu}{\tau} \equiv \frac{2\tau_c}{\tau} = \frac{\omega\chi}{\sin\alpha}. \tag{3.1.19}$$

The lowest possible value of the ratio in the right-hand side is attained at $\omega = 0$. Expanding χ and α in this limit shows that it is equal to $2(1 + k^2)^{3/2}$. Thus, the front is stable in the entire domain $\tau_c < \tau$, $\tau_c < 1$.

3.1.3 Propagating Fronts

Propagating fronts exist at $\tau < \tau_c$, and traveling instability precedes the zigzag instability at $\tau < 1$, when the long-range variable is sufficiently inertial and its distribution lags behind the displacement of the sharp front. By analogy with magnetic domain walls, stationary and propagating fronts have been labeled, respectively, as *Ising* and *Bloch* fronts, and the transition between them as the Ising–Bloch bifurcation (Coullet et al, 1990). The term has taken roots, although the specific structure of magnetic domain walls is not reproduced in equations of the reaction-diffusion type, including the FN model, and magnetic Bloch, as well as Ising, walls are stationary. The term "traveling instability" is more suggestive and precise.

The long-range variable should verify the equation in the comoving frame $z = x - Ct$, where $C = c/\tau$.

$$\begin{aligned} Cv'(z) + v''(z) - v - \nu - \mu = 0 \quad &\text{at} \quad z < 0, \\ Cv'(z) + v''(z) - v - \nu + \mu = 0 \quad &\text{at} \quad z > 0. \end{aligned} \tag{3.1.20}$$

The general solution that satisfies the matching conditions at the short-scale discontinuity at $z = 0$ and does not diverge at $z \to \pm\infty$ is

$$v = \begin{cases} \mu\left[\left(1 + s^{-1}\right) e^{\frac{1}{2}(s-1)Cz} - 1\right] - \nu & \text{at } z \leqslant 0 \\ \mu\left[\left(s^{-1} - 1\right) e^{-\frac{1}{2}(1+s)Cz} + 1\right] - \nu & \text{at } z \geqslant 0, \end{cases} \tag{3.1.21}$$

where $s^2 = 1 + (2/C)^2$. The value of v on the front, $v(0) = -\nu + \mu/s$, must equal $C\tau/b$. This gives the equation for C,

$$\frac{C}{\sqrt{4 + C^2}} = \frac{1}{\mu}\left(\nu + \frac{C\tau}{b}\right) \equiv \frac{\nu}{\mu} + \frac{C\tau}{2\tau_c}. \tag{3.1.22}$$

Forgetting for the moment the orders of magnitude of the parameters compatible with the weakly biased model, we can observe that the left-hand side of this equation is an S-shaped curve (Fig. 3.2), which may have more than one intersection with the straight line representing the right-hand side. In the symmetric case $\nu = 0$ (Fig. 3.2a), besides the stationary solution $C = 0$, there are two symmetric nontrivial solutions

$$C = \pm\sqrt{(b\mu/\tau)^2 - 4} \equiv \pm 2\sqrt{(\tau_c/\tau)^2 - 1} \tag{3.1.23}$$

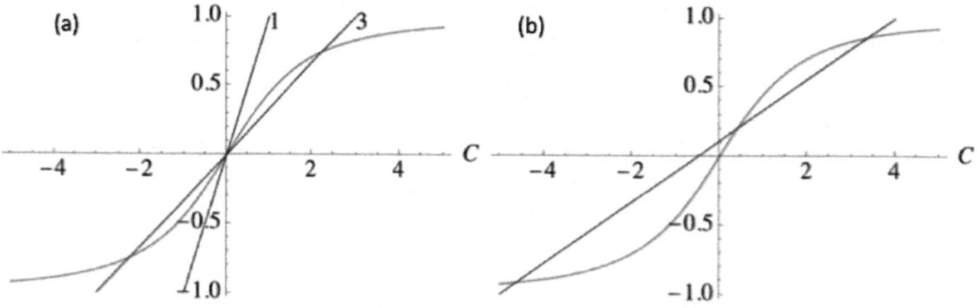

Fig. 3.2. The S-shaped curve in (3.1.22). (a) The symmetric case ($\nu = 0$); the number of solutions is shown at the respective straight lines. (b) The asymmetric case ($\nu > 0$)

corresponding to fronts propagating in opposite directions. These solutions appear when τ drops below the critical value $\tau_c = b\mu/2$, which has been identified in Sect. 3.1.2 as the point where the dispersion relation (3.1.14) acquires a positive root at $k = 0$. This is a pitchfork bifurcation, which can exist only when the two propagation directions are symmetric[2]. In the asymmetric case $\nu \neq 0$, as the straight line shifts, the unstable stationary solutions is replaced by an unstable front propagating in the "wrong" direction (positive at $\nu > 0$ and *vice versa*). A pair of solutions, one stable and another unstable, disappear at a saddle-node bifurcation. The bifurcation condition is determined by tangency, and can be derived from (3.1.22) in an implicit form

$$\frac{\tau}{\tau_c} = (1 + C^2/4)^{-3/2}, \qquad \frac{\nu}{\mu} = C(1 + 4/C^2)^{-3/2}. \tag{3.1.24}$$

The relation between the ratios τ/τ_c and ν/μ at the bifurcation locus is drawn in Fig. 3.3a. The front velocity increases with ν along this curve, and the critical value of τ required for the bifurcation of a pair of fronts going in a "wrong" direction decreases with growing bias. Qualitatively, this picture fits the bifurcations in the strongly biased FN model computed numerically by Hagberg and Meron (1994), but solving (3.1.20) analytically is impossible in this case due to the nonlinear dependence of quasistationary levels $u_s(v)$ in (3.1.3), which makes (3.1.4) nonlinear as well through the term $\mu u_s(v)$.

We have to recall at this stage that in the weakly biased FN system, $C \sim O(1)$ is possible only at $\tau \sim O(\epsilon)$. If other parameters are of $O(1)$, the tangency condition cannot be met. Otherwise, $C \sim O(\epsilon)$, and (3.1.22) is satisfied only at $\nu \sim O(\epsilon)$. The expressions (3.1.23), (3.1.24) remain in force only in the limit $\mu, \nu, \tau \sim O(\epsilon)$, $C \sim O(1)$, so that the bifurcation diagram fitting the strongly biased model is reproduced under these conditions in qualitative way. Therefore we restrict to this limit to investigate the zigzag and oscillatory instabilities of propagating fronts, presuming that a qualitative semblance will persist.

The spectral perturbation equation (3.1.12) is transformed in the comoving frame to

$$C\psi'(z) + \psi''(z) - q^2\psi = 2\mu\chi(k,\lambda)\delta(z)\,, \tag{3.1.25}$$

leading to

$$\psi(0) = -\frac{\mu}{\sqrt{q^2 + C^2/4}}\chi(k,\lambda)\,. \tag{3.1.26}$$

Another change stems from the term $v_s'(0)$ in (3.1.9), which is now expressed as $v_s'(0) = 2\mu/\sqrt{4 - C^2}$. This leads, upon using these results in (3.1.9), to the implicit dispersion relation modifying (3.1.14) to

$$\lambda = -k^2 + 2\tau_c\left(\frac{1}{\sqrt{1 + C^2/4}} - \frac{1}{\sqrt{q^2 + C^2/4}}\right). \tag{3.1.27}$$

The translational invariance remains in place, since $q = 1$ at vanishing k and λ. As before, expanding (3.1.27) at $\lambda \sim k^2 \ll 1$ yields in the leading order

[2] It is also known as a "drift pitchfork bifurcation"

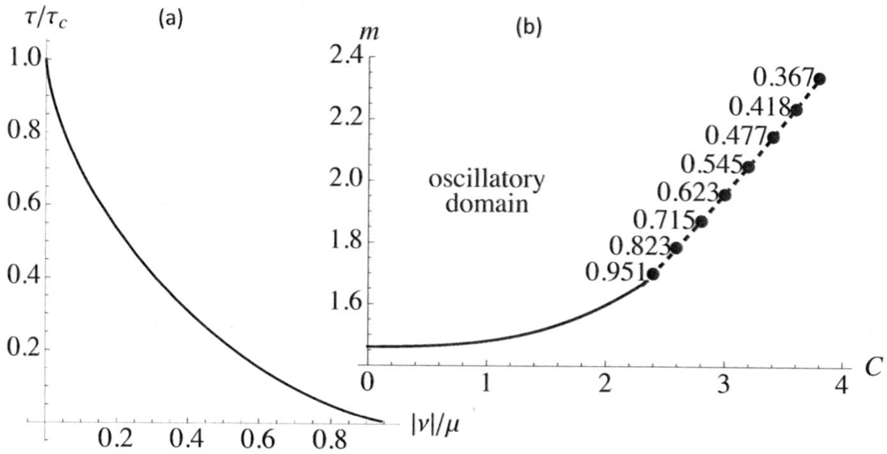

Fig. 3.3. (a) The locus of bifurcation of a pair of fronts with ν and C of the same sign coinciding with the locus of the zigzag instability. (b) The locus of oscillatory instability. The solid curve shows the instability limit at $\theta = \tau/\tau_c = 1$. The numbers at the dashed line indicate the maximum θ at the respective velocity C.

$$\lambda = -k^2 \frac{\tau_c - (1 + C^2/4)^{3/2}}{\tau_c/\tau - (1 + C^2/4)^{3/2}}, \tag{3.1.28}$$

which reduces to (3.1.16) at $C = 0$. The numerator is negative with the adopted scaling, and instability sets on when the denominator is positive. The instability limit coincides with the tangency condition (3.1.24). When a pair of solutions emerges at a saddle-node bifurcation, one of them must be unstable, and the above result demonstrates that in this case the instability of the slower front is undulatory. The dispersion relation (3.1.16) is also applicable with the common scaling $\tau \sim O(1)$, $C \sim O(\epsilon)$, in which case the zigzag instability would set on when the *numerator* is negative, only slightly shifting the threshold of a stationary front.

Oscillatory instability is tested, as in Sect. 3.1.2, by setting in (3.1.27) $\lambda = \mathrm{i}\omega$ to obtain after separating the real and imaginary parts

$$k^2 = 2\tau_c \left(\frac{1}{\sqrt{1 + C^2/4}} - \frac{\cos\alpha}{\chi} \right) + \mathrm{i} \left(\frac{2\tau_c \sin\alpha}{\chi} - \omega \right), \tag{3.1.29}$$

$$\chi = [(C^2/4 + 1 + k^2)^2 + (\omega/\tau)^2]^{1/4}, \tag{3.1.30}$$

$$\alpha = \frac{1}{2} \arg[C^2/4 + 1 + k^2 + \mathrm{i}\omega/\tau)] = \frac{1}{2} \arctan \frac{\omega/\tau}{(C^2/4 + 1 + k^2)}.$$

The condition of vanishing imaginary part of (3.1.29) has the same form as (3.1.19) with χ and α defined by (3.1.30), but with $\tau \sim O(\epsilon)$ expanding in the limit $\omega \to 0$ does not work. Assuming $\omega \ll \tau$ leads to $\tau \sim O(1)$; therefore ω and τ must be of the same order of magnitude, and instability would set on at $\omega \neq 0$.

The real part of (3.1.29) is balanced at $k^2 \sim O(\epsilon)$, so that it can be neglected in (3.1.30), presenting it in the leading order as

$$\alpha = \frac{1}{2}\arctan\frac{\omega}{\tau\chi_0}, \qquad \chi_0 = [S + (\omega/\tau)^2]^{1/4}, \qquad S = (C^2/4+1)^2. \quad (3.1.31)$$

The imaginary part takes a visually simple form $\chi_0 = 2m\sin\alpha$, where $m = 2\tau_c/\omega$. It must vanish at the instability locus. The explicit form of the resulting equation, which can be solved to obtain the dependence of m on the ratio $\theta = \tau/\tau_c$ and propagation velocity, is

$$\left[S + \left(\frac{2}{m\theta}\right)^2\right]^{1/4} = \frac{m}{\sqrt{2}}\sqrt{1 - \left[1 + \frac{4}{(m\theta)^2 S}\right]^{-1/2}}, \qquad (3.1.32)$$

We are interested in the largest value of θ allowing for acceptable solutions of this equation. The computation shows that the maximum value $\theta = 1$ is attained up to $C \approx 2.3$ and further decreases with C, as shown in Fig. 3.3b, so that increasing velocity has a stabilizing effect. The frequency at the bifurcation locus can be extracted from m. The wavenumber is determined by combining the expressions for real and imaginary parts, $\sin\alpha = \frac{1}{2}\chi_0/m$, $\omega\cos\alpha = k^2\omega\chi_0/m$ to obtain

$$k^2 = \omega\sqrt{m^2[S + (\omega/\tau)^2]^{-1/2} - 1} = \omega\sqrt{m^2[S + 4(m/\theta)^{-2}]^{-1/2} - 1}. \quad (3.1.33)$$

One can check that the expression under the radical is always positive under conditions when solutions of (3.1.32) exist, so that the limits indicated above remain in force.

3.1.4 Stationary Bands

A more interesting structure is a solitary band of width $2L$ bound by two fronts of opposite polarity. For the down state inside the band, the profile of v should verify

$$\begin{aligned} v''(x) - v - \nu - \mu = 0 \quad &\text{at} \quad |x| < L, \\ v''(x) - v - \nu + \mu = 0 \quad &\text{at} \quad |x| > L. \end{aligned} \qquad (3.1.34)$$

There are now two symmetric sets of matching conditions at $x = \pm L$, and the solution is

$$v = \begin{cases} \mu(2\mathrm{e}^{-L}\cosh x - 1) - \nu & \text{at } |x| \leqslant L \\ \mu(1 - 2\mathrm{e}^{-|x|}\sinh L) - \nu & \text{at } |x| \geqslant L. \end{cases} \qquad (3.1.35)$$

The fronts are stationary when $v(\pm L) = \mu\mathrm{e}^{-2L} - \nu = 0$, which gives $L = \ln\sqrt{\mu/\nu}$. The solutions exist only at $\mu > \nu > 0$.

In the same way as for solitary fronts, we can write equations of motion for infinitesimal displacements $\zeta^{\pm}$ of the kink at $x = L$ and the antikink at $x = -L$ that confine a solitary band, coupled to the perturbation of the slow variable $\widetilde{v}(x, y, t)$. Take note that both the sign of the bias term in the eikonal equation (3.1.5) and the effect of front displacement are reversed when applied to the antikink. Using the expression for the stationary fluxes at $x = \pm L$ following from (3.1.35), we obtain

$$\zeta_t^{\pm} = \zeta_{yy}^{\pm} b \left[\pm\widetilde{v}(\pm L, y, t) + \mu(1 - \mathrm{e}^{-2L})\zeta^{\pm} \right]. \tag{3.1.36}$$

The equation for the perturbation field is

$$\tau^{-1}\partial_t \widetilde{v} = \nabla^2 \widetilde{v} - \widetilde{v}^{\pm} - 2\mu\zeta^{+}(y,t)\delta(x - L) + 2\mu\zeta^{-}(y,t)\delta(x + L). \tag{3.1.37}$$

It is advantageous to consider the symmetric and antisymmetric combinations of displacements, $\xi^{\pm} = \frac{1}{2}(\zeta^{+} \mp \zeta^{-})$, that may lead to the *varicose* or *zigzag* instabilities, respectively. It is clear from (3.1.36) that $\xi^{\pm}$ are coupled to the symmetric and antisymmetric parts $\widetilde{v}^{\pm}$ of the perturbation field $\widetilde{v}$:

$$\xi_t^{\pm} = \xi_{yy}^{\pm} + b \left[\widetilde{v}^{\pm}(L, y, t) + \mu(1 - \mathrm{e}^{-2L})\xi^{\pm} \right]. \tag{3.1.38}$$

It is sufficient to define the equations for both perturbation fields derived from (3.1.37) in the half-plane $x > 0$, continuing the solutions to $x < 0$ by symmetry:

$$\tau^{-1}\partial_t \widetilde{v}^{\pm} = \nabla^2 \widetilde{v}^{\pm} - \widetilde{v}^{\pm} - 2\mu\xi^{\pm}(y,t)\delta(x - L). \tag{3.1.39}$$

Setting as before $\widetilde{v}^{\pm} = \psi^{\pm}\mathrm{e}^{\lambda t + iky}$, $\xi^{\pm} = \chi^{\pm}\mathrm{e}^{\lambda t + iky}$ leads to the equations for the spectral amplitudes

$$\psi_{\pm}''(x) - q^2\psi_{\pm} = 2\mu\chi_{\pm}(k, \lambda)\delta(x - L). \tag{3.1.40}$$

The equations are identical, and the solutions differ by the boundary conditions at $x = 0$: $\psi_{+}' = 0$ and $\psi_{-}(0) = 0$:

$$\psi^{+}(x) = \begin{cases} -2\mu\chi^{+}q^{-1}\mathrm{e}^{-qL}\cosh qx & \text{at } x \leqslant L, \\ -2\mu\chi^{+}q^{-1}\mathrm{e}^{-q(L-x)}\cosh qL & \text{at } x \geqslant L, \end{cases}$$
$$\psi^{-}(x) = \begin{cases} -2\mu\chi^{-}q^{-1}\sinh qx & \text{at } x \leqslant L, \\ -2\mu\chi^{-}q^{-1}\mathrm{e}^{-q(L-x)}\sinh qL & \text{at } x \geqslant L. \end{cases} \tag{3.1.41}$$

Further derivations require only the values of $\psi_{\pm}(L)$. Using $\psi^{\pm}(L)$ in the Fourier expansion of (3.1.38) yields the dispersion relation for the varicose and zigzag branches:

$$\lambda_{\pm} = -k^2 + b\mu \left[1 - \mathrm{e}^{-2L} - q^{-1}\left(1 \pm \mathrm{e}^{-2qL}\right) \right]. \tag{3.1.42}$$

Vanishing λ_{-} at $k = 0$ reflects the translational symmetry of the band. Expanding (3.1.42) at $\lambda_{-} \sim k^2 \ll 1$ yields in the leading order:

$$\lambda_{-} = k^2\tau \frac{2 - b\mu + b\mu(1 + 2L)\mathrm{e}^{-2L}}{b\mu - 2\tau - b\mu(1 + 2L)\mathrm{e}^{-2L}}, \tag{3.1.43}$$

As for a single front, the stationary band is stable to a traveling transition, when the denominator is negative, and it is required for stability that the numerator be negative as well:

$$\frac{2}{b\mu} \equiv \tau_c^{-1} > 1 - (1+2L)\mathrm{e}^{-2L} = 1 - \left(1 + \ln\frac{\mu}{\nu}\right)\frac{\nu}{\mu} \tag{3.1.44}$$

The latter expression follows from the stationarity condition $\mathrm{e}^{-2L} = \nu/\mu$. In the limit $L \to \infty$, $\nu \to 0$, the instability condition of a single front $b\mu < 2$ is recovered. Within the range $b\mu < 1/\sqrt{3}$, instability is possible at $L < 0.9697$. As L decreases, the remaining domain covers a wider parametric region. Beyond the marginal instability locus, the maximum growth rate is observed at a finite wavelength.

However, the stable domain is encroached by the oscillatory instability. It is analyzed in the same way as for a single front, but with different results. Setting in (3.1.42) $\lambda = \mathrm{i}\omega\tau$ and separating the real and imaginary parts yields two equations for k^2 and ω, where the positive sign corresponds to symmetric (varicose) and the negative sign, to antisymmetric (zigzag) instabilities:

$$k^2 = b\mu\left\{1 - \mathrm{e}^{-2L} - \chi^{-1}\left[\cos\alpha \pm \mathrm{e}^{-2L\chi\cos\alpha}\sin\alpha\cos(\alpha + 2L\chi)\right]\right\}, \tag{3.1.45}$$

$$\omega\tau = b\mu\chi^{-1}\sin\alpha\left[1 \pm \mathrm{e}^{-2L\chi\cos\alpha}\sin(\alpha + 2L\chi)\right]. \tag{3.1.46}$$

The notation follows (3.1.17), which is recovered at $L \to \infty$. In the asymmetric case, the lowest possible value of the ratio μ/τ satisfying (3.1.46) is again attained at $\omega \to 0$, and is expanded at $\omega \ll 1$ as

$$\frac{b\mu}{2\tau} \equiv \frac{\tau_c}{\tau} = 2(1+k^2)^{3/2} + \omega\sqrt{1+k^2}\left[1 \pm \mathrm{e}^{-2\sqrt{1+k^2}}\sin\left(2\sqrt{1+k^2}L\right)\right]. \tag{3.1.47}$$

The coefficient at ω is always positive, so that, as in the case of a single front, the lowest value of the right-hand side is 2. However, the right-hand part of (3.1.45) is *negative* in a range of small ω, k that shrinks as L increases. This, in turn, raises the minimum level of μ/τ satisfying (3.1.46), and therefore instability in the region stable to monotonic perturbation is possible. In the symmetric case, the expansion

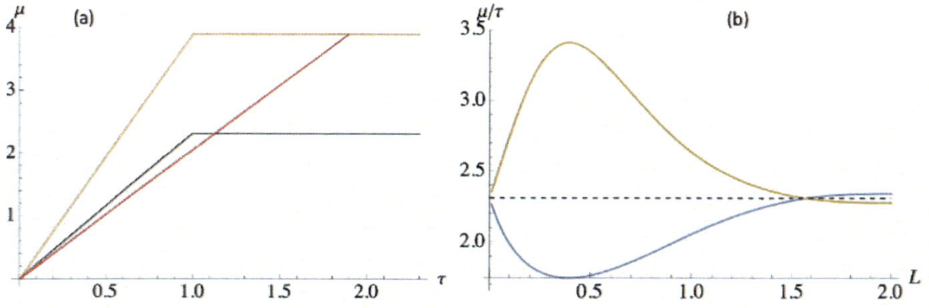

Fig. 3.4. (a): The stable parametric domain of a single front, below the black lines, compared to that of a band with the half-thickness $L = 1$. The limit set by the monotonic instability only is shown by the orange lines. The zigzag instability cuts off the triangle above the red line. (b): The lower limit of the ratio μ/τ for the zigzag (blue curve) and varicose (orange curve) instability as a function of the half-thickness L. The dashed line shows the respective value for a single front

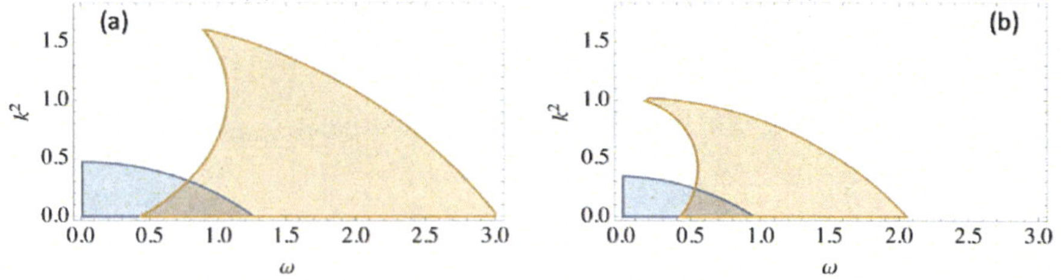

Fig. 3.5. The range of possible wavenumbers and frequencies for the zigzag (a) and varicose (b) instability at $L = 1$ (shaded blue) and $L = 1/2$ (shaded orange)

near $\omega = 0$ does not show a definite result, but numerical minimization shows that also in this case the minimum is observed at zero wavenumber and small ω.

As shown in Fig. 3.4b, the antisymmetric (zigzag) is more dangerous than the symmetric (varicose) instability, unless at larger L when both are already close to the stability limit of a single front. The locus of the traveling bifurcation

$$\frac{\tau}{\tau_c} > 1 - (1 + 2L)e^{-2L} = 1 - \left(1 + \ln\frac{\mu}{\nu}\right)\frac{\nu}{\mu}, \tag{3.1.48}$$

always lies above the level $\tau_c/\tau = 1$. The range of possible wavenumbers and frequencies, shown in Fig. 3.5, grows and shifts to higher ω as L decreases, and is wider for the antisymmetric (zigzag) that for the symmetric (varicose) instability. Beyond the marginal instability locus, the maximum growth rate is observed at a finite wavelength. The zigzag instability is apt either to destroy the band turning it into an array of spots or to trigger spreading out into a labyrinthine pattern – but this can be only determined by full numerical simulation in 2D.

3.1.5 Migration-Enhanced Structures

A system similar to the FN model can be constructed as a nonequilibrium extension of the phase field model described in Sect. 2.3.5 (Hildebrand et al, 1999). The equations for the activator and the inhibitor are constructed in a way similar to (2.3.44) with addition of nongradient terms, which may describe, for example, a nonequilibrium chemical reaction. This way of constructing a system capable of generating nonequilibrium patterns has an advantage of being directly related to thermodynamics. As a consequence of equilibrium relations between the two species, the diffusional fluxes are supplemented by migration, which enhances the gradient of the long-range inhibitor across the front of the short-range activator.

The variational equations are constructed treating the activator as a nonconserved and the inhibitor as a conserved field:

$$\frac{\partial u}{\partial t} = -\gamma^{-1}\frac{\delta \mathcal{F}}{\delta u}, \quad \frac{\partial v}{\partial t} = \nabla \cdot \left[\Gamma(v)\nabla\frac{\delta \mathcal{F}}{\delta v}\right] + g(u, v), \tag{3.1.49}$$

where γ^{-1} (presumed to be constant) and $\Gamma(v)$ are kinetic coefficients. The energy functional $\mathcal{F}$ is written in the form

$$\mathcal{F} = \int \left[V(u,v) + W(v) + \tfrac{1}{2}\epsilon^2|\nabla u|^2\right] \mathrm{d}\mathbf{x}. \tag{3.1.50}$$

The potential $V(u, v)$ is such that $f(u, v) = -\partial V/\partial u$ in (3.1.3). The v-dependent part of the potential, as well as the dependence of mobility on v, should be chosen to obtain the correct form of the diffusional term in (3.1.2):

$$W(v) = \frac{(v - v^-)\ln(v - v^-) + (v^+ - v)\ln(v^+ - v)}{\beta(v^+ - v^-)}, \tag{3.1.51}$$

$$\Gamma(v) = \beta(v - v^-)(v^+ - v), \tag{3.1.52}$$

where $v^\pm$ bound the allowed range of v.

With this choice of functions, the equation of u retains the form (3.1.3) and can be written, to emphasize its gradient structure, as

$$\partial_t u = \epsilon^2\nabla^2 u - \partial_u V. \tag{3.1.53}$$

The equation of v takes the form

$$\partial_t v = \nabla^2 v + \beta\nabla\cdot\left[(v - v^-)(v^+ - v)\nabla\frac{\partial V}{\partial v}\right] + g(u, v). \tag{3.1.54}$$

The additional migration term, which causes transport of the inhibitor under the influence of the activator gradient, is dominant near the activator front and causes a jump of v across the interface. In the front region, the short-scale coordinate normal to the front should be used in (3.1.54). This renders the last term negligible ($\propto \epsilon^2$). For a stationary straight-line front, (3.1.54) reduces then to

$$v''(x) = -\beta\frac{\mathrm{d}}{\mathrm{d}x}\left[(v - v^-)(v^+ - v)\left(\frac{\partial^2 V}{\partial u\partial v}\frac{\mathrm{d}u}{\mathrm{d}x} + \frac{\partial^2 V}{\partial v^2}\frac{\mathrm{d}v}{\mathrm{d}x}\right)\right]. \tag{3.1.55}$$

As an example, we take the particular form of $f(u, v)$ in (3.1.6), which is obtained from the potential

$$V(u, v) = \tfrac{1}{4}(1 - u^2)^2 + uv(1 - \tfrac{1}{3}u^2). \tag{3.1.56}$$

Setting also $v^\pm = \pm 1$ brings (3.1.6) to the form

$$v''(x) = -\beta\frac{\mathrm{d}}{\mathrm{d}x}\left[(1 - u^2)^2\frac{\mathrm{d}u}{\mathrm{d}x}\right]. \tag{3.1.57}$$

This equation can be integrated twice to obtain the relation between u and v:

$$v = v_0 - jx - \beta u\left[1 - \tfrac{2}{3}u^2 + \tfrac{1}{5}u^4\right], \tag{3.1.58}$$

where v_0, j are integration constants; the latter determines the flux through the interface, which may be caused by the nonequilibrium part of (3.1.54) negligible in the vicinity of the front. The flux should be measurable on the outer (long) scale and therefore must be small, so that the second term in (3.1.58) should be negligible within the interfacial region where u switches between its two equilibrium values. Taking into account that, on the short scale, $u \to \pm 1$ at $x \to \infty$, this gives the difference between the equilibrium values of v on the two sides of the interface $2\Delta = \frac{16}{15}\beta$. A stationary straight-line front is a solution of (3.1.53) when the values of the potential (3.1.56) are equal on both sides of the front. By symmetry, this requires v to be an odd function of u, so that $v_0 = 0$.

In the outer region, the migration term in (3.1.54) is negligible, and the equation of v reverts to the common form (3.1.2). Choosing again a linear function $g(u, v)$ as in (3.1.4), we obtain the same equations for the stationary inhibitor field as in Sect. 3.1.2; the only difference is that v is not required now to be continuous across the interface, but has to jump by 2Δ. In particular, the solution for a solitary band with the half-width L analogous to (3.1.35) is obtained in the form

$$v = \begin{cases} 2(\mu + \Delta)\mathrm{e}^{-L} \cosh x - \mu - \nu & \text{at } |x| \leqslant L \\ 2(\mu + \Delta)\mathrm{e}^{-|x|} \sinh L) + \mu - \nu & \text{at } |x| \geqslant L. \end{cases} \tag{3.1.59}$$

The flux $j = -v'(\pm L) = \pm 2(\mu + \Delta)(1 - \mathrm{e}^{-2L})$ is continuous across the interface. The stationarity condition, following from the condition $v(L+0) + v(L-0) = 0$, is $L = \ln \sqrt{(\mu + \Delta)/\nu}$. Although the structure of solutions remains formally similar, comparing the expressions for the stationary band width with and without migration shows that the size of stationary structures can be significantly affected by the migration term.

3.1.6 Propagating Bands

The profile of v in a traveling band is computed by solving the equations in the comoving frame, which coincide with (3.1.20), the only difference being that, as in (3.1.34), the down state is localized at $|x| < L$ and the up state, at $|x| > L$. The solution satisfying the matching conditions at both fronts is

$$v = \begin{cases} \mu \left[1 - (1 + -s^{-1})\varPhi_- \left(\mathrm{e}^{-\frac{1}{2}(s-1)CL} - 1\right)\right] - \nu & \text{at } z \leqslant -L \\ \mu \left[(1 + s^{-1})\varPhi_- + (1 - s^{-1})\varPhi_+ - 1\right] - \nu & \text{at } |z| \leqslant L \\ \mu \left[1 - (1 - s^{-1})\varPhi_+ \left(\mathrm{e}^{\frac{1}{2}(s+1)CL} - 1\right)\right] - \nu & \text{at } z \geqslant L \end{cases} \tag{3.1.60}$$

where $s^2 = 1 + (2/C)^2$, $\varPhi_\pm = \mathrm{e}^{-\frac{1}{2}(s\pm 1)C(L\pm z)}$. The values of v on the kink at $x = L$ and the antikink at $x = -L$ must verify $v(\pm L) = \pm C\tau/b$, which translates into the relations defining both velocity and width of a propagating band:

$$\pm s^{-1} + (1 \pm s^{-1})\mathrm{e}^{-(1\pm s)CL} = \frac{\nu}{\mu} \pm \frac{C\tau}{b\mu}. \tag{3.1.61}$$

This system cannot be solved analytically, but is easily resolved with respect to the parameters by presenting it in the form

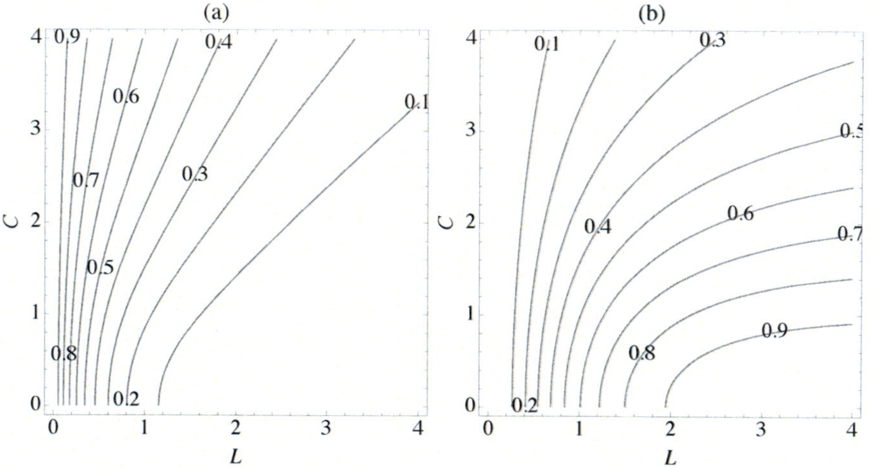

Fig. 3.6. (a) The contour plot of $V_+(L, C)$. The numbers at the curves indicate levels of ν/μ. (b) The contour plot of $V_-(L, C)$. The numbers at the curves indicate levels of τ/τ_c. Both plots continue symmetrically into the half-plane $C < 0$

$$V_+ \equiv \frac{1}{2}\left[(1+s^{-1})\mathrm{e}^{-(1+s)CL} + (1-s^{-1})\mathrm{e}^{-(1-s)CL}\right] = \frac{\nu}{\mu}, \tag{3.1.62}$$

$$V_- \equiv C^{-1}\left[\frac{2}{s} + (1+s^{-1})\mathrm{e}^{-(1+s)CL} - (1-s^{-1})\mathrm{e}^{-(1-s)CL}\right] = \frac{2\tau}{b\mu} \equiv \frac{\tau}{\tau_c},$$

These relations can be solved graphically by reading the values of the parameters ν/μ, τ/τ_c from the plots of $V_\pm$ in Fig. 3.6. The plots continue symmetrically into the half-plane $C < 0$; therefore, at any L there are symmetric solutions propagating in opposite directions. We can also notice that at relatively large L, when the traveling bifurcation precedes the oscillatory one, ν is small, as is velocity, and τ is close to τ_c. In this limit, the dependence of either ν/μ or $1 - \tau/\tau_c$ on C is parabolic. The profile $v(x)$ in a moving band is asymmetric; an example is shown in Fig. 3.7.

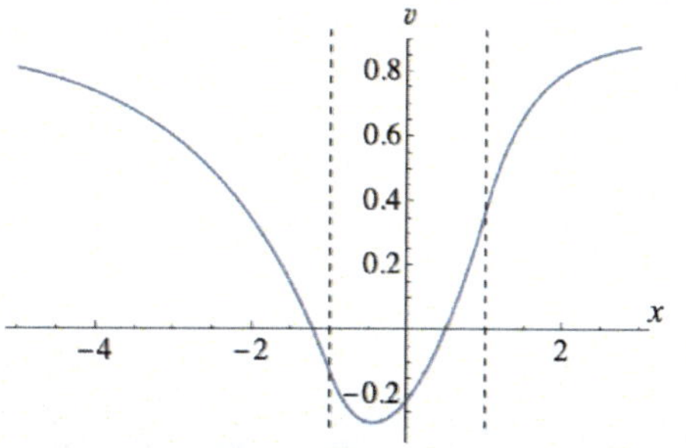

Fig. 3.7. A typical profile of the long-range variable a band propagating to the right. Parameters: $C = L = \mu = 1$, $\nu = 0.1$

The dispersion relation determining stability of moving bands is constructed and analyzed in the same way as before, but become more involved. The spectral perturbation equation is similar to (3.1.25), with two antisymmetric sources at $z = \pm L$:

$$C\psi'(z) + \psi''(z) - q^2\psi = 2\mu[\chi_+(k,\lambda)\delta(z-L) - \chi_-(k,\lambda)\delta(z+L), \tag{3.1.63}$$

Solving it yields the Fourier amplitudes of $v(x)$ at the kink and antikink

$$\psi(\pm L) = -\frac{2\mu}{\sqrt{4q^2 + C^2}} \left[1 - \mathrm{e}^{\mp(1\pm s)CL}\right] \chi_\pm(k,\lambda). \tag{3.1.64}$$

Using this in (3.1.36) together with the expression for the derivative of the stationary field derived from (3.1.60) leads to the dispersion relations for the varicose and zigzag branches:

$$\lambda_\pm = -k^2 + 2b\mu \left[(4 + C^2)^{-1/2} \left(1 - \mathrm{e}^{-L\left(\sqrt{4+C^2}\pm C\right)}\right) \right.$$
$$\left. -(4q^2 + C^2)^{-1/2} \left(1 \pm \mathrm{e}^{-L\left(\sqrt{4q^2+C^2}\pm C\right)}\right)\right]. \tag{3.1.65}$$

The single-front expression (3.1.27) is recovered at $L \to \infty$, and so does λ_- when the signs of L, C are inverted, while setting $C = 0$ brings it back to (3.1.42).

Short of dealing with a complicated expression obtained when λ_- is expanded at $\lambda_- \sim k^2 \ll 1$, the effect of propagation on the most dangerous zigzag instability can be tested near the traveling bifurcation threshold when C is small. This leads to the correction $\widetilde{\lambda}_-$ to the value (3.1.43) defining instability of a stationary band

$$\widetilde{\lambda}_- = \frac{2k^2\tau(1-\tau)L(2L-1)C}{b\mu\left[2L - 1 + (1 - \tau/\tau_c)\mathrm{e}^{-2L}\right)^2}. \tag{3.1.66}$$

Notably, the velocity-dependent correction depends on the sign of C. Since $\tau < 1$ when traveling bifurcation is possible, the coefficient at C is positive, unless at very small L when a stationary band is already unstable. Thus, propagation is destabilizing

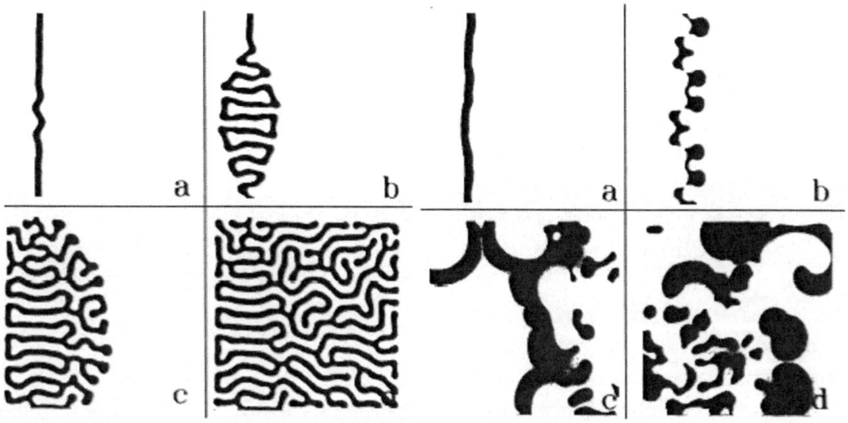

Fig. 3.8. Left: labyrinthine pattern developing from a solitary stripe following the zigzag instability. Right: spiral turbulence developing from a propagating stripe. The panels a–d show development of the patterns (Hagberg and Meron, 1994; reproduced by permission, ©American Physical Society)

at $C > 0$ when the band moves in the direction of a steeper gradient (see Fig. 3.6). This is in accordance with computations by Hagberg and Meron (1994) who observed that the instability of the leading front is stronger. Their simulations (based on the phenomenological approach described in Sect. 3.2.5) have shown that the zigzag instability of both stationary and moving bands is never saturated, but develops in unlikely ways. In the former case, instability either destroys the band turning it into an array of spots or triggers spreading out into a labyrinthine pattern, as shown in the left panel of Fig. 3.8. A propagating stripe evolves in a different way: it breaks into disjoint pieces (Fig. 3.8, right), which subsequently develop into a complex spatiotemporal pattern.

3.2 Curvilinear Structures

3.2.1 Solitary Disk and Sphere

In 2D, the front should acquire a circular shape due to the surface tension, and the stationary solution depends only on the long-range radial coordinate r. The stationary outer equation is

$$v_{rr} + r^{-1} v_r - v - \nu \pm \mu = 0. \tag{3.2.1}$$

The stationary solution for a disk with $u < 0$ immersed in a continuum with $u > 0$, satisfying the continuity and smoothness conditions at the front located at $r = L$, is

$$v = \begin{cases} \mu[2LK_1(L)I_0(r) - 1] - \nu & \text{at} \quad r \leqslant L \\ \mu[1 - 2LI_1(L)K_0(r)] - \nu & \text{at} \quad r \geqslant L, \end{cases} \tag{3.2.2}$$

where $I_n,\ K_n$ are modified Bessel functions. Using the identity $K_1(L)I_0(L) + K_0(L)I_1(L) = L^{-1}$, the common value of $v(L)$ can be expressed as

$$v(L) = \mu L\, [K_1(L)\, I_0(L) - K_0(L)\, I_1(L)] - \nu. \tag{3.2.3}$$

Using $v(L)$ and $\kappa = L^{-1}$ in the eikonal equation (3.1.5) yields

$$c\tau + L^{-1} + b\nu = b\mu L\, [K_1(L)\, I_0(L) - K_0(L)\, I_1(L)]\,. \tag{3.2.4}$$

A convenient form of the stationarity condition $c = 0$ is

$$\mu^{-1}(b^{-1} + L\nu) = L^2\, [K_1(L)\, I_0(L) - K_0(L)\, I_1(L)]\,. \tag{3.2.5}$$

The right-hand side of (3.2.5), denoted as $F(L)$, has a peak value $F_m \approx 0.556$ at $L \approx 1.60$ and relaxes asymptotically to 1/2 at $L \to \infty$ (Fig. 3.9). The equation can be solved graphically by finding an intersection of this universal function with straight lines starting from the point $(b\mu)^{-1}$, $L = 0$ and inclined by the slope ν. We expect the solution to be stable to perturbations preserving circular symmetry when the slope of the straight line exceeds $F'(L)$, so that, as the radius grows, the shrinking action due to the surface tension decreases slower than spreading driven by

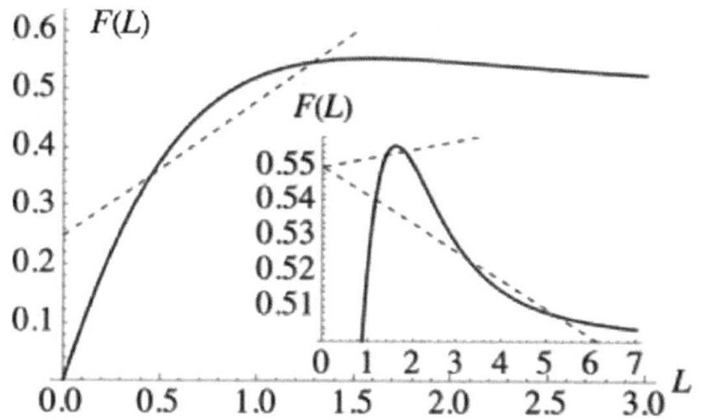

Fig. 3.9. Plot of the function $F(L)$ defined by (3.2.5). The solutions are obtained as intersection points with straight lines presenting the left-hand side of this equation at $b = 3/\sqrt{2}$. Inset: the range $1/2 < F(L) < F_m$ showing solutions at $\nu < 0$

the slow variable, and vice versa[3]. The natural expectation is a solution at $\nu > 0$ when spreading of the lower state is counteracted by curvature. However such a solution exists only at sufficiently high values of μ satisfying $(b\mu)^{-1} < F_m$, i.e., $\mu \gtrsim 0.8478$. Counterintuitively, solutions with $\nu < 0$ exist above this limit, as demonstrated by the blowup in the inset of Fig. 3.9. In this case, the above mentioned stability condition is never satisfied on the ascending branch of $F(L)$, which is convex, and can be only met on the descending branch. A pair of solutions, one of which is stable to perturbations preserving the circular symmetry, bifurcates when the straight line in Fig. 3.9a is tangent to $F(L)$.

The existence region in the parametric plane μ, ν is bounded by the axis $\nu = 0$, $\mu > 2/b \approx 0.9428$ and the bifurcation line that can be obtained by differentiating (3.2.4) with respect to L:

$$L^{-2} + 2b\mu L\,[K_1(L)\,I_1(L) - K_0(L)\,I_0(L)] = 0. \tag{3.2.6}$$

The existence boundary can be drawn as a parametric plot of the functions $\mu(L)$, $\nu(L)$ verifying (3.2.5), (3.2.6), shown by curve C in Fig. 3.10). It merges tangentially to the axis $\nu = 0$ forming the lower boundary, but penetrates into the half-plane $\nu < 0$ as a narrow tongue terminating in a cusp at $\mu \approx 0.7849$, $\nu \approx -0.0195$. The upper branch of the existence boundary corresponds to $L \lesssim 2.2937$ and the short lower segment to larger radii. Combining the cusped plot in Fig. 3.10 with a mirror plot drawn for a disk with $u > 0$ immersed in a continuum with $u < 0$, we see that both may coexist within a narrow region at very small values of $|\nu|$ and μ slightly below $2/b$. This feature is specific to 2D and is lost in both one and three dimensions.

In 3D, solving the long-scale equation in spherical coordinates yields

$$v = \begin{cases} \mu[2(1+L)\mathrm{e}^{-L} r^{-1} \sinh r - 1] - \nu & \text{at } r \leqslant L \\ \mu[1 - 2r^{-1}\mathrm{e}^{-r}(L\cosh L - \sinh L)] - \nu & \text{at } r \geqslant L, \end{cases} \tag{3.2.7}$$

$$v(L) = \mu[L^{-1} - \mathrm{e}^{-2L}(1 + L^{-1})] - \nu. \tag{3.2.8}$$

The stationarity condition $bv(L) = \kappa = 2/L$ in the form analogous to (3.2.5) is

$$\mu^{-1}(2/b + L\nu) = 1 - L(1+L)\mathrm{e}^{-2L}. \tag{3.2.9}$$

[3] This solution may be still subject to symmetry-breaking or oscillatory instabilities

The right-hand side of this equation is a monotonically increasing function of L, approaching unity at $L \to \infty$. The possible range of solutions, determined, analogous to the 2D case, by the tangency condition, is restricted to $\nu > 0$, but extends to a wider range of μ than in 2D, $\mu > b^{-1} \approx 0.4714$.

A regular assembly of spots in 2D or spheres or cylinders in 3D may form a pattern resembling a crystalline structure where symmetric solitary objects serve the role of "atoms." This analogy is not precise, since the environment of such an "atom" in the crystal is not isotropic, and its symmetric shape must be distorted to some degree[4]. Nevertheless, both experiment and computations often show patterns (typically, hexagonal in 2D) closely resembling a crystalline arrangement of almost symmetric spots, as seen, for example, in two left panels of Fig. 1.10. Moreover, one often observes a variety of "crystalline splinters" comprising finite clusters of spots, which are a telltale sign of coexistence of a pattern with a stable stationary state (see Sect. 4.5.4).

3.2.2 Splitting and Bending Instabilities

Stability of a stationary solitary disk is investigated in cylindrical coordinates (r, ϕ) by testing perturbations $\zeta(\phi, t) \ll L$. The curvature of the front is expressed as $\kappa = L^{-1}(1 - \zeta - \zeta_{\phi\phi})$. Expanding, as usual, the long-range variable in the vicinity of the unperturbed position brings the expression for velocity (3.1.5) to the form

$$\zeta_t = L^{-2}\left(\zeta_{\phi\phi} + \zeta\right) + b\left[L^{-1}\widetilde{v}(0, \phi, t) + v_s'(0)\zeta\right], \tag{3.2.10}$$

where $v_s'(0) = 2\mu L I_1(L) K_1(L)$ is the stationary flux at the front following from (3.2.2). The equation for the perturbation field $\widetilde{v}$ is

$$\tau^{-1}\partial_t \widetilde{v} = \nabla^2 \widetilde{v} - \widetilde{v} - 2\mu L \zeta(\phi, t)\delta(r - L). \tag{3.2.11}$$

We are looking for a solution in the form

$$\zeta(\phi, t) = \chi_n \mathrm{e}^{\lambda_n t + in\phi}, \quad \widetilde{v}(r, \phi, t) = \psi(r; n)\mathrm{e}^{\lambda_n t + in\phi}. \tag{3.2.12}$$

The only difference from (3.1.11) is that an arbitrary wavenumber k is replaced by an integer n. The equation of ψ is

$$\psi_{rr} + \frac{1}{r}\psi_r - \left(q^2 + \frac{n^2}{r^2}\right)\psi = 2\mu L \chi_n \delta(r - L), \tag{3.2.13}$$

where $q^2 = 1 + \lambda/\tau$. The solution of this equation is

$$\psi(r) = \begin{cases} -2\mu L^2 \chi_n q^{-1} K_n(qL) I_n(qr) & \text{at } r \leqslant L, \\ -2\mu L^2 \chi_n q^{-1} I_n(qL) K_n(qr) & \text{at } r \geqslant L. \end{cases} \tag{3.2.14}$$

Using (3.2.12) and (3.2.14) in (3.2.10) yields the dispersion relation

$$\lambda_n = -L^{-2}(n^2 - 1) + 2b\mu L[I_1(L)K_1(L) - q^{-1} I_n(qL) K_n(qL)]. \tag{3.2.15}$$

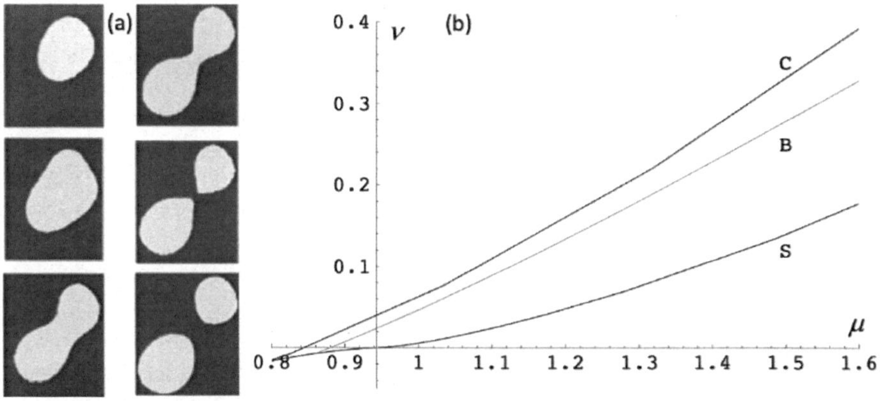

Fig. 3.10. (a) Splitting of a solitary spot (experimental sequence by Li et al, 1996; reproduced by permission, ©American Institute of Physics). (b) Existence boundary (C) and loci of splitting (S) instability of a solitary disk and bending (B) instability of an infinite cylinder

As expected, λ_n vanishes at $n = 1$, which corresponds to a shift of the disk without deformation. As n is a discrete parameter, expansion near the neutrally stable state is unavailable, but stability can be checked numerically by finding the conditions when the right-hand side of (3.2.15) vanishes at $\lambda_n = 0$. This shows that, of all angular modes, the dipole mode $n = 2$, deforming the disk into an ellipse, is the most dangerous one, causing instability at the lowest value of μ at any L. Beyond the instability threshold, there is no nonlinear saturation, and the disk splits into two spots (Fig. 3.10a). This may set off a continuing "multiplication" process (Reynolds et al, 1994), eventually generating a pattern filling the entire plane. The locus of splitting instability shown by the curve S in Fig. 3.10b starts at the point $\nu = 0$, $\mu = 2/b \approx 0.9428$, which corresponds to the limit $L \to \infty$. The stable domain in the parametric plane μ, ν is restricted to the area between this curve and the existence boundary, cutting off the domain of smaller ν but retaining the counterintuitive solutions with large radii at $\nu < 0$.

Extending the solitary disk solution to 3D, we can consider the stability of a cylindrical domain $r < L$ infinitely extended along the z-axis. It is subject, in addition to planar modes with $k = 0$, $q = 1$, to distortions along the z-axis. Using the expression for the mean curvature of a slightly distorted cylindrical surface, $\kappa = L^{-1}(1 - \zeta - \zeta_{\phi\phi}) + \zeta_{zz}$, setting $\widetilde{v} = \psi_n(r)\mathrm{e}^{\lambda t + in\phi + ikz}$, $\zeta = \chi_n \mathrm{e}^{\lambda t + in\phi + ikz}$, and solving the resulting equations as before, we arrive at the dispersion relation similar to (3.2.15), but with $q^2 = 1 + \lambda/\tau + k^2$ and an added term due to axial perturbations:

$$\lambda = L^{-2}(1 - n^2) - k^2 + 2b\mu\left[K_1(L)\,I_1(L) - q^{-1}K_n(qL)\,I_n(qL)\right]. \quad (3.2.16)$$

The most dangerous mode, which can lead to bending instability, is the translational mode $n = 1$, which is neutral at $k = 0$. In the infinitely long cylinder, the marginal

[4] This is true, incidentally, also for real atoms in real crystals

stability condition for this mode is vanishing of the derivative $\mathrm{d}\lambda/\mathrm{d}(k^2)$ at $k = 0$, $n = 1$. Differentiating (3.2.16) we express this condition as

$$(b\mu)^{-1} = L^2[I_1(L)K_0(L) - I_0(L)K_1(L)] + 3LI_1(L)K_1(L). \tag{3.2.17}$$

The marginal stability curve B in the parametric plane μ, ν shown in Fig. 3.10b lies above the locus of splitting instability, so that it always prevails over planar distortion modes, but one can check numerically when both thresholds equalize at a finite k as the cylinder shortens.

3.2.3 Traveling and Oscillatory Instabilities

Similar to rectilinear structures, a disk may also suffer dynamic instabilities: traveling and oscillatory, which both depend on the inertial parameter τ. Traveling instability can be detected, as in Sect. 3.1.2, by finding a positive eigenvalue of (3.2.15) at $n = 1$, which may exist, besides the trivial solution indicating translational invariance, at sufficiently small τ. In this case, this task is far more challenging but can be handled numerically by defining $\Lambda = \lambda/\tau$, $\theta = \tau/\tau_c = 2\tau/(b\mu)$ and rewriting (3.2.15) at $n = 1$ as

$$\theta = \frac{4L}{\Lambda}\left[I_1(L)K_1(L) - \frac{I_1(L\sqrt{1+\Lambda})K_1(L\sqrt{1+\Lambda})}{\sqrt{1+\Lambda}}\right]. \tag{3.2.18}$$

The right-hand side is maximal at $\Lambda = 0$. This gives the largest value of the ratio θ that defines the limit of traveling instability:

$$\theta_T = 2L[LK_2(L) - LI_2(L)K_1(L) - I_1(L)K_1(L)]. \tag{3.2.19}$$

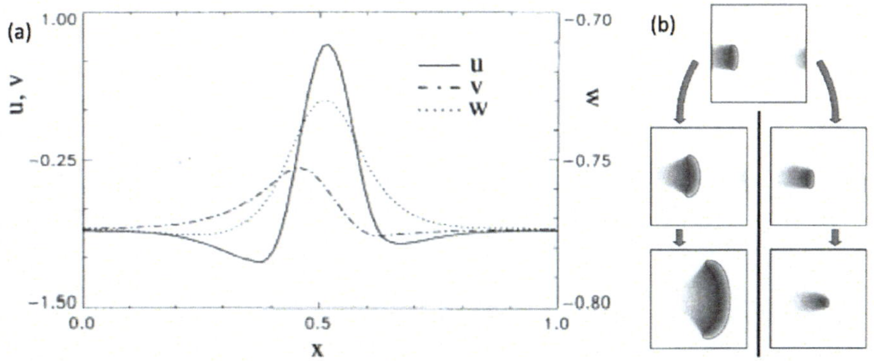

Fig. 3.11. (a) Evolution of a traveling spot (Schenk et al, 1997; reproduced by permission, ©The American Physical Society) Traveling spot in the three-variable model. (b) Typical profiles of the variables u, v, w along the direction of motion are shown by the solid, dot-dashed , and dotted lines, respectively (Or-Guil et al, 1998; reproduced by permission, ©The American Physical Society)

This function, drawn in Fig. 3.12b, increases monotonically with L dropping to zero at $L = 0$ and approaching 2 at $L \to \infty$, indicating that, at any given μ, traveling instability requires the inhibitor to be less inertial as the size of the disk decreases. As a moving disk loses its circular shape, a propagating solution cannot be constructed analytically. Numerical computations (Krischer and Mikhailov, 1994; Schenk et al, 1997) carried out at finite values of ϵ showed (Fig. 3.11b) that a moving disk either shrinks and disappears or elongates in the direction normal to the direction of motion, bends at the ends, and eventually turns into a spiral pattern (see Sect. 3.3).

Spreading or shrinkage can be suppressed by adding interaction with a global variable in a finite domain (Krischer and Mikhailov, 1994). This idea was refined by Or-Guil et al (1998), who replaced a rapidly reacting global variable by an additional fast long-range inhibitor w which, like the principal inhibitor v controlling the front propagation speed, is linearly coupled with the activator u, but serves a different purpose of stabilizing the shape of the spot. Typical profiles of the variables along the direction of motion are shown in Fig. 3.11a. The shift between the distribution of u and v is responsible for propagation; the distribution of w is broad and symmetric, and serves to contain lateral growth or shrinkage.

An oscillatory, or breathing, instability occurs when a pair of complex conjugate eigenvalues crosses the imaginary axis, signaling bifurcation to a pulsating disk. This instability is more difficult to detect analytically. The locus of the breathing instability of a solitary disk is obtained by setting in the dispersion relation (3.2.15) $\lambda = \mathrm{i}\omega\tau$ and $n = 0$, bringing it to the form

$$\mathrm{i}\omega\tau = L^{-2} + 2b\mu F(L,\omega) \equiv L^{-2} + \tau_c F(L,\omega), \tag{3.2.20}$$

$$F(L,\omega) = 4L\left[I_1(L)K_1(L) - (1+\mathrm{i}\omega)^{-1/2} I_0\left(L\sqrt{1+\mathrm{i}\omega}\right) K_0\left(L\sqrt{1+\mathrm{i}\omega}\right)\right].$$

This relation can be resolved with respect to the parameters as

$$\tau_c = -\frac{1}{L^2 \mathrm{Re}\, F(L,\omega)}, \qquad \tau = -\frac{\mathrm{Im}\, F(L,\omega)}{\omega L^2 \mathrm{Re}\, F(L,\omega)}, \qquad \theta = \frac{\mathrm{Im}\, F(L,\omega)}{\omega}. \tag{3.2.21}$$

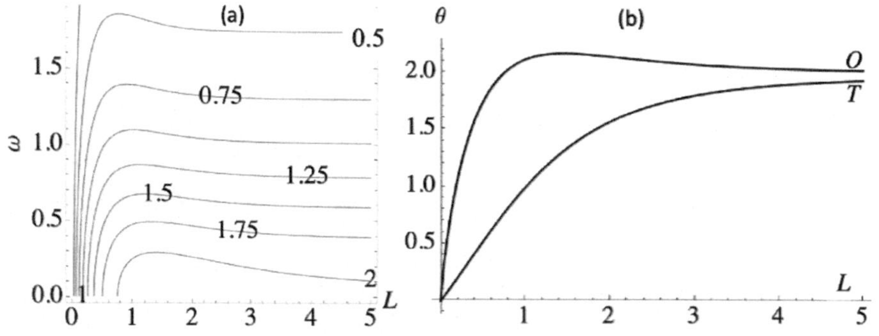

Fig. 3.12. (a) The dependence of the oscillatory instability limit of the ratio $\theta = \tau/\tau_c$ (shown at respective curves) on the disk radius L and frequency ω. (b) The limits of traveling (T) and oscillatory instabilities as function of the disk radius L

This, together with the stationarity condition (3.2.5), defines the locus of marginal breathing instability as a surface in the parametric space μ, ν, τ parametrized by the disk radius L and frequency ω.

Test computations shown in Fig. 3.12a shows that at any L the instability limit of the of ratio $\theta = \tau/\tau_c$ increases with decreasing frequency. At $\omega \ll 1$, a Taylor series can be used, leading to the instability limit

$$\theta_O = \tfrac{1}{2}L[I_0(L)K_0(L) - LI_1(L)K_0(L) + LI_0(L)K_1(L)]. \tag{3.2.22}$$

As shown in Fig. 3.12b, this limit lies consistently above that given by (3.2.19); hence, breathing instability is always more dangerous than traveling instability. Pulsating spots have been observed experimentally by Haim et al (1996).

3.2.4 Quasistationary Dynamics

The stationary form of the long-scale equation (3.1.4) may be applicable to dynamical problems when the mobility ratio τ is very large. Under these conditions, the dynamics of the long-range variable is so fast that it follows quasistationarily the front displacements. In the leading order, the time derivative in (3.1.4) falls out, and it reduces to a stationary form

$$\nabla^2 v - v - \nu + \mu u = 0. \tag{3.2.23}$$

This equation can be resolved with the help of an appropriate Green's function $\mathcal{G}(\mathbf{r})$ as

$$v(\mathbf{x}) = -\nu + \mu \int \mathcal{G}(\mathbf{x} - \boldsymbol{\xi}) u(\boldsymbol{\xi})\, \mathrm{d}^2\boldsymbol{\xi}. \tag{3.2.24}$$

In 2D, the Green's function $\mathcal{G}$ of (3.2.23) is a modified Bessel function dependent on the distance between the two points $r = |\mathbf{r}| = |\mathbf{x} - \boldsymbol{\xi}|$:

$$\mathcal{G}(r) = (2\pi)^{-1} K_0(r). \tag{3.2.25}$$

Using (3.2.24) in (3.1.3) brings it to the variational form $\partial_t u = -\delta\mathcal{F}/\delta u$ (2.1.2) with the energy functional

$$\begin{aligned}\mathcal{F} = &\int \left[\tfrac{1}{2}|\nabla u|^2 + \epsilon^{-1}\nu u + \tfrac{1}{4}\epsilon^{-2}(1-u^2)^2\right] \mathrm{d}\mathbf{x} \\ &+ \tfrac{1}{2}\epsilon^{-1}\mu \int\!\!\int u(\mathbf{x})\mathcal{G}(\mathbf{x}-\boldsymbol{\xi})u(\boldsymbol{\xi})\, \mathrm{d}\mathbf{x}\mathrm{d}\boldsymbol{\xi}.\end{aligned} \tag{3.2.26}$$

If $\mu > 0$, the nonlocal term is negative when $u(\mathbf{x})$ and $u(\boldsymbol{\xi})$ have opposite signs. This may compensate the distortion energy and make a nonuniform (patterned) state energetically favorable.

Since the front displacement is fully determined by local values of v, it is possible to derive a closed nonlocal equation of the front motion by transforming (3.2.24) to a boundary integral (Petrich and Goldstein, 1994). Since $u = \pm 1 + O(\epsilon)$, (3.2.23) becomes

$$\nabla^2 v - v - \nu \pm \mu = 0. \tag{3.2.27}$$

The general solution of this equation can be presented as

$$v(\mathbf{x}) = \mu - \nu - 2\mu \int_S \mathcal{G}(\mathbf{x} - \boldsymbol{\xi})\, \mathrm{d}\boldsymbol{\xi}, \tag{3.2.28}$$

where the integration is carried out over the area S occupied by the down state. This integral can be transformed into a contour integral along a boundary Γ with the help of the Gauss theorem. To avoid divergent expressions, the contour should exclude the point $\mathbf{x} = \boldsymbol{\xi}$. Clearly, excluding an infinitesimal circle around this point does not affect the integral (3.2.28), since its kernel (3.2.25) is only logarithmically divergent ($K_0(r) \asymp \ln r$ at $r \to 0$). Replacing $\mathcal{G}(\mathbf{r}) = \nabla^2 \mathcal{G}(\mathbf{r})$ ($\mathbf{r} \neq 0$), we transform the integral in (3.2.28) as

$$\int_S \mathcal{G}(\mathbf{x} - \boldsymbol{\xi})\, \mathrm{d}^2\boldsymbol{\xi} = -\int_S \nabla_{\boldsymbol{\xi}} \cdot \mathbf{H}(\mathbf{x} - \boldsymbol{\xi}) \mathrm{d}^2\boldsymbol{\xi} = -\oint_{\Gamma'} \mathbf{n}(s) \cdot \mathbf{H}(|\mathbf{x} - \boldsymbol{\xi}(s)|) \mathrm{d}s, \tag{3.2.29}$$

where $\mathbf{H}(\mathbf{r}) = \nabla \mathcal{G}(\mathbf{r})$, and $\mathbf{n}$ is the normal to the contour Γ' drawn to exclude the singularity. The vector Green's function $\mathbf{H}$ corresponding to the kernel (3.2.25) is computed as

$$\mathbf{H}(\mathbf{r}) = -\frac{\mathbf{r}}{2\pi r} K_1(r). \tag{3.2.30}$$

If $\mathbf{x}$ is a boundary point, the contour Γ' consists of the boundary Γ cut at this point and closed by an infinitesimally small semicircle about $\mathbf{x}$. The integral over the semicircle equals $1/2$ and cancels with the first term in (3.2.28). Defining the external normal to Γ as the tangent vector $\mathbf{t} = \mathbf{x}'(s)$ rotated clockwise by $\pi/2$, the required value of the long-range variable on the spot boundary (parametrized by the arc length s or σ) is expressed, using the 2D cross product $\times$, as

$$v(s) = -\nu - 2\mu \oint_\Gamma \mathbf{H}(\mathbf{x}(s) - \mathbf{x}(\sigma)) \times \mathbf{x}'(\sigma) \mathrm{d}\sigma. \tag{3.2.31}$$

The integral can be evaluated for the boundary points and used in (3.1.5) to obtain the local front velocity. The resulting expression depends nonlocally on the instantaneous front position. As an exercise, the stationary disk solution obtained in Sect. 3.2.1 can be reproduced using in (3.1.5) the radius vector $\mathbf{r} = L\{\cos\phi - 1, \sin\phi\}$ between the points on the circle of radius L separated by the angle ϕ and its length $r = 2L\sin(\phi/2)$ and evaluating the integral

$$\begin{aligned} v + \nu &= \frac{\mu L}{\pi} \int_0^\pi \sin\frac{\phi}{2} K_1\left(L \sin\frac{\phi}{2}\right) \mathrm{d}\phi = -\frac{\mu L}{\pi} \frac{\mathrm{d}}{\mathrm{d}L} \int_0^\pi K_0\left(L\sin\frac{\phi}{2}\right) \mathrm{d}\phi \\ &= -\mu L \frac{\mathrm{d}}{\mathrm{d}L} [I_0(L) K_0(L)] = \mu L [K_1(L)\, I_0(L) - K_0(L)\, I_1(L)]. \end{aligned} \tag{3.2.32}$$

The result is identical to (3.2.3).

The boundary integral method can be used to evolve the interface in time by discretizing the boundary and translating the points with the prescribed velocity along the normal $\mathbf{n}$. Practical computations using this method are not easy, as they require adding and removing points in the course of evolution and watching for possible self-intersections of the evolving boundary, but the substantial advantage is that one only needs to keep in memory a 1D array of values of v at the front, rather than the entire 2D field.

3.2.5 Phenomenological Velocity–Curvature Relation

It is advantageous to derive closed evolution equations for fronts that would enable direct computation of front displacements without a need to solve a higher dimensional evolution equation for the long-scale field. Such a closed equation, derived above for the case of quasistationary dynamics, has a nonlocal form not treatable analytically and not very convenient for numerical computation. Trying to use a history-dependent kernel for a general dynamic problem with a linear long-scale field is still less practical. An alternative is to derive an approximate *local* equation valid under restricted conditions and faithfully reproducing, at least in a qualitative sense, essential features of the front motion, including both geometric factors and action of the long-scale field.

A simple relation of this kind has been suggested by Hagberg and Meron (1994). The starting point is the long-scale equation (3.1.4) rewritten in the aligned coordinate frame (Sect. 2.2.1). This adds both the advective term $\tau^{-1}cv'(x)$, as in (3.1.20), and the term $\kappa v'(x)$, similar to the respective term in the equation of the short-scale variable that is responsible for the curvature dependence in the eikonal equation (3.1.5), which has exactly the same form as the advective term. Repeating the derivation in Sect. 3.1.3, we require the value of v on the front, computed now as $v(0) = -\nu + 2\mu(C+\kappa)/s$, where $C = c/\tau$ and $s^2 = 1 + \frac{1}{4}(C+\kappa)^2$, to be equal to $(c+\kappa)/b$. In this way, we arrive at (3.1.22) with C replaced by $C+\kappa$ on the left-hand side and $C\tau$ by $C\tau+\kappa$ on the right-hand side:

$$\frac{C+\kappa}{\sqrt{4+(C+\kappa)^2}} = \frac{1}{\mu}\left(\nu + \frac{C\tau+\kappa}{b}\right). \tag{3.2.33}$$

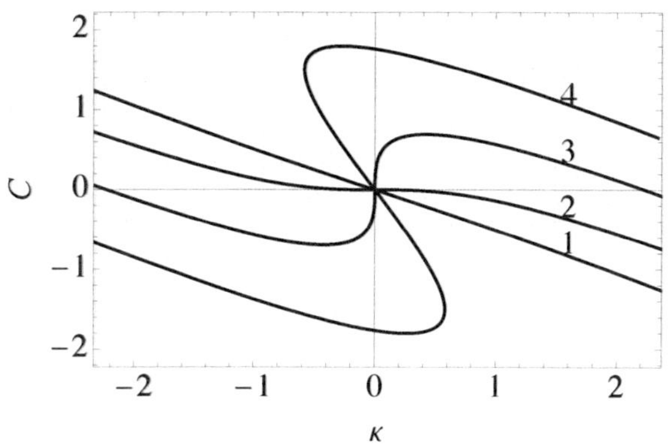

Fig. 3.13. Dependence of the velocity on the curvature at $\tau = 3/2$, $\nu = 0$ and several values of $b\mu$ (shown by the numbers at the curves). Segments with $\mathrm{d}C/\mathrm{d}\kappa > 0$ appear at $b\mu/2 = \tau_c > 1$ and multivalued S-shaped curves at $\tau_c > \tau$

The velocity-curvature relation defined in this way becomes S-shaped at sufficiently large values of μ, as shown in Fig. 3.13a. The loci of traveling and zigzag instabilities are obtained by differentiating (3.2.33). The limit of traveling instability is defined by the derivative with respect to C at $C = \kappa = 0$, yielding $\frac{1}{2}b\mu \equiv \tau_c = \tau$. The limit of zigzag instability is defined by the derivative with respect to κ at $\kappa = 0$, resulting in $\tau_c = 1$ for a stationary front. Both results coincide with those derived in Sect. 3.1.2. For a propagating front, the locus of zigzag instability is $\tau_c = (1 + C^2/4)^{3/2}$, coinciding with that corresponding to vanishing numerator of (3.1.28).

The relation (3.2.33) does not result from a rational expansion in a small parameter; it is not restricted to a weakly biased model but cannot be precise. One can see this, for example, by computing with its help the radius $L = 1/\kappa$ of a stationary disk. The result is much simpler than in Sect. 3.2.1 and gives the same existence region, $0 < \nu < \mu$, as for a solitary band, rather than a cusped region given by the precise computation. The reason is that curvature, entering as a small correction in the short-scale equation, is not small on the extended scale of the long-range variable. The dependence of the long-range field on the interface geometry is *nonlocal* and cannot be expressed precisely by any local relation; neither can it account for interaction of kinks and antikinks responsible for peculiar features of solitary bands. Distinctions in location of bifurcation points for different shapes (band, disk, etc.) cannot be discerned, since global shape dependence is missing.

Nevertheless, (3.2.33) is useful as a simple phenomenological formula that captures in a qualitative way both traveling and zigzag instabilities and reproduces thereby the essential features of front dynamics for both slow and fast long-range field. Oscillatory instability is not directly reproduced in this way, but Haim et al (1996) attributed it to transitions between the two branches of the multivalued velocity–curvature relations depicted by an S-shaped curve. This approach enables computations in large domains, such as those presented in Fig. 3.8. Beyond the traveling bifurcation, the dependence of velocity on curvature, which can be obtained by resolving (3.2.33) numerically, becomes S-shaped (Fig. 3.13).

A simplified form of (3.2.33) incorporating asymmetry can be obtained through the expansion near the traveling bifurcation. The simplest formula retaining a multivalued velocity–curvature dependence can be obtained by setting in (3.2.33)

$$C = \epsilon\widetilde{C}, \quad \kappa = \epsilon^3\widetilde{\kappa}, \quad \tau = \tau_c(1 - \epsilon^2\widetilde{\tau}), \quad \nu = \mu\epsilon^3\widetilde{\nu} \tag{3.2.34}$$

and expanding in $\epsilon \ll 1$. This yields in the third order a cubic equation

$$\tfrac{1}{2}\left(\tau_c^{-1} - 1\right)\widetilde{\kappa} = -\widetilde{\nu} + \tfrac{1}{2}\widetilde{\tau}C - 2^{-4}C^3. \tag{3.2.35}$$

At $\tau < 1$, when the traveling bifurcation precedes the zigzag instability, the coefficient at $\widetilde{\kappa}$ is positive.

A longer but more convincing way to derive this formula is to expand (3.1.4), (3.1.5) for a weakly curved front in the vicinity of the traveling bifurcation. A sequence of equations is obtained for functions v_i in the expansion of the long-range variable $v = v_0 + \epsilon v_1 + \cdots$ written in the comoving frame. Taking (3.2.34) as a scaling guideline, we assume $C = O(\epsilon)$, $\nu = O(\epsilon^3)$, $\kappa = O(\epsilon^3)$. The respective scaling of transverse derivatives $\partial_y = O(\epsilon^3)$ eliminates them in the relevant orders of the expansion. Then the first-order equation is

$$v_1''(x) - v_1 + Cv_0'(x) = 0, \tag{3.2.36}$$

where v_0 verifies (3.1.7) with $\nu = 0$. The solution satisfying the matching conditions at the front located at $x = 0$ is

$$v_1 = \begin{cases} \frac{1}{2}C\mu(1+x)\mathrm{e}^{x} & \text{at } x \leqslant 0, \\ \frac{1}{2}C\mu(1-x)\mathrm{e}^{-x} & \text{at } x \geqslant 0. \end{cases} \tag{3.2.37}$$

The first-order eikonal equation is simply $c = C\tau = bv_1(0) = Cb\mu/2$. This equation is verified trivially at any C at the traveling bifurcation point $\tau_0 = b\mu/2$. Fixing τ at this value, we proceed to the second-order equation

$$v_2''(x) - v_2 + Cv_1'(x) = 0, \tag{3.2.38}$$

solved by

$$v_2 = \begin{cases} -\frac{1}{8}C^2\mu x(1-x)\mathrm{e}^{x} & \text{at } x \leqslant 0 \\ -\frac{1}{8}C^2\mu x(1+x)\mathrm{e}^{-x} & \text{at } x \geqslant 0. \end{cases} \tag{3.2.39}$$

Since $v_2(0) = 0$, the eikonal equation is trivially verified also in the second order. Finally, in the third order, adding also an $O(\epsilon^3)$ bias and curvature and $O(\epsilon^2)$ deviation of τ from the bifurcation value, we have

$$v_3''(x) - v_3 + Cv_2'(x) + \kappa v_0'(x) - \nu = 0, \tag{3.2.40}$$

solved by

$$v_3 = \begin{cases} \frac{1}{2}\mu\mathrm{e}^{x}\left[\kappa(1+x) - \frac{1}{24}C^3(3 - 3x + x^3)\right] - \nu & \text{at } x \leqslant 0 \\ \frac{1}{2}\mu\mathrm{e}^{-x}\left[\kappa(1-x) - \frac{1}{24}C^3(3 + 3x - x^3)\right] - \nu & \text{at } x \geqslant 0. \end{cases} \tag{3.2.41}$$

This leads to the third-order eikonal equation identical to (3.2.35). This local velocity–curvature relation is especially useful since it can be combined with intrinsic equations of the front motion (Sect. 3.3.1). It will be used for construction of rotating spiral bands in Sect. 3.3.4.

3.2.6 Long-Scale Evolution Equations

A straight-line propagating front (or a planar front in 3D) is neutrally stable to translations along the propagation direction. By continuity, the dynamics of transverse distortions is slow when their wavelength is large compared to the $O(1)$ scale characteristic to the inhibitor field. The dynamic equations on extended spatial and temporal scales can be obtained as in Sect. 2.4.5 through a regular expansion procedure, leading to a universal form independent of the underlying system.

It is most convenient to work in the aligned frame. We assume that the curvature is scaled as $O(\sqrt{\varepsilon})$ when measured on the inhibitor scale[5]. The matching extended time variable is scaled as $\partial_t = O(\varepsilon)$. The inhibitor field is expanded in the scale ratio ε: $v = v_0 + \varepsilon v_1 + \cdots$, and the normal velocity (measured in the inhibitor units) is expanded as $C = C_0 + \varepsilon C_1 + \cdots$.

The zero-order equation in the moving frame is stationary and one-dimensional, and coincides with (3.1.20). We denote its solution given by (3.1.21) as $v_0(x)$. The first-order equation is

$$v_1''(x) - v_1 + C_0 v_1'(x) + (C_1 + \kappa) v_0'(x) = 0. \tag{3.2.42}$$

Solving this equation subject to the matching conditions at $x = 0$, we obtain the value of v_1 at the front:

$$v_1(0) = \tfrac{1}{2}\mu(1 + \tfrac{1}{4}C_0^2)^{3/2}\,(C_1 + \kappa). \tag{3.2.43}$$

The correction to the normal velocity C_1 is defined by the first-order eikonal equation obtained by expanding (3.1.5) rescaled to the inhibitor time units:

$$\tau C_1 = b v_1(0) - \kappa. \tag{3.2.44}$$

Using here (3.2.43) yields

$$\left[\frac{2\tau}{b\mu}\left(1 + \tfrac{1}{4}C_0^2\right)^{3/2} - 1\right] C_1 = -\left[\frac{2}{b\mu}\left(1 + \tfrac{1}{4}C_0^2\right)^{3/2} - 1\right]\kappa. \tag{3.2.45}$$

This equation is well posed when the bracketed coefficients are both positive. Then it has the same form as (2.2.12) and can be reduced in a Cartesian frame to the Burgers equation as in Sect. 2.2.3. The coefficient at C_1 vanishes at the traveling bifurcation; further expansion near this point has been carried out in the preceding subsection. Vanishing coefficient at κ marks the limit of the long-scale zigzag instability.

The expansion can be continued to the next order to describe slow dynamics of long-scale distortions near this point. For this aim, we fix the parameter μ at the instability limit, allowing for a small deviation:

$$\mu = 2b^{-1}\left(1 + \tfrac{1}{4}C_0^2\right)^{3/2} + \varepsilon\mu_1. \tag{3.2.46}$$

Then $C_1 = 0$, and a higher-order correction C_2 has to be computed in the next order. The second-order equation should be written using the Laplacian in the aligned frame (2.2.4) or (2.2.7), expanded to the second order:

$$v_2''(x) - v_2 + \kappa v_1'(x) + (C_2 - x\kappa^2) v_0'(x) + \nabla_\perp^2 v_1 = 0. \tag{3.2.47}$$

In 3D, κ^2 should be replaced here by $\kappa_1^2 + \kappa_2^2$. We need to use here the solution of (3.2.42). To avoid cumbersome expressions, we write this solution at the Maxwell construction ($C_0 = 0$):

[5] Recall that curvature was assumed to be small on a shorter activator scale when the eikonal equation (3.1.5) was derived. The notation for a small parameter is modified here in order to emphasize that it has nothing to do with the scale ratio ϵ used so far

$$v_1 = \begin{cases} b^{-1}\mathrm{e}^{x}(1+x)\kappa + \mu_1(\mathrm{e}^{x}-1) & \text{at } x \leqslant 0 \\ b^{-1}\mathrm{e}^{-x}(1-x)\kappa + \mu_1(1-\mathrm{e}^{-x}) & \text{at } x \geqslant 0. \end{cases} \tag{3.2.48}$$

Since we only need to compute the value of v_2 at the front, some rules verified by direct computation can be set to save the effort of solving this equation (subject to usual matching conditions). An inhomogeneity of the form $Q\mathrm{e}^{x}$ at $x < 0$, $Q\mathrm{e}^{-x}$ at $x > 0$ contributes the constant $Q/2$ (which may depend on the transverse coordinates and slow time) to the value $v_2(0)$ entering the eikonal equation. An inhomogeneity of the same form but with different signs in the positive and negative domains gives no contribution. The contribution of an inhomogeneity of the form $-Qx\mathrm{e}^{x}$ at $x < 0$, $Qx\mathrm{e}^{-x}$ at $x > 0$ is $Q/4$, while the same inhomogeneity with common signs in the two domains gives no contribution.

Applying these rules, one can read the value of $v_2(0)$ directly from (3.2.48) and use the result in the second-order eikonal equation

$$\tau C_2 = b v_2(0) - \kappa. \tag{3.2.49}$$

The result turns out to be dependent only on first-order terms in the expansion of the Laplacian in the aligned frame:

$$(\tau - 1)C_2 = \tfrac{1}{2}\mu_1\kappa + \tfrac{1}{4}\nabla_\perp^2\kappa. \tag{3.2.50}$$

The coefficient at C_2 is positive at $\tau > 1$ when the zigzag instability precludes traveling bifurcation. The first term on the right-hand side is destabilizing at $\mu_1 > 0$. Perturbations are stabilized by the last term that suppresses short-scale corrugations.

This equation is transformed to the Kuramoto–Sivashinsky (KS) equation (1.1.5) by presenting the front position, as in Sect. 2.2.3, in the Cartesian frame $(X, \mathbf{Y})$ propagating with the speed C_0 as $X = \zeta(\mathbf{Y}, t)$ with $\nabla_\perp\zeta \ll 1$ and using approximate expressions for the curvature $\kappa = -\nabla_\perp^2\zeta$ and the displacement speed along the X axis:

$$\zeta_t = \left(1 + \epsilon|\nabla_\perp\zeta|^2\right)^{-1/2}\left(C_0 + \epsilon^2 C_2\right) - C_0 \approx -\frac{1}{2}\epsilon C_0|\nabla_\perp\zeta|^2 + \epsilon^2 C_2. \tag{3.2.51}$$

The nonlinear advective term turns out to be dominant, unless either the basic velocity C_0 or the displacement are small. After rescaling $C_0 \to \epsilon C_1$, $\partial_t \to \epsilon^2\,\partial_T$, plugging (3.2.50) in (3.2.51) yields the KS equation balanced on an extended time scale:

$$\zeta_t - \frac{1}{2}C_1|\nabla_\perp\zeta|^2 + \tfrac{1}{2}\mu_1(\tau-1)^{-1}\nabla_\perp^2\zeta + \tfrac{1}{4}(\tau-1)^{-1}\nabla_\perp^4\zeta = 0. \tag{3.2.52}$$

The original derivation of the KS equation for an unstable front in the FitzHugh–Nagumo system (Kuramoto and Tsuzuki, 1976) carried out in the "laboratory" rather than aligned frame is far more lengthy, as is an earlier derivation by Sivashinsky (1977) in the context of combustion fronts. We shall return to the KS equation in Sect. 5.1.3 where it appears as a long-scale evolution equation of phase instabilities.

3.3 Locally Induced Motion

3.3.1 Intrinsic Equations of Motion

Kinematics of a planar curve can be described without reference to coordinates, in terms of its intrinsic characteristics: metrics and curvature. The starting point is the 2D version of the Frenet–Serret equations

$$\mathbf{X}_s = \mathbf{l}, \quad \mathbf{l}_s = \kappa \mathbf{n}, \quad \mathbf{n}_s = -\kappa \mathbf{l}. \tag{3.3.1}$$

The two orthogonal unit vectors, tangent $\mathbf{l}$ and normal $\mathbf{n}$, define at any point of the curve the *Frenet dihedron*. The curvature defines locally the rotation of the Frenet dihedron and therefore determines the shape of a smooth curve. If the function $\kappa(s)$ is known, the curve can be reconstructed, in principle, by integrating (3.3.1), although this task is technically difficult due to the implicit form of these equations. For practical computations, (3.3.1) should be transformed to a more convenient form involving a minimal number of variables.

Suppose that the curve is set into motion in such a way that each point has a velocity $\mathbf{v}(s,t)$. For the time being, we allow $\mathbf{v}$ to be an arbitrary vector function but require it to be continuously dependent on s. In order to obtain an evolution equation of a curve in terms of its intrinsic parameters, it is advantageous to use projections of $\mathbf{v}$ on the Frenet dihedron; thus, we assume

$$\mathbf{v} = \mathbf{X}_t = \chi \mathbf{l} - c\mathbf{n}. \tag{3.3.2}$$

The sign of the last term follows the sign convention of Sect. 2.2.1.

Since the arc length, generally, changes as the curve moves, the derivatives with respect to t and s do not commute. It might be therefore more convenient to choose a constant parametrization y, such that $\partial_t \partial_y = \partial_y \partial_t$. One can use for this purpose, for example, the arc length at a fixed moment of time $y = s(t_0)$; this parametrization can be viewed just as a Lagrangian tag carried by each point as it moves. The derivative $g = \partial s / \partial y$ defines the metric of the curve in the chosen parametrization. Then the commutator between the derivatives with respect to t and s is computed as

$$\partial_t \partial_s - \partial_s \partial_t = \partial_t (g^{-1} \partial_y) - g^{-1} \partial_y \partial_t = -g^{-2} g_t \partial_y = -g^{-1} g_t \partial_s. \tag{3.3.3}$$

Using this commutation relation, we obtain, by differentiating the first Frenet–Serret equation with respect to time and replacing the derivatives with respect to s with the help of other equations, the evolution equation of the tangent vector $\mathbf{l}$:

$$\begin{aligned} \mathbf{l}_t &= \partial_t \partial_s \mathbf{X} = \partial_s \partial_t \mathbf{X} - g^{-1} g_t \partial_s \mathbf{X} = \partial_s \mathbf{v} - g^{-1} g_t \mathbf{l} \\ &= (\chi_s - g^{-1} g_t + \kappa c)\mathbf{l} + (\kappa \chi - c_s)\mathbf{n}. \end{aligned} \tag{3.3.4}$$

Since $\mathbf{l}$ is a unit vector, the projection $\mathbf{l} \cdot \mathbf{l}_t$ should vanish. This yields the metric continuity equation

$$g_t = g(\chi_s + \kappa c). \tag{3.3.5}$$

Equation (3.3.4) thus reduces to

$$\mathbf{l}_t = (\kappa\chi - c_s)\mathbf{n}. \tag{3.3.6}$$

In the same manner, differentiating the second Frenet–Serret equation with respect to time and using (3.3.5) and (3.3.6) yields

$$\begin{aligned}\kappa\mathbf{n}_t &= -(\kappa_t + g^{-1}g_t)\mathbf{n} + \partial_s((c_s + \kappa\chi)\mathbf{n}) \\ &= -(\kappa_t + \kappa^2 c + c_{ss} - \kappa_s\chi)\mathbf{n} - (\kappa\chi - c_s)\mathbf{l}.\end{aligned} \tag{3.3.7}$$

Again, requiring the projection $\mathbf{n} \cdot \mathbf{n}_t$ to vanish gives the evolution equation of the curvature:

$$\kappa_t = -\kappa^2 c - c_{ss} + \kappa_s\chi. \tag{3.3.8}$$

The tangential velocity χ is just a gauge variable that expresses an arbitrary reparametrization of the evolving curve. If the normal velocity component is expressed as a function of curvature κ, the system of two evolution equations (3.3.5), (3.3.8) for three variables κ, g, and χ can be closed by imposing an arbitrary gauge condition. Two gauges are most useful: the *comoving* gauge $\chi = 0$ and the *isometric* gauge with time-independent metric g.

Since the arc length is not conserved in the comoving gauge, it cannot be used as an independent variable, and the intrinsic equations of locally induced motion should be rewritten using a constant parametrization y related to s by the metric function, $\mathrm{d}s = g\,\mathrm{d}y$. Using the local equation of the front motion (2.2.12) with $c_0 = \text{const}$, the intrinsic equations in the comoving gauge (3.3.5), (3.3.8) are written as

$$g_t = g\kappa(c_0 - D\kappa), \tag{3.3.9}$$

$$\kappa_t = g^{-1}\partial_y\left(g^{-1}\kappa_y\right) - \kappa^2(c_0 - D\kappa). \tag{3.3.10}$$

Clearly, at the Maxwell construction ($c_0 = 0$) the curve shrinks as a result of motion along the normal, resulting in the decrease of energy roughly proportional to the length of the boundary. The line tension can be viewed, in fact, as the physical cause of the normal motion. If $\kappa = R^{-1} = \text{const}$, both (3.3.9) and (3.3.10) yield $\kappa_t = D\kappa^3$ or $R_t = -D/R$, integrating to $R = \sqrt{2D(t - t_0)}$, where t_0 is the moment of collapse. Of course, the theory becomes inapplicable at the ultimate brief stage of collapse when the curvature radius becomes comparable to the width of the front.

The diffusive character or normal motion becomes apparent in the limit of very small curvatures when the nonlinear terms in (3.3.10) can be neglected. As the rate of change of the metric is quadratic in κ, it can be considered constant in the leading order, and (3.3.10) reduces to the diffusion equation $\kappa_t = \kappa_{yy}$. This indicates the tendency of relaxation to circular shape. Conversely, at large curvatures, when the nonlinear terms prevail, the evolution is catastrophic, leading to rapid local collapse of warped segments.

3.3.2 Steadily Rotating Spiral

Intrinsic equations of motion provide the most convenient means to study the front dynamics whenever a local velocity–curvature relation is available. We shall apply it now to construct a *rotating spiral*. Its development can be understood qualitatively starting from a narrow semi-infinite propagating band, or a "finger" (see Sect. 3.5.2). Near the end, the propagation speed tends to be lower due to the curvature effects. As these segments are left behind, the band is bending. If the end point does not retreat in spite of the high local curvature, the overall length of the band keeps increasing when it curves while propagating outward as in the left sequence of Fig. 3.11b. The resulting structure is a rotating spiral. In a similar fashion, a pair of counter-rotating spirals would grow from a segment of a finite length or out of any defect in a propagating band.

The isometric gauge is most suitable for identifying a self-similar shape of an open-ended curve that remains invariant as it moves. The metric equation (3.3.5) turns in this case into the equation of the tangential velocity

$$\chi_s + \kappa c = 0. \tag{3.3.11}$$

The curvature equation (3.3.8) can be rewritten then as a conservation equation

$$\kappa_t + q_s = 0, \tag{3.3.12}$$

where the flux q is defined as

$$q = c_s - \kappa\chi. \tag{3.3.13}$$

For an invariant curve, the flux q must be constant.

An invariant curve rotating with a constant frequency ω can be constructed using the polar frame (r, ϕ). In order to preserve the shape defined by some function $\phi = \Phi(r)$, the tangential velocity χ should be fixed to ensure that each point be translated strictly in the angular direction while it propagates along the normal. The tangent and normal unit vectors are

$$\mathbf{l} = \frac{\widehat{\mathbf{r}} + r\Phi'(r)\,\widehat{\boldsymbol{\phi}}}{\sqrt{1 + [r\Phi'(r)]^2}}, \qquad \mathbf{n} = \frac{r\Phi'(r)\,\widehat{\mathbf{r}} - \widehat{\boldsymbol{\phi}}}{\sqrt{1 + [r\Phi'(r)]^2}}, \tag{3.3.14}$$

where $\widehat{\mathbf{r}}, \widehat{\boldsymbol{\phi}}$ are unit vectors in the radial and angular directions. The direction of the normal is chosen to be opposite to the positive rotation direction. The radial and angular velocities are expressed now as

$$\begin{aligned} c_r &= (\chi\mathbf{l} - c\mathbf{n}) \cdot \widehat{\mathbf{r}} = \frac{\chi - cr\Phi'(r)}{\sqrt{1 + [r\Phi'(r)]^2}} = 0, \\ c_\phi &= (\chi\mathbf{l} - c\mathbf{n}) \cdot \widehat{\boldsymbol{\phi}} = \frac{\chi r\Phi'(r) + c}{\sqrt{1 + [r\Phi'(r)]^2}} = \omega r. \end{aligned} \tag{3.3.15}$$

These conditions are resolved as

$$\Phi'(r) = \frac{\chi}{cr}, \quad \chi = -\sqrt{\omega^2 r^2 - c^2}. \tag{3.3.16}$$

The choice of the sign of χ ensures that the front is convex, i.e., propagates against the direction of the normal.

The radial coordinate and the arc length are related by

$$\mathrm{d}s/\mathrm{d}r = \sqrt{1+[r\Phi'(r)]^2} = \sqrt{1+(\chi/c)^2} = \pm\omega r/c. \tag{3.3.17}$$

This can be used to differentiate the relation $\chi^2 + c^2 = \omega^2 r^2$ with respect to the arc length. Substituting there the derivatives of χ and c given by (3.3.11), (3.3.13) yields $q = \pm\omega$. This relation holds for any steadily rotating spiral.

The simplest case is propagation with a constant normal velocity c_0 (Wiener and Rosenblueth, 1946). Clearly, (3.3.16) becomes inapplicable within an inner circle with the radius $R = c_0/\omega$. Thus, the spiral can only rotate around a central hole of the radius R. This relation can also be interpreted in the opposite sense: the frequency $\omega = c_0/R$ is fixed by the radius of the central hole. By (3.3.16), the tangential velocity vanishes on the inner circle, and the spiral kicks off normally to the circle.

The actual shape of the spiral is readily obtained by integrating (3.3.16) starting from an arbitrary angular position ϕ_0 on the hole perimeter:

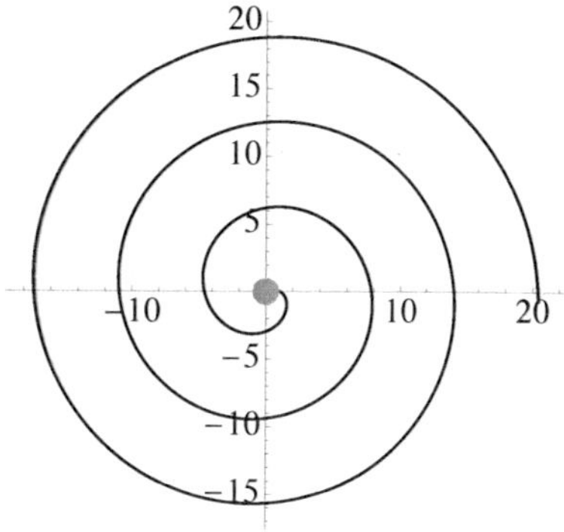

Fig. 3.14. A Wiener–Rosenblueth spiral. The central hole is shaded

$$\Phi(r) = \phi_0 - \left[\sqrt{(r/R)^2-1} + \arctan\sqrt{(r/R)^2-1}\right]. \tag{3.3.18}$$

Asymptotically at large r, this is an Archimedean spiral with the pitch $2\pi R$ equal to the circumference of the central hole, as shown in Fig. 3.14. The unsettling feature of this solution is a singularity at the free end. The curvature $\kappa \approx \Phi''(r)$ diverges at $r \to R$, as can be already seen by inspecting (3.3.16).

One should also keep in mind that computing a single open-ended curve is not sufficient for constructing a realistic spiral pattern, since the front propagation velocity must change sign to create a rotating spiral band. The desired pattern can be imitated by adding an identical curve shifted by a different angle ϕ_0. This construction is quite artificial, since the two curves become totally unrelated when bailed out of direct contact by the central hole.

3.3.3 Propagation into a Quiescent State

A more realistic local model including a curvature dependence for a curve moving around a circular obstacle is based on the eikonal equation (3.1.5). The value of the long-range variable v defining the propagation speed of a straight-line excitation front is, generally, unknown and coupling it to the front position makes the problem severely nonlocal. Postponing the analysis of spiral waves coupled to the inhibitor field till the next subsection, we consider here a simple case when $v = v_0$ and hence $c_0 = bv_0$ is constant everywhere along the front. This may happen in an excitable system where an excitation pulse propagates into a "quiescent" state corresponding to an only HSS. An additional requirement imposed on a spiral excitation pulse is that the excited region should be narrow and the time interval between subsequent passages of the pulse through any point should far exceed the inhibitor relaxation time, so that the system would have sufficient time to relax back to the quiescent state after the passage of a preceding excitation.

For any local velocity–curvature relation resolved with respect to κ, combining (3.3.11) and (3.3.13) yields

$$c\frac{\mathrm{d}c}{\mathrm{d}\chi} = -\chi - \frac{q}{\kappa(c)}. \tag{3.3.19}$$

Another form of this equation, which is more directly related to geometry, can be obtained if the normal velocity is replaced with the help of the second invariance condition (3.3.16) by the rescaled squared radial coordinate $\rho = \omega^2 r^2$. It is convenient to use the latter as the independent variable:

$$\frac{\mathrm{d}\chi}{\mathrm{d}\rho} = -\frac{1}{2q}\kappa\left(\sqrt{\rho - \chi^2}\right). \tag{3.3.20}$$

The basic local velocity–curvature relation is given by the eikonal equation. We shall use here its original dimensional form (2.2.12). Substituting it in (3.3.19) yields a closed equation for $c(\chi)$, which is further reduced by using c_0 as a velocity scale to

$$c\frac{\mathrm{d}c}{\mathrm{d}\chi} = -\chi - \frac{\Omega}{1-c}, \qquad \Omega = \frac{qD}{c_0^2}. \tag{3.3.21}$$

We expect the curvature to decrease, and the normal velocity to grow along the curve parametrized by the arc length s. The tangential velocity must vanish at the free end $s = 0$, where the curvature should be maximal and the normal velocity minimal. Since at the free end $c'(s) > 0$, while $\chi = 0$, the flux q is positive and equal to the rotation frequency ω, according to the result of the preceding subsection. As the flux remains constant along the curve, χ must be negative at $s \to \infty$ where $c'(s)$ vanishes. The dependence $c(\chi)$ can be obtained therefore by integrating (3.3.21) from $\chi = 0$ in the negative direction, starting from some initial condition $c(0) = \eta,\ 0 < \eta < 1$. The rescaled normal velocity must approach unity asymptotically at $\chi \to -\infty$. This happens only at a certain unique value of the rescaled frequency $\Omega = \Omega_c(\eta)$. This value is found iteratively with high precision as a border between trajectories $c(\chi)$ turning up to $c = 1$ or down to $c = 0$ with increasing $|\chi|$; the result is shown in Fig. 3.15a.

It is very difficult to continue integration to large $|\chi|$, and the curve should be matched there to the asymptotic solution

$$c \asymp 1 - \Omega/|\chi| + O(\chi^{-4}) \quad \text{at} \quad |\chi| \to \infty. \tag{3.3.22}$$

The computed curve $c(\chi)$ can be translated to the radial velocity dependence with the help of the second invariance condition (3.3.16) and the dependence $\Omega_c(\eta)$,

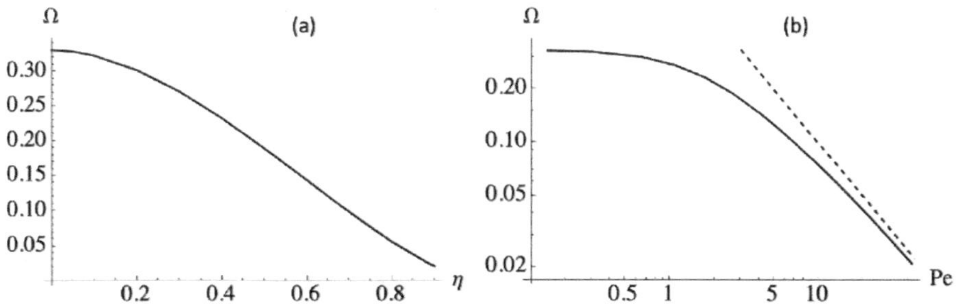

Fig. 3.15. (a) The relation between the rescaled rotation frequency Ω and the velocity at the origin η. (b) The dependence of the rescaled rotation frequency on the Péclet number obtained as the integrability condition of (3.3.21). The dashed line shows the WR limit

transformed into the dependence of the rotation frequency on the hole radius R. Reverting to the dimensional variables, we observe that, for kinematic reasons, $c = \omega R$ on the hole circumference, i.e., $\eta = \Omega R/c_0$. The relation between the frequency and the hole radius is expressed therefore as $\text{Pe} = \eta/\Omega_c(\eta)$, where $\text{Pe} = Rc_0/D$ is the Péclet number based on the hole radius. The numerical dependence is shown in Fig. 3.15b. The curve Ω(Pe) approaches the WR limit $\Omega = 1/\text{Pe}$ asymptotically at $\text{Pe} \to \infty$. The other limiting value is $\Omega \approx 0.331$ at Pe, $\eta \to 0$. The actual shape can be restored by integrating (3.3.16). It differ from the WR spiral only close to the hole circumference, and the overall difference in form is hardly detectable visually.

Clearly, what is missing in this kinematic description is the reverse transition to quiescent state. Mikhailov et al (1994) interpreted the kinematic solution as a narrow excitation zone, without distinguishing between the dynamics of its front and back sides. Such a narrow zone can, indeed, be formed under certain conditions; however, the kinematic theory excludes the crucial end region of the excitation band. The vanishing tangential velocity at the free end of the spiral, which is arbitrarily imposed as a boundary condition in kinematic theory, is actually fixed by the local curvature at the end point. More sophisticated treatment is needed to relate both the local curvature and the spiral core radius (replacing an artificial hole) to physical parameters of an excitable system (see Sect. 3.5.3). The kinematic theory of Mikhailov et al (1994), though not mathematically precise, correctly predicted a general form of this relationship, which has been later derived through rational expansion in the large core limit by Hakim and Karma (1999).

3.3.4 Spiral Band near Traveling Bifurcation

In the plane lacking a contrived hole, the front propagation velocity must change sign to create a rotating spiral band. Both the front and back side can approach asymptotically the Archimedean spiral form with the curvature vanishing at large distances when the function $\kappa(c)$ has two stable roots with opposite signs. The simplest local velocity–curvature relation suitable for generating such a structure is the cubic form (3.2.35) derived in the vicinity of the traveling bifurcation.

Using the symmetric rescaled version of (3.2.35), $\kappa = c - c^3$, we rewrite (3.3.20) as

$$\frac{d\chi}{d\rho} = -\frac{1}{2q}\sqrt{\rho - \chi^2}\,(1 - \rho + \chi^2). \tag{3.3.23}$$

It is convenient to integrate this equation backward starting from the large ρ asymptotics that can be expressed as a series

$$|\chi| = \sqrt{\rho} - \frac{1}{2\sqrt{\rho}} - \frac{q}{2\rho} - \frac{1-2q^2}{8\rho^{3/2}} + \cdots . \tag{3.3.24}$$

The propagation speed $c = \sqrt{\rho - \chi^2}$ should be less than unity (its asymptotic value at zero curvature). One can see from the above expansion that this condition holds when the flux q is negative, and, using the general relation between the flux and frequency from Sect. 3.3.2, we set $q = -\omega$. The frequency should be chosen in such a way that the tangential velocity vanishes at the spiral's end point. If this is achieved, the solution can be matched smoothly with the antisymmetric solution that has the normal velocity of the opposite sign and corresponds to the reverse transition front. For the symmetric case considered here, χ vanishes together with c and κ at $\rho = 0$ if one chooses $\omega \approx 0.267$. The normal and tangential velocities and curvature are plotted in Fig. 3.16a.

The actual form of the curve obtained by integrating the first invariance condition (3.3.16) is shown in Fig. 3.16b. Asymptotically at large r, it is again an Archimedean spiral, since the normal velocity approaches a constant value. The two spiral branches joined at the origin, where the normal velocity vanishes, divide the plane into two regions of identical shape occupied by the up and down states. The apparent tip of the spiral band is located at $\sqrt{\rho} \approx 0.62$, where the curvature is maximal. In an asymmetric case ($\nu \neq 0$), both branches of the spiral band should be integrated separately starting from an asymptotic expansion at large ρ where c approaches either of the two stable values with opposite signs. The frequency chosen to match both branches at the location of the unstable zero of the velocity–curvature relation.

The "local induction" approximation gives the best chance to construct the rotating spiral almost analytically, as we did above. Near the traveling bifurcation, this is,

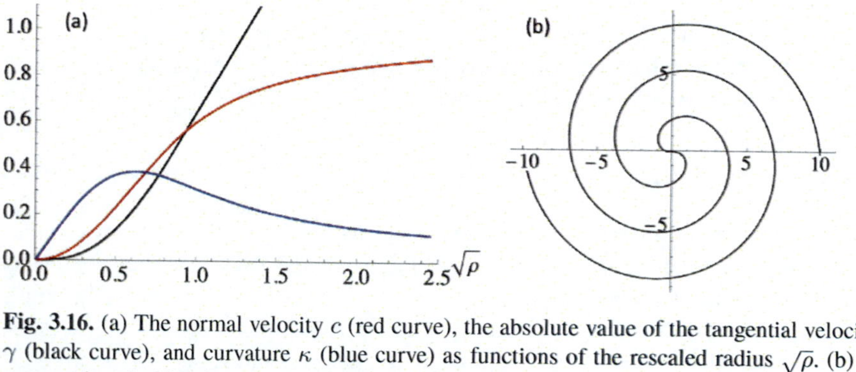

Fig. 3.16. (a) The normal velocity c (red curve), the absolute value of the tangential velocity γ (black curve), and curvature κ (blue curve) as functions of the rescaled radius $\sqrt{\rho}$. (b) A symmetric spiral band near a traveling bifurcation

indeed, a rational approximation based on our choice of extra-long spatial scales. Since not only the curvature but also the spiral pitch will be in this case extremely long when rescaled back to physical units, nonlocal effects due to the influence of faraway spiral segments should be negligible as well. The approximation may, however, break down when more subtle effects, such as instabilities, are investigated.

Returning to the kinematic equation (3.3.10), we can observe that the region near the center is apt to be unstable, since close to the unstable zero of the velocity–curvature relation the propagation speed *increases* with curvature and (3.3.10) becomes ill-posed. This is related to the phenomenon of spiral tip meandering observed in numerous experiments and computations (more on this in Sect. 3.5.4). The instability due to the local negative diffusivity may either saturate nonlinearly upon acceleration or be advected by normal propagation to stable outward segments and decay there. Formally, the instability develops most rapidly on the shortest wavelength, but the approximation clearly breaks down on shorter scales. Short-scale instabilities are saturated nonlinearly when the implicit local velocity–curvature relation (3.2.33) is used in its full form, as in computations by Hagberg and Meron (1994), or replaced by a form cubic in *both* velocity and curvature. The latter form can be derived by bifurcation expansion near a higher codimension parametric point $\mu = 2/b$, $\tau = 1$, where the traveling and zigzag instabilities are degenerate and both C and κ can be scaled as $O(\epsilon)$ (Sect. 3.2.5). However, nonlocal effects mediated by the inhibitor field are likely to become at shorter scales at least as strong as nonlinear curvature effects, and should play an important role in the development or stabilization of emerging short-scale modulations.

3.4 Advective Limit

3.4.1 Inertial Scaling

The scaling of Sect. 3.1.1 is not suitable for the analysis of oscillatory or excitable systems, since strong coupling of the two variables in the FN system is necessary to eliminate one or both fixed points. Typical structures associated with oscillatory or excitable dynamics are *propagative*: excitation pulses and wave trains, which can be viewed as excitation spikes or oscillations of dynamical systems translated from the temporal to spatial domain. A wave train in an oscillatory system would combine alternating transitions from the lower to the upper state and back (Fig. 3.17a). For an excitable system (Fig. 3.17b), both wave trains and solitary spikes are possible.

Unlike stationary or slowly evolving patterns where the characteristic length scale is set by the diffusional range of the long-scale variable, the wavelength of a propagating pattern is tied to the propagation speed and remains finite even when the inhibitor is nondiffusive. Accordingly, it is appropriate to redefine the long scale on the basis of the characteristic relaxation time of the inhibitor γ_2 and the characteristic propagation speed of the activator front $c^* = \sqrt{D_1/\gamma_1}$. Using this "advective" length unit, $L^* = \gamma_2 c^* = \gamma_2 \sqrt{D_1/\gamma_1}$, we bring (3.1.1) to the dimensionless form

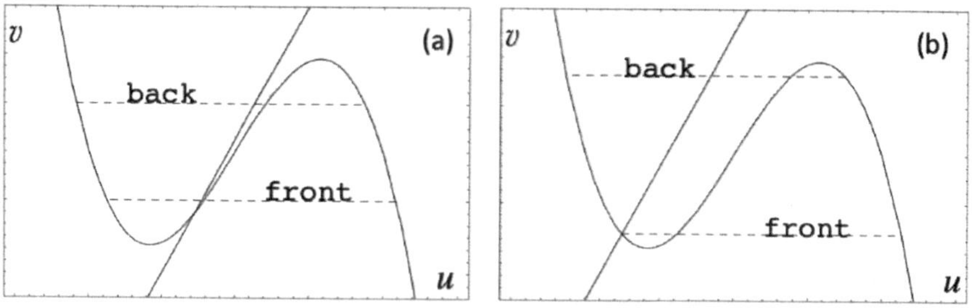

Fig. 3.17. Null isoclines of an oscillatory (a) and an excitable (b) system; dashed lines show transition fronts of the short-scale variable in a wave train (a) or a single excitation pulse (b)

$$\gamma \partial_t u = \gamma^2 \nabla^2 u + f(u, v), \tag{3.4.1}$$
$$\partial_t v = \delta^2 \nabla^2 v + g(u, v), \tag{3.4.2}$$

where $\delta = \gamma/\epsilon$. The "inner" scale of the transitional layer, where the system switches between the two alternative activator states, $u = u_g^{\pm}$, is now set exclusively by the capacitance ratio γ independently of diffusivities and, provided $\gamma \ll 1$, remains small even when the inhibitor is *less* diffusive than the activator and ϵ (as defined in Sect. 3.1.1) is not small. The parameters can be chosen in such a way that $\delta \ll 1$, so that inhibitor's diffusion is negligible. In terms of the original parameters of (3.1.1), this requires $\gamma \ll D_1/D_2$. Under these conditions, the inhibitor diffusion can be neglected, reducing (3.1.2) to

$$\partial_t v = g(u, v). \tag{3.4.3}$$

Although this equation contains no mechanism for healing discontinuities in v, the inhibitor field should remain smooth in the course of evolution, barring freaky initial conditions or strongly localized perturbations.

As before, (3.4.1) is replaced on the $O(1)$ length scale by the eikonal equation (2.2.12) or (3.1.5), which is rewritten in the adopted scaling as

$$c = c_0 - \gamma\kappa, \tag{3.4.4}$$

where $c_0 = c(v_0)$ is the speed of the straight-line front governed by the local value of the controlling variable v_0. This formula is applicable only when the curvature radius far exceeds the inner scale equal to γ in the adopted dimensionless units. The front motion slows down when v_0 is close to the "stall" value at the Maxwell construction v^*. Under these conditions, c_0 may become of the same order of magnitude as the speed of curvature-driven motion. If both v_0 and c_0 are of $O(\gamma^n)$, the two terms in (3.4.4) are balanced at $\kappa = O(\gamma^{n-1})$.

The characteristic dimensionless time τ of evolution of the controlling variable required for switching between the values of v at the "front" and the "back" sides of a propagating wave train or excitation pulse is reduced likewise to $O(\gamma^n)$. Accordingly, the dimensionless wavelength or the length of the pulse, $\lambda = c_0\tau$, becomes of $O(\gamma^{2n})$. The curvature radius should be of the same order of magnitude as $\lambda^{-1} \propto$

γ^{-2n}. Comparing this with the above estimate, we see that the orders of magnitude match when $n = 1/3$. This is known as *Fife scaling* (Fife, 1985). The curvature $\kappa = O(\gamma^{-2/3})$ is still less than the inverse interface width γ^{-1}, and therefore the sharp interface limit leading to (3.1.5) remains valid.

After rescaling

$$\begin{aligned}
&\partial_t \to \gamma^{-1/3}\partial_t, \quad \partial_x \to \gamma^{-2/3}\partial_x, \quad c \to \gamma^{1/3}c, \\
&v \to v^* + \gamma^{1/3}v, \quad \kappa \to \gamma^{-2/3}\kappa,
\end{aligned} \tag{3.4.5}$$

the eikonal equation (3.4.4) reverts to the form (3.1.5), which does not depend on the small parameter γ and only contains an $O(1)$ model-dependent numerical parameter $b = O(1)$, while (3.4.2) becomes

$$\partial_t v = \delta^2\gamma^{-1}\nabla^2 v + g_\pm(v^*) + \gamma^{1/3} g'_\pm(v^*)v \ , \tag{3.4.6}$$

where $g^\pm(v) = g(u^\pm(v), v)$. The diffusional term remains negligible after rescaling when $\delta^2 \ll \gamma$, i.e. $\gamma \ll \epsilon^2$ or, reverting to the original parameters, $D_2 \ll D_1$. As the last term is small, (3.4.6) reduces then to (3.4.3) in the leading order.

3.4.2 Dispersion Relation for Wave Trains

An undistorted excitation pulse or wave train propagating with a constant speed c_0 along the x-axis is defined in the advective limit by (3.4.3) rewritten in the comoving frame:

$$c_0 v_x + g^\pm(v) = 0, \qquad g^\pm(v) = g(u^\pm(v), v). \tag{3.4.7}$$

This equation is, generally, nonlinear even with a linear function $g(u, v)$, in view of the nonlinear dependence of quasistationary activator levels on v. This technical difficulty is usually removed by fixing $f(u, v)$ in such a way that $u^\pm(v) = \text{const}$. A convenient choice is (3.1.6), leading to $u_s^\pm = \pm 1$, with only the unstable intermediate state $u_s^0 = v$ being v-dependent. The respective phase portrait is shown in Fig. 3.18. The propagation speed, following from (2.1.13), is $c_0 = bv_0$ with $b = \sqrt{2}$. With this

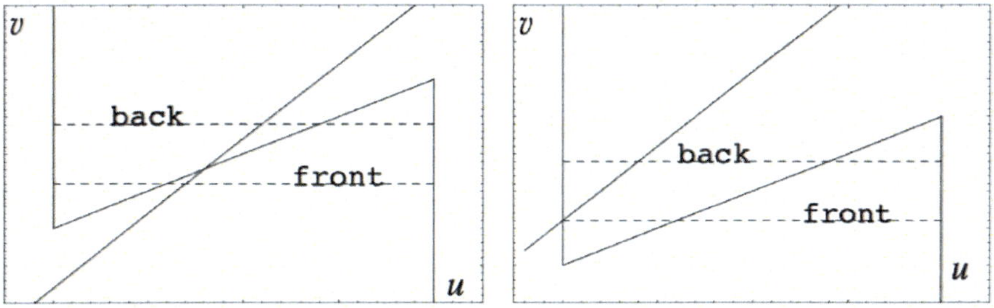

Fig. 3.18. Null isoclines of an oscillatory (a) and an excitable (b) system with the piecewise-linear function (3.1.6); dashed lines show transition fronts of the short-scale variable in a wave train (a) or a single excitation pulse (b)

assignment, and a linear $g(u, v)$ in its usual form, (3.4.7) becomes linear and can be easily solved analytically:

$$c_0 v_x - v - \nu \pm \mu = 0. \tag{3.4.8}$$

Equation (3.4.8) admits a continuum of periodic solutions parameterized by the velocity c_0. The dependence between the velocity and wavelength L, i.e. a dispersion relation, can be obtained by fixing the values of v at the transition fronts. Qualitatively similar solutions for nonlinear systems with phase portraits of the type shown in Fig. 3.17 can be routinely computed numerically.

Consider an oscillatory system with the dynamical phase portrait shown in Fig. 3.18a. The only possible regular pattern in this system is a wave train combining alternating transitions from the lower to the upper state and back, as shown by the dashed line in this figure. To conform to the common usage, we shall call the upper state "excited." On its advancing edge, called the "front" side and located, say, at $x = nL,\ n = \ldots, -1, 0, 1, \ldots$, the propagation speed, by our usual sign convention (Sect. 2.1.2), is negative. The reverse transition to a "quiescent" lower state takes place at the "back" side located at $x = nL \mp L_\pm$, where the propagation speed is positive; $L_\pm$ are the widths of the excited and quiescent regions.

Clearly, the values of v_0 at the front and back edges should be of the opposite sign and equal by absolute value. The solution of (3.4.8) satisfying the boundary conditions $v(0) = -v_0 = -c_0/b$ is (within a single period)

$$v = \begin{cases} \mu - \nu - (\mu - \nu + c_0/b)e^{x/c_0} & \text{at} \ -L_+ \leqslant x \leqslant 0, \\ -(\mu + \nu) + (\mu + \nu - c_0/b)e^{x/c_0} & \text{at} \ \ 0 \leqslant x \leqslant L_-. \end{cases} \tag{3.4.9}$$

The widths of the excited and quiescent regions $L_\pm$ should be obtained from the boundary conditions $v(-L_+) = v(L_-) = v_0$. This gives the relation between $L_\pm$ and the propagation speed c_0 in the implicit form

$$\mu = \frac{c_0}{b} \frac{e^{L/c_0} - 1}{(e^{L_+/c_0} - 1)(e^{L_-/c_0} - 1)}, \qquad \nu = \frac{c_0}{b} \frac{e^{L_+/c_0} - e^{L_-/c_0}}{(e^{L_+/c_0} - 1)(e^{L_-/c_0} - 1)}. \tag{3.4.10}$$

The values of $L_\pm$ at particular levels of the parameters μ, ν can be read from Fig. 3.19a. In the symmetric case $\nu = 0$ and $L_\pm = L/2$, one can obtain an explicit dispersion relation

$$L = 4c_0 \operatorname{arc\,coth}(b\mu/c_0). \tag{3.4.11}$$

The period diverges at $c_0 \to b\mu$ when the trajectory approaches a heteroclinic orbit passing through the corners of the zigzag null isocline in Fig. 3.18. At $c_0 \to 0$, (3.4.11) yields formally a vanishing period; the formula is, however, inapplicable at small velocities, since diffusion has to be taken into account.

The periodic solutions constructed in this way are stable to uncorrelated displacements of individual fronts, since any advance would bring it to an unrelaxed zone where the value of v is unfavorable, and thus cause deceleration. Conversely, any delay will shift the front further down the relaxation tail of the preceding transition and cause acceleration, restoring the original wavelength. Eckhaus instability limiting the

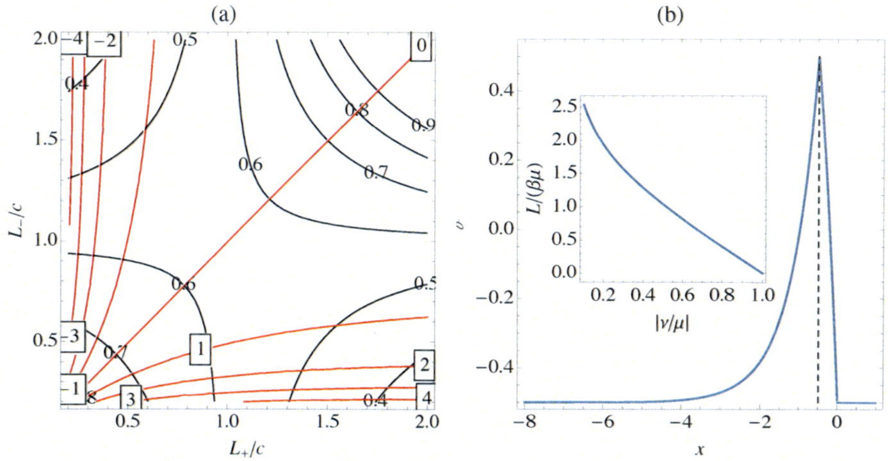

Fig. 3.19. (a): Resolving (3.4.10). Black and red contours show the values of $\beta\mu/c$, $\beta\nu/c$, respectively, as functions of $L_\pm/c$. The labels of $\beta\nu/c$ contours are framed. (b): A profile of v for a typical propagating pulse in an excitable system. Inset: the dependence of the scaled width of the excited zone on the ratio ν/μ

minimal possible wavelength is observed, however, at finite values of γ. The instability threshold can be only obtained numerically (see, e.g., Bär and Or-Guil, 1999). The fastest-growing modes correspond to an alternating compression and expansion of subsequent pulses.

Similar patterns can be constructed for an excitable system with the dynamical phase portrait shown in Fig. 3.18b. The only limitation is that the front edge should lie above the fixed point. However, generating such a pattern will require special initial conditions. It could be sustained, in a more or less realistic setting, on a large ring, where excitation waves of a finite period would chase one another. A more natural structure for this system is a single excitation pulse propagating into the quiescent state $v = v_s$. The speed of the excitation front is $c_0 = c(v_s)$. The reverse transition in the relaxation tail may be delayed, in the absence of perturbations, to the turning point of the positive branch in Fig. 3.18b, but the back edge forming there would move faster and catch up to the level of v where both speeds equalize.

The profile satisfying the linear equation (3.4.8) should be computed with $c_0 = \sqrt{2}v_s$, $v_s = -(\nu + \mu)$. An excitable state on the lower branch with $-1 < v_s < 0$ exists at $-\mu < \nu < 1 - \mu$. The solution in the comoving frame with the front edge at the origin and the back edge at $-L$, satisfying the boundary condition $v(-L) = -v(0) = \nu + \mu$, is

$$v = \begin{cases} -(\mu+\nu) & \text{at } x \geqslant 0 \\ \mu - \nu - 2\mu \exp\left[\frac{x}{b(\mu+\nu)}\right] & \text{at } -L \leqslant x \leqslant 0 \\ -(\mu+\nu) - 2\mu(1+\frac{\mu}{\nu}) \exp\left[\frac{x}{b(\mu+\nu)}\right] & \text{at } x \leqslant -L \end{cases} \quad (3.4.12)$$

$$L = -b(\mu+\nu)\ln|\nu/\mu|. \quad (3.4.13)$$

A typical profile is shown in Fig. 3.18b.

3.4.3 Chaotic Wave Trains

Replacing a generic system with the phase portrait in Fig. 3.17 by a system with a zigzag phase portrait as in Fig. 3.18 has its price, as it eliminates oscillatory approach to the stationary state in the wake of a propagating excitation pulse. Oscillatory tails affect the interaction between successive pulses in a nontrivial way, which can lead to spatiotemporal complexity of the wave train structure (Elphick et al, 1988).

We return to the general FN system in the advective limit (3.4.1), (3.4.3). After rescaling the coordinate $x \to \gamma x$, the system is rewritten in the comoving frame as

$$cu_x + u_{xx} + f(u,v) = 0, \quad (3.4.14)$$

$$c\gamma^{-1} v_x + g(u,v) = 0. \quad (3.4.15)$$

Suppose the system is excitable, having a single stationary state (u_s, v_s). Linearizing (3.4.14) and (3.4.15) near this solution leads to a cubic equation for the eigenvalues λ, which characterize the exponential tail of the pulse. In a certain parametric region, this equation has a pair of complex conjugate roots, $\lambda = \lambda_r \pm i\lambda_i$, in addition to a real root λ_s, with λ_r and λ_s having opposite signs. Under these conditions, the fixed point (u_s, v_s) of (3.4.14), (3.4.15), viewed as a dynamical system with the coordinate x playing the role of time, is a *saddle focus*.

Trajectories passing near this point tend to generate a *chaotic* attractor, known as "Shilnikov's snake" (Shilnikov, 1965 – see, e.g., Pismen, 2021a). This indicates that a train of pulses with oscillatory tails may form a chaotic sequence. Elphick et al (1988) described this sequence in terms of interaction of well-separated pulses centered at $x = X_j$ (Fig. 3.20). Because of the exponential decay toward the fixed point both ahead and behind each pulse, it is sufficient to take into account interaction between the nearest neighbors only. For a train propagating in the negative direction, $\lambda_r < 0$ and $\lambda_s > 0$, and each pulse is influenced by the oscillatory tail of the preceding pulse located at $x = X_{j-1}$ and the exponentially weak precursor of the

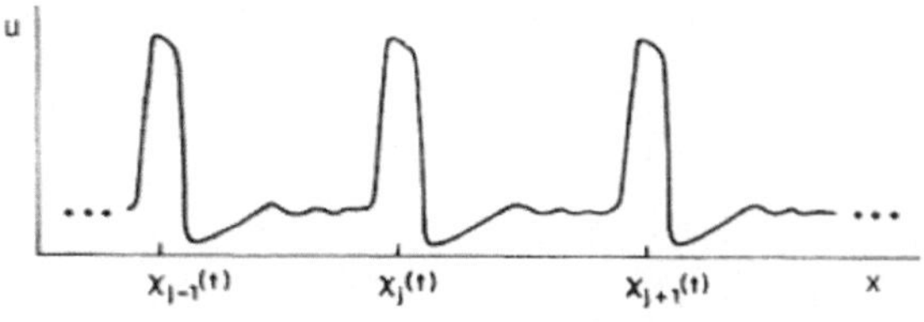

Fig. 3.20. Schematic illustration of a train of pulses with oscillatory tails propagating to the left (Elphick et al, 1988; reproduced by permission, ©The American Physical Society)

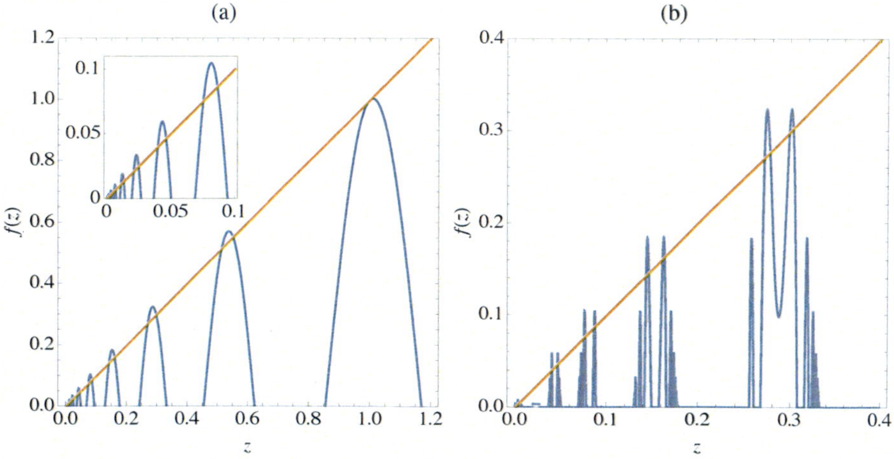

Fig. 3.21. (a): The pattern map (3.4.18) with $r = 0.9\,\omega = 10,\ \delta = \theta = 0$. Inset: blowup near the origin. (b): An iterated pattern map

trailing pulse located at $x = X_{j+1}$. The corrections to the propagation speed due to these interactions can be computed by applying a solvability condition of perturbation equations. In this way, one arrives at the sequence of equations

$$\dot{X}_j = a_+ \mathrm{e}^{-\lambda_s (X_{j+1}-X_j)} + a_- \mathrm{e}^{\lambda_r (X_j - X_{j-1})} \cos\left[\lambda_i (X_j - X_{j-1}) - \theta\right], \quad (3.4.16)$$

where the coefficients $a_\pm$ and θ can be related to the parameters of the original system by computing the above-mentioned solvability condition explicitly.

For a steadily propagating train, $\dot{X}_j = \Delta c = \text{const}$, where Δc is the speed increment relative to the comoving frame. Then (3.4.16) can be converted into a *pattern map* analogous to a Poincaré map of a dynamical system. Denoting

$$Z_j = a_+ \mathrm{e}^{-\lambda_s (X_j - X_{j-1})}, \quad a = -\frac{a_-}{a_+}, \quad r = \frac{\lambda_r}{\lambda_s}, \quad \omega = \frac{\lambda_i}{\lambda_s}, \quad \delta = \frac{\Delta c}{a_+}, \quad (3.4.17)$$

we present the pattern map in the form

$$Z_{j+1} = \delta + a Z_j^r \cos\left(\omega \ln Z_j - \theta\right) \equiv f(Z_j). \quad (3.4.18)$$

The fixed points of this map are solutions with equal spacing of the pulses. As the parameters change, these solutions appear or annihilate in pairs (one stable and one unstable) at saddle–node bifurcations. The fixed points accumulate at $Z \to 0$, which corresponds to the infinite spacing (see Fig. 3.21a).

The map (3.4.18) cannot be iterated to obtain chaotic sequences in the same way as Poincaré maps of dynamical systems are iterated, since it contains nonphysical segments, with $f(Z) < 0$ (not shown in Fig. 3.21). However, it is possible to obtain solutions with unequal spacing as fixed points of iterated maps, e.g., period-two sequences are fixed points of the iterated map in Fig. 3.21b. Numerical simulations

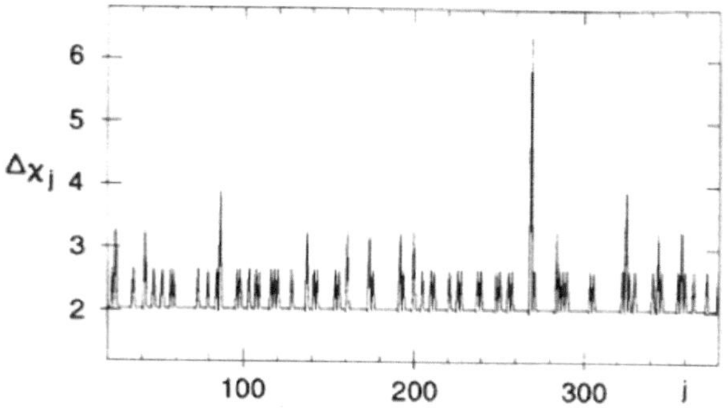

Fig. 3.22. Spacings between subsequent pulses $\Delta X_j = X_j - X_{j-1}$ (Elphick et al, 1988; reproduced by permission, ©The American Physical Society)

of (3.4.16) starting from an unstable equally spaced solution showed the tendency of the wave train to break into groups of pulses separated by "defects" with very large spacing (Elphick et al, 1988). A sequence of spacings in a train consisting of several hundred pulses is shown in Fig. 3.22.

3.5 Rotating Spiral Waves

3.5.1 Advective Limit for a Rotating Spiral

We have seen in Sect. 3.3 that purely kinematic, or local, theories of spiral waves are unable to describe both front and back sides of a propagating excitation band connected by the spiral tip, unless a sign-changing velocity–curvature relation is used as in Sect. 3.3.4. Nonlocal theory involving direct computation of the inhibitor field is far more difficult, but becomes tractable when the inhibitor's diffusion is neglected in the advective limit.

The inhibitor field $v(r, \varphi - \omega t)$ of a rotating spiral should be stationary in a polar coordinate frame $(r, \phi = \varphi - \omega t)$ rotating with some frequency ω, so far indeterminate. Equation (3.4.7) in its usual linear form, rewritten in the rotating polar frame, is similar to (3.4.8):

$$-\omega v_\phi = -v - \nu \pm \mu. \tag{3.5.1}$$

The values of v at the front and back sides of the "excited" region situated at $\phi = \Phi_\pm(r)$ should be fixed to ensure propagation of both curves without change of form with a constant angular velocity. Integrating (3.5.1) yields the respective values

$$v_+ = -\nu - \mu \frac{1 + e^{2\pi/\omega} - 2e^{\Delta/\omega}}{e^{2\pi/\omega} - 1}, \quad v_- = -\nu + \mu \frac{1 + e^{2\pi/\omega} - 2e^{(2\pi - \Delta)/\omega}}{e^{2\pi/\omega} - 1}, \tag{3.5.2}$$

where $\Delta = \Phi_+ - \Phi_-$.

Using the kinematic equations is, generally, not practical, since the geometrical separation of the two fronts, Δ, cannot be eliminated. One can compute instead the shape of both fronts directly after substituting into the eikonal equation the normal propagation speed obtained by eliminating χ from (3.3.16),

$$c = \pm \frac{\omega r}{\sqrt{1 + [r\Phi'(r)]^2}}, \tag{3.5.3}$$

together with the expression for curvature in polar coordinates

$$\kappa(\phi) = \frac{2\Phi'(r) + r^2[\Phi'(r)]^3 + r\Phi''(r)}{\left(1 + [r\Phi'(r)]^2\right)^{3/2}}, \tag{3.5.4}$$

and the values of $v_\pm$ given by (3.5.2). Although the Fife scaling (Sect. 3.4.1) is most suitable for analyzing the motion of curved fronts in the advective limit, we shall keep using the eikonal equation in its original form (2.2.12) as in Sect. 3.3.3; the diffusivity D should be set equal to γ in the inertial and to unity in the Fife scaling. The resulting coupled equations for $\Phi_\pm$ are

$$\frac{\omega r}{D} = \frac{b v_\pm}{D}\left(1 + [r\Phi'_\pm(r)]^2\right)^{1/2} + \Phi'_\pm(r) + \frac{\mathrm{d}_r[r\Phi'_\pm(r)]}{1 + [r\Phi'_\pm(r)]^2}\,. \tag{3.5.5}$$

This system, coupled to (3.5.2), was integrated numerically by Pelcè and Sun (1991). Fine-tuning of the rotation frequency and the tip radius where the two curves meet is needed to achieve proper convergence to the asymptotic Archimedean shape. The solution is singular at the spiral tip where the advective approximation breaks down.

A special case is a symmetric spiral, which can be constructed with the help of intrinsic equations of motion even when no local equation of motion is available (Keener, 1994). In the symmetric case ($\nu = 0$), the bands of the up and down state should be equal, so that the values $v_\pm$ at the branches where either state advances should be of the opposite sign and have the same absolute value v_0 independent of r. The respective velocity of a straight-line front is

$$c_0 = b|v_\pm| = b\mu \tanh\frac{\pi}{2\omega}. \tag{3.5.6}$$

Thus, we have a local relation between the curvature and propagation speed, which is the same as in Sect. 3.3.3, but with the maximum speed c_0 dependent on frequency.

As before, the spiral's local velocity and eventually its shape can be found by integrating (3.3.19) or (3.3.20). The difference is that either equation is now applicable on both symmetric branches separated by π. According to our usual sign convention, both propagation speed and curvature change sign when the advancing states interchange, which leaves the eikonal equation invariant. Another substantial difference is that the solution should be extended to the origin, leaving no hole. By symmetry, the tangential as well as normal velocity should vanish at the origin. The solution for both branches is therefore identical to the limit R, $\eta \to 0$ in Sect. 3.3.3, and is attained at the same unique value $|\Omega| = \omega D/c_0^2 \approx 0.331$.

In view of (3.5.6), the frequency can be obtained as the root of

$$\omega \tanh^{-2}\frac{\pi}{2\omega} = 0.331\frac{(b\mu)^2}{D}. \tag{3.5.7}$$

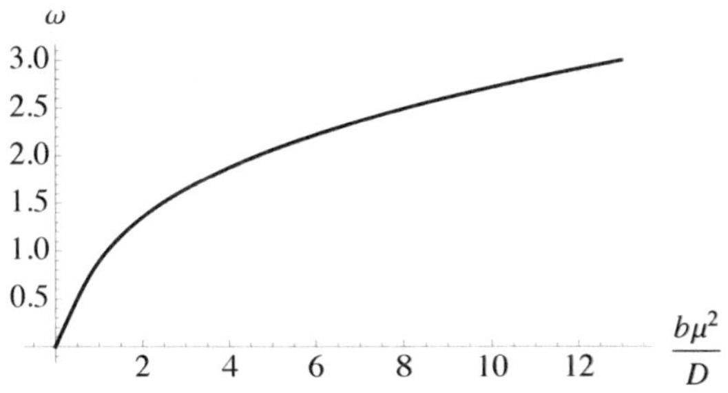

Fig. 3.23. Dependence of the rotation frequency ω on the parameter μ for a symmetric spiral

The left-hand side of this equation is a monotonically increasing function of ω; hence, there is a unique solution with the frequency monotonically increasing with growing μ (Fig. 3.23).

The troubling feature of the solution obtained in this way, as well as of the above numerical solution, is a discontinuity of v at the origin where the two branches with $v = v_{\pm} = \pm v_0$ of the opposite sign come together. This can be mended only by taking into account the inhibitor's diffusion (Keener, 1994). The value of v on the spiral branches turns out then to be dependent on the radial coordinate; this dependence should be taken into account at least in the vicinity of the origin.

3.5.2 Propagating Finger

If a straight-line excitation pulse propagating along the x-axis in the plane is truncated at $y = 0$, the curved tip of the remaining semi-infinite "finger" will lag behind faraway segments continuing to propagate with an unchanged speed. Generally, such a finger would either grow or retract, but a special solution corresponding to a steadily translating "critical finger" can be obtained by adjusting the parameters of the inhibitor equation (3.4.3). This solution plays a special role in the dynamics of a spiral tip to be considered in the next subsection, and forms an essential part of the so far most comprehensive analytical theory of rotating spiral waves (Hakim and Karma, 1999).

We shall use the Cartesian frame with the origin at the tip and coordinates x and y directed along and across the propagation direction, respectively, and define the front and back sides of the finger by some function $x = X_{\pm}(y)$. The front side propagates into the stationary state of the inhibitor field, and the normal velocity is fully determined by local geometry. The shape of the interface can be therefore computed with the help of intrinsic equations of Sect. 3.3.

To preserve a stationary shape, the tangential velocity χ should be fixed to ensure that each point be translated strictly along the x-axis while it propagates along the normal. The tangent and normal unit vectors are

$$\mathbf{l} = \frac{\widehat{\mathbf{y}} + X'(y)\,\widehat{\mathbf{x}}}{\sqrt{1 + [X'(y)]^2}}, \qquad \mathbf{n} = \frac{X'(y)\,\widehat{\mathbf{y}} - \widehat{\mathbf{x}}}{\sqrt{1 + [X'(y)]^2}}, \tag{3.5.8}$$

where $\widehat{\mathbf{x}}$ and $\widehat{\mathbf{y}}$ are unit vectors in the respective directions. The velocity components along the axes are now expressed as

$$c_y = (\chi \mathbf{l} - c\mathbf{n}) \cdot \widehat{\mathbf{y}} = \frac{\chi - cX'(y)}{\sqrt{1 + [X'(y)]^2}} = 0,$$
$$c_x = (\chi \mathbf{l} - c\mathbf{n}) \cdot \widehat{\mathbf{x}} = \frac{\chi X'(y) + c}{\sqrt{1 + [X'(y)]^2}} = c_0, \tag{3.5.9}$$

where c_0 is the propagation speed, which should coincide by its absolute value with the speed of a straight-line propagating band approached at $y \to \infty$. This implies

$$X'(y) = \chi/c, \quad c^2 = c_0^2 - \chi^2. \tag{3.5.10}$$

Using the latter relation in (3.3.19) we see that the flux q must vanish, which, in view of (3.3.13), yields the equation defining the change of the normal velocity along the arc length s:

$$c'(s) = \kappa\sqrt{c_0^2 - c^2}. \tag{3.5.11}$$

In accordance with the eikonal equation, which we take again in its basic form (2.2.12), the curvature is $\kappa = (c_0 - c)/D$. Using this in (3.5.11) yields a closed equation for the normal velocity, which can be easily integrated to

$$\frac{c}{c_0} = \frac{c_0 s/D\,(2 + c_0 s/D)}{2 + 2c_0 s/D + (c_0 s/D)^2}. \tag{3.5.12}$$

The origin $s = 0$ is placed at the tip where the normal velocity vanishes and the curvature reaches the maximum value $\kappa_t = c_0/D$; thus, the tip radius is $R_t = D/c_0$. At $s \to \infty$, where the front is flat, $c \to c_0$ and the curvature vanishes.

The actual shape of the front can be restored with the help of (3.5.10) supplemented by the metric relation $\mathrm{d}s = \sqrt{1 + [X_+'(y)]^2}\mathrm{d}y$:

$$\frac{\mathrm{d}y}{\mathrm{d}s} = \frac{c}{c_0}, \qquad \frac{\mathrm{d}X_+}{\mathrm{d}s} = \sqrt{1 - \frac{c^2}{c_0^2}}. \tag{3.5.13}$$

It is convenient to combine these equations with (3.5.11):

$$\frac{\mathrm{d}y}{\mathrm{d}c} = \frac{Dc}{c_0(c_0 - c)\sqrt{c_0^2 - c^2}}, \qquad \frac{\mathrm{d}X_+}{\mathrm{d}c} = \frac{D}{c_0(c_0 - c)}. \tag{3.5.14}$$

These equations are integrated to obtain the function $X_+(y)$ in the parametric form

$$y(c) = \frac{D}{c_0}\left(\sqrt{\frac{c_0 + c}{c_0 - c}} - \arcsin\frac{c}{c_0} - 1\right), \quad X_+(c) = -\frac{D}{c_0}\ln\left(1 - \frac{c}{c_0}\right). \tag{3.5.15}$$

The normal velocity at the back side depends on the local inhibitor's level, which, in turn, is dependent on the local finger thickness. This dependence can be obtained, as in the preceding subsection, by solving (3.4.8). However, since the finger is presumed to be narrow and the shift of v relative to its stationary value v_s (lying close to the stall value v^*) to be small, the solution can be linearized by approximating the change of v across the finger by a linear function of the separation,

$$\Delta v = g^+(v^*)(X_+ - X_-)/c_0. \tag{3.5.16}$$

This approximation works for the general form (3.4.3), and only the proportionality coefficient is dependent on a particular form of the inhibitor equation. Accordingly, the normal velocity of a straight-line front corresponding to the local value of v at the back side is expressed as $c = c_0 - \beta(X_+ - X_-)$. The parameter $\beta = bg^+(v^*)/c_0$ (where b is the derivative $\mathrm{d}c/\mathrm{d}v$ computed at the stall value) serves in this context as a measure of *excitability* of the system, which increases as β decreases.

In view of the position dependence, intrinsic equations of motion become insufficient, and the equation for $X_-(y)$ should be obtained by using explicit expressions for the curvature and normal velocity in the Cartesian frame in the eikonal equation (3.1.5). Since this equation depends on $X_+(y)$, which is defined by the above analytic solution only implicitly, it is more convenient to solve the equations of both interfaces simultaneously. The system to be solved is

$$\frac{D}{c_0}\frac{\mathrm{d}^2 X_+}{\mathrm{d}y^2} = 1 + \left(\frac{\mathrm{d}X_+}{\mathrm{d}y}\right)^2 - \left[1 + \left(\frac{\mathrm{d}X_+}{\mathrm{d}y}\right)^2\right]^{3/2}, \tag{3.5.17}$$

$$\frac{D}{c_0}\frac{\mathrm{d}^2 X_-}{\mathrm{d}y^2} = 1 + \left(\frac{\mathrm{d}X_-}{\mathrm{d}y}\right)^2 + \left[1 - \frac{\beta}{c_0}(X_+ - X_-)\right]\left[1 + \left(\frac{\mathrm{d}X_-}{\mathrm{d}y}\right)^2\right]^{3/2}.$$

The solution of this equation should satisfy the asymptotic condition $X_+ - X_- = 2/\beta$ at $y \to \infty$, since the propagation velocity far from the tip must be equal (by its absolute value) to c_0. On the other end, the curve $X_-(y)$ should match near $y = 0$ the front curve $X_+(y) \asymp \sqrt{2Dy/c_0}$. The problem is overdetermined, and the solution exists only at a particular value of the excitability parameter, $\beta = \beta_c$. When (3.5.17) with some test value of β is solved numerically starting from the square root asymptotics at $y \to 0$, the curve $X_-(y)$ turns away either upwards or downwards from the desired large y asymptotics, respectively, at $\beta < \beta_c$ or $\beta > \beta_c$. The value $\beta_c/c_0 = 0.535326\ldots$ can be found iteratively. The numerical solution is plotted in Fig. 3.24. The solution is weakly nonanalytic at $y = 0$, since the subleading terms in the asymptotics of $X_\pm(y)$ at $y \to 0$ differ: $X_+(y) \asymp \sqrt{2Dy/c_0} + \frac{1}{3}$, while $X_-(y) \asymp \sqrt{2Dy/c_0} + \frac{1}{3}(1 - 2\beta/c_0)$. As in the preceding subsection, this is caused by a breakdown of the advective approximation at the spiral tip. The mismatch can be healed by taking into account the inhibitor diffusion near the tip (Hakim and Karma, 1999).

A retracting finger solution can be constructed by allowing the propagation speed along the x axis c_t to be *larger* than the normal speed of the straight-line front c_0. The finger should be then inclined relative to the y axis, so that asymptotically at $y \to \infty$, $X'_\pm(y) \asymp \sqrt{(c_t/c_0)^2 - 1}$. The propagation speed c_t should be used instead of c_0 in (3.5.10) and (3.5.13). Accordingly, (3.5.11) is rewritten as

$$c'(s) = \frac{c_0 - c}{D}\sqrt{c_t^2 - c^2}, \tag{3.5.18}$$

and (3.5.14) takes the form

$$\frac{\mathrm{d}y}{\mathrm{d}c} = \frac{Dc}{c_t(c_0 - c)\sqrt{c_t^2 - c^2}}, \qquad \frac{\mathrm{d}X_+}{\mathrm{d}c} = \frac{D}{c_t(c_0 - c)}, \tag{3.5.19}$$

which integrates to the parametric equation of the front interface:

$$y(c) = \frac{D}{c_t}\left[\frac{c_0}{\sqrt{c_t^2 - c_0^2}} \ln \frac{c_0\left(c_t^2 - cc_0 + \sqrt{(c_t^2 - c_0^2)(c_t^2 - c^2)}\right)}{c_t(c_0 - c)\left(1 + \sqrt{c_t^2 - c_0^2}\right)} - \arcsin\frac{c}{c_t}\right],$$
$$X_+(c) = -\frac{D}{c_t}\ln\left(1 - \frac{c}{c_0}\right). \tag{3.5.20}$$

For simultaneous computation of the front and back interface, one has to integrate the equations generalizing (3.5.17):

$$\frac{D}{c_0}\frac{\mathrm{d}^2 X_+}{\mathrm{d}y^2} = \frac{c_t}{c_0}\left[1 + \left(\frac{\mathrm{d}X_+}{\mathrm{d}y}\right)^2\right] - \left[1 + \left(\frac{\mathrm{d}X_+}{\mathrm{d}y}\right)^2\right]^{3/2}, \tag{3.5.21}$$
$$\frac{D}{c_0}\frac{\mathrm{d}^2 X_-}{\mathrm{d}y^2} = \frac{c_t}{c_0}\left[1 + \left(\frac{\mathrm{d}X_-}{\mathrm{d}y}\right)^2\right] + \left[1 - \frac{\beta}{c_t}(X_+ - X_-)\right]\left[1 + \left(\frac{\mathrm{d}X_-}{\mathrm{d}y}\right)^2\right]^{3/2}.$$

Retracting solutions exist at $\beta > \beta_c$; the dependence of the deviation $\beta - \beta_c$ on the ratio c_t/c_0 can be obtained as before by adjusting this parameter to arrive at the desired asymptotics when (3.5.21) is integrated numerically. Near $\beta = \beta_c$, the dependence is linear:

$$\frac{c_t}{c_0} = 1 + \frac{\beta - \beta_c}{k}. \tag{3.5.22}$$

Hakim and Karma (1999) found the numerical value $k \approx 0.630$ by expanding the solutions of (3.5.21) near the critical finger and applying a solvability condition of the first-order equation. This result is applicable as well at $\beta < \beta_c$ when the finger is *advancing*, and will be further used in Sect. 3.5.3 for the analysis of a rotating spiral band.

At c_t/c_0 slightly deviating from unity, the tip of the retracting finger closely approximates the tip of the critical finger, as seen in Fig. 3.24.

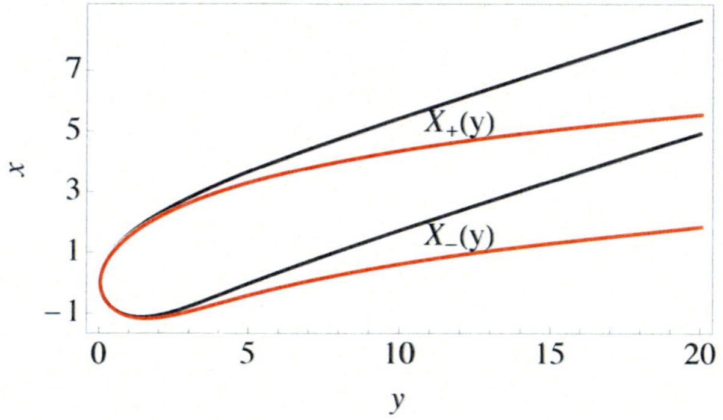

Fig. 3.24. The critical finger obtained by integrating (3.5.17) with $\beta_c/c_0 = 0.535326$ (red line) and a retracting finger satisfying (3.5.21) with $c_t/c_0 = 1.05$, $\beta/c_0 = 0.566907$ (black line). The coordinates are scaled by D/c_0

3.5.3 Slender Spiral Band

Analytical theory for a slender spiral with a large core radius was constructed by Hakim and Karma (1999). They analyzed the spiral shape by decomposing it into three overlapping regions where different approximations can be applied. Close to the tip, the curvature of the tip trajectory can be neglected, and the shape of the excited region can be obtained by perturbing the critical finger solution described in Sect. 3.5.2. This yields a linear relation between the tip propagation velocity c_t and deviation from the critical excitability $\Delta\beta = \beta - \beta_c$. Steadily rotating spirals exist when $\Delta\beta$ is *negative*. The shape of the critical finger in the inner region near the tip remains virtually unchanged as long as $|\Delta\beta|$ is small.

Far from the tip, the interfacial curvature is negligible, and the shape of the interface can be found by setting in the invariance conditions (3.3.16) $c = c_0(v)$, the propagation speed of the straight-line front. The resulting equation is then the same as in the WR limit (Sect. 3.3.2), and the outer solution is given by (3.3.18), with R replaced by c_0/ω. This solution, as we know, breaks down at $r \to R$. The inner radius R, which was interpreted as the radius of a circular obstacle (hole) in Sect. 3.3.2, has now a different meaning of the *core radius*, and is defined as the radial position of the tip.

The inner and outer approximations do not overlap, but should be matched in an intermediate region where the interface is almost normal to the circle of rotation and its curvature is small. Recalling that the spiral end was normal to the hole circumference in the kinematic theory of Sect. 3.3.3, we observe that the radius of the "equivalent hole" should fit, in fact, the radial position of this intermediate region rather than the tip position, and the rotation frequency is determined by the normal propagation speed in this region.

The relation between the tip translation velocity c_t, the core radius R, and the tip curvature $\kappa_t = 1/R_t$ can be obtained by balancing the corrections to the short-scale and long-scale approximations in this region. The small parameter of the problem is the inverse of the Péclet number, Pe $= Rc_0/D$. In order to find out a correct scaling in terms of $\varepsilon = 1/\text{Pe}$, we express tentatively the radial and angular variables as

$$r = R(1 + \varepsilon^m \widetilde{y}), \qquad \Phi = \varepsilon^n \widetilde{X}, \tag{3.5.23}$$

with so far indefinite exponents $m, n > 0$, and expand (3.5.5) in powers of ε. Retaining the leading terms only, we have

$$\frac{c_t - c_0}{c_0} + \varepsilon^n \widetilde{y} = \frac{\varepsilon^{2(m-n)}}{2}\left(\frac{\mathrm{d}\widetilde{X}}{\mathrm{d}\widetilde{y}}\right)^2 + \varepsilon^{m-2n+1}\frac{\mathrm{d}^2\widetilde{X}}{\mathrm{d}\widetilde{y}^2}. \tag{3.5.24}$$

All terms of this equation will be of the same order of magnitude if one chooses $m = 1, n = 2/3$, and sets

$$c_t/c_0 - 1 = a\,\varepsilon^{2/3}, \tag{3.5.25}$$

where a is an $O(1)$ constant. Transforming now the variables $\widetilde{y} = 2^{1/3}\eta - a$, $\widetilde{X}(\widetilde{y}) = 2\ln\xi(\eta)$ reduces (3.5.24) to the Airy equation $\xi''(\eta) = \eta\xi(\eta)$. Of the two Airy functions solving this equation, one has to choose the one decreasing at $\eta \to \infty$, $\mathrm{Ai}(\eta)$. The solution is therefore[6]

$$\widetilde{X}(\widetilde{y}) = 2\ln \mathrm{Ai}(2^{-1/3}(a+\widetilde{y})). \tag{3.5.26}$$

The asymptotics of this solution at $\widetilde{y} \to \infty$, $\widetilde{X}(\widetilde{y}) \asymp -\frac{1}{3}(2\widetilde{y})^{3/2}$ matches the asymptotics of the outer solution (3.3.18) at $r \to R$, as one can see by expanding the latter near this point and rewriting the result in the rescaled variables (3.5.23).

At $\widetilde{y} \to 0$, (3.5.26) should match the asymptotics of the finger solution (3.5.15) at $y \to \infty$,

$$X_+(y) \asymp \frac{2D}{c_0} \ln \frac{c_0 y}{\sqrt{2}D}. \tag{3.5.27}$$

Since the argument of the logarithm in (3.5.26) has to be linear in $\widetilde{y}$ to enable matching, $2^{-1/3}a$ must be a zero of the Airy function; this must be the largest zero, $\eta_0 \approx -2.3381$, to ensure smooth connection to asymptotics at $\widetilde{y} \to \infty$.

The solution for the back interface $X_-(y)$ is virtually identical in the intermediate and far regions to the solution for the front interface. Since in the tip region the curvature of the tip trajectory is negligible, the linear relation (3.5.22) remains valid and can be combined with (3.5.25) to obtain the dependence on excitability of both the ratio of the tip radius $R_t = D/c_0$ to the core radius R and the rotation frequency

$$\varepsilon = \frac{R_t}{R} = \frac{\omega D}{c_0^2} = \frac{1}{\sqrt{2}}\left(\frac{\beta_c - \beta}{k|\eta_0|}\right)^{3/2} = \left(\frac{\beta_c - \beta}{1.856}\right)^{3/2}. \tag{3.5.28}$$

The theory has to be further corrected by taking into account that the front interface of the rotating spiral band propagates not right into the quiescent state $v = v_s$ but rather into the tail of the preceding pulse where the relaxation of the controlling variable is still less than complete. The correction is necessary when the resulting velocity decrement is of the same order of magnitude as $c_t/c_0 - 1 \propto \varepsilon^{2/3}$. The deviation of the controlling variable is computed by integrating (3.4.3) rewritten in the rotating polar frame. The result can be written precisely using the special linear form (3.5.1), but, since the shift Δv across the narrow excitation band is always small in a weakly excitable system, the general form (3.4.3) can be linearized as well, and the only relevant constant characterizing the relaxation dynamics is the relaxation time $\tau = 1/g'_-(v_s)$. Neglecting the width of a slender excited band, the residual deviation is

$$v(y) - v_s = \Delta v(y) \exp\left(-\frac{2\pi}{\omega\tau}\right) \approx \frac{\beta_c}{b} \exp\left(-\frac{2\pi R}{c_0 \tau}\right). \tag{3.5.29}$$

The latter expression is written taking into account that the shape of the perturbed band is close to that of the critical finger.

The respective small velocity decrement, $b(v - v_s)$, can be treated as a weak perturbation of the critical finger in the same way as $c_t/c_0 - 1$ was treated before.

---

[6] The other Airy function $\mathrm{Bi}(\eta)$, diverging at $\eta \to \infty$, would describe a spiral rotating in the negative direction (clockwise).

The relation (3.5.25) is modified in such a way that c_t is compared not to the velocity of a single excitation pulse but to the velocity of a wave train with the wavelength $2\pi R$, which coincides in the leading order with the asymptotic wavelength of the spiral:

$$\frac{c_t}{c_0} - 1 + 2\exp\left(-\frac{2\pi R}{c_0 \tau}\right) = 2^{1/3}|\eta_0|\,\varepsilon^{2/3}. \tag{3.5.30}$$

The relation modifying (3.5.28) obtained by perturbing the critical finger is

$$\beta_c - \beta = k_1 \varepsilon^{2/3} + k_2 \exp\left(-\frac{2\pi R}{c_0 \tau}\right), \tag{3.5.31}$$

with $k_1 = 2^{1/3}k|\eta_0| \approx 1.856$ and $k_2 \approx 2.262$. The solvability condition applied in this derivation effectively averages the influence of the residual deviation $\Delta v(y)$ over the tip region.

3.5.4 Tip Meandering

Interaction between successive excitation bands forming a rotating spiral wave may destabilize steady rotation. The instability arises at the spiral core and causes unsteady motion of the spiral tip. This phenomenon of *tip meandering*, observed both in experiment and numerical simulations, can be understood by extending the above large-core expansion to unsteady motion (Hakim and Karma, 1999).

We continue to use the inverse Péclet number ε as the small parameter of the problem, and suppose that deviations of both the tip position and the rotation radius from their common stationary value, which we now denote as R_0, are of the same order of magnitude as the tip curvature radius, i.e. $R - R_0 = O(\varepsilon)$. Likewise, the instantaneous tip velocity c_t should differ from its steady-state value, now denoted as $c_t^0 = \omega R_0$, by $O(\varepsilon)$. Thus, we set

$$R = R_0 + \varepsilon \widetilde{R}, \qquad c_t = c_t^0 + \varepsilon \widetilde{c}_t. \tag{3.5.32}$$

In the leading order, the instantaneous rotation radius and tip velocity are related by (3.5.30) with the Péclet number based on R rather than R_0:

$$\frac{c_t}{c_0} - 1 + 2\exp\left(-\frac{2\pi R_0}{c_0 \tau}\right) = 2^{1/3}|\eta_0|\left(\frac{\varepsilon R_0}{R}\right)^{2/3}. \tag{3.5.33}$$

Unsteady rotational motion or the spiral tip is most conveniently characterized by the tip velocity c_t and the tip rotation rate c_t/r, which is proportional to the vector product of the tip velocity c_t and acceleration $\dot{c}_t$ (the dot denotes the time derivative).

The instantaneous tip position can be presented in polar coordinates as

$$r = R[1 + \varepsilon\rho(t)], \qquad \phi = \omega t + \varepsilon\psi(t). \tag{3.5.34}$$

It is advantageous to use the complex representation of the tip position, written in the polar form as $z = r\mathrm{e}^{\mathrm{i}\phi}$, and its time derivatives. Then we have

$$c_t = |\dot{z}|, \qquad c_t/R = \mathrm{Im}\left(\ddot{z}\dot{\bar{z}}/|\dot{z}|^2\right), \tag{3.5.35}$$

where the bar denotes the complex conjugate. The right-hand sides of (3.5.35) are expanded to the first order in ε as

$$|\dot{z}| = R[\omega + \varepsilon(\dot{\psi} + \omega\rho)], \qquad \mathrm{Im}\left(\ddot{z}\dot{\bar{z}}/|\dot{z}|^2\right) = \omega + \varepsilon(\dot{\psi} + \omega^{-1}\ddot{\rho}). \tag{3.5.36}$$

Combining (3.5.32) and (3.5.36) cancels the zero-order terms, while the first-order terms yield the dynamic equations for ψ, ρ:

$$\dot{\psi} = R^{-1}\widetilde{c}_t - \omega\rho, \qquad \omega^{-1}\ddot{\rho} + \dot{\psi} = R^{-1}(\widetilde{c}_t - \omega\widetilde{R}). \tag{3.5.37}$$

Both equations combine to a forced harmonic equation for the radial displacement ρ:

$$\ddot{\rho} + \omega^2\rho = \omega^2\widetilde{R}/R. \tag{3.5.38}$$

The perturbation of the rotation radius can be expressed through the velocity perturbation with the help of a simplified relation (3.5.33) where the change of the stationary radius due to incomplete relaxation is neglected:

$$\frac{\widetilde{R}}{R} = \frac{3\varepsilon^{2/3}}{2^{4/3}|\eta_0|}\,\frac{\widetilde{c}_t}{c_0}. \tag{3.5.39}$$

Replacing also time by the angular coordinate brings (3.5.38) to the form

$$\frac{\mathrm{d}^2\rho}{\mathrm{d}\phi^2} + \rho = \frac{3\varepsilon^{2/3}}{2^{4/3}|\eta_0|}\,\frac{\widetilde{c}_t}{c_0}. \tag{3.5.40}$$

This equation, though looking innocuous, is in fact highly nonlocal because of the dependence of $\widetilde{c}_t$ on perturbations of the inhibitor field caused by perturbations of excitation pulses that had passed the same location in the past. Fortunately, in view of exponential decay of perturbations, it is sufficient to take into account only one immediately preceding pulse. Expressing $\widetilde{c}_t$ through the latter's position is the next task necessary for closing (3.5.40).

Since the origin of the coordinate y in the critical finger solution that serves as the basis for the perturbations is located at the instantaneous tip position, the front of the propagating pulse feels at any given moment t, or at any given angular position of the tip ϕ, the residual value of the inhibitor that corresponds to its level v_- at the back of the preceding pulse at the point $\widehat{y} = y + \varepsilon[\rho(\phi) - \rho(\phi - 2\pi)]$. The latter value, in its turn, depends on the level of v at the front of the preceding pulse at the same location and the local thickness of this pulse, $\Xi = X_+ - X_-$, as defined by (3.5.16). This yields the relation between the inhibitor values at the front of the two successive pulses generalizing the stationary relation (3.5.29):

$$v(y;\phi) - v_s = \left[v(\widehat{y};\phi - \pi) + \frac{\beta}{b}\,\Xi(\widehat{y};\phi - \pi) - v_s\right]\exp\left(-\frac{2\pi R}{c_0\tau}\right). \tag{3.5.41}$$

The respective small velocity decrement, $\widetilde{c}_t(y;\phi) = b[v(y;\phi) - v_s]$, can be treated as a weak perturbation of the critical finger in the same way as $c_t/c_0 - 1$ was treated

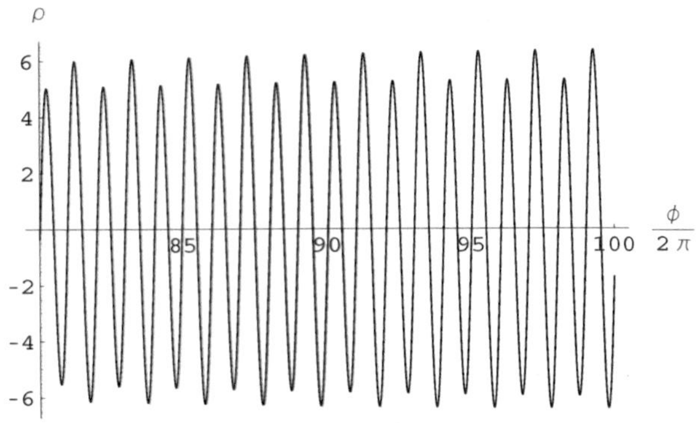

Fig. 3.25. Angular dependence of the deviation of the tip position $\rho(\phi)$ attained after a number of revolutions. Parameters: $a = 0.2,\ m = 0.5$

before. A tedious but straightforward computation described in detail in the original publication by Hakim and Karma (1999) uses the solvability condition to express $\widetilde{c}_t$ as a numerical function $F(\varDelta)$ of the difference between perturbations of the tip positions of the two successive pulses $\varDelta = \rho(\phi) - \rho(\phi - 2\pi)$. At large values of its argument, which bring either tip into interaction with the asymptotic (Archimedean) tail of its counterpart, this function should approach a constant value, while $F(0)$ must vanish. The function is reasonably well approximated by a shifted hyperbolic tangent:

$$F(\varDelta) = \tanh(\varDelta - \alpha) + \tanh\alpha. \tag{3.5.42}$$

A qualitatively truthful picture of meandering transition and motion of the spiral in the supercritical region can be reconstructed by integrating the differential delay equation obtained by using (3.5.42) in (3.5.40):

$$\frac{\mathrm{d}^2\rho}{\mathrm{d}\phi^2} + \rho = mF[\rho(\phi) - \rho(\phi - 2\pi)]. \tag{3.5.43}$$

Beyond the instability threshold, the tip trajectory can be obtained by integrating (3.5.43) numerically. An example of the angular dependence of the deviation of the tip position is shown in Fig. 3.25. One can clearly see here a period-two orbit attained after a number of revolutions.

As the parameter m increases, the trivial solution $\rho = 0$ loses stability. The instability threshold is detected in a standard way by linearizing (3.5.43) and seeking for solutions in the form $\rho \propto \mathrm{e}^{\lambda\phi}$. This leads to the eigenvalue problem

$$1 + \lambda^2 = m\left(1 - \mathrm{e}^{-2\pi\lambda}\right). \tag{3.5.44}$$

At any m, this equation is verified by $\lambda = \pm\mathrm{i}$. The respective eigenmodes correspond to simple translation of the spiral in the angular direction. Other solutions of (3.5.44) depend on the parameter m. At $m \ll 1$, the right-hand side of (3.5.44) may become of $O(1)$ only when the real part of λ is negative, and therefore the trivial solution must be stable. As m increases, the roots of (3.5.44) move in the complex plane, and stability is lost when one of them crosses the imaginary axis. At the instability threshold $m = m_c$, $\lambda = \mathrm{i}\Omega$ is purely imaginary. Substituting this in (3.5.44) and separating the real and imaginary part yields the pair of equations defining both the critical value m_c and the incipient meandering frequency Ω:

$$1 - \Omega^2 = m_c\,(1 - \cos 2\pi\Omega) \qquad m_c \sin 2\pi\Omega = 0. \tag{3.5.45}$$

The second equation requires $\Omega = n/2$ to be half-integer. The first equation reduces then to $1 - (n/2)^2 = m_c\,[1 - (-1)^n]$. For $m > 0$, the only solution of this equation is $n = \pm 1,\ m_c = 3/8$. Thus, meandering sets on at this value of m with a frequency exactly twice less than the basic rotation frequency, indicating period doubling of the spiral rotation.

3.5.5 Phenomenology of Complex Spiral Motion

Generally, meandering instability can be detected only numerically. For the FN model (3.4.1), (3.4.2) with the nonlinear function (3.1.6) and linear $g(u, v) = -v - \nu + \mu u$, the computations were carried out by Barkley (1992). By symmetry, the spiral wave solution in the infinite plane has three neutral eigenmodes: one with zero eigenvalue, corresponding to infinitesimal rotations of the entire spiral, and two with imaginary eigenvalues, $\pm\mathrm{i}\omega$, corresponding to translations in the plane. Meandering instability arises as a result of a Hopf bifurcation when another isolated pair of complex conjugate eigenvalues crosses the imaginary axis. The imaginary part of these eigenvalues, which we denote as $\widehat{\omega}$, determines the frequency of the meandering motion near the bifurcation point. The eigenmode responsible for meandering instability has the form of an attenuating wave propagating away from the zero curvature point at the spiral tip; the perturbation amplitude is maximal at the point of maximum curvature.

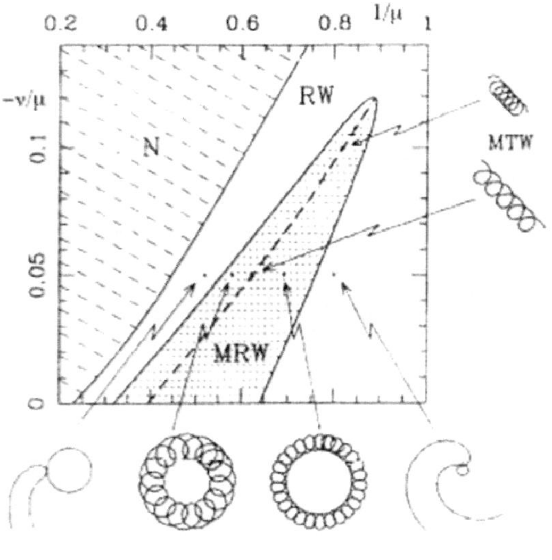

Fig. 3.26. Phase diagram for spiral-wave dynamics, showing the regions containing no spiral waves (N), stable rotating waves (RW), and modulated rotating waves (MRW). The latter two regions are separated by the locus of a supercritical Hopf bifurcation. The dashed locus of modulated traveling waves (MTW) emerges from the resonant point on the Hopf curve and separates MRW states with inward (left) and outward (right) petals of spiral tip trajectories. Schematic tip paths are shown for six chosen parametric points (Barkley, 1994; reproduced by permission, ©The American Physical Society)

Meandering instability in the system studied by Barkley (1992, 1994) is supercritical and saturates at finite amplitude, creating various meandering patterns determined by the ratio of the basic rotation frequency ω and the meandering frequency $\widehat{\omega}$, which changes along the Hopf bifurcation locus in the parametric plane. A typical phase diagram in the parametric plane μ, ν for $\gamma = 0.02$ is shown in Fig. 3.26. This diagram can be reproduced qualitatively by an ODE model containing five equations for the frequency, coordinates, and the velocity of the spiral tip (Barkley, 1994).

Following the meandering transition, steadily rotating waves (RW) with a circular tip trajectory give way to *modulated* rotating waves (MRW). The latter are *quasiperiodic* solutions, which are seen as periodic in the frame rotating with the basic rotation frequency ω. The tip trajectories form "flowers" with either inward (for $\widehat{\omega} < \omega$) or outward (for $\widehat{\omega} > \omega$) petals. This causes modulation of the entire spiral wave structure, creating a "superspiral" seen in Fig. 3.27. At the boundary between these two types of motion, emerging from the resonance point $\widehat{\omega} = \omega$, the flowers radii diverge, and the tip translates on the average in a certain direction; this intermediate structure can be characterized as a modulated *traveling* wave. More complex "hypermeandering" motion may arise at secondary bifurcations (Winfree, 1991).

Besides meandering instability involving an isolated eigenvalue, spiral instability can originate in a continuous band limiting the range of stable wavelengths of plane waves. Since the wave emanated by the spiral core approaches a plane wave as its curvature decreases with distance, this instability is likely to arise in outlying spiral arms. The Eckhaus instability arising at short wavelengths never saturates, but results in nucleation of a pair of counterrotating spirals. In the final state, the remaining spiral fragment is surrounded by a sea of vortex turbulence, as seen in Fig. 3.28d. Similar distinction between core and far-field instabilities in spiral solutions of the CGL equation will be discussed in Sects. 5.1.2 and 5.3.4.

A nontrivial interaction between both types of instabilities was suggested by Bär and Eiswirth (1993) and further studied by Bär and Or-Guil (1999) through simulations using a model with a somewhat more complicated function $g(u, v)$. Tip meandering may destabilize outlying spiral arms through a large variation of the period caused by the Doppler effect due to the tip motion. The wave trains are

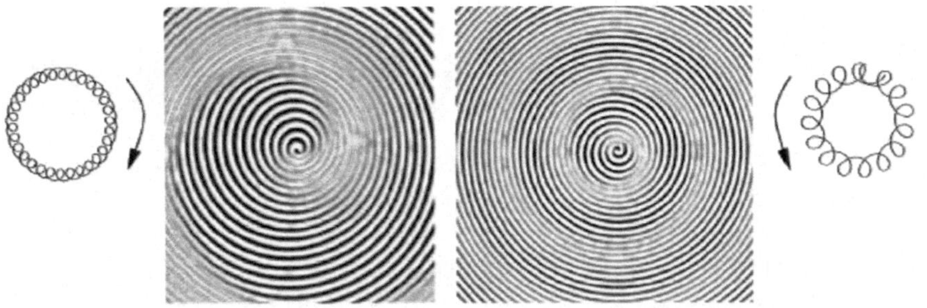

Fig. 3.27. Superspiral images and magnified spiral tip trajectories for inwardly (left) and outwardly (right) meandering spirals (Sandstede and Scheel, 2001; reproduced by permission, ©The American Physical Society)

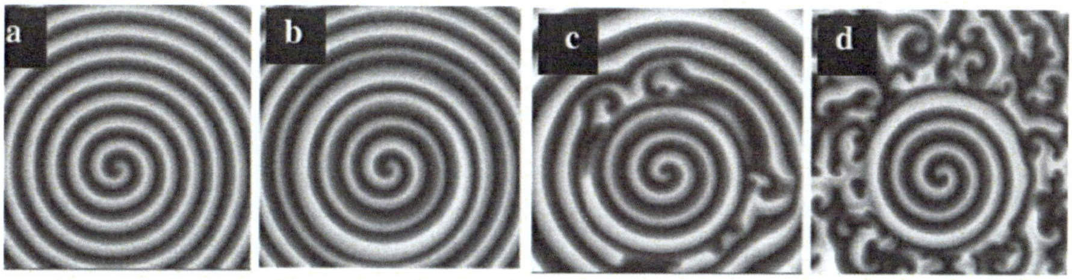

Fig. 3.28. Spiral breakup in an oscillatory system. (a) Initial symmetric spiral. (b) Development of modulational instability. (c) Spiral breakup and nucleation of vortex pairs away from the tip. (d) The final state with a remaining spiral fragment surrounded by a sea of vortex turbulence (Bär and Or-Guil, 1999; reproduced by permission, ©The American Physical Society)

compressed in the direction of drift and expanded in the opposite direction, as seen in Fig. 3.27 or Fig. 3.28b. This may take the wavelength out of the stable range, initiating Eckhaus instability, which breaks the wave front through nucleation of vortex pairs as seen in Fig. 3.28c. The vortex turbulence of Fig. 3.28d may be the final result of this process; otherwise, it may develop directly as a result of a far-field Eckhaus instability. Bär and Or-Guil (1999) argued that the Doppler mechanism is characteristic to excitable systems, while direct transition to spatiotemporal turbulence away from the core is prevalent in oscillatory systems.

It should be noted that natural spiral wave patterns observed in numerous experiments as well as in model computations starting from random initial conditions are spatially disordered even under conditions when isolated spirals are stable. Typical multispiral patterns consist of domains of uneven sizes filled by rotating spirals and separated by shocks where the waves emanating from different centers collide (more on this in Sect. 5.4.5). Interest to minor details of spiral motion has been sustained for decades by the supposed role of spiral excitations in heart fibrillation. The notion of the heart tissue as a particular example of an excitable medium was powerful enough to persuade funding agencies to support studies based on universal models and lacking any physiological detail.

3.5.6 Scroll Waves

The 3D extension of the spiral wave is a *rotating scroll wave* that has been observed in experiments with oscillatory chemical reactions (Winfree, 1987). The core filament of a scroll wave is a *line* vortex. A scroll wave with a straight-line core directed along the z-axis has identical spiral waves in each cross-section. Even then, the 3D extension can be nontrivial if the spiral phases are given a *phase twist*, i.e. are shifted along the z-axis. A *curved* core filament may be also closed into a ring or even form knots. A stable scroll structure evolves to decrease the filament's curvature (Keener, 1988). This kind of dynamics is similar to curvature-driven motion of 2D fronts (Sect. 3.3.1), but may be reversed when the filament is unstable, as we shall see below.

A steady scroll wave with a constant twist τ is a time-independent solution of applicable reaction-diffusion equations rewritten in rotating cylindrical coordinates

r, $\phi = \varphi - \omega t - \tau z$, z. In the advective limit, the equations stemming from (3.4.1) and (3.4.3) are written after rescaling length as

$$\partial_t u = \left[\omega \partial_\phi + (\partial_z - \tau \partial_\phi)^2 + \nabla_\perp^2\right] u + \gamma^{-1} f(u, v), \tag{3.5.46}$$

$$\partial_t v = \omega v_\phi + g(u, v), \tag{3.5.47}$$

where $\nabla_\perp^2$ is the Laplacian in the plane transverse to z. The stationary fields $u(r, \phi)$ and $v(r, \phi)$ are also z-independent and are obtained by solving a 2D problem. For $\gamma \ll 1$, (3.5.46) can be replaced, as for a line front in the plane, by the eikonal equation (3.4.4), where the curvature κ should be now interpreted as twice the mean curvature of the activator front, in accordance with the expression (2.2.7) for the Laplacian in the aligned frame. Otherwise, the full system (3.5.46), (3.5.47) should be solved numerically. In either case, the rotation frequency ω should be determined as a function of the imposed twist and other parameters by solving a nonlinear eigenvalue problem.

Detailed stability analysis of scroll waves was carried out by Henry and Hakim (2002). Since the linearized problem contains no explicit z-dependence, taking a Fourier transform eliminates the axial direction, so that the perturbation equations are purely two-dimensional but dependent on the axial wavenumber k. Thus, all 2D modes expand into k-dependent spectral bands. For zero twist, the eigenvalues are even functions of k. Owing to the translational symmetry along the z axis, the most dangerous 3D perturbation modes are the long-scale modes in the spectral bands corresponding to the neutrally stable 2D modes specified in the preceding subsection.

Instabilities may occur both in the meander band, which corresponds to z-dependent deformations of the 2D spiral meandering modes, and in the translation band, which corresponds to z-dependent translations of the 2D spiral in different transverse planes. Henry and Hakim (2002) found that 3D meandering instability is most dangerous in the outward-meandering region to the right of the dashed line in Fig. 3.26, so that a scroll wave destabilizes through z-dependent meandering under conditions when a 2D spiral wave is stable. On the contrary, in the inward-meandering region on the left, meandering starts in the 2D mode. The most dangerous instability

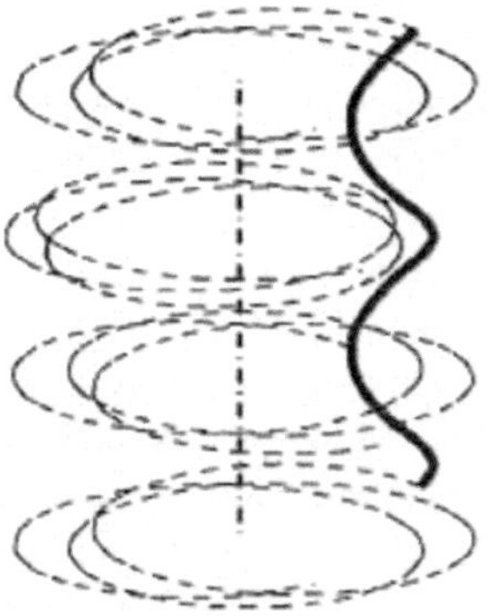

Fig. 3.29. View of a restabilized core filament in the weakly nonlinear regime. The instantaneous helical filament shown by the bold line. The dashed lines are the trajectories of the tip of the spiral in four regularly spaced horizontal planes. The dashed-dotted line is the axis around which the axis of the helical filament rotates (Henry and Hakim, 2002; reproduced by permission, ©The American Physical Society)

originates in this region in the translation band, causing spontaneous expansion and bending of the core filament. Instability in the meander band generally restabilizes as a distorted scroll wave with the twisted rotating core (Fig. 3.29).

Instability in the translation band causes spontaneous bending of the scroll axis. This instability may occur also in the absence of twist. It is sometimes interpreted as "negative surface tension," like that causing transverse instability of line fronts in the local approximation of Sect. 3.2.5. As a matter of fact, this effect is nonlocal and is caused by interaction of different wave fronts mediated by the long-range inhibitor field. In contrast to the meandering instability, the bending instability does not saturate on the nonlinear level, but gives rise to a scroll wave with a continuously extending core (Fig. 3.30a). This leads to a turbulent state visualized as a tangle of breaking wave fronts (Fig. 3.30b). Bending instability of the core filament is more readily analyzed using the CGL model (see Sect. 5.5).

For a nonzero twist, the inversion symmetry along the z axis is lost, and it only remains true that the eigenvalues $\lambda(\pm k)$ are complex conjugate. Also, a neutrally stable eigenmode should combine translation along the z axis with rotation, and therefore, as follows directly from (3.5.46), it corresponds to $k = \pm\tau$ rather than $k = 0$. These values remain extrema on the translation bands, and determine a finite marginal axial wavelength of a "sproing" instability which occurs in the translation

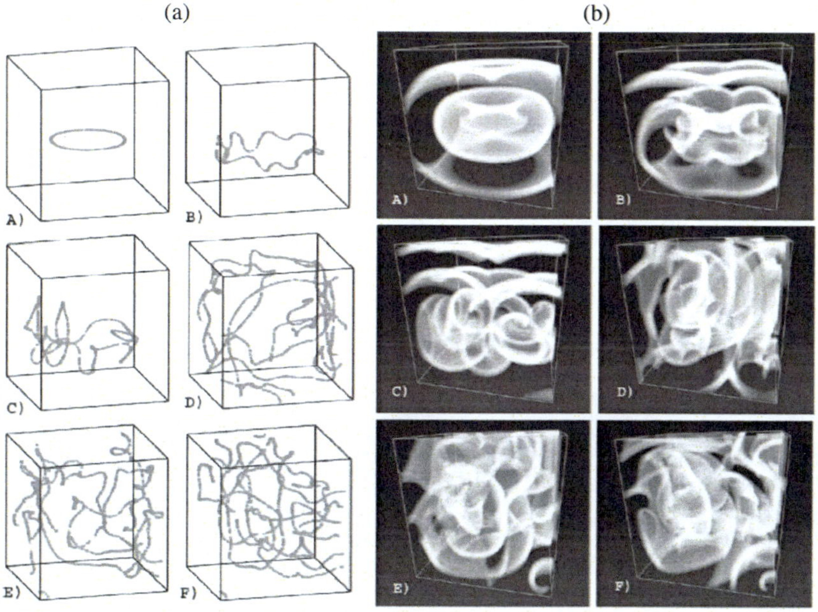

Fig. 3.30. Development of 3D turbulence due to extension of the filament's core and breakup of scroll waves. (a) Snapshots of the core filament, starting from a closed loop. (b) Respective snapshots of wave patterns showing a semitransparent visualization of the activator fronts (Alonso et al, 2004; reproduced by permission, ©The American Physical Society)

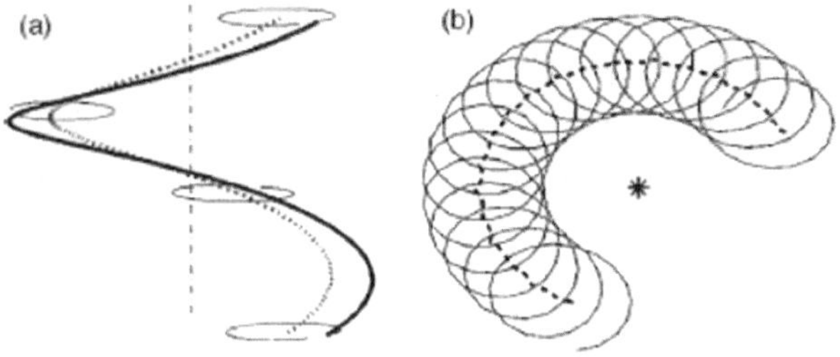

Fig. 3.31. A restabilized helical filament in the weakly nonlinear regime. (a) The bold solid line represents the instantaneous helical filament, the dotted line, the mean filament, and the thin solid lines are trajectories of the tip of the spiral in four equally spaced horizontal planes. (b) The thin solid line is the trajectory of the spiral tip in a horizontal plane and the bold dashed line is the trajectory of the mean filament in that plane. The mean filament rotates clockwise while the fast rotation of the spiral tip is counterclockwise. The position of the axis of the helix is marked by the asterisk (Henry and Hakim, 2002; reproduced by permission, ©The American Physical Society)

band when twist exceeds a certain threshold. Unlike bending instability in the absence of twist, nonlinear development of this instability results in a restabilized helical wave, as seen in Fig. 3.31. When several unstable modes are present, simulations show core filaments taking a more complex shape, which is found to travel as a steady nonlinear wave in the vertical direction (Henry and Hakim, 2002).

4 Irregular Stationary Patterns

4.1 Striped Patterns

4.1.1 Stationary Solutions of Amplitude Equations

The real Ginzburg–Landau equation (RGL) (1.1.7) is the simplest pattern-forming system derived in Sect. 1.3.3 as the amplitude equation describing long-scale modulations near a cusp bifurcation point. The wavenumber dependence of the leading eigenvalue of a linearized system follows in this case the curve 1 in Fig. 1.2. The complex amplitude is actually a shorthand for two real variables and can be presented in different forms. We used in (1.1.11) the symmetric form where the two variables are u and its complex conjugate $\overline{u}$, but it is usually advantageous to work with the polar form $u = \rho e^{i\theta}$ where the two variables are the modulus $\rho = |u|$ and the phase $\theta = \frac{i}{2}\ln(\overline{u}/u)$. Both RGL and the corresponding Lagrangian (1.1.11) are invariant to shifting the phase of the complex amplitude by a constant increment. This invariance reflects the symmetry of the regular pattern to translation along its wave vector.

Setting in (1.1.7) $u = \rho e^{i\theta}$, removing the phase factor, and separating the real and imaginary parts we obtain

$$\partial_t \rho = \nabla^2 \rho + (1 - |\nabla\theta|^2 - \rho^2)\rho, \tag{4.1.1}$$

$$\partial_t \theta = \nabla^2 \theta + \frac{2}{\rho}\nabla\rho \cdot \nabla\theta. \tag{4.1.2}$$

These equations can be also obtained directly as $\partial_t \rho = -\delta\mathcal{L}/\delta\rho,\ \partial_t\theta = -\rho^{-2}\delta\mathcal{L}/\delta\theta$ by varying the Lagrangian (1.1.11) recast as.

$$\mathcal{L} = \tfrac{1}{2}\left[|\nabla\rho|^2 + \rho^2|\nabla\theta|^2 + \tfrac{1}{2}(1-\rho^2)^2\right]. \tag{4.1.3}$$

Evident stationary solutions of (4.1.1), (4.1.2) are periodic structures

$$\theta = \mathbf{q}\cdot\mathbf{x}, \quad \rho = \rho_0 = \sqrt{1-q^2}, \quad q = |\mathbf{q}|. \tag{4.1.4}$$

L. Pismen, *Patterns and Interfaces in Dissipative Dynamics*, Springer Series in Synergetics,
https://doi.org/10.1007/978-3-031-29579-9_5

Keeping in mind the origin of the RGL equation as an amplitude equation, we recognize $\epsilon\mathbf{q}$ as a wave vector of a long-scale modulation of a basic stationary solution at the bifurcation point.

Other stationary solutions can be detected by rewriting the phase equation (4.1.2) in the form of a conservation law

$$\partial_t \theta = \rho^{-2} \nabla \cdot \mathbf{j}, \quad \mathbf{j} = \rho^2 \nabla \theta. \tag{4.1.5}$$

The phase is stationary when $\mathbf{j} = \text{const}$. Using this in (4.1.1) reduces the stationary equation of the real amplitude to a single diffusion equation

$$\nabla^2 \rho - V'(\rho) = 0, \quad V(\rho) = -\tfrac{1}{2}\left[\rho^2 - \tfrac{1}{2}\rho^4 + (j/\rho)^2\right], \tag{4.1.6}$$

where $j = |\mathbf{j}|$. The potential $V(\rho)$ is monotonic at $j \geq 2/3^{3/2}$, but at lower values of j it has a minimum and a maximum (Fig. 4.1a), which correspond to two stationary solutions. Replacing $\rho = \rho_0$, $\mathbf{j} = \rho_0 \mathbf{q}$, one can see that the minimum coincides exactly with the periodic solution (4.1.4); however, the minima disappear at $q^2 = 1/3$, $\rho_0^2 = 2/3$, i.e., $j = 2/3^{3/2}$. We shall see in the following subsection that this coincides with the Eckhaus instability limit of periodic solutions.

One-dimensional *inhomogeneous* solutions of (4.1.6) correspond to *quasiperiodic* structures. They can be obtained in the same way as in Sect. 2.1.3. The solutions exist at the values of V lying between the minimum and the maximum of $V(\rho)$ in Fig. 4.1. The lower bound of this interval corresponds to a homoclinic orbit approaching at $x \to \pm\infty$ a stable periodic structure (4.1.4). These solutions constitute a family of *dark solitons* parametrized by the asymptotic wavenumber q and can be presented as orbits in the phase plane $\rho, p = \rho'(x)$ surrounding the unstable homogeneous stationary state. They are given in the implicit form by (2.1.23), but can be expressed here explicitly:

$$\rho^2(x) = 2q^2 + (1 - 3q^2)\tanh^2\left[x\sqrt{(1-3q^2)/2}\right]. \tag{4.1.7}$$

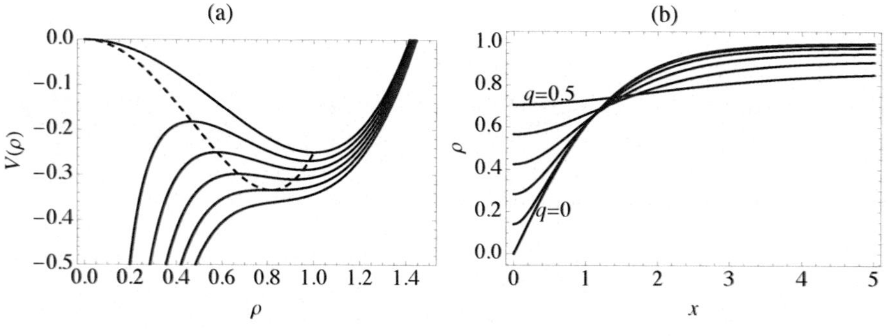

Fig. 4.1. (a) Plots of the potential (4.1.6) with $|j|^2$ varying from 0 (upper curve) to $5/27$ with the increment $1/27$. The dashed line is the locus of the extrema. (b) Dark solitons (4.1.7); q varies from 0 to 0.5 with the increment 0.1

The amplitude profile spreads out and becomes more shallow with increasing q (Fig. 4.1b). The simplest representative of this family is the solution with $j = 0$, $V = -1/2$, $q = 0$, which approaches at $x \to \pm\infty$ the basic state $\rho = 1$, $q = 0$. It is identical to the stationary front solution (2.1.21), which can be written now as $\rho(x) = \tanh |x/\sqrt{2}|$. The phase turns by π at the singular point $x = 0$ where the real amplitude vanishes and the phase is indefinite. Unlike vortices in 2D (Sect. 4.2), this singularity is *not* topological, and the solution, which is not protected by conservation of a topological charge, is not expected to be stable.

According to (4.1.3), (4.1.4), the energy density of periodic solutions $\mathcal{L} = \frac{1}{2}q^2\left(1 - \frac{1}{2}q^2\right)$ increases with q and exceeds the energy of the uniform state. Nevertheless, they correspond to *local* energy minima and are stable to infinitesimal perturbations as long as the wavenumber q is sufficiently small. One can ask how it may be possible that the solutions belonging to a continuous branch are all stable. The reason is that periodic solutions on the infinite line cannot be deformed one into the other by infinitesimal perturbations even when their wavenumbers are infinitesimally close, since the difference between them accumulates at long distances[1]. One can also observe that stable solutions should be separated by unstable solutions marking the boundaries of their attraction basins. Those are the quasiperiodic and homoclinic solutions found above. Though unstable, they play an important role in dynamics, as they correspond to saddle points of wavelength-changing transitions accompanied by disappearance of one or more periods (rolls, or stripes) constituting the pattern (Kramer and Zimmermann, 1985).

Periodic stationary solutions of the Newell–Whitehead–Segel (NWS) equation (1.3.34) applicable in the vicinity of a stationary symmetry breaking bifurcation (named after Rayleigh or Turing in different contexts) have the same form (4.1.4) with q defined by

$$q = \epsilon^{-1}\left(|\mathbf{k}_0 + \epsilon\mathbf{q}| - k_0\right) = q_x + \frac{q_y^2}{2k_0}\,. \tag{4.1.8}$$

Although the polar representation of the NWS equation is quite clumsy, this becomes clear by comparing (4.1.8) with the relation

$$\heartsuit \mathrm{e}^{\mathrm{i}\epsilon\mathbf{q}\cdot\mathbf{x}} = \mathrm{i}\epsilon\left(q_x + \frac{q_y^2}{2k_0}\right)\mathrm{e}^{\mathrm{i}\epsilon\mathbf{q}\cdot\mathbf{x}}. \tag{4.1.9}$$

This certifies that the NWS equation accounts precisely for the equivalence of all structures with identical wavenumbers, independently of the direction of the wave vector. Quasiperiodic solutions of RGL apply to NWS as well, provided the flux $\mathbf{j}$ is directed along the basic wave vector $\mathbf{k}_0$.

4.1.2 Stability of Stationary Solutions

The study of stability of a stationary striped pattern with a spatial period L, $u_0(x) = u_0(x + L)$, is a Floquet–Bloch problem, which is solved by presenting

[1] On a finite interval, the paradox disappears, since the wavenumbers become discrete and no longer form a continuous band

an infinitesimally small perturbation $\widetilde{u}(x,y,t) = u(x,y,t) - u_0(x)$ in the form

$$\widetilde{u}(x,t) = U e^{\lambda(\mathbf{K})t + i\mathbf{K}\cdot\mathbf{x}} w(x), \tag{4.1.10}$$

where $w(x)$ is L-periodic. The specific feature of this representation is that the basic periodicity is extracted from the perturbation. The component K_x of the perturbation wave vector $\mathbf{K}$ along the x-axis should lie within the limits $0 \leqslant K_x < 2\pi/L$ (the first Brillouin zone of the striped structure). As in the standard case of homogeneous states (Sect. 1.1.3), the stability condition is Re $\lambda(\mathbf{K}) \leqslant 0$ for all $\mathbf{K}$.

In a spatially homogeneous infinitely extended system, necessarily exists a neutrally stable (Goldstone) mode $\widetilde{u}(x) = u_0'(x)$. This is a consequence of translational symmetry, implying neutral stability to phase shifts, $u_0(x) \to u_0(x+\phi)$ with $\phi =$ const. Most commonly, the most dangerous perturbation modes have a vanishing wavenumber and can be viewed as long-scale distortions of neutrally stable translational modes. Loss of stability at $\mathbf{K} \neq 0$, which is far less probable, may cause formation of more complicated structures, e.g., period doubling when q_x is twice the wavenumber of the basic structure k. A distorted 2D structure would form when stability is lost to a perturbation mode with $\mathbf{q}$ noncollinear with the wave vector of the basic structure[2]. Most commonly, instabilities of this kind do not saturate but lead to a spatially chaotic (turbulent) pattern.

Both RGL and NWS equations being gradient systems, their periodic stationary solutions exhibit only the simplest kinds of instabilities. According to the general Floquet–Bloch method, stability of (4.1.4) is checked by presenting the perturbed solution in the form

$$u = e^{i\mathbf{q}\cdot\mathbf{x}} \left[\sqrt{1-q^2} + U(\mathbf{K}) \exp[\lambda(\mathbf{K})t + i\,\mathbf{K}\cdot\mathbf{x}\right], \tag{4.1.11}$$

where the basic periodicity, expressed by the factor $e^{i\mathbf{q}\cdot\mathbf{x}}$, is extracted from the perturbation. Linearizing the RGL equation (1.1.7) leads to the eigenvalue problem

$$\lambda \begin{pmatrix} U(\mathbf{K}) \\ \overline{U}(-\mathbf{K}) \end{pmatrix} = -\begin{pmatrix} (1-q^2) + 2\mathbf{K}\cdot\mathbf{q} + K^2 & 1-q^2 \\ 1-q^2 & (1-q^2) - 2\mathbf{K}\cdot\mathbf{q} + K^2 \end{pmatrix} \cdot \begin{pmatrix} U(\mathbf{K}) \\ \overline{U}(-\mathbf{K}) \end{pmatrix}. \tag{4.1.12}$$

The determinant of the above matrix is $K^2\{K^2 + 2[1 - q^2(1 + 2\cos^2\chi)]\}$, where $K = |\mathbf{K}|$ and χ is the angle between $\mathbf{K}$ and $\mathbf{q}$. As the determinant increases with K, the instability first sets on in the *long-scale* mode. The most dangerous perturbations are directed along the wave vector of the stationary pattern $\mathbf{q}$. The determinant is negative at small K, indicating long-scale instability, at $q > 1/\sqrt{3}$. This is the *Eckhaus* instability limit confining the range of stable wavenumbers.

Stability analysis of periodic solutions of the NWS equation is carried out in the same way. In spite of the complicated form of the $\heartsuit$ operator, the algebra is made easy by using (4.1.9), (4.1.8), and (4.1.11). Then (4.1.12) is rewritten as

[2] It should not be mixed up with a "sideband" instability leading to excitation of a primary mode with a differently directed wave vector, which can be treated in the framework of amplitude equations without long-scale spatial dependence as in Sect. 1.5.1

$$\lambda \begin{pmatrix} U(\mathbf{K}) \\ \overline{U}(-\mathbf{K}) \end{pmatrix} = \begin{pmatrix} 1-2(1-q^2)+K_+ & -(1-q^2) \\ -(1-q^2) & 1-2(1-q^2)+K_- \end{pmatrix} \cdot \begin{pmatrix} U(\mathbf{K}) \\ \overline{U}(-\mathbf{K}) \end{pmatrix}, \tag{4.1.13}$$

where $K_\pm = \epsilon^{-1}|\mathbf{k}_0 + \mathbf{q} \pm \mathbf{K}| - k_0$. It is sufficient to assume $\mathbf{q}$ to be parallel to the basic wave vector $\mathbf{k}_0$ and to consider separately perturbations with $\mathbf{K}$ parallel and normal to this direction. For longitudinal perturbations, $K_\pm = q \pm K$, and the limit of Eckhaus instability is the same as in the RGL case. For transverse perturbations, (4.1.8) yields $K_\pm = q + K^2/(2k_0)$, and the determinant of the matrix in (4.1.13) is approximated at $K \ll 1$ as $2(q/k_0)K^2(1-q^2)$. It is negative at $q < 0$, which sets the limit of the long-scale *zigzag* instability. This implies that *all* patterns with the wavelength longer than the basic wavelength $2\pi/k_0$ are unstable to long-scale transverse perturbations.

Since both Eckhaus and zigzag instability occur at $K \to 0$, their thresholds can be obtained in a simpler way by using the phase dynamics approach, as described in the following. Recalling the basic scaling of the amplitude equation, where a parametric deviation from the bifurcation point is scaled as ϵ^2 and the spatial scale of distortions as ϵ, we observe that the band of stable wavenumbers scales as the square root of supercriticality. The effective band width observed in experiments and simulations is more narrow, since selection of wavelengths close to the optimal one at the "band center" is facilitated by motion of defects in disordered patterns (see Sect. 4.2.1).

4.1.3 Universal Form of Phase Equations

The notion of a phase variable that appears in the polar form of the amplitude equations can be extended to large-amplitude striped patterns, which should be predominant away from the symmetry-breaking bifurcation point. A striped pattern is defined as an ordered spatially periodic structure with translational symmetry broken along a certain direction. Although the basic structure cannot be presented in a simple harmonic form, it can be characterized *locally* by the wave vector $\mathbf{k}$. The phase variable θ is introduced in such a way that $\mathbf{k} = \nabla\theta$. In an isotropic system, the direction of $\mathbf{k}$ can be arbitrary, and different orientations may prevail in different far removed locations of a spatially extended system. As a rule, stable patterns can exist within a certain range of wavenumbers. Thus, besides changes of orientation, the wavelength of the striped pattern can vary between different locations.

A *regular* pattern with $\mathbf{k} =$ const and $k = |\mathbf{k}|$ lying within the stable range should be stationary, but a pattern arising naturally in an extended system is likely to be *distorted*. Distorted structures are, generally, nonstationary. It is also useful therefore to define the frequency $\omega = -\partial_t\theta$. The wave vector and frequency are related by the integrability condition

$$\partial_t \mathbf{k} + \nabla\omega = 0. \tag{4.1.14}$$

A universal macroscopic description of weakly distorted and slowly evolving striped patterns, independent of either their short-scale structure or the nature of underlying equations, is provided by the *phase dynamics* formalism. The idea of phase dynamics has been first put forward in the context of convective patterns by Pomeau and Manneville (1979), and formulated in a more general way by Cross and Newell (1984). The distortion can be treated as *weak* when the wave vector $\mathbf{k}$ is slowly varying on an extended scale far exceeding the wavelength of the underlying structure: $\mathbf{k} = \mathbf{k}(\mathbf{X}, T)$, where $\mathbf{X} = \epsilon\mathbf{x}$ and $T = \epsilon t$ are extended spatial and temporal variables, respectively, and ϵ is a small parameter equal to the ratio of the prevailing wavelength to a characteristic macroscopic scale of pattern distortions.

Equations of phase dynamics that describe evolution of a weakly distorted pattern on an extended time scale can be constructed by setting in original equations $\partial_t = \epsilon^2\partial_T$, $\nabla = \epsilon\widehat{\nabla}$ (where $\widehat{\nabla}$ is the gradient with respect to the extended spatial variables), and expanding in powers of the scale ratio ϵ. Introducing also the extended phase variable $\Theta = \epsilon\theta$ and the rescaled frequency $\Omega = \omega/\epsilon$, one can write

$$\mathbf{k} = \nabla\theta = \widehat{\nabla}\Theta, \quad \Omega = -\epsilon^{-1}\partial_t\theta = -\partial_T\Theta, \quad \partial_T\mathbf{k} + \widehat{\nabla}\Omega = 0. \tag{4.1.15}$$

If all dependent variables in the original system are scalars, the general form of the phase equation is determined by scaling and symmetry considerations alone. The frequency is a scalar and may depend only on scalars formed with the help of $\mathbf{k}$ and $\widehat{\nabla}$. In the leading order, the only possible combinations are $\mathbf{n}\cdot\widehat{\nabla}k$ and $\widehat{\nabla}\cdot\mathbf{k}$, where $\mathbf{n} = \mathbf{k}/k$ is the unit vector normal to a line $\theta =$ const. Thus, the most general form of a rotationally invariant phase equation should be

$$-\Omega = D_1\mathbf{n}\cdot\widehat{\nabla}k + D_2\widehat{\nabla}\cdot\mathbf{k}, \tag{4.1.16}$$

where D_1, D_2 are phase diffusivities that both depend on a particular underlying problem and are, generally, functions of k. This equation can also be rewritten through the extended phase variable Θ:

$$\partial_T\Theta = D_1(\mathbf{n}\cdot\widehat{\nabla})^2\Theta + D_2\widehat{\nabla}^2\Theta. \tag{4.1.17}$$

Another equivalent form is

$$-\Omega = D_{\parallel}\mathbf{n}\cdot\widehat{\nabla}k + kD_{\perp}\widehat{\nabla}\cdot\mathbf{n}, \tag{4.1.18}$$

where $D_{\parallel} = D_1 + D_2$ and $D_{\perp} = D_2$ are, respectively, the longitudinal and transverse phase diffusivities.

The phase equation (4.1.17) is, in fact, strongly nonlinear due to the dependence of both the diffusivities and the direction of the unit vector $\mathbf{n}$ on the local phase gradient. It can be linearized, yielding an anisotropic diffusion equation, only when deviations from a prevailing wave vector $\mathbf{k} = \mathbf{k}_0$ are arbitrary small. If the X- and Y-axes are drawn along and across $\mathbf{k}_0$, respectively, (4.1.18) is reduced to

$$\partial_T\Theta = D_{\parallel}(k_0)\partial_X^2\Theta + D_{\perp}(k_0)\partial_Y^2\Theta, \tag{4.1.19}$$

where the longitudinal and transverse phase diffusivities $D_{\parallel}$, $D_{\perp}$ both depend on $k_0 = |\mathbf{k}_0|$. The pattern with the wavenumber k_0 is stable to long-scale perturbations when both phase diffusivities are positive. Vanishing $D_{\parallel}$ corresponds to the Eckhaus instability and vanishing $D_{\perp}$ to the zigzag instability.

4.1.4 Variational Formulation

Cross and Newell (1984) presented the phase equation as

$$C(k)\partial_T\Theta = \widehat{\nabla} \cdot [\mathbf{k}B(k)]. \tag{4.1.20}$$

This form makes transparent the *gradient structure* of the phase equation, which can be presented in the form

$$C(k)\partial_T\Theta = -\delta\mathcal{F}/\delta\Theta, \qquad \mathcal{F} = \int \mathcal{L}\,\mathrm{d}\mathbf{x}, \tag{4.1.21}$$

with the Lagrangian

$$\mathcal{L}(k) = \int_{k_0}^{k} kB(k)\mathrm{d}k = \tfrac{1}{2}\int_{k_0^2}^{k^2} B(k^2)\mathrm{d}(k^2). \tag{4.1.22}$$

In this approximation, energetic costs are totally determined by deviations from the optimal wavenumber k_0. Keeping in mind that variations of $\mathbf{k}$ are restricted by (4.1.15), the variation of (4.1.22) is computed as

$$\delta\mathcal{L} = kB(k)\delta k = B(k)\mathbf{k}\cdot\delta\mathbf{k} = -\widehat{\nabla} \cdot [\mathbf{k}B(k)]\delta\Theta. \tag{4.1.23}$$

Hence, the Euler–Lagrange equation derived from (4.1.22) reduces to (4.1.20) or (4.1.18), where the longitudinal and transverse phase diffusivities are related to the functions $B(k)$, $C(k)$ as

$$D_{\|} = \frac{1}{C(k)}\frac{\mathrm{d}(kB)}{\mathrm{d}k} = \frac{1}{C(k)}\frac{\mathrm{d}^2\mathcal{L}}{\mathrm{d}k^2}, \quad D_{\perp} = \frac{B(k)}{C(k)}. \tag{4.1.24}$$

Longitudinal stability of the pattern, dependent on the sign of $D_{\|}$, is determined by the sign of the *second derivative* of the energy density with respect to the wavenumber. This can be understood qualitatively in the following way. A deformation of the striped pattern changing locally its wavelength causes a local *compression* on the one side and local *dilation* on the other side. This causes an *increase* of the overall energy, indicating *stability*, when the dependence $\mathcal{L}(k)$ is concave, but a *decrease*, indicating *instability*, when the dependence is convex, as seen in Fig. 4.2.

For a gradient system, the phase equation can be deduced directly from the variational formula (4.1.21), assuming that the entire spatial structure of the pattern can be expressed through the phase variable. A simple example is given by the RGL equation (1.1.7). For the time being, we can forget that the amplitude u is a complex envelope of an underlying pattern, and identify $\mathbf{q}$ with $\mathbf{k} = \nabla\theta$. Stationary solutions of (1.1.7) are given by (4.1.4). Neglecting in the leading order $\nabla q \ll q^2$, the Lagrangian (4.1.3) is expressed through q only:

$$\mathcal{L}(q) = \tfrac{1}{2}q^2(1 - \tfrac{1}{2}q^2). \tag{4.1.25}$$

A Lagrangian can be expressed in a way independent of the choice of a representation of the order parameter or, in geometric language, of coordinates u^i in the order

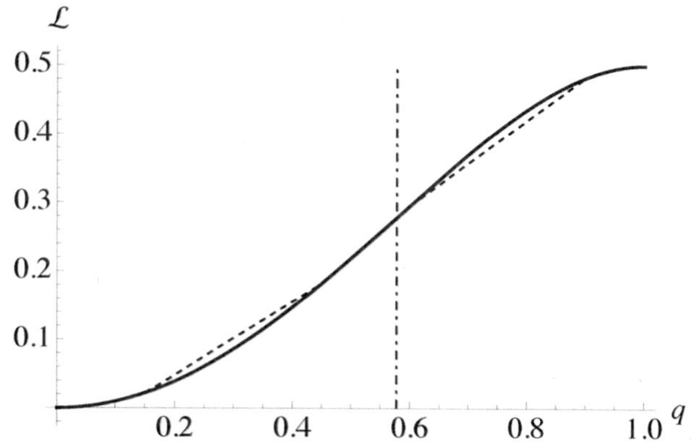

Fig. 4.2. Plot of the dependence of the energy density on the wavenumber $\mathcal{L}(k)$ (4.1.25) containing a concave and a convex segments to the left and to the right, respectively, of the dashed-dotted line corresponding to the Eckhaus instability limit. The dashed lines indicate gain or loss of energy in the respective regions

parameter space. This is done by introducing the metric g_{ij} of the order parameter space that defines an infinitesimal interval ds as

$$\mathrm{d}s^2 = g_{ij}\mathrm{d}u^i\mathrm{d}u^j \tag{4.1.26}$$

(summation over repeated upper and lower indices is presumed). The differential term possessing the required symmetry properties can be written as the scalar product $g_{ij}\nabla u^i \cdot \nabla u^j$. Then the RGL Lagrangian is written as

$$\mathcal{L}_{\mathrm{RGL}} = \tfrac{1}{2}\left[g_{ij}\nabla u^i \cdot \nabla u^j + \tfrac{1}{2}(1-|u|^2)^2\right]. \tag{4.1.27}$$

The NWS Lagrangian is written in the same way, with ∇ replaced by $\heartsuit$ and its complex conjugate. The gradient dynamic equations are derived in a usual way by varying the energy integral $\mathcal{F} = \int \mathcal{L}\,\mathrm{d}\mathbf{x}$, and contain the inverse g^{ij} of g_{ij}:

$$u^i_t = -g^{ij}\delta\mathcal{F}/\delta u^j. \tag{4.1.28}$$

The coefficient $C(q) = \rho^2 = 1 - q^2$ is found by comparing (4.1.21) to (4.1.28) and identifying it with the phase component of the metric tensor g_{ij} (i.e. inverse of g^{ij}). Comparing (4.1.25) to (4.1.22), where we replace now k by q, yields

$$B(q) = 2\,\mathrm{d}\mathcal{L}/\mathrm{d}(q^2) = 1 - q^2, \tag{4.1.29}$$

and, according to (4.1.24),

$$D_{\parallel} = \frac{1}{C(q)}\frac{\mathrm{d}(qB)}{\mathrm{d}q} = \frac{1-3q^2}{1-q^2}, \quad D_{\perp} = 1. \tag{4.1.30}$$

Thus, $q^2 = 1/3$ is the Eckhaus instability limit, in accordance with the results of Sect. 4.1.2.

The phase equation following from the NWS Lagrangian (1.3.35) *does not* reduce to the form (4.1.20) due to the dependence on transverse derivatives of $\mathbf{k}$ generating fourth-order derivatives in the Euler–Lagrange equation. The correct leading-order expression for the NWS Lagrangian, obtained by expressing the complex amplitude in (1.3.35) in the polar form and making use of (4.1.8), (4.1.4), is

$$\mathcal{L} = \tfrac{1}{2}q^2(1 - \tfrac{1}{2}q^2) + \tfrac{1}{8}(1 - q^2)(\partial q_y/\partial y)^2, \tag{4.1.31}$$

where q is defined by (4.1.8). The phase equation can be obtained in two different forms, depending on whether it is expressed through the phase θ of the amplitude $u = \rho e^{i\theta}$ or through the total phase of the underlying pattern, which we denote now as ϑ. The relation between the variations of q and θ is obtained in both cases from the two alternative forms of (4.1.8):

$$q = \epsilon^{-1}\left(|\nabla\vartheta| - k_0\right) = \theta_x + \frac{\theta_y^2}{2k_0}. \tag{4.1.32}$$

The last expression leads to a very complicated nonlinear fourth-order equation[3]. Its awkward structure is, in essence, a consequence of the noncovariance of the NWS equation.

Using the total phase yields a more transparent expression (see also Sect. 4.1.6). As follows from (4.1.8), (4.1.32), the variations of q and ϑ are related as

$$\delta q = |\mathbf{k}_0 + \epsilon\mathbf{q}|^{-1}(\mathbf{k}_0 + \epsilon\mathbf{q}) \cdot \delta\mathbf{q} = \mathbf{n} \cdot \delta(\nabla\vartheta), \tag{4.1.33}$$

where $\mathbf{n}$ is the normal vector directed along the corrected wave vector $\mathbf{k}_0 + \epsilon\mathbf{q}$. If the last term in (4.1.31) is omitted, the variation $\delta\mathcal{L}$ retains the form (4.1.23):

$$\delta\mathcal{L} = (\mathrm{d}\mathcal{L}/\mathrm{d}q)\,\delta q = -\widehat{\nabla} \cdot [\mathbf{n}qB(q)]\,\delta\vartheta. \tag{4.1.34}$$

Hence, the Euler–Lagrange equation retains the form (4.1.20) with k replaced by q and B, C dependent on q only:

$$C(q)\vartheta_T = \widehat{\nabla} \cdot [\mathbf{n}qB(q)]\,. \tag{4.1.35}$$

The expression for $D_\parallel$ in (4.1.30) defining the Eckhaus instability limit remains without change. The expression for the transverse diffusivity $D_\perp$ remains without change as well, but the factor q, which may be negative, replaces nonnegative k in (4.1.18). This signals *zigzag instability* at $q < 0$, which destabilizes patterns with the wavelength larger than optimal. Bringing back the last term in (4.1.31) generates higher order terms stabilizing the zigzag instability at shorter wavelengths and finite amplitudes, but does not affect the instability threshold.

One should keep in mind that when an amplitude equation is taken as the "original" system, one should assume that the phase changes on the scale exceeding the spatial scale of this equation (already extended compared to the scale of the basic structure). Only under this condition, the real amplitudes are slaved to the phase, and the Lagrangian can be presented as above in the leading order as a function of q only.

---

[3] A simplified form of this phase equation valid near the band center will appear in Sect. 4.2.3

4.1.5 Phase Dynamics of a Nongradient System

A phase equation with the gradient structure (4.1.21) may be also valid when the original system is nongradient; then the gradient structure would be destroyed in higher order approximations. Consider a general reaction–diffusion system (1.2.5) with a constant diffusivity matrix $\mathbf{D}$:

$$\partial_t \mathbf{u} = \mathbf{D}\nabla^2 \mathbf{u} + \mathbf{f}(\mathbf{u}). \tag{4.1.36}$$

Suppose that stationary spatially periodic solutions of (4.1.36) can be written in the form of a 2π-periodic function $\mathbf{u} = \mathbf{u}_0(\theta; k)$ parametrized by the phase variable θ instead of a spatial coordinate and dependent on the wavenumber k. The function $\mathbf{u}_0(\theta; k)$ verifies the stationary equation

$$\mathbf{D}(k^2)\mathbf{u}^0_{\theta\theta} + \mathbf{f}(\mathbf{u}_0) = 0. \tag{4.1.37}$$

This function must be even, $\mathbf{u}_0(\theta; k) = \mathbf{u}_0(-\theta; k)$, due to the symmetry to inversion of the wave vector $\mathbf{k}$ in an isotropic system.

A weakly distorted and slowly evolving pattern is sought for as an expansion

$$\mathbf{u}(\mathbf{X}, T) = \mathbf{u}^0(\theta(\mathbf{X}, T); k(\mathbf{X}, T)) + \epsilon \mathbf{u}^1(\theta, \mathbf{X}, T) + \cdots , \tag{4.1.38}$$

where the small parameter ϵ is defined, as in Sect. 4.1.3, as the ratio of the prevailing wavelength to a characteristic scale of long-scale distortions. The expansion uses the relations

$$\begin{aligned} \partial_t \mathbf{u}^0 &= \epsilon \mathbf{u}^0_\theta \Omega + O(\epsilon^2), \qquad \nabla \mathbf{u}^0 = \mathbf{k}\mathbf{u}^0_\theta + \epsilon \mathbf{u}^0_k \widehat{\nabla} k, \\ \nabla^2 \mathbf{u}^0 &= \mathbf{k}\mathbf{u}^0_\theta + \epsilon \left(\mathbf{u}^0_\theta \widehat{\nabla} \cdot \mathbf{k} + 2\mathbf{u}^0_{k\theta} \mathbf{k} \cdot \widehat{\nabla} k \right) + O(\epsilon^2). \end{aligned} \tag{4.1.39}$$

The first-order equation is a linear inhomogeneous equation of a general form

$$\mathbf{L}\mathbf{u}_1 + \boldsymbol{\Psi}(\mathbf{u}_0) = 0 \tag{4.1.40}$$

containing the linear operator

$$\mathbf{L} = \mathbf{D}k^2 \partial_\theta^2 + \mathbf{f_u}, \tag{4.1.41}$$

where $\mathbf{f_u}$ is the Jacobi matrix of the function $\mathbf{f}(\mathbf{u}_0)$ evaluated at $\mathbf{u} = \mathbf{u}^0(\theta)$, and the inhomogeneity is

$$\boldsymbol{\Psi} = \mathbf{D} \left(\mathbf{u}^0_\theta \widehat{\nabla} \cdot \mathbf{k} + 2\mathbf{u}^0_{k\theta} \mathbf{k} \cdot \widehat{\nabla} k \right) - \mathbf{u}^0_\theta \Omega. \tag{4.1.42}$$

The operator $\mathbf{L}$ has an eigenfunction with zero eigenvalue (Goldstone mode) $\mathbf{u}^0_\theta$ which corresponds to the translational symmetry of the periodic solution. Generally, this operator is not self-adjoint, and one has to compute the eigenfunction $\varphi^\dagger(\theta)$ of the adjoint operator

$$\mathbf{L}^\dagger = \mathbf{D}^\dagger k^2 \partial_\theta^2 + \mathbf{f_u}^\dagger, \tag{4.1.43}$$

containing transposed matrices $\mathbf{D}^\dagger$ and $\mathbf{f_u}^\dagger$.The adjoint eigenfunction should also be even in θ. The solvability condition of (4.1.40) is

$$\int_0^{2\pi} \varphi^\dagger(\theta)\boldsymbol{\Psi}\mathrm{d}\theta = 0. \tag{4.1.44}$$

Using here (4.1.42) yields the phase equation (4.1.16) with the phase diffisivities

$$D_1 = 2k \int_0^{2\pi} \varphi^\dagger(\theta)\mathbf{D}\mathbf{u}_{k\theta}^0\mathrm{d}\theta \bigg/ \int_0^{2\pi} \varphi^\dagger(\theta)\mathbf{u}_\theta^0\mathrm{d}\theta\,,$$
$$D_2 = \int_0^{2\pi} \varphi^\dagger(\theta)\mathbf{D}\mathbf{u}_\theta^0\mathrm{d}\theta \bigg/ \int_0^{2\pi} \varphi^\dagger(\theta)\mathbf{u}_\theta^0\mathrm{d}\theta\,. \tag{4.1.45}$$

4.1.6 Covariant Phase-Amplitude Equation

The reader may have noticed that the phase dynamics approach, when applied to a striped pattern in an isotropic system, yields symmetric covariant expressions, whereas the amplitude equation at a symmetry-breaking bifurcation point (1.3.35) is noncovariant and contains a higher order derivative. One can ask what is the root of this contradiction and whether it might be resolved. The expansion leading to the phase equation is essentially simpler than that leading to the amplitude equation, even though computation of the integrals (4.1.45) may be technically difficult. Indeed, phase dynamics involves only one dependent variable – phase – while in the amplitude expansion the real amplitude is an additional dynamical variable not slaved to the phase.

This, however, has nothing to do with the loss of covariance, which is caused, essentially, by extraction of the basic structure, making it necessary to introduce the operator $\heartsuit$ dependent on local orientation. In the following, we derive an alternative amplitude equation using a procedure similar to derivation of the phase dynamics equation in the preceding subsections but applicable near a bifurcation point where the dependence on the phase can be expressed in a simple harmonic form but the real amplitude is not slaved to the phase. The covariance will be retained at the price of keeping as a dynamic variable the *total* phase of the emerging pattern, rather than an *envelope* phase as in the NWS equation.

Since the form of the amplitude equation should be universal, it can be obtained starting from any underlying system. We start therefore with a simple example taking as the original system the SH equation (1.1.3) with $\mu = \epsilon^2$, $q = 1$:

$$\partial_t u = -(1+\nabla^2)^2 u + u(\epsilon^2 - u^2). \tag{4.1.46}$$

A striped pattern with the wavenumber $k_0 = 1$ bifurcates at $\epsilon = 0$. At small ϵ, we are looking for an expansion in the form

$$u = \epsilon\left(\rho e^{\mathrm{i}\vartheta} + \text{c.c.}\right) + \epsilon^2 u_2 + \cdots\,. \tag{4.1.47}$$

We assume that the phase obeys the relations

$$\nabla\vartheta = \mathbf{k}, \quad |\mathbf{k}| = k_0 + \tfrac{1}{2}\epsilon q. \tag{4.1.48}$$

Both the real amplitude ρ and the wavenumber mismatch q are allowed to change on an extended scale, so that $\nabla = O(\epsilon)$ *except* when applied to the phase itself.

The expansion is made easy by using the relation

$$(1+\nabla^2)\rho e^{i\vartheta} = \epsilon\left(-q + 2i\mathbf{k}\cdot\nabla + i\nabla\cdot\mathbf{k}\right)\rho e^{i\vartheta} + O(\epsilon^2). \tag{4.1.49}$$

There are no terms of $O(\epsilon)$ in the expansion of the right-hand side of (4.1.46); hence, ∂_t scales as ϵ^2. The second-order equation can be written, after compounding the operator (4.1.49) and separating the real and imaginary part, in the form

$$\begin{aligned}\partial_t\rho = &\left[1 - q^2 - \rho^2 + (\nabla\cdot\mathbf{k})^2 + 2\mathbf{k}\cdot\nabla(\nabla\cdot\mathbf{k})\right]\rho \\ &+ 4(\nabla\cdot\mathbf{k})\,\mathbf{k}\cdot\nabla\rho + 2(\mathbf{k}\cdot\nabla)^2\rho,\end{aligned} \tag{4.1.50}$$

$$\partial_t\vartheta = 2\rho^{-2}\nabla\cdot\left(\mathbf{k}q\rho^2\right). \tag{4.1.51}$$

The phase equation (4.1.35) immediately follows from (4.1.51) when $\rho = 1 - q^2$ is slaved to the phase. Unfortunately, (4.1.50) has a very complicated form ill-suited for numerical simulations, and the short-scale phase variation precludes using a sparse grid fit to a slow change of the wave vector and real amplitude. The phase can be excluded by applying the gradient operator to (4.1.51):

$$\partial_t\mathbf{k} = 2\nabla[\rho^{-2}\nabla\cdot\left(\mathbf{k}q\rho^2\right)]. \tag{4.1.52}$$

This vector equation preserves the compatibility condition $\nabla\times\mathbf{k} = 0$ and can be solved on a long-scale grid together with (4.1.50).

Having three equations replacing a single scalar SH equation that served as the basis for the derivation might be too high a price for universality and covariance. The SH equation remains the most handy tool for qualitative numerical modeling of striped patterns. An alternative equation, derived in an ad hoc way, was suggested by Qian and Mazenko (2004). This model equation, called a "nonlinear phase model" by the authors, can be written in the form of a vector Cahn–Hilliard equation for the gradient of the local wavenumber $\mathbf{Q} = \nabla k$:

$$\mathbf{Q}_t = -\nabla\left\{\nabla\cdot\left[\nabla^2\mathbf{Q} + (1 - |\mathbf{Q}|^2)\right]\right\}. \tag{4.1.53}$$

This equation has a gradient structure

$$\mathbf{Q} = -\nabla\left(\nabla\cdot\frac{\delta\mathcal{F}}{\delta\mathbf{Q}}\right), \quad \mathcal{F} = \tfrac{1}{2}\int\left[(\nabla\cdot\mathbf{Q})^2 + \tfrac{1}{2}(1 - |\mathbf{Q}|^2)^2\right]d\mathbf{x}. \tag{4.1.54}$$

The algebraic term encourages relaxation to the "optimal" wavenumber $|\mathbf{Q}| = 1$, so that this model, like SH, generates in the long run an ordered pattern with the optimal wavelength, unless restricted by boundary conditions, e.g., the condition requiring the wave vector to be parallel to the walls. Numerical simulations using the model (4.1.54) produce, however, at intermediate stages distorted striped patterns rather different from those generated by the SH model and less similar to natural patterns observed in experiment.

4.2 Defects in Striped Patterns

4.2.1 Natural Patterns

Natural patterns seen both in experiment and simulations have various *defects*: dislocations, disclinations, and domain walls. Proliferation of defects in striped patterns is, in essence, a consequence of rotational symmetry of the system. Different orientations of stripes may be chosen at different locations, either randomly or under influence of boundary conditions or local inhomogeneities. The discrepancies of local orientations can be reconciled through the formation of disclinations and domain walls, while dislocations reconcile discrepancies of local wavelengths. Different kinds of defects are seen in Fig. 4.3 (Bowman and Newell, 1998).

Both dislocations and disclinations are *codimension two* topological defects localized at a point in the plane and characterized by circulation around a contour surrounding this point. A *dislocation* is a topological singularity of the phase field. The total change of phase around any contour surrounding the singularity must be a multiple of 2π:

$$\oint \mathrm{d}\theta = 2\pi N. \tag{4.2.1}$$

The integer N is called a *topological charge*. The phase cannot be defined globally as a continuous univalued function when dislocations are present. This does not cause any nonphysical discontinuities, inasmuch as physically relevant quantities – wave vector and frequency – are well defined everywhere except the defect locations.

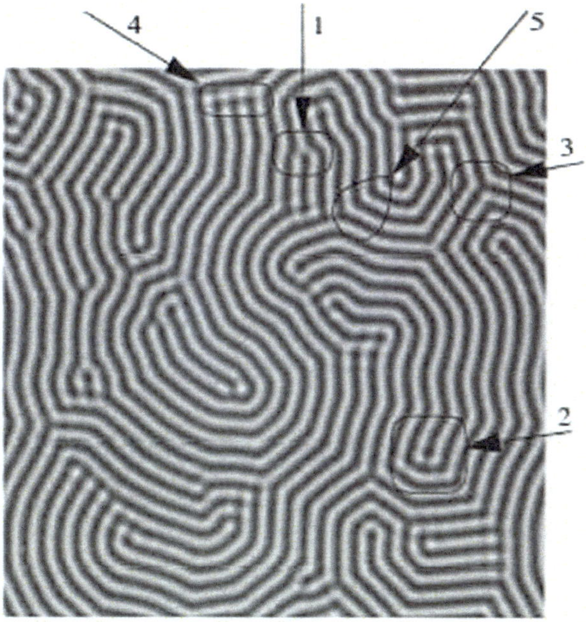

Fig. 4.3. Various forms of pattern defects. *1* – dislocation, *2* – concave disclination, *3* – convex disclination, *4* – amplitude domain wall, *5* – phase domain wall (Bowman and Newell, 1998; reproduced by permission. ©The American Physical Society)

A *disclination* involves rotation of the orientation of the underlying pattern. If one follows the direction of stripes continuously around a disclination, it rotates through an angle of $\pm\pi$ (more on this in Sect. 4.2.5). The rotation is positive for a *convex* and negative for a *concave* disclination. Topological charges of disclinations as well as of dislocations add up defining total circulation of either direction or phase along a contour surrounding several defects. Topological charges are *conserved* when the pattern evolves in time.

Codimension one defects (lines in 2D) are *domain walls* which separate regions with different prevailing orientations of the wave vector. One can distinguish between *phase walls*, which are lines where the phase is continuous, but the pattern's wave vector changes abruptly, and *amplitude walls* where the phase is discontinuous and the amplitude vanishes. Point and line defects are related to each other in the overall texture of the pattern, as phase walls meet and terminate at disclinations, while amplitude walls disintegrate at a close look into dislocation chains. The phase theory is applicable in each domain, but full equations have to be reintroduced at domain walls, as well as at disclinations and dislocation cores. Manneville and Pomeau (1983) and later Passot and Newell (1994) and Newell et al (1996) described disclinations and domain walls as *singular* solutions of equations of phase dynamics. This gives a qualitatively correct structure of the far field of singularities, but is still inadequate for a *dynamic* description that would hinge upon the structure of defect cores.

Both point and line defects, though being elements of *disorder*, contribute to *ordering* of the pattern, as they facilitate selection of an optimal wavelength. As we have seen in Sect. 4.1.2, an ordered pattern is stable within a range of wavenumber deviations scaling as the square root of parametric deviation from the bifurcation point. When dislocations are present, their motion will eliminate patches with the wavelength deviating from the optimal one (see Sect. 4.3.1). Likewise, if a domain wall separates patches with different wavelength, it will move in the direction reducing the overall energy of the pattern, thereby helping to eliminate nonoptimal domains.

4.2.2 Phase Field of a Dislocation

Dislocations occur in patterns of almost parallel stripes and therefore are most amenable to analysis. Consider a pattern with a constant prevailing wavenumber k_0 described by the phase diffusion equation (4.1.19). This anisotropic equation can be transformed, by rescaling the coordinates and reverting from the extended phase variable Θ to the actual phase θ, into an *isotropic* diffusion equation

$$\partial_T \theta = \widehat{\nabla}^2 \theta. \tag{4.2.2}$$

This equation is verified by a stationary symmetric vortex solution centered at the origin, written in the polar coordinates as $\theta = N\phi$, where N should be integer to avoid multivaluedness of physical fields dependent on the phase modulo 2π. Since (4.2.2) is a linear equation, vortex solutions centered at any point can be additively combined as $\theta(\mathbf{x}) = \sum \theta_i$, where θ_i obeys the circulation condition (4.2.1) on a contour surrounding the ith defect. A naive solution corresponding to an assembly of static defects is

$$\theta(\mathbf{x}) = \sum \phi_i(\mathbf{x} - \mathbf{x}_i), \tag{4.2.3}$$

where ϕ_i is the angle counted around the position of a particular defect located at $\mathbf{x} = \mathbf{x}_i(t)$. All phases can be counted from arbitrary reference angles, as the phase is defined only up to a constant. This solution can be combined as well with any constant-gradient solution, say, $\theta = qX$.

A dislocation in the underlying striped pattern corresponds to a *vortex* in the envelope field with the complex amplitude u. Adding the vortex solution with the unit charge to the underlying pattern $\theta = k_0 x = \epsilon^{-1} k_0 X$ yields the phase field of a dislocation, such as seen in Fig. 4.4a. Dislocations with multiple charges (Fig. 4.4b) are also possible, but are apt to be unstable, and tend to break up into singly charged dislocations, such as seen in Fig. 4.4c, while keeping the total charge conserved. Pairs of dislocations of opposite charge, as in Figs. 4.4d, e, may annihilate restoring a nonsingular phase field. The precise location of a dislocation in a natural pattern is revealed when the short-scale structure is removed by a spectral shift. The dislocations are located at the intersections of the level lines Re $u = 0$ and Im $u = 0$, as shown in Fig. 4.4f, which corresponds to the pair of dislocations in Fig. 4.4d. At these points, the real amplitude of the envelope field vanishes, and its phase becomes indefinite.

The superposition (4.2.3) appears to be a stationary solution that, indeed, satisfies both (4.2.2) and the circulation conditions (4.2.1) for each defect – and this is for an arbitrary distribution of defects, subject only to the restriction that all mutual distances should be large. The trouble lies, of course, in the phase singularity at $\mathbf{x} = \mathbf{x}_i$. Near the defect core, the long-scale approximation breaks down, and the full system (4.1.1), (4.1.2) should be solved. This system, being nonlinear, does not obey

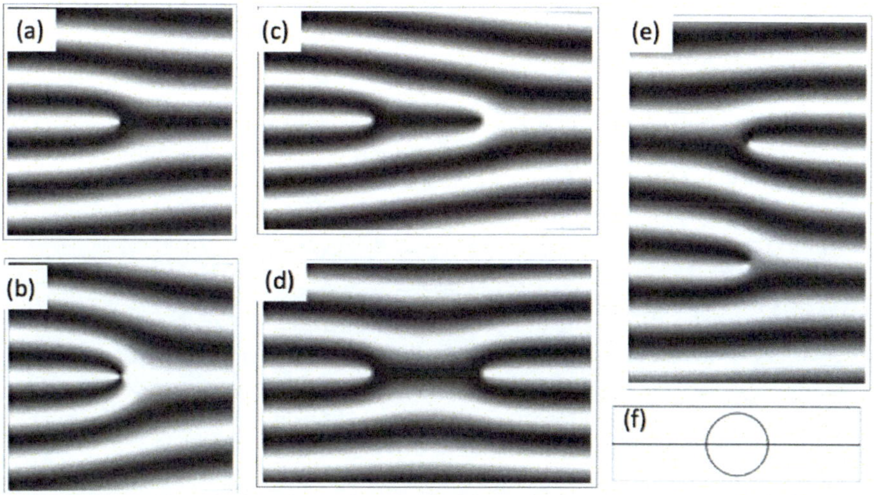

Fig. 4.4. Dislocations with a single (a) and double (b) charge and pairs of like (c) and unlike (d, e) dislocations. Shading shows levels of the real part of the complex order parameter (amplitude). (f) Loci of vanishing real and imaginary part of the complex order parameter for the pair of dislocations in (d) with the basic stripe structure removed

the superposition principle implied in (4.2.3). The full solution can be approximated as the product of solutions corresponding to isolated defects at $\mathbf{x} = \mathbf{x}_i$:

$$u(\mathbf{x}) = \prod u_i(\mathbf{x} - \mathbf{x}_i) = \prod \rho_0(|\mathbf{x} - \mathbf{x}_i|) \exp\left[\mathrm{i} \sum \phi_i(\mathbf{x} - \mathbf{x}_i)\right] . \tag{4.2.4}$$

This is not an exact solution of the amplitude equation, since the circular symmetry of an isolated defect is broken by phase and amplitude gradients induced by other defects. Phase gradients turn out to be the main source of an effective force setting defects into motion. We shall return to dynamics of dislocations in Sect. 4.3.

4.2.3 Dislocations in NWS Equation

In a striped pattern with a nearly optimal wavelength described by the NWS equation (1.3.35), the simple symmetric solution $\theta = N\phi$ does not hold, and obtaining its equivalent (the phase field of a single static dislocation) requires some ingenuity. If a change of the real amplitude is neglected, while the characteristic scale of phase distortions is assumed to be large even on the extended NWS scale, (1.3.35) reduces to the phase equation dependent on long-scaled variables $X = \epsilon x$, $Y = \sqrt{\epsilon} y$, $T = \epsilon t$. Unlike (4.2.2), this equation, obtained by applying the operator $\partial_T - \heartsuit^2$ to $u = \rho \exp[\mathrm{i}\theta(X, Y, T)]$ and collecting the imaginary part[4], is *nonlinear*:

$$\partial_T \theta = \partial_X^2 \theta - \tfrac{1}{4} \partial_X^4 \theta + \partial_X \theta_X \partial_Y^2 \theta + 2\partial_{XY}^2 \theta \partial_Y \theta_Y + \tfrac{3}{2} (\partial_Y \theta)^2 \partial_Y^2 \theta. \tag{4.2.5}$$

Meiron and Newell (1985) showed that a stationary singular solution of (4.2.5) corresponding to a quiescent defect placed at the origin has a self-similar form, i.e., can be presented as a function of a single coordinate $\eta = Y/\sqrt{2|X|}$ that satisfies the equation

$$\theta'''' = 4\eta^2 \theta'' + 12\eta\theta' - 12\eta\theta'\theta'' - 8\theta'^2 + 6\theta'^2\theta''. \tag{4.2.6}$$

The boundary conditions are stipulated by the requirements that the derivatives of $\theta(\eta)$ vanish at $\eta \to \pm\infty$, and the total change of θ measured around a contour enclosing the defect be equal to $2\pi N$, where N is the topological charge. The solution with the charge $+1$ is obtained by setting $\theta(-X, Y) = -\theta(X, Y)$ and imposing the boundary conditions (at $X > 0$)

$$\theta(-\infty) = 0, \quad \theta(\infty) = \pi. \tag{4.2.7}$$

Nepomnyashchy and Pismen (1991) showed that the self-similar solution verifying (4.2.6) is a representative of a wider class of solutions of the phase equation (4.2.5) that also satisfy a linearizable equation of the Burgers type. As a corollary, the self-similar solution can be presented in a remarkably simple analytical form. One can check by differentiation that all solutions of the equation

$$\pm \partial_Y^2 \theta = 2\partial_X \theta + (\partial_Y \theta_Y)^2 \tag{4.2.8}$$

[4] The real part of this expression contributes to the dependence of ρ on the derivatives or θ in the far field where the real amplitude is slaved to the phase, $\rho^2 = 1 - \mathrm{Re}\,[\heartsuit^2 \exp[\mathrm{i}\theta(X, Y, T)]$

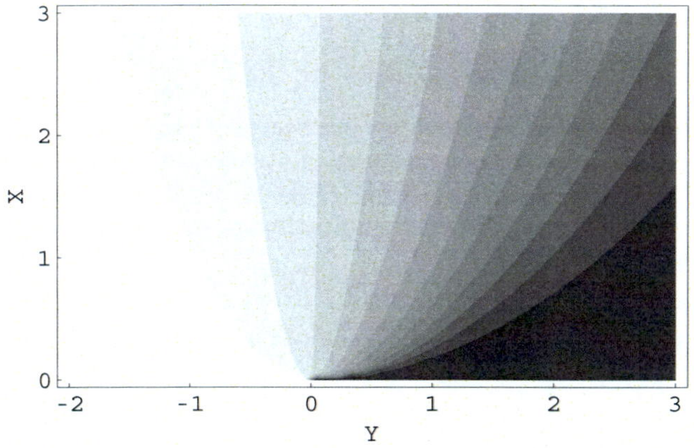

Fig. 4.5. Phase field of a dislocation in the NWS equation shown in gray scale

also verify (4.2.5). In each of the two domains separated by the axis $X = 0$, (4.2.8) is recognized as the integral form of the Burgers equation with the coordinate $\pm X$ playing the role of time. The expression in parentheses on the right-hand side of (4.2.8) represents the wavenumber shift relative to the optimal value $k = 1$.

Equation (4.2.8) can be linearized in a standard way through the Hopf–Cole transformation $\theta = \mp \ln f$, yielding

$$\partial_Y^2 f = \pm 2\partial_X f. \tag{4.2.9}$$

The upper and lower signs refer, respectively, to $X > 0$ and $X < 0$. Equation (4.2.9) can be integrated in the positive and negative half-planes with "initial" conditions fixed at the axis $X = 0$. The self-similar solution to (4.2.9) $f(\eta)$ satisfies

$$f'' + 2\eta f' = 0 \tag{4.2.10}$$

with the boundary conditions [(cf. (4.2.7)]

$$f(-\infty) = 1, \quad f(\infty) = \mathrm{e}^{-\pi}. \tag{4.2.11}$$

After transforming back to the phase variable we obtain

$$\theta = -\mathrm{sign}(x) \ln \left[\frac{1 + \mathrm{e}^{-\pi}}{2} - \frac{1 - \mathrm{e}^{-\pi}}{2} \mathrm{erf}(\eta) \right]. \tag{4.2.12}$$

The asymptotics of $\theta'(\eta)$ at $\eta \to \pm\infty$ is $\theta' \asymp C_\pm \mathrm{e}^{-z^2}$, where

$$C_+ = \frac{\mathrm{e}^\pi - 1}{\sqrt{\pi}} \approx 12.5, \quad C_- = \frac{1 - \mathrm{e}^{-\pi}}{\sqrt{\pi}} \approx 0.54.$$

The difference between both asymptotic values reflects the strong asymmetry of the solution (see Fig. 4.5).

Of course, solutions of (4.2.8) do not exhaust all possible stationary solutions of the higher-order equation (4.2.5). Since (4.2.8) is solved as an "initial value" problem in X, one cannot satisfy boundary conditions that might be imposed in a finite domain (say, in a band parallel to the Y-axis). Self-similarity breaks down in finite domains as well. Equation (4.2.12) gives the analytical form of the self-similar singular solution corresponding to a solitary quiescent defect in a pattern with the optimal wavenumber. This solution is fundamental for pattern-forming systems characterized by a nonvanishing optimal wavenumber, playing the same role of a "basic nontrivial state" as the symmetric static solution $\Theta = \phi$ of (4.2.2).

Linearizability of (4.2.8) implies a superposition principle. One can construct a solution of (4.2.8) corresponding to an arbitrary array of defects centered at points $Y = Y_i$ along the axis $X = 0$ by superimposing solutions to (4.2.9). The solution, being a generalization of (4.2.12), is

$$\theta = -\mathrm{sign}(X) \ln \frac{1}{2} \left[\left(\mathrm{e}^{-\theta_0} + \mathrm{e}^{-\theta_N}\right) + \sum_{i=1}^{N} \left(\mathrm{e}^{-\theta_i} - \mathrm{e}^{-\theta_{i-1}}\right) \mathrm{erf}\left(\frac{Y - Y_i}{\sqrt{2|X|}}\right)\right], \tag{4.2.13}$$

where $\theta_i - \theta_{i-1} = \pm\pi$ for defects of unit charge.

The existence of this stationary solution is contrary to our expectation that the defects, if initially arranged at some intervals along the X-axis, must start climbing under the influence of mutual interaction. This means that interaction of defects, at least in this linear configuration, vanishes in the phase diffusion approximation – exactly the way it does in the RGL equation. As in the latter case, the phase equation alone cannot fully account for the defect dynamics, even though the nonlinear phase equation (4.2.5) does not look as innocuous as (4.2.2).

4.2.4 Dislocation Core

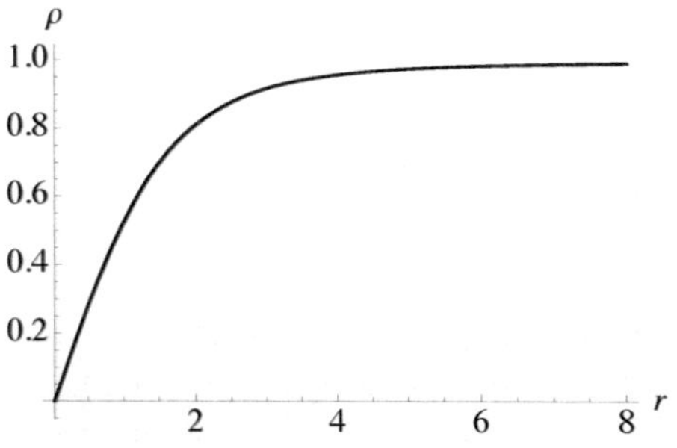

Fig. 4.6. The approximate dependence of the real amplitude on scaled radial distance from a point defect with unit charge

The phase description of dislocations fails at short distances where phase gradients become strong. The structure of the dislocation core is model-specific and has to be computed by solving full underlying equations. As a simple example, we consider the RGL equation (1.1.7).

A stationary vortex solution is obtained by rewriting (4.1.1) and (4.1.2) in polar coordinates r, ϕ and using the *ansatz* $\rho = \rho_0(r)$, $\theta = N\phi$. Then (4.1.2) is satisfied automatically, while (4.1.1) becomes

$$\rho_0'' + r^{-1}\rho_0' + (1 - (N/r)^2 - \rho_0^2)\rho_0 = 0, \tag{4.2.14}$$

subject to the boundary conditions $\rho_0(0) = 0$, $\rho_0(\infty) = 1$. Vanishing of the real amplitude at the origin is necessary to eliminate the divergence due to the phase singularity. Indeed, it can easily be seen that $\rho_0(r) \propto r^{|N|}$ at $r \to 0$.

Equation (4.2.14) cannot be solved analytically, but the solution is easily found numerically (Ginzburg and Pitaevskii, 1958). The solution corresponding to a vortex of unit charge ($N = \pm 1$) is shown in Fig. 4.6. A useful analytical form of $\rho_0(r)$ is a Padé approximant

$$\rho_0^2(r) = \frac{r^2(0.34 + 0.07r^2)}{1 + 0.41r^2 + 0.07r^4}. \tag{4.2.15}$$

This function has correct asymptotics both at the origin and at infinity, and approximates the exact numerical solution fairly well elsewhere.

The energy can be computed by plugging the numerical function $\rho_0(r)$ in (4.1.3). We shall first compute the "potential" energy due to the last term. Since this integral is converging, the integration limit may be extended to infinity. The integration can be done analytically, making use of (4.2.14) and integrating by parts:

$$\begin{aligned} \mathcal{I}_1 &= \tfrac{1}{2}\int_0^\infty r(1-\rho_0^2)^2\,\mathrm{d}r = \int_0^\infty r^2(1-\rho_0^2)\rho_0\rho_0'(r)\,\mathrm{d}r \\ &= -\int_0^\infty \left[r\rho_0'(r)\frac{d(r\rho_0')}{\mathrm{d}r} - N^2\rho_0\rho_0'(r)\right]\mathrm{d}r = \tfrac{1}{2}N^2. \end{aligned} \tag{4.2.16}$$

The remaining distortion energy is computed most conveniently by extracting from the integrand the function $N^2 r^{-1} H(r-1)$, where $H(x)$ is the Heaviside function: $H(x) = 1$ at $x > 0$ and $H(x) = 0$ otherwise. This function, integrating to $N^2 \ln L$, compensates the logarithmic divergence. After the divergence has been extracted, integration may be extended to infinity:

$$\mathcal{I}_0 = \int_0^\infty \left\{ r[\rho_0'(r)]^2 + N^2 r^{-1}[\rho_0^2 - H(r-1)]\right\}\mathrm{d}r = -N^2 \ln a_0. \tag{4.2.17}$$

The result is expressed through a constant a_0 dependent on the core structure. The numerical value for $|N| = 1$ is $a_0 \approx 1.126$. The total energy is

$$\mathcal{F} = \pi(\mathcal{I}_0 + \mathcal{I}_1 + N^2 \ln L) = \pi N^2 \ln \frac{L\sqrt{\mathrm{e}}}{a_0}. \tag{4.2.18}$$

Since the energy is proportional to N^2, only vortices with the unit topological charge are expected to be stable[5]. The phase field, due to its weak decay, gives the largest overall contribution to the energy integral and is solely responsible for the divergence at large distances. The total energy of the vortex within a circle with the radius L diverges logarithmically with the characteristic linear dimension L of the medium. This divergence is not dangerous by itself, since a sole defect in the infinite plane is unphysical. However, it is a disquieting circumstance that may lead to paradoxes when one approaches a perturbation problem without due caution (see Sect. 4.3).

[5] The energy argument still does not exclude metastable multiply charged vortices that might be stable to infinitesimal perturbations

4.2.5 Disclinations

The description of *disclinations* poses more difficult problems. On a naive level, the wave vector may be treated as a *director*, i.e., an "arrowless vector." This is specific to stationary patterns, which are invariant to inversion $\mathbf{k} \to -\mathbf{k}$. In this way, the pattern is viewed as an oriented medium similar to a nematic liquid crystal (more on this in Sect. 6.2.1). A more precise mathematical form of the order parameter characterizing this medium in 2D is a traceless symmetric matrix $\mathbf{n} \otimes \mathbf{n} - \frac{1}{2}\mathbf{I}$, where $\mathbf{I}$ is the 2D unity matrix and $\mathbf{n}$ is a unit vector directed along $\mathbf{k}$. Clearly, inversion of $\mathbf{n}$ does not affect $\mathbf{u}$, while using such a mathematically ambiguous term as an "arrowless vector" is avoided. Topological singularities of the director field in 2D are disclinations with integer or half-integer circulation (Mermin, 1979; Kléman and Lavrentovich, 2003). The latter are allowed in view of the inversion symmetry, and, moreover, are generic, since they have a lower energy. Disclinations with positive and negative half-integer circulation (respectively, convex and concave) can be seen in Figs. 4.3, 4.7.

This approach is too simplistic, as it does not take into account that a striped pattern, unlike a director field, is characterized locally by the phase, and not only by its gradient – the wave vector. If singularities of the phase, as well as of the wave vector field, are taken into account, a striped pattern is recognized as a medium with a very complex topology that cannot be described in the framework of standard homotopy theory (Mermin, 1979). The difference between a director field and a striped pattern is clearly seen when the applicable energy functionals are compared: for the former, the Lagrangian depends on the derivatives of the director only, while for the latter, the energy strongly depends on the absolute value of $\mathbf{k}$.

Disclinations in natural patterns, as well as in anisotropic fluids, may be necessitated by boundary conditions. For example, a common boundary condition in convection patterns is orthogonality of the stripes to the bounding wall, as observed both experimentally and numerically (see e.g. Fig. 4.7) and justified theoretically by

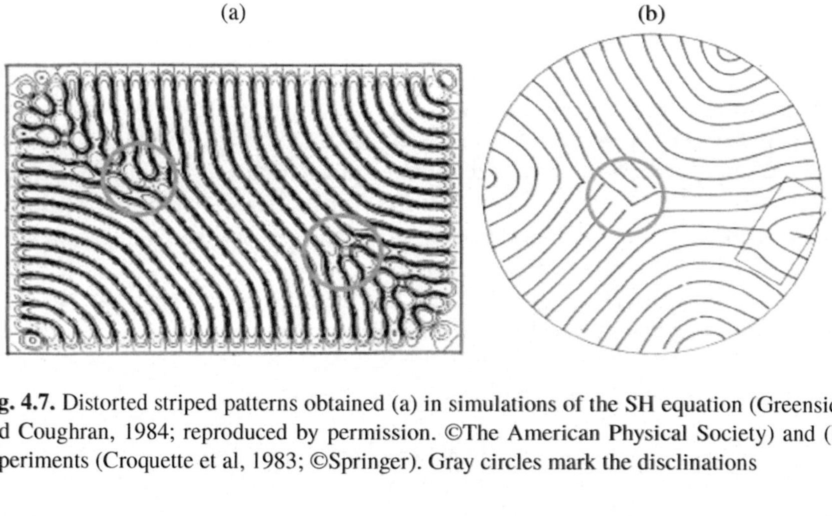

Fig. 4.7. Distorted striped patterns obtained (a) in simulations of the SH equation (Greenside and Coughran, 1984; reproduced by permission. ©The American Physical Society) and (b) experiments (Croquette et al, 1983; ©Springer). Gray circles mark the disclinations

Zaleski et al (1984). This means that the wave vector parallel to the walls should rotate through 2π along the perimeter. In Fig. 4.7a, this rotation is compensated by two concave disclinations (marked by gray circles). One can also see in this figure domain walls going from the upper-left and lower-right corners toward the two disclinations. In Fig. 4.7b, the three target sources on the border rotate the wave vector by π, so that a single disclination is sufficient to compensate the remaining rotation. Dislocations, such as the one within the rectangular marker in the figure, do not affect the rotation of the wave vector.

4.2.6 Domain Walls

A change of orientation across a domain wall in patterns generated in simulations and experiments is usually effected on a length comparable with the prevailing wavelength of the pattern. Nevertheless, using the amplitude equation may give a qualitatively correct picture of the transition. Malomed et al (1990) described the structure of a domain wall by solving coupled NWS equations for striped patterns with two alternative orientations. One can expect that a stationary solution exists only when the wavelengths are equal on both sides of the wall; otherwise, the wall would propagate in the direction decreasing the overall energy of the pattern. It turns out that an even stronger restriction is true, and both wavelengths should be *optimal*.

The derivative with respect to the coordinate z normal to the wall should scale as $O(\epsilon)$, in the same way as the derivative along the wave vector (i.e., as the square root of the parametric deviation from the symmetry-breaking bifurcation point), unless the angles $\alpha_\pm$ between this direction and the directions of the respective wave vectors are close to $\pi/2$. Under these conditions, the fourth-order derivatives in $\heartsuit^2$ can be neglected, and the amplitude equations reduce to two coupled nonlinear diffusion equations with the amplitude diffusivities $\cos^2\alpha_\pm$. The stationary equations for the two amplitudes $u_\pm$ satisfy the equation

$$\cos^2\alpha_\pm u_\pm''(z) + u_\pm\left(1 - |u_\pm|^2 - \nu|u_\mp|^2\right) = 0. \tag{4.2.19}$$

Transforming to the polar form $u_\pm = \rho_\pm e^{i\theta_\pm}$, removing the phase factor, and separating the real and imaginary parts yields the coupled equations for the real amplitudes and phases,

$$\cos^2\alpha_\pm\left(\rho_\pm''(z) - \rho_\pm K_\pm^2\right) + \rho_\pm\left(1 - \rho_\pm^2 - \nu\rho_\mp^2\right) = 0, \tag{4.2.20}$$

$$\frac{\cos^2\alpha_\pm}{\rho_\pm^2}\frac{d}{dz}\left(\rho_\pm^2 K_\pm\right) = 0, \tag{4.2.21}$$

where $K_\pm = \theta_\pm'(z)$. It follows from the last equation that $\rho_\pm^2 K_\pm = \text{const}$, and since $\rho_\pm(\mp\infty) = 0$, $K_\pm$ must vanish. This means that the wavelength on both sides is the optimal one, i.e., corresponds to the "band center"; thus, the domain wall strongly selects a unique wavenumber from the allowed stable range.

We see that only real amplitudes $\rho_\pm$, defined by (4.2.20) with $K = 0$, change across the domain wall interpolating from zero on one side to unity on the other

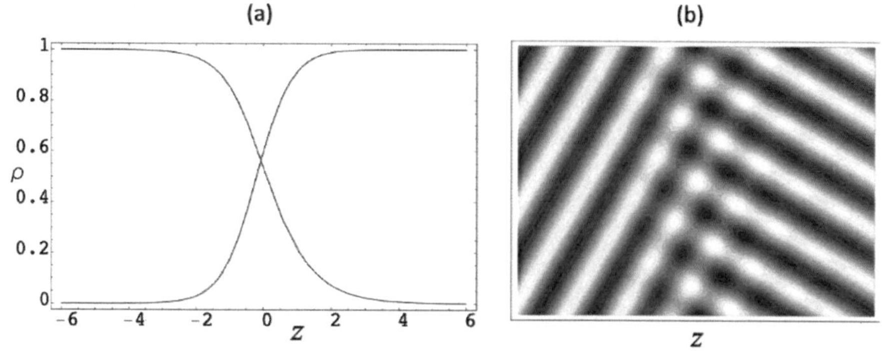

Fig. 4.8. Change of real amplitudes across a domain wall (a) and the respective reconstructed pattern (b)

side; thus, it is an amplitude wall in terms of Sect. 4.2.1. The boundary conditions are $\rho_{\pm}(\pm\infty) = 1$, $\rho_{\pm}(\mp\infty) = 0$. Thc coefficient ν should be larger than unity to discourage coexistence of alternatively directed modes (see Sect. 1.5.1); in particular, for the SH model $\nu = 2$.

The result of a numerical computation, presenting a wall between domains with **k** turning from $\alpha_- = -\pi/6$ to $\alpha_+ = \pi/3$ relative to the normal to the wall, is shown in Fig. 4.8a; Fig. 4.8b presents the reconstructed pattern near the wall. The two modes appear to be mixed near the wall where a patch of a rectangular pattern is observed. This is at variance with the natural patterns seen in Figs. 4.3 and 4.7 where the transition between alternative orientations is abrupt, and points out to limitations of the amplitude representation.

The approximation breaks down when at least on one side of the wall $|\alpha_{\pm} - \pi/2| = O(\epsilon)$; then the fourth-order derivatives along the direction of the stripes should be taken into account. In a particular case when the stripes are parallel to the wall on the one side and normal on the other side, the stationary equations for real amplitudes are

$$\rho_+''(z) + \rho_+ \left(1 - \rho_+^2 - \nu\rho_-^2\right) = 0, \tag{4.2.22}$$

$$-\left(\frac{\epsilon}{2k_0}\right)^2 \rho_-''''(z) + \rho_- \left(1 - \rho_-^2 - \nu\rho_+^2\right) = 0, \tag{4.2.23}$$

subject to the boundary conditions $\rho_{\pm}(\pm\infty) = 1$, $\rho_{\pm}(\mp\infty) = 0$. An approximate solution of these equations is obtained in the following way (Manneville and Pomeau, 1983). The fourth-order derivative in (4.2.23) can be neglected everywhere except a narrow $O(\sqrt{\epsilon})$ boundary layer near some location $z = z_0$. At $z > z_0$, one can set $\rho_- = 0$, and the solution of (4.2.22) is simply $\rho_+ = \tanh\left(z/\sqrt{2}\right)$ [cf. (2.1.21)]. At $z < z_0$, $\rho_-^2 = 1 - \rho_+^2$ can be eliminated from (4.2.23), bringing (4.2.22) to the form

$$\rho_+''(z) - (\nu - 1)\rho_+ + \left(\nu^2 - 1\right)\rho_+^3 = 0, \tag{4.2.24}$$

which is solved by $\rho_+ = \sqrt{2/(\nu-1)}\,\mathrm{sech}\left[\sqrt{\nu-1}(z-z_1)\right]$. The elimination makes sense only when $\rho_+^2 < 1/\nu$. Thus, the two solutions can be matched by requiring

$$\tanh\left(z_0/\sqrt{2}\right) = \mathrm{sech}\left[\sqrt{\nu-1}(z-z_1)\right] = \nu^{-1/2}, \tag{4.2.25}$$

which fixes both z_0 and z_1. The solution exists only at $\nu > 1$, since otherwise excitation of an alternative mode is preferred in each domain, and a quadratic pattern is formed. The solution obtained in this way is not smooth at $z = z_0$, but this can be repaired by restoring the fourth-order derivative in (4.2.23) within the boundary layer without affecting the solutions in both outer regions in a considerable way.

If the change of orientation across the domain wall is small, it can be incorporated in the phase. Then a single amplitude is sufficient for the description of the entire pattern including the domain wall, being treated now as a phase wall (Sect. 4.2.1). No stationary solutions exist in this case in an infinite region, as the pattern oriented in a slightly different way at $z \to \pm\infty$ tends to straighten up with time; the last stages of this process can be described by solving a self-similar asymptotic form of the amplitude equation (Malomed et al, 1990).

An exception is the case when the stripes are almost normal to a weak domain wall. Then one can find a stationary solution verifying the phase equation (4.2.6) and, more specifically, the Burgers equation (4.2.8). Directing the x-axis along the domain wall and replacing the transverse coordinate by z, this equation can be rewritten as

$$\mathrm{d}K/\mathrm{d}z = \pm(2C + K^2), \tag{4.2.26}$$

where K is the z-component of the wave vector and C is a correction to the x-component, assumed to be constant. This constant has to be set to $C = -q^2/2$ to ensure that the pattern is at the band center far from the domain wall. A solution with K approaching a constant asymptotic value at $z \to \pm\infty$ is obtained when the negative sign is chosen in (4.2.26):

$$K = q\tanh qx. \tag{4.2.27}$$

The existence of this solution is a consequence of the marginal zigzag instability at the band center (Sects. 4.1.2, and 4.1.4), which allows a weak bend in the pattern to remain stationary rather than straighten up as it does when the pattern is stable.

4.3 Motion of Dislocations

4.3.1 Phase Field of a Moving Defect

A dislocation *climbing* across the direction of the wave vector of an underlying pattern effects a change of the wavenumber over an extended region. Indeed, the phase gradient $\nabla\theta$ due to the dislocation adds up to the prevailing wave vector $\mathbf{k}_0$ on the one side of the vortex and is subtracted on the other side. The *Peach–Köhler force* driving the dislocation is due to the wavenumber mismatch, i.e., a deviation from the optimal wavenumber (the "band center"). In terms of the envelope field, it corresponds to a nonvanishing constant wave vector $\mathbf{q}$ in (4.1.4). When a number of dislocations are present, the action of their phase fields is superimposed on the the prevailing wave vector.

The first attempts to understand dynamics of dislocations in striped patterns (Siggia and Zippelius, 1981; Pomeau et al, 1983; Tesauro and Cross, 1986) concentrated upon a more difficult problem of patterns in *isotropic* systems near a critical point, which are described by the NWS equation (1.3.35). The analysis was based on a linearized version of the far field equation derived from (1.3.35), and on the naive phase field solution (4.2.3). We have seen in Sect. 4.2.3 that the actual situation is far more complex, and even the core structure of a static NWS vortex is not known to this day[6]. In addition, patterns in isotropic systems are apt to contain disclinations and domain walls, and are strongly influenced by lateral boundaries, which makes controlled measurement of defect motion all but impossible. Later on, attention focused on patterns in *anisotropic* systems, bolstered by precision experiments on electroconvection in nematics. The governing equations reduce in this case to the much simpler RGL model (1.1.7). The analytical theory to be described further in this section is restricted to this case.

It is clear that different approximations should be used in the far and near field, as has been already done when evaluating the energy of an isolated dislocation in Sect. 4.2.4. In both regions, approximate solutions can be obtained by means of a rational expansion assuming that defects are at mutual distances far exceeding the core size (the "healing length") and are slowly moving. These solutions should be matched in an intermediate region, large compared to the core size but small compared to the defect separation or other relevant macroscopic scale. In this region, both near and far field solutions must hold, and their simplified asymptotic forms should apply (Neu, 1990; Pismen and Rodriguez, 1990).

Roughly speaking, the mechanism which sets the defects into motion is the incompatibility of the symmetries in the far and near fields. The far field equation (4.2.2) is invariant to the gauge transformation $\theta \to \theta + \mathbf{k} \cdot \mathbf{x}$, i.e., adding an arbitrary constant phase gradient $\mathbf{k}$; thus, it seems not to be affected altogether by an imposed phase gradient that has to set the defect into motion. The far field equation has solutions corresponding to steady motion of the defect with an arbitrary speed v. Finally, being linear, it obeys a superposition principle, so that defects do not interact at all in this approximation. It is clear, of course, that taking the core into consideration is essential. The nonlinear RGL equation (1.1.7) lacks the symmetries of the far field equation. On the other hand, the circular symmetry of the defect core is lost in the far field and is distorted by the action of the driving force.

A dislocation moving under the action of a constant force is expected to propagate with a constant speed. The direction of motion breaks the symmetry of the basic solution $\theta = \phi$, so that the phase field of static and moving dislocations might be qualitatively different. We shall suppose that both the driving force due to an externally imposed phase gradient $\mathbf{k} = \epsilon \mathbf{K}$ and the propagation speed $\mathbf{v} = \epsilon \mathbf{V}$ are

[6] Written so in the first edition, but, according to my search, still remains true

small. Under these conditions, changes of the real amplitude are negligible outside the dislocation core, and the phase field can be computed as a stationary solution of (4.2.2) in a comoving coordinate frame. Without loss of generality, we assume that the motion is along the Y-axis, and the defect is at the origin. The far field equation (4.2.2) is written in the comoving frame as

$$V\partial_Y\theta + \partial_X^2\theta + \partial_Y^2\theta = 0, \tag{4.3.1}$$

subject to the circulation condition (4.2.1) along any contour surrounding the origin. This equation is solved by introducing a *dual* function related to θ as follows (Dubois-Violette et al, 1983):

$$\partial_X\theta = -N(\partial_Y\Phi + V\Phi), \qquad \partial_Y\theta = N\partial_X\Phi. \tag{4.3.2}$$

The integrability condition for Φ is (4.3.1), while the integrability condition for θ combined with the circulation condition (4.2.1) results in the equation of Φ containing a point source at the defect location:

$$V\partial_Y\Phi + \partial_X^2\Phi + \partial_Y^2\Phi = 2\pi\delta(\mathbf{X}). \tag{4.3.3}$$

Unlike the phase, the dual function is univalued. The solution of (4.3.3) is expressed through a modified Bessel function:

$$\Phi = -\exp\left(-\frac{VY}{2}\right)K_0\left(\frac{VR}{2}\right), \tag{4.3.4}$$

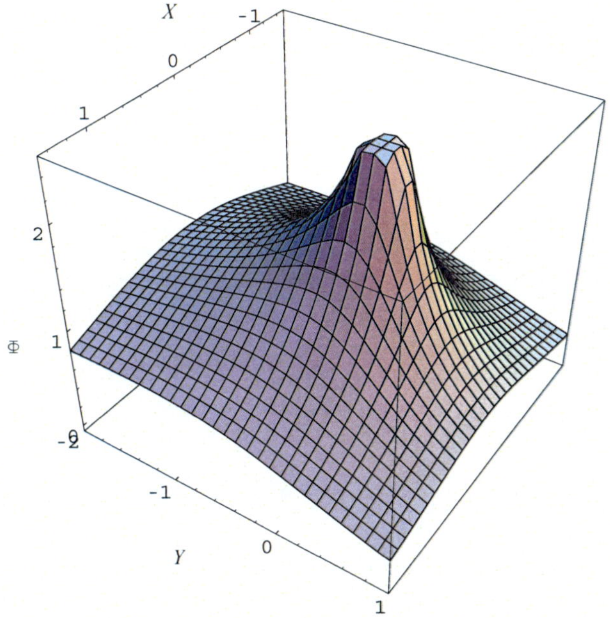

Fig. 4.9. The rescaled dual function of a moving defect. Take note of the fast drop-off in the direction of motion

where $R = \epsilon r$ is the extended radial coordinate.

The limit of the dual function at $V \to 0$ and R fixed is, up to a constant, $\Phi = \ln R$; this is a circularly symmetric function, which is the dual of the polar angle. The symmetry is broken in a singular way by any nonvanishing velocity. Indeed, (4.3.4) is scale-invariant and can be made independent of velocity by rescaling the coordinate $\mathbf{X} \to \mathbf{X}/V$. The rescaled function is drawn in Fig. 4.9. Its strong asymmetry is evident. The function falls off exponentially everywhere except the tail along the negative Y-axis; the steepest fall-off is in the direction of motion. Asymptotically at $R \to \infty$,

$$\Phi \asymp -\sqrt{\frac{\pi}{VR}} \exp\left[-\frac{VR}{2}(1+\sin\phi)\right]. \tag{4.3.5}$$

The components of the phase gradient are read directly from (4.3.2). The resulting expressions are scale-invariant and can be rewritten in short-scale variables without change:

$$\begin{aligned}
\theta_x &= \frac{Nv}{2} \exp\left(-\frac{vr}{2}\sin\phi\right)\left[K_0\left(\frac{vr}{2}\right) - \sin\phi\, K_1\left(\frac{vr}{2}\right)\right], \\
\theta_y &= \frac{Nv}{2} \exp\left(-\frac{vr}{2}\sin\phi\right)\cos\phi\, K_1\left(\frac{vr}{2}\right).
\end{aligned} \tag{4.3.6}$$

4.3.2 Dissipation Integral and Peach–Köhler Force

A velocity estimate can be obtained by comparing two equivalent expressions for the energy dissipation rate computed in a comoving frame, which both contain the unknown defect velocity (Pomeau et al., 1983; Bodenschatz et al., 1988). Consider a single defect steadily moving along the y-axis. After transforming $y \to y - vt$, ∂_t is replaced by $-v\partial_y$:

$$\begin{aligned}
\partial_t \mathcal{F} &= \int \left(\frac{\delta \mathcal{F}}{\delta u}\partial_t u + \text{c.c.}\right) \mathrm{d}\mathbf{x} \\
&= -2\int \partial_t u \partial_t \overline{u}\, \mathrm{d}\mathbf{x} = -2v^2 \int \partial_y u \partial_y \overline{u}\, \mathrm{d}\mathbf{x} \equiv -2v^2 \mathcal{I}\,,
\end{aligned} \tag{4.3.7}$$

where $\mathcal{I}$, by definition, is a *dissipation integral.*

Another way to write the same expression is to transform to the comoving frame at an earlier stage:

$$\begin{aligned}
\partial_t \mathcal{F} &= -v\int \left(\frac{\delta \mathcal{F}}{\delta u}\partial_y u + \text{c.c.}\right) \mathrm{d}\mathbf{x} \\
&= -v\int \left(\partial_y \overline{u}\left[\nabla^2 u + (1-|u|^2)u\right] + \text{c.c.}\right) \mathrm{d}\mathbf{x} \equiv -vF.
\end{aligned} \tag{4.3.8}$$

Comparing (4.3.8) with (4.3.7) defines the velocity, as yet unknown, as the ratio of the two integrals:

$$v = F/2\mathcal{I}. \tag{4.3.9}$$

Analogous to familiar equations of classical mechanics, F is interpreted as a driving force. This force is thermodynamic by its nature, as it originates in the energy integral and is called a *Peach–Köhler force*. By the same analogy, the dissipation integral in the denominator is interpreted as a *friction factor*.

The integral (4.3.7) is mostly accumulated in the far field, and the main contribution comes from the phase gradient. Thus, it is crucial whether a static (circularly symmetric) or a propagating solution defined by (4.3.6) is used in the computation. A traditional approach was to take the symmetric solution $\theta = N\phi$; however, since it leads to a diverging integral, it is necessary to assume a long-scale cut-off length equal to a typical distance between defects L (Bray, 1995). Then the integral in (4.3.7) is $\propto N^2 \ln L$. A more quantitative result can be obtained by taking into account the fall-off of the amplitude in the defect core to eliminate the logarithmic divergence at $r \to 0$. Using the numerical function $\rho_0(r)$ computed in Sect. 4.2.4 yields

$$\mathcal{I} = \int_0^{2\pi} \mathrm{d}\phi \int_0^L \{r[\rho_0'(r)]^2 \sin^2\phi + N^2 r^{-1} \rho_0^2(r) \cos^2\phi\}\, \mathrm{d}r = \pi N^2 \ln\frac{L}{a_0}. \tag{4.3.10}$$

This reduces to the radial integral (4.2.17) after angular integration, so that we obtain the same numerical constant $a_0 \approx 1.126$ for $|N| = 1$.

If the motion is caused by the interaction of two defects at a mutual distance L, one can express the driving force directly as the derivative of the total energy with respect to L taken with the negative sign:

$$\partial_t \mathcal{F} = \frac{\partial \mathcal{F}}{\partial L}\frac{\partial L}{\partial t} \equiv -Fv. \tag{4.3.11}$$

A rough estimate of the energy of an interacting pair can be obtained using the static solution and assuming the cutoff length equal to a typical defect separation L (Bray, 1995):

$$F = \frac{\partial \mathcal{F}}{\partial L} = \frac{\pi N^2}{L}. \tag{4.3.12}$$

Using (4.3.10), the velocity estimate is

$$v = \frac{1}{2L\ln(L/a_0)}. \tag{4.3.13}$$

The numerical factors should not, of course, be taken too seriously.

In view of the result of the preceding subsection, the phase field of a moving defect loses circular symmetry. Using the propagating, rather than stationary, far field solution in the dissipation integral appears to be logical in view of the assumption of stationary propagation implied in (4.3.7). However, this assumption is expected to be valid *quantitatively* only when the driving force is also constant. This is not realistic in an ensemble of moving and interacting defects, but can be realized when the background solution has the form (4.1.4). In this case, there is a prevailing nonvanishing wave vector $\mathbf{k}$ that corresponds to a constant phase gradient acting upon a defect. This setup is common in the context of moving dislocations in a striped pattern.

The defect moves in the direction normal to $\mathbf{k}$ in such a way that $k = |\mathbf{k}|$ ahead of it is larger than that behind, and therefore both the wavenumber and energy decrease as a result. As the y-axis is taken as the direction of motion, the direction of $\mathbf{k}$ can be taken as the x-axis. Let $u = U\mathrm{e}^{\mathrm{i}kx}$ where $U = \sqrt{1-k^2}\mathrm{e}^{\mathrm{i}\phi}$ asymptotically at large distances. This corresponds to (4.1.4) with added phase circulation. Equation (4.3.8) now becomes

$$\partial_t \mathcal{F} = -v \int_{-\infty}^{\infty}\int_{-\infty}^{\infty} \left(\partial_y \overline{U}\left[\nabla^2 + (1 + 2\mathrm{i}k\partial_x - k^2 - |U|^2)\right] U + \text{c.c.}\right) \mathrm{d}x\, \mathrm{d}y. \tag{4.3.14}$$

All terms, except the one containing the imaginary unit, can be presented as total derivatives and vanish upon integration. The remaining integral, defining the Peach–Köhler force, is reduced by integrating by parts to a contour integral:

$$\begin{aligned}\mathcal{F} &= 2\mathrm{i}k \int_{-\infty}^{\infty}\int_{-\infty}^{\infty} \left(\partial_y \overline{U}\partial_x U - \text{c.c.}\right) \mathrm{d}x\, \mathrm{d}y \\ &= 2\mathrm{i}k \left(\int_{-\infty}^{\infty} \mathrm{d}x\, \overline{U}\partial_x U\Big|_{y=-\infty}^{y=\infty} - \int_{-\infty}^{\infty} \mathrm{d}y\, \overline{U}\partial_y U\Big|_{x=-\infty}^{x=\infty}\right).\end{aligned} \tag{4.3.15}$$

The two terms on the right-hand side combine to a closed contour integral containing the derivative of the phase along the contour. Since the latter increases by $2\pi N$ around the contour enclosing the defect, we finally obtain

$$\mathcal{F} = 2k(1-k^2)\oint \frac{\partial\theta}{\partial s}\mathrm{d}s = 4\pi k N(1-k^2). \tag{4.3.16}$$

It remains to compute the dissipation integral. Plugging the propagating solution into (4.3.7) makes a lot of difference, as the long-distance divergence of the energy integral *disappears* at any $v \neq 0$. The far field contribution to the dissipation integral is computed by assuming in (4.3.7) $\rho = 1$ outside a circle of a radius $1 \ll r_0 \ll v^{-1}$, which is small on the extended scale. Denoting in (4.3.6) $z = vr/2$ yields

$$\mathcal{I} = N^2 \int_{\frac{1}{2}vr_0}^{\infty} zK_1^2(z)\mathrm{d}z \int_0^{2\pi} \mathrm{e}^{-2z\sin\phi}\cos^2\phi\, \mathrm{d}\phi = \pi N^2 \int_{\frac{1}{2}vr_0}^{\infty} I_1(2z)K_1^2(z)\mathrm{d}z. \tag{4.3.17}$$

The integrand in the last expression decays asymptotically as $z^{-3/2}$ at $z \to \infty$, and therefore the integral is converging. Due to the logarithmic divergence at $z \to 0$, $\mathcal{I} \propto \ln(1/v)$. A quantitative result is obtained by computing the inner part of the dissipation integral with due account for the fall-off of the amplitude in the defect core (Bodenschatz et al, 1988). This is done as in (4.3.10) with the cutoff radius L replaced by r_0. The sum of the inner and outer integrals is independent of the cutoff radius, provided it is large compared to the size of the defect core but small on the extended outer scale. The result is

$$\mathcal{I} = \pi \ln \frac{3.29}{v}. \tag{4.3.18}$$

Using (4.3.16) and (4.3.18) in (4.3.9) yields the mobility relation connecting the wavenumber mismatch with the drift velocity:

$$\frac{Nv}{2}\ln\frac{3.29}{v} = k(1-k^2). \tag{4.3.19}$$

This result is, in fact, exact for the motion in a constant phase gradient. This is confirmed by a more rigorous computation based on the method of matched asymptotic expansion in the next subsection. On the other hand, the "quasistatic" estimate (4.3.13) is mere qualitative and does not account for charges of defects and directions of motion in a proper way. The logarithmic divergence of the integral (4.3.10) was taken by some authors to imply a logarithmic decrease of mobility with the size of the system. This conclusion is flawed because the far field solution depends in a singular way on the defect velocity, so that the dissipation integral (4.3.17) evaluated in the comoving frame converges even when the velocity is arbitrarily small. The problem of far field divergences is overcome by applying an asymptotic method based on the matching of solutions in the inner (in the vicinity of the defect) and outer (far field) regions, as explained in detail in the next subsection. The dependence on the size of the system comes to be far more subtle and can be related to the influence of the shape of a planar region on the Green's function of the heat equation (1.1.7) or to the action of images of defects.

Nevertheless, (4.3.10) and (4.3.18) are not contradictory in a qualitative way. Both give two alternative estimates of the dissipation integral: one, provided by a nonvanishing velocity, and the other, by either the size of the region or a characteristic distance between defects. In a realistic situation, when there are many moving defects in an extended region, the lowest of the two bounds might be most relevant. The velocities computed with the help of (4.3.13) and (4.3.19) are likely to be close one to another, since (neglecting logarithmic corrections) we expect both the typical defect velocity and the background wavenumber to be inversely proportional to a typical distance between defects.

4.3.3 Matched Asymptotic Expansions

The dislocation core, like its far field, is distorted when the defect is set into motion. As long as the velocity is small, this effect can be treated perturbatively. Matching the perturbed core solution to the far field is the most precise way to determine the drift velocity.

Consider a single defect steadily moving along the y-axis with a slow speed $v = \epsilon V$, as yet unknown. The RGL equation (1.1.7) is rewritten in a comoving coordinate frame centered at the defect as

$$\nabla^2 u + (1 - |u|^2)u = -v\partial_y u. \tag{4.3.20}$$

The term containing the velocity is seen as a small correction, so that the solution can be sought for in the form of an expansion in either a dummy small parameter ϵ, to be further connected to an externally imposed weak phase gradient, or in v itself:

$$u = u_0 + v u_1 + \cdots . \tag{4.3.21}$$

The zero-order equation is

$$\nabla^2 u_0 + (1 - |u_0|^2) u_0 = 0, \tag{4.3.22}$$

satisfied by the symmetric static defect solution $u_0 = \rho_0(r) e^{iN\phi}$.

The first-order equation can be rewritten in a compact form

$$\mathcal{H}(u_1, \overline{u}_1) + \Psi(\mathbf{x}) = 0, \tag{4.3.23}$$

where $\Psi(\mathbf{x}) = \partial_y u_0$ is the inhomogeneity and $\mathcal{H}$ is the linear operator

$$\mathcal{H}(u_1, \overline{u}_1) \equiv \nabla^2 u_1 + (1 - 2|u_0|^2) u_1 - u_0^2 \overline{u}_1. \tag{4.3.24}$$

It is not necessary to actually solve (4.3.23); it suffices to obtain the *solvability condition* of this equation. The operator $\mathcal{H}$ is self-adjoint and has two eigenfunctions $\varphi(\mathbf{x})$ with zero eigenvalue that correspond to two translational degrees of freedom in the plane. These eigenfunctions are just the two vector components of ∇u_0; indeed, taking the gradient of (4.3.22) yields $\mathcal{H}(\nabla u_0, \nabla \overline{u}_0) = 0$.

The well-known *Fredholm alternative* states that if a *homogeneous* linear equation has a nontrivial solution, the respective *inhomogeneous* equation is solvable only if the inhomogeneity is orthogonal to the eigenfunction with the zero eigenvalue. If the Fredholm alternative is applied to (4.3.23) in the infinite plane, an apparent result is that it should have no solution: indeed, its right-hand side is certainly not orthogonal to itself, i.e., to the y-component of the eigenfunction ∇u_0. This conclusion is erroneous since integration in the infinite plane has no sense whatsoever: first, because the integrals involving the static solution u_0 diverge and, second, because this solution is inapplicable in the far field, as the phase of a moving defect is strongly asymmetric at distances $r = O(v^{-1})$ – see Sect. 4.3.1.

The correct way to obtain the solvability condition is to apply the Fredholm alternative *in a circle* of a radius r_0 large compared to the core size but small on the far field scale, i.e., $1 \ll r_0 \ll v^{-1}$. The resulting condition involves both area and contour integrals. It is formally derived by multiplying (4.3.23) by the eigenfunction $\varphi(\mathbf{x})$ and integrating over a circle or, generally, any finite region. Since (4.3.23) is a shorthand for two complex conjugate equations of u_1 and $\overline{u}_1$, one has to add as well the complex conjugate, and the resulting condition will be real. The differential term in $\mathcal{H}$ is transformed by twice integrating by parts:

$$\int \overline{\varphi} \nabla^2 u_1 \, d\mathbf{x} = \int u_1 \nabla^2 \overline{\varphi} \, d\mathbf{x} + \oint (\overline{\varphi} \mathbf{n} \cdot \nabla u_1 - u_1 \mathbf{n} \cdot \nabla \overline{\varphi}) \, ds, \tag{4.3.25}$$

where $\mathbf{n}$ is the outward normal to the integration contour, and s is the arc length. After the algebraic terms in $\mathcal{H}$ are added, the area integrand (excluding the inhomogeneous part) reduces to $u_1 \mathcal{H}(\overline{\varphi}, \varphi)$+c.c., and vanishes identically since φ is an eigenfunction. Restricting now to integration over a circle, the resulting solvability condition is

$$\mathrm{Re}\left\{\int_0^{r_0} r\,\mathrm{d}r \int_0^{2\pi} \overline{\varphi}\Psi(r,\phi)\,\mathrm{d}\phi + r_0 \int_0^{2\pi} (\overline{\varphi}\partial_r u_1 - u_1\partial_r\overline{\varphi})_{r=r_0}\,\mathrm{d}\phi\right\} = 0. \tag{4.3.26}$$

With the particular inhomogeneity,

$$\Psi = \partial_y u_0 = \left[\rho_0'(r)\sin\phi + \mathrm{i}Nr^{-1}\rho_0(r)\cos\phi\right]\mathrm{e}^{\mathrm{i}N\phi}, \tag{4.3.27}$$

the left-hand side of (4.3.26) does not vanish upon angular integration only when φ is the y-component of ∇u_0 given by the same expression (4.3.27).

The contour integral depends on the asymptotics of the first-order solution u_1. Assuming $r_0 = O(v^{-1/2})$, we have $\rho_0(r_0) = 1 - O(v)$. It follows that the contour integral can be expressed, to the leading order $O(v)$, through the phase field only, and is computed using the inner limit of the outer solution (4.3.6) with an arbitrary constant gradient $k = vK$ added to θ_x:

$$\begin{aligned}\theta_x &= -\frac{N}{r}\sin\phi - \frac{Nv}{2}\left(\ln\frac{vr}{4} + \gamma - \sin^2\phi\right) + vK,\\ \theta_y &= \frac{N}{r}\cos\phi - \frac{Nv}{2}\cos\phi\sin\phi,\end{aligned} \tag{4.3.28}$$

where $\gamma \approx 0.577$ is the Euler constant. The phase itself can be reconstituted from this asymptotic expression in the form containing a dipole component which contributes to the contour integral:

$$\theta = N\phi + r\chi(r)\cos\phi, \quad \chi(r) = -\frac{Nv}{2}\ln\left(\frac{vr}{4}\mathrm{e}^{\gamma-1-2K/N}\right). \tag{4.3.29}$$

Using the asymptotics of (4.3.27),

$$\partial_y u_0 \approx \mathrm{i}\theta_y \mathrm{e}^{\mathrm{i}N\phi} = \mathrm{i}Nr^{-1}\cos\phi\,\mathrm{e}^{\mathrm{i}N\phi}, \tag{4.3.30}$$

the contour integral is computed now as $N\pi[N - 2\chi(r_0)]$. The area integral in (4.3.26) is evaluated as in (4.3.10), yielding $\pi N^2\ln(r_0/a_0)$. When the area and contour integrals are added up, the auxiliary radius r_0 falls out, and the solvability condition takes the form

$$\ln\left(\frac{a_0 v}{4}\mathrm{e}^{\gamma-\frac{1}{2}}\right) = \frac{2K}{N}. \tag{4.3.31}$$

This expression can be rewritten in the form of a mobility relation resolved with respect to the weak driving phase gradient $k = vK$:

$$k = -\frac{Nv}{2}\ln\frac{v_0}{v}, \quad v_0 = \frac{4}{a_0}\mathrm{e}^{-\gamma+\frac{1}{2}} \approx 3.29. \tag{4.3.32}$$

This weakly nonlinear dependence is shown in Fig. 4.10. The numerical value relates to $|N| = 1$. The result coincides (at $k \ll 1$) with (4.3.19), though evaluation of the integral (4.3.17) has been circumvented. Note that the analytical form of the mobility relation is universal, and only the constant a_0, hence, v_0, depends on the numerical integration in the core region.

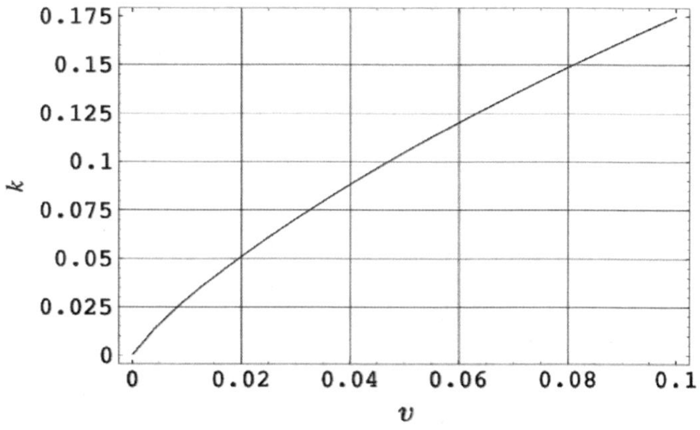

Fig. 4.10. Relation between the phase gradient k and the defect velocity v

The motion is in the direction orthogonal to the imposed phase gradient. Since v must be small, the logarithm is always positive, and the direction of motion is turned clockwise or anticlockwise through $\pi/2$ with respect to the direction of the phase gradient when the defect is, respectively, positively or negatively charged. The vector form of (4.3.32) can be written with the help of the operator of rotation through the right angle $\mathcal{J}$:

$$\mathbf{k} = \frac{N}{2} \ln \frac{v_0}{|\mathbf{v}|} \mathcal{J}\mathbf{v}; \quad \mathcal{J} = \begin{pmatrix} 0 & -1 \\ 1 & 0 \end{pmatrix}. \tag{4.3.33}$$

Since $v_0 = O(1)$, while $|\mathbf{v}| = O(\epsilon)$, the argument of the logarithm is *large*. Thus, the final result turns out to be weakly dependent on the ratio of the characteristic scales of problem.

4.3.4 Interaction of Dislocations

In a disordered pattern, dislocations themselves serve as sources of phase gradients acting on other defects. Pick any defect, and place the origin of the polar coordinate system (r, ϕ) at its instantaneous position. Take tentatively (4.2.4) with the phase given by (4.2.3) and the real amplitude $\rho_0(r)$ by the circularly symmetric stationary solution satisfying (4.2.14). Near the origin, the only significant contribution is due to the "resident" defect; all other amplitudes can be presented by asymptotic expressions, and the corrections are of $O(L^{-2})$. Neglecting these terms and taking the phase of the resident defect to be equal to the polar angle, the trial solution becomes

$$u(\mathbf{x}) = \rho_0(r) \exp\left[\mathrm{i}\left(\phi + \sum \theta_i(\mathbf{x} - \mathbf{x}_i)\right)\right], \tag{4.3.34}$$

where the summation is now over phases θ_i of all defects except the resident one. Using this in (4.1.1) and (4.1.2) leaves $O(L^{-1})$ residual terms due to the phase gradients $\nabla\theta_i$. The combined gradient can be identified with an externally imposed gradient considered in Sect. 4.3.3, and the mobility relation (4.3.33) can be applied to compute the respective defect velocity. When the phase gradient is created by the

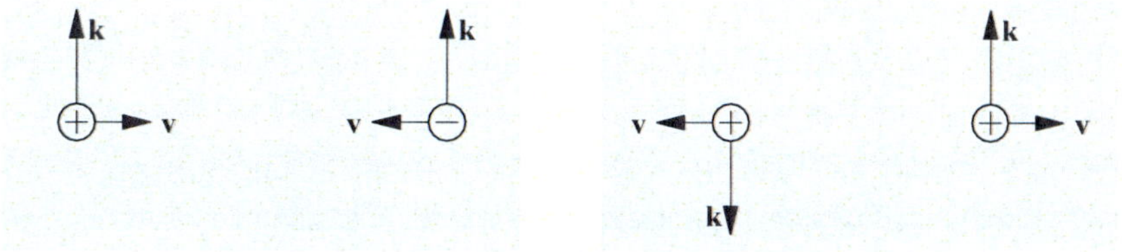

Fig. 4.11. Attracting and repelling defect pairs

phase field of another defect, the result is, as expected, repulsion of likely charged defects and attraction of oppositely charged ones (see Fig. 4.11).

The motion of an ensemble of interacting defects can be followed then by computing the instantaneous phase gradient at the defect location due to the action of all other defects, moving each defect with the speed determined by the mobility relation, recomputing the gradients, etc. The algorithm would fail only when two defects (say, mutually attracting defects of the opposite sign) closely approach each other, and the weak field theory becomes inapplicable. This would be, however, only a very brief stage, resulting in annihilation of the colliding defects, while all others would not be considerably disturbed. In this way, the field problem is reduced in effect to a dynamical *particle–field* problem with a finite number of degrees of freedom.

Unfortunately, this approach is not precise due to the lack of a universally applicable (history-independent) mobility relation. Equation (4.3.33), strictly speaking, applies only to the motion in a constant phase gradient. Indeed, it relies on the near field asymptotics of a particular far field solution (4.3.6). The correct form of the asymptotics, (4.3.29), must be retained for any background phase field slowly changing in time and space. The numerical constants in the argument of the logarithm in (4.3.29) might, however, be different. The phase field of a moving defect (4.3.6) is velocity-dependent, and, as the velocity changes, either by its absolute value or direction, it cannot adjust instantaneously to a new quasistationary distribution. Therefore, the phase gradient induced by each defect at the core of another one depends, strictly speaking, on the entire past history of accelerations. Since the characteristic time scale of the phase field evolution is the same as that of the defect displacement, one would need to solve the *nonstationary* phase equation (4.2.2) simultaneously with the equations of motion of defects.

A rough estimate can be based on a "self-consistent" approximation that neglects these complications (Pismen and Rodriguez, 1990). It is assumed that each defect carries the quasistationary phase field corresponding to its instantaneous velocity. Under this assumption, the dependence of the drift velocity on the instantaneous separation of a pair of dislocations is obtained in the following way. First, we compute the value of the phase gradient due to one of the defects at the core of the other one with the help of (4.3.6). The velocity is then given by the mobility relation (4.3.32) or (4.3.33). For oppositely charged defects moving toward each other along the y-axis, the phase gradient generated by one of the defects at the core of the other is $\theta_x(r, \frac{\pi}{2})$, while for like-charged defects, moving apart, it is $\theta_x(r, -\frac{\pi}{2})$. Quasistationary velocities are computed then as solutions of the equation

$$\ln \frac{v_0}{v} = \mathrm{e}^{\pm vr/2} \left[K_0(vr/2) \pm K_1(vr/2)\right], \qquad (4.3.35)$$

where the positive sign should be taken for the repelling and the negative sign for the attracting pair of defects of unit charge. A stable quasistationary solution exists only beyond a certain separation; at shorter distances velocities are large, and the entire approximation scheme breaks down. At larger separations, two branches of solutions exist, of which the lower one, corresponding to velocities decreasing with separation, is physical. Quasistationary velocities in both cases are shown in Fig. 4.12. The velocities of mutually repelling defects are much larger, at a comparable distance, than those of attracting ones, due to a slower fall-off of the phase gradient behind a moving defect.

The self-consistent solution should be viewed as an estimate rather than as a rational approximation, although it is expected to work well when the separation is very large even on the extended scale. An opposite, less congruous, assumption would be to use instead of (4.3.6) a "solid body" solution (4.2.3) neglecting the velocity-dependent deformation of the phase field entirely. The truth may lie between these two extremes. As the vortex accelerates, the phase readjusts fast near the core but is retarded in far outlying regions. Thus, for example, when a circularly symmetric vortex at rest is suddenly accelerated, its phase field will evolve to the asymmetric stationary form (4.3.6) only gradually, spreading the deformation from the center outward and fully readjusting only asymptotically at $t \to \infty$. The solid body and self-consistent velocities correspond to the vortex motion in the respective limits of infinite and zero acceleration.

A better way to follow the evolution of the phase field is to integrate the phase equation (4.2.2) subject to the circulation condition (4.2.1). Since the phase equation is linear, its solution can be presented as a superposition of phase fields θ_i generated by different defects and subject to the respective circulation conditions. For the latter fields, it appears to be possible to write down solutions corresponding to arbitrary

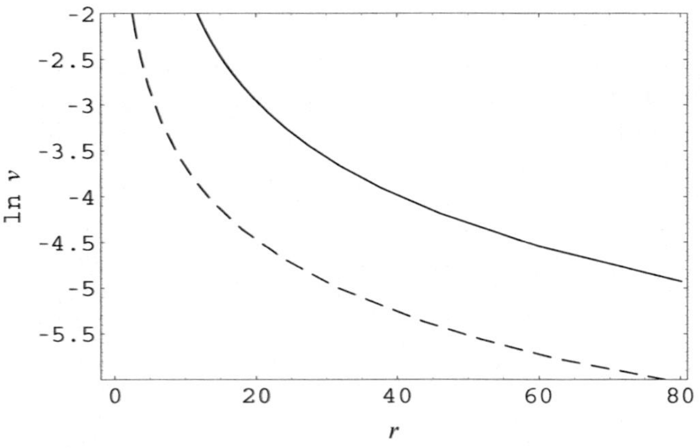

Fig. 4.12. Quasistationary velocities vs. separation for attracting (dashed line) and repelling (solid line) defects

vortex trajectories with the help of an appropriate Green's function. A difficulty arises, however, due to the multivaluedness of θ. A possible way out is to introduce, as in Sect. 4.3.1, a univalued function dual to θ which would obey the usual heat equation with a point source located at a mobile defect. This has been done consistently only for the case of rectilinear motion of a pair of interacting defects (Rodriguez et al, 1991).

As before, the motion can be assumed, without loss of generality, to be along the y-axis. Then the dual function $\Phi(X, Y, T)$ is defined as

$$\partial_X \theta = N\left(\partial_Y \Phi - \int^Y \partial_T \Phi \mathrm{d}Y\right), \quad -\partial_Y \theta = N \partial_X \Phi. \tag{4.3.36}$$

One can see that the equation

$$\partial_X^2 \Phi + \partial_Y^2 \Phi_- \partial_T \Phi = 2\pi\delta(X)\delta(Y) \tag{4.3.37}$$

both serves as the integrability condition for θ in (4.3.36) and ensures that the circulation condition (4.2.1) be satisfied. The dual function can be expressed using the Green's function of (4.3.37) as

$$\Phi(X, Y, T) = \frac{1}{2}\int_0^t \frac{1}{T-\tau} \exp\left[-\frac{[Y + \eta(T) - \eta(\tau)]^2 + X^2}{4(T-\tau)}\right] \mathrm{d}\tau, \tag{4.3.38}$$

where $\eta(T)$ is an instantaneous position of the vortex moving along the axis $X = 0$. The phase gradient computed from (4.3.38), (4.3.36) is further used in the mobility relation (4.3.32) to obtain the velocities and subsequent positions of the vortices.

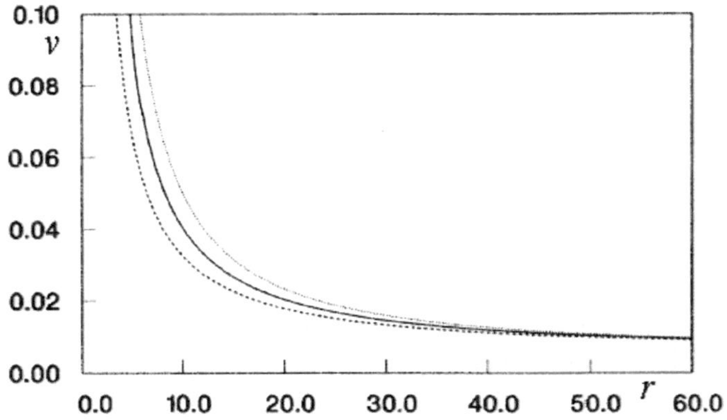

Fig. 4.13. The dependence of the velocity on the distance for a pair of interacting defects of opposite sign, starting at a certain distance from a self-consistent phase field, and eventually annihilating (Rodriguez et al, 1991, ©The American Physical Society). The solid line shows the computed $v(r)$ curve. It is flanked by the dotted lines showing the self-consistent approximation (below) and the "solid body" approximation with the initial phase unchanged (above)

Figure 4.13 (Rodriguez et al, 1991) shows the results of computations utilizing the above procedure for the case of a pair of unlikely charged defects moving toward each other in an unbounded domain from some initial distance and eventually annihilating at the point $Y = 0$. Due to the symmetry, it is sufficient to consider a single defect together with its image in a reflecting wall placed at $Y = 0$. Ideally, the interacting defects should be nucleated at an infinite separation distance. For a practical numerical calculation the integration must start from a finite separation, and hence there is some uncertainty in the choice of initial conditions. The initial phase distribution in the computation shown in Fig. 4.13 corresponds to the self-consistent solution at the initial separation. The velocity dependence on the instantaneous separation, $v(r)$, is shown by the solid line in the figure. For comparison, the dashed line in the same plot shows the quasistationary self-consistent dependence (4.3.35) and the dotted line, a "solid body" solution computed under the assumption that one of the vortices moves in the unchanging phase field of the other. The numerically computed velocity lies between these two extreems.

The vortices accelerate steeply at close approach, and the computations have to stop shortly before the moment of annihilation, when the basic assumptions of slow motion and weak phase gradients are no longer valid. The computations are in a reasonable agreement with experiments carried out on weakly distorted electroconvection patterns in nematics, where the basic alignment of the pattern was ensured by treated solid boundaries (Braun and Steinberg, 1991), but experimental precision is not sufficient to resolve the scaling factor of the logarithmic nonlinearity.

4.3.5 Motion in a Supercriticality Ramp

Although the phase field is the dominant cause of vortex motion in extended homogeneous systems with low density of defects, it is also possible to set vortices into motion by affecting the background amplitude. In the framework of the RGL model, this can be done in the simplest way by making the linear growth rate position-dependent. Then the RGL Lagrangian (1.1.10) is rewritten as

$$\mathcal{L} = \frac{1}{2}\left[\nabla\overline{u}\cdot\nabla u + \frac{1}{2}\left(\mu(\mathbf{x}) - |u|^2\right)^2\right], \tag{4.3.39}$$

and the evolution equation (1.1.7) becomes

$$u_t = \nabla^2 u + (\mu(\mathbf{x}) - |u|^2)u. \tag{4.3.40}$$

The supercriticality parameter μ may be position-dependent due to either local inhomogeneities or a long-range ramp induced by boundary conditions. Since the energy of the vortex is lower where μ is small, stable configurations should correspond to vortices pinned at local minima of $\mu(\mathbf{x})$. It is natural to expect that, neglecting vortex interactions, the gradient dynamics would be reduced to a vortex drift down the gradient of $\mu(\mathbf{x})$. The eventual effect of inhomogeneities might be to suppress the annihilation of vortices at some stage of the coarsening process, thereby freezing a disordered "vortex glass" state.

The case when $\mu(\mathbf{x})$ varies on a characteristic scale far exceeding the size of the vortex core (the healing length) and the vortex drift is slow is treated perturbatively as above. The local landscape in the core region is expanded as

$$\mu(\mathbf{x}) = \mu_0(1 + \epsilon\mathbf{M}\cdot\mathbf{x} + \ldots), \tag{4.3.41}$$

where both $\mu_0 = \mu(\mathbf{x}_0)$ and its logarithmic gradient $\mathbf{m} \equiv \epsilon\mathbf{M} = \nabla \ln \mu(\mathbf{x}_0)$ are measured at the vortex vocation.

We set $\mathbf{v} = \epsilon\mathbf{V}$ and expand in the dummy small parameter ϵ. In order to bring the core expansion equations to the same form as in the preceding subsection, it is convenient to eliminate μ_0 by rescaling both the order parameter u and the temporal and spatial variables as $u \to \mu_0^{1/2}u,\ r \to \mu_0^{-1/2}r,\ t \to \mu_0^{-1}t$. Then the rescaled zero-order vortex solution retains the form $u_0 = \mathrm{e}^{\mathrm{i}\phi}\rho_0(r)$, where the function $\rho_0(r)$ verifies (4.2.14). The first-order equation has the general form (4.3.23) with the inhomogeneity

$$\Psi(\mathbf{x}) = \mathbf{V}\cdot\nabla u_0 + (\mathbf{M}\cdot\mathbf{x})u_0. \tag{4.3.42}$$

The solvability condition (4.3.26) is obtained exactly as before, but the area integral includes now a new term proportional to $\mathbf{M}$. If V, M are the respective projections on the y-axis, the area integral corresponding to the translational eigenfunction along the y-axis is

$$\begin{aligned}\pi V\int_0^{r_0}\left[r(\rho_0')^2 + N^2r^{-1}\rho_0^2\right]\mathrm{d}r + \pi M\int_0^{r_0} r^2\rho_0\rho_0'\,\mathrm{d}r \\ \asymp \pi\left(V\ln\frac{r_0}{a_0} + M\ln\frac{r_0}{a_m}\right).\end{aligned} \tag{4.3.43}$$

Both integrals depend logarithmically on the upper limit r_0 at large distances but with different constants; for a vortex of unit charge $a_m \approx 0.67$. This numerical discrepancy cannot be disregarded, and necessitates matching with the far field, as first-order distortions of the vortex core are not suppressed, in spite of the similarity of the asymptotics of the two integrals.

The simplest large-scale setup is a single vortex moving along a supercriticality ramp. The amplitude in the far field is no longer constant, but it remains a slaved variable, so that $\rho = \sqrt{\mu}$ in leading order. Due to the variability, an extra term $\mu^{-1}\widehat{\nabla}\mu\cdot\widehat{\nabla}\theta = \mathbf{M}\cdot\widehat{\nabla}\theta$ is added to the far field equation (4.2.2). Clearly, $\mathbf{V}$ and $\mathbf{M}$ are collinear, and we can assume that both are directed along the Y-axis. Then the stationary far field equation in the comoving frame becomes

$$(V + M)\partial_Y\theta + \partial_X^2\theta + \partial_Y^2\theta = 0. \tag{4.3.44}$$

This equation has the same form as (4.3.1), so, assuming there is no extrinsic phase gradient, we can write immediately the required inner asymptotics modifying (4.3.29):

$$\theta = N\phi - \tfrac{1}{2}\epsilon N(V + M)r\cos\phi\,\ln\left(\tfrac{1}{4}\epsilon(V + M)r\mathrm{e}^{\gamma-\frac{1}{2}}\right). \tag{4.3.45}$$

Using (4.3.43), (4.3.45), the solvability condition (4.3.26) becomes

$$-(\mathbf{v}+\nabla\mu)\ln\left(\tfrac{1}{4}|\mathbf{v}+\nabla\mu|e^{\gamma-\frac{1}{2}}\right)=\mathbf{v}\ln a_0+\nabla\mu\ln a_m, \tag{4.3.46}$$

where we have rescaled to the short-scale units and restored the vector notation. Note that, as long as $a_m \neq a_0$, the relation between the velocity and the local supercriticality gradient is *nonlinear*, although, at small $|\mathbf{v}|, |\nabla\mu|$ when this formula is applicable, the deviation from the "naive" linear dependence $\mathbf{v} = -\nabla\mu$ is small.

One can combine motion under the action of both amplitude and phase gradients. The mobility relation including the local phase gradient is just a merger of (4.3.33) and (4.3.46):

$$\mathbf{k}=\frac{N}{2}\left[\ln\frac{v_0}{|\mathbf{v}+\nabla\mu|}\mathcal{J}\mathbf{v}+\ln\frac{v_m}{|\mathbf{v}+\nabla\mu|}\nabla\times\mu\right], \tag{4.3.47}$$

where $v_m = (4/a_m)\exp(-\gamma+\frac{1}{2}) \approx 5.55$.

Even when no extrinsic inhomogeneities are present, vortices themselves may, in effect, serve as a source of variable supercriticality in their neighborhood. If the real amplitude is presented by a product of vortex amplitudes as in (4.2.4), the action of other vortices effectively reduces the linear growth rate at the vortex location. This should cause *mutual attraction* of all vortices, irrespectively of charge. The topological (charge-dependent) interaction, mediated by the phase field, is, however, stronger, since the phase gradient decays at large distances as L^{-1}, compared with an L^{-2} decay of amplitude inhomogeneities. Topological repulsion of like vortices can be compensated by nontopological attraction, producing an unstable equilibrium at some critical separation. However, this configuration, which corresponds to an incipient break-up of a double-charged vortex, would be attained at distances where the perturbation theory is no longer applicable.

4.4 Propagation of Pattern and Pinning

4.4.1 Wavelength Selection in a Propagating Pattern

When a stable pattern advances into an unstable homogeneous state, the problem of selection of the propagation speed considered in Sect. 2.1.4 is augmented by the problem of selection of the wavelength of the emerging pattern. As a rule, there is a one-parametric family of patterned states parametrized by wavelength, which expands into a two-parametric family of propagating solutions. With all this, we expect that a unique propagation speed and a unique wavelength will be selected dynamically.

The simplest nontrivial example of selection of a patterned state gives the Swift–Hohenberg model (1.1.3), which we shall write here in the 1D form

$$\partial_t u = -(1+\partial_x^2)^2 u + \mu u - u^3. \tag{4.4.1}$$

We assume, as in Sect. 2.1.4, that selection is determined by the leading edge of the propagating pattern and use the dispersion relation computed according to (2.1.34), (2.1.35):

$$\omega(k) = \mathrm{i}[\mu - (1 - k^2)^2]. \tag{4.4.2}$$

The group velocity is

$$\widehat{c} = \omega'(k) = 4\mathrm{i}k(1 - k^2). \tag{4.4.3}$$

Using this expression, we find that the condition keeping $\widehat{c}$ real has a nontrivial solution

$$k_r = \sqrt{1 + 3k_i^2}. \tag{4.4.4}$$

The indices r, i denote here, as in Sect. 2.1.4, the real and imaginary parts, respectively.

The function $C(k_i) = \omega_i/k_i - \widehat{c}_r(k)$ defined by (2.1.44) and computed with the help of this relationship is monotonically decreasing, and vanishes at

$$k_i = \frac{1}{2\sqrt{3}}\left(\sqrt{1 + 6\mu} - 1\right)^{1/2}. \tag{4.4.5}$$

This defines the slope of the marginally stable amplitude profile. The propagation speed is now evaluated as

$$c = \widehat{c} = 8k_i(1 + 4k_i^2) = \frac{4}{3\sqrt{3}}\left(2 + \sqrt{1 + 6\mu}\right)\left(\sqrt{1 + 6\mu} - 1\right)^{1/2}. \tag{4.4.6}$$

The wavenumber k_r selected at the leading edge is not identical to the wavenumber of the full-grown pattern formed behind the front. The latter can be computed assuming that the number of wave crests does not change in the wake of the propagating front (Ben Jacob et al, 1985). The frequency of the emerging wave is determined by ω_r, which is Doppler-shifted in the moving frame to $\widehat{\omega} = \omega_r - ck_r$, and the wave crests are convected into the interior of the advancing pattern with the speed c. Therefore, the actual wavenumber of the stationary pattern left behind the propagating front must be

$$k^* = -\frac{\widehat{\omega}}{c} = k_r - \frac{\omega_r}{c} = \frac{3\left(3 + \sqrt{1 + 6\mu}\right)^{3/2}}{8\left(2 + \sqrt{1 + 6\mu}\right)}. \tag{4.4.7}$$

It is remarkable that the selected wavenumber coincides with the optimal value minimizing the energy functional of the SH model only in the limit $\mu \to 0$.

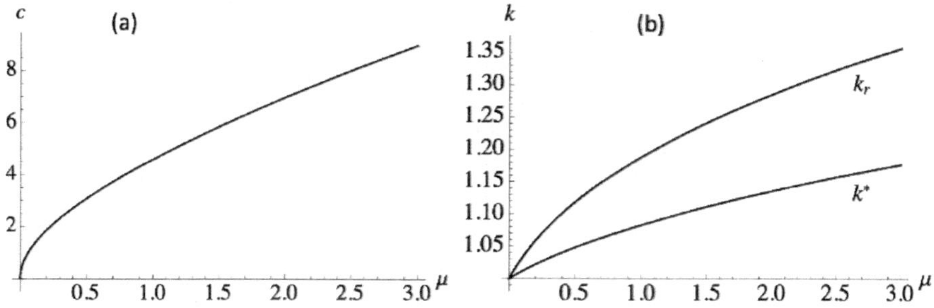

Fig. 4.14. Dependence of (a) the propagation velocity and (b) the wavenumber at the leading edge k_r and the selected wavenumber k^* on the supercriticality parameter μ

The dependence of k_r and k^* and c on μ is shown in Fig. 4.14. When the bifurcation point $\mu = 0$ is approached, these values vary as

$$k_i \approx \frac{1}{2}\sqrt{\mu}, \quad k_r \approx 1 + \frac{3}{8}\mu, \quad k^* \approx 1 + \frac{1}{8}\mu, \quad c \approx 4\sqrt{\mu}. \tag{4.4.8}$$

These predictions were confirmed with high precision by direct computations (Ben Jacob et al, 1985). An example of the front advancement is seen in Fig. 4.15a.

In spite of the success achieved in this and other simple cases, selection at the leading edge may have only a limited influence on the properties of the full-grown pattern, since nonlinear interaction behind the front may modify the geometry and affect the propagation speed. One can observe that, while the linear spreading dynamics of the SH equation is similar to that of the Cahn–Hilliard equation in Sect. 2.3.3, the dynamical nonlinear behavior of the patterns forming behind the front, which is determined by nonlinear interactions, is very different in the two cases. The pattern formed in the SH equation does not coarsen as in Fig. 2.12, even though the selected wavelength is not optimal. However, it may happen also in this case that the selected pattern is unstable, as seen, for example, in Fig. 4.15b. This particular computation was carried out for the SH equation modified by including a symmetry-breaking quadratic term:

$$\partial_t u = -(1 + \nabla^2)^2 u + u(\mu - u^2) + \nu u^2. \tag{4.4.9}$$

The simulation results show that another front establishing a homogeneous nontrivial stationary state is formed behind the front of the advancing pattern.

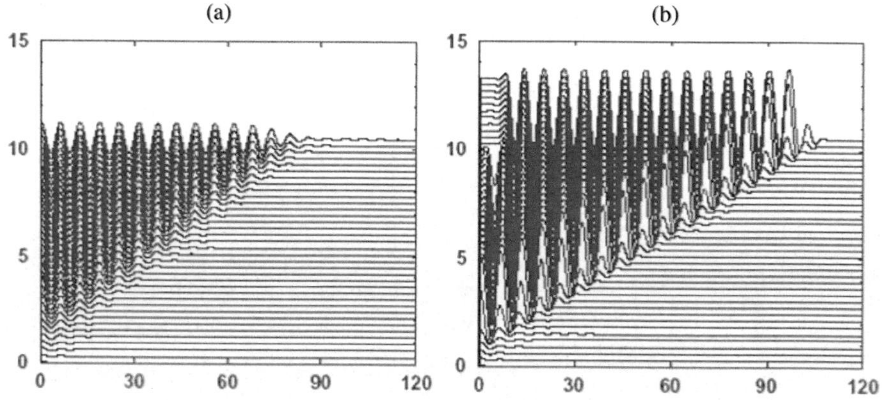

Fig. 4.15. Space-time plot of simulations results for pulled fronts in the SH equation with $\mu = 0.4$ (a) and the modified equation (4.4.9) with $\mu = 0.4$, $\nu = 3$ (b). The lines denote the profiles of u at successive time intervals and are shifted upward relative to each other (van Saarloos, 2003; reproduced by permission, ©Elsevier)

4.4.2 Self-Induced Pinning

In the case when a stable homogeneous solution coexists with a stable periodic pattern, Pomeau (1984, 1986) has described the mechanism of robust existence of stable stationary fronts between the two states. The motion of this front is affected by the *discrete* structure of the pattern, which causes *self-induced pinning* hindering the retreat of a metastable state. There are two *depinning* transitions, corresponding to "crystallization" or "melting" of the pattern, shown schematically by the thick lines in Fig. 4.16. To the right of the crystallization threshold C, the pattern advances by a periodic nucleation process which creates new elementary cells at the interface, while to the left of the melting limit M, the pattern recedes as elementary cells at the interface are destroyed.

A variety of metastable stationary structures including a different number of cells can be observed within the interval between the melting transition M and the single-cell spreading limit 1-C lying to the right of the crystallization threshold. The minimal number of cells in a sustainable stationary structure increases as the melting threshold is approached; the corresponding melting limits, starting from the single-cell melting limit 1-M, are shown schematically by thin vertical lines in Fig. 4.16. The respective crystallization thresholds for clusters of finite size lie to the right of C approaching this point as the cluster size increases.

Pinned structures may include a single-cell ("soliton") solution, a finite patterned inclusion sandwiched between semi-infinite domains occupied by a uniform state, or a semi-infinite pattern, coexisting with a uniform state. A soliton can be viewed as an island where one of the uniform states is approached, immersed in the infinite region occupied by an alternative state; the basic soliton "atom" is equivalent to a single period of the pattern immersed in a uniform state. A patterned inclusion can be also placed asymmetrically between the regions occupied by alternative uniform states. A variety of localized structures were reviewed by Knobloch (2015).

A system suitable for the study of pinned structures should be at least of the fourth order, and the simplest choice is the standard 1D SH equation (4.4.1), although somewhat more complicated equations were used in numerical computations: the modified SH equation (4.4.9) including a quadratic term (Hilali et al, 1995) and the quintic equation (Sakaguchi and Brand, 1996). All metastable pinned structures mentioned above were detected in these numerical experiments.

Fig. 4.16. A scheme of depinning transitions showing crystallization (C) and melting (M) thresholds for an infinite cluster, as well as the corresponding limits for clusters of different sizes, terminating in single-cell limits 1-C, 1-M

A small-amplitude striped pattern with the wavenumber $k = 1$ first appears in the SH equation at the symmetry-breaking bifurcation point $\mu = 0$. As the parameter μ further increases, the available wavenumber band widens, until it reaches the limiting value $k = 0$. This signals the appearance of a pair of nontrivial uniform states, $u = \pm\sqrt{\mu - 1}$. The two symmetric states are stable to infinitesimal perturbations at $\mu > 3/2$. Since the SH equation is a gradient system, one can use the expression for the energy functional

$$\mathcal{F} = \int \tfrac{1}{2} \left\{ [(1 + \nabla^2)u]^2 - \mu u^2 + \tfrac{1}{2} u^4 \right\} \mathrm{dx}. \tag{4.4.10}$$

to compute the energy of stationary solutions obtained in numerical computations, and characterize them as *metastable* states. The energy of the uniform state is higher than that of the regular pattern with the optimal wavenumber at $\mu < \mu_M \approx 5.7$, and all above solutions containing uniform stretches should be metastable as well.

In spite of the difference between the energies of the uniform and patterned state, the one-dimensional interface between them remains immobile at moderate and large values μ, and pinned fronts, as well as multiple localized states, remain stable to infinitesimal perturbations. The general scheme of crystallization and melting thresholds is the same as in Fig. 4.16, with μ increasing from the right to the left. Numerical computation (Aranson et al, 2000) shows that the depinning transition for a single soliton (point 1-C) occurs at $\mu \approx 1.74$. The depinning threshold increases with the size of the patterned cluster, rapidly converging to the limiting crystallization threshold for the semi-infinite pattern $\mu = \mu_c \approx 1.7574$ (point C).

Self-induced pinning impeding propagation of a pattern into a metastable uniform state ("crystallization") at moderate μ or the reverse "melting" process at $\mu \gg 1$ can be attributed to the oscillatory character of the asymptotic perturbations of the uniform state, as demonstrated by the phase space analysis based on the correspondence between spatial structures and trajectories of a dynamical system. A stationary pattern-forming equation, e.g., the SH equation rewritten as

$$(1 - \mu)u + u^3 + 2u''(x) + u''''(x) = 0, \tag{4.4.11}$$

can be viewed as a dynamical system with the coordinate playing the role of time. A regular pattern corresponds to a periodic orbit, a symmetric patterned inclusion to a homoclinic orbit, an asymmetric inclusion between two different homogeneous stationary states (HSS) – to a heteroclinic orbit, and a pinned front of a semi-infinite cluster – to a heteroclinic connection between a fixed point and a periodic orbit. Stable and unstable trajectories near hyperbolic points of higher-order dynamical systems may behave in a highly irregular way, leading to an infinite number of intersections. In the context of 1D patterns, this is translated to *snaking* trajectories that correspond to pinned periodic inclusions (Coullet et al, 2000).

The spatial inversion symmetry is translated to reversibility of the dynamical system (4.4.11), which can be transformed to a nondissipative equation of motion of two coupled particles. The conserved quantity is

$$\mathcal{H} = \tfrac{1}{4}(1 - \mu + u^2)^2 + u'(x)^2 + u'(x)u'''(x) - \tfrac{1}{2}u''(x)^2. \tag{4.4.12}$$

Both homoclinic and heteroclinic orbits lie on the 3D hypersurface $\mathcal{H} = 0$ in the 4D phase space of (4.4.11) spanned by u, $u'(x)$, $u''(x)$, $u'''(x)$. At $\mu > 3/2$, when the fixed points are dynamically stable as solutions of (4.4.1), they are double foci when viewed as solutions of the dynamical system (4.4.11). The two pairs of complex eigenvalues obtained by linearizing (4.4.11) near the fixed points have real parts of the opposite sign:

$$k = \pm\sqrt{\sqrt{\tfrac{1}{2}(\mu-1)} - \tfrac{1}{2}} \pm \mathrm{i}\sqrt{\sqrt{\tfrac{1}{2}(\mu-1)} + \tfrac{1}{2}}. \qquad (4.4.13)$$

A homoclinic orbit exits the fixed point along its unstable manifold spanned by the pair of eigenvalues with positive real parts and returns along its stable manifold spanned by the pair of eigenvalues with negative real parts. Both stable and unstable manifolds are two-dimensional, and, since they are confined to the 3D manifold $\mathcal{H} = 0$, they have to intersect generically along a line. Due to inversion symmetry, the homoclinic trajectories have to intersect the plane $u''(x) = u'''(x) = 0$; the stable and unstable manifolds are mirror images of each other with respect to this plane.

A heteroclinic orbit connects the unstable manifold of one fixed point with the stable manifold of its counterpart. This trajectory is antisymmetric and has to intersect the plane $u = u''(x) = 0$. Heteroclinic orbits owe their existence to the symmetry of (4.4.1) or (4.4.11) to the inversion of u and disappear when this symmetry is broken as in (4.4.9).

A periodic orbit must intersect the plane $u''(x) = u'''(x) = 0$ exactly twice. At any chosen value of $\mu > 0$, there is a continuum of periodic orbits with periods varying within a certain range. This family of orbits can be parametrized by the value of $\mathcal{H}$. Since $\mathcal{H}$ is conserved, only a single orbit corresponding to $\mathcal{H} = 0$ can connect to a fixed point. Thus, a pinned front, like a pulled front in Sect. 4.4.1, selects a certain wavelength of the pattern, which does not need to coincide with the optimal wavelength minimizing (4.4.10).

Genesis of pinned states can be understood (Coullet et al, 2000) by constructing a Poincaré map in the phase space of (4.4.11). The appropriate Poincaré section is the

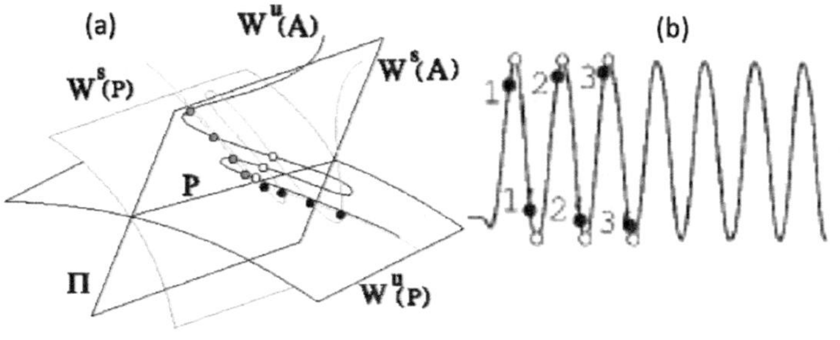

Fig. 4.17. (a) Phase portrait of (4.4.11) under conditions when pinned fronts exist. The stable and unstable manifolds are shown by the gray and black lines, respectively; see the text for explanations. (b) Correspondence between the marked points in (a) and elementary cells of pinned fronts (Coullet et al, 2000; reproduced by permission, ©The American Physical Society)

plane $u'(x) = u'''(x) = 0$, denoted as Π in Fig. 4.17a, which is crossed by all periodic trajectories. The family of periodic orbits parametrized by $\mathcal{H}$ is projected onto the line P in this plane. Each periodic orbit, under conditions when it is dynamically stable as a solution of (4.4.11), has a one-dimensional stable and a one-dimensional unstable manifolds. The stable and unstable manifolds of the family of periodic orbits span the surfaces $W^s(P)$ and $W^u(P)$ symmetrical with respect to Π (Fig. 4.17)a. The stable and unstable manifolds of the fixed points are also symmetrical with respect to Π. Trajectories $W^s(A)$ and $W^u(A)$ lying in these manifolds intersect, respectively, $W^u(P)$ and $W^s(P)$ at the points marked by filled circles in Fig. 4.17a when a pinned front exists. The intersections of these curves with Π are marked in Fig. 4.17a by empty circles; the corresponding points on a localized structures are shown in Fig. 4.17b.

4.4.3 Crystallization Kinetics

The front propagation process can be described in terms of periodic nucleation events triggered by explosive growth of the localized zero-eigenvalue mode of the corresponding linear problem (Aranson et al, 2000). Fig. 4.18a shows propagation of a semi-infinite pattern into a uniform stable state at $\mu_c = 1.75$. The front propagation takes the form of well separated in time periodic nucleation events creating new "atoms" of the "crystalline" state at the front. Between each successive nucleation events, the solution remains close to the stationary semi-infinite pattern found at μ_c. This process resembles crystallization in equilibrium solids, with the important distinction that the new "atoms" are created directly from the metastable "vacuum" state. The time between consecutive nucleation events diverges as μ approaches the pinning threshold. The average front speed c can be approximated as $c = c_0\sqrt{\mu_c - \mu}$ with $c_0 = 2.292$.

At $|\delta| = \mu - \mu_c > 0$, the immobile front solution $u_0(x, \mu)$ is linearly stable. Numerical stability analysis shows that there exists an exponentially decaying eigenfunction $\tilde{u}(x, \mu)$ localized at the front with a negative eigenvalue λ_s, which approaches zero as $|\delta|^{1/2}$. In addition, there is also an unstable front solution and a corresponding eigenfunction with a positive eigenvalue λ_u which also approaches zero as $|\delta|^{1/2}$, as shown in the inset of Fig. 4.18b. At the pinning threshold $\mu < \mu_c$, these two solutions collide and disappear via a saddle-node bifurcation, and the front solution becomes nonstationary. Nevertheless, the numerical study suggests that at $|\delta| \ll 1$ the solution remains close to the stationary front solution at the pinning threshold all the time except short intervals when a new cell nucleates.

The front propagation can be analyzed perturbatively near the critical value $\mu = \mu_c$. At $|\delta| \ll 1$, the front solution can be presented in the form $u(x, t) = u_0(x, \mu_c) + |\delta|^{1/2} u_1(x, t)$, where $u_0(x)$ is the stationary front solution at $\delta = 0$ and $u_1(x, t)$ is a small correction. This solution is uniformly valid at small positive δ; it is also valid during quasistationary phases between nucleation events for small negative δ. Plugging this *ansatz* in (1.1.3), we obtain in the leading order

$$\partial_t u_1 = L[u_0]u_1 - 3|\delta|^{1/2}\left[u_0 u_1^2 - \text{sign}(\delta)U_0\right], \tag{4.4.14}$$

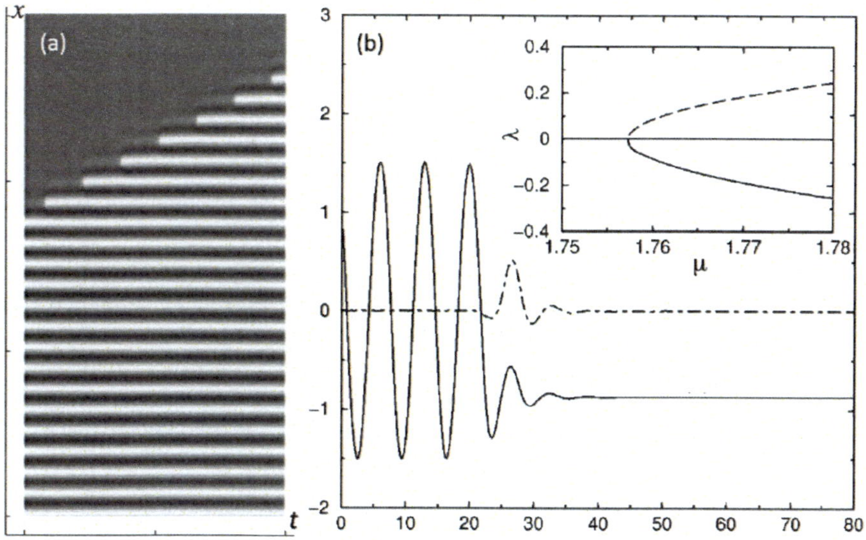

Fig. 4.18. (a) Space-time plot of the stick-and-slip propagation of a large cluster (approximating a semi-infinite pattern) into the metastable uniform phase at $\mu = 1.757$. The horizontal axis is spatial coordinate, and the vertical axis is time (only a part of the cluster near the advancing boundary is shown). (b) A semi-infinite pattern $u_0(x)$ coexisting with a uniform state at $\mu - \mu_c$ (solid line) and the perturbation mode $u_1(x)$ (dashed-dotted line). Inset: the stable (solid line) and unstable (dashed line) eigenvalues as functions of μ. (Aranson et al, 2000, ©The American Physical Society)

where $L[u_0] = \mu_c - 3u_0^2 - (1+\nabla^2)^2$ is the linearized SH operator at μ_c. At $\delta = 0$, (4.4.14) yields the linear equation $\partial_t u_1 = L[u_0]u_1$, which is always verified by the stationary translational mode $u_0'(x)$. In addition, the linearized operator $L[u_0]$ has a localized neutral eigenfunction u_1 shown together with the stationary front solution $u_0(x, \mu_c)$ in Fig. 4.18b. Since all other eigenmodes have negative eigenvalues, the evolution of the system close to the bifurcation point can be reduced to single-mode dynamics near a saddle-node bifurcation point. Therefore, in the lowest order we can write $u(x,t) = u_0(x) + |\delta|^{1/2} a(t) u_1(x)$, with an amplitude $a(t)$.

Close to the bifurcation threshold, $\partial_t u_1 = a'(t)U_1$ can be treated as perturbation, and in the second order we derive

$$L[u_0]u_2 = a'(t)u_1 + |\delta|^{1/2}\left[3a^2 u_0 u_1^2 - \text{sign}(\delta)u_0\right]. \tag{4.4.15}$$

This equation has a bounded solution if its right-hand side is orthogonal to the zero mode u_1 of the operator L. This leads to the solvability condition defining the dynamic equation of the amplitude a:

$$\alpha a'(t) = |\delta|^{1/2}\left[\text{sign}(\delta)\beta - \gamma a^2\right], \tag{4.4.16}$$

$$\alpha = \int_{-\infty}^{\infty} u_1^2 \mathrm{d}x, \quad \beta = \int_{-\infty}^{\infty} u_1 u_0 \mathrm{d}x, \quad \gamma = 3\int_{-\infty}^{\infty} u_1^3 u_0 \mathrm{d}x.$$

At $\delta > 0$, $a(t)$ reaches a stationary amplitude $a_0 = (\beta/\gamma)^{1/2}$. This value corresponds to the difference between the stable front solutions at μ and μ_c. At small *negative* δ, (4.4.16) describes *explosive* growth of the amplitude, which passes from $-\infty$ to ∞ in a finite time $\tau_e = \pi\alpha/(|\delta|\beta\gamma)^{1/2}$. This explosion time gives an upper bound for the period between the nucleation events, after which the whole process repeats. The front speed is computed as $c = \Lambda/\tau_e$, where Λ is the asymptotic spatial period of the pattern selected by the nucleation process, and τ_e is the time interval between nucleation events. The calculations give the value $c = 1.8|\delta|^{1/2}$. This is in a qualitative agreement with the results of the numerical simulations, but the prefactor 1.8 is noticeably lower than the corresponding value $c_0 = 2.29$ obtained by numerical simulation of the SH equation. This discrepancy may be caused by distortions behind the moving front triggered by nucleation events. These distortions may effectively provoke a consequent nucleation event by creating an initial perturbation of the zero mode $u_1(x)$ thereby leading to an increase of the front velocity. This effect, apparently, goes beyond what a perturbation theory can capture.

The above results cannot be trivially extended to a regular 2D periodic structure (striped pattern) selected by the SH equation (1.1.3). By analogy with 1D dynamics, one could expect a depinning mechanism via local nucleation of new stripes of the patterned state, which would subsequently spread out sidewise. However, the actual picture seen in computations (Hagberg et al, 2006) is completely different. It is initiated by a zigzag instability of the pattern followed by nucleation of convex–concave disclination pairs, with the convex disclinations moving toward the uniform

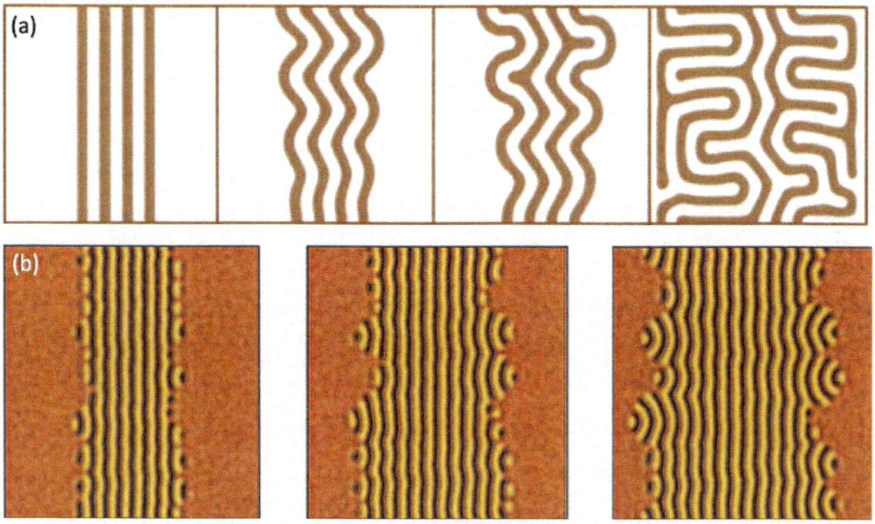

Fig. 4.19. Depinning of a striped pattern in 2D. (a) Simulation of the SH equation at $\mu = 4$ (Hagberg et al, 2006, ©Elsevier). (b) Simulation of the quintic equation (4.4.17) at $\mu = 0.07$, $\nu = 0.7$, $q = 0.5$ with added noise (Clerc et al, 2022, reproduced by permission, ©Elsevier)

state, as seen in Fig. 4.19a. This generates stripes extending in the normal direction, turning eventually the original boundary into a domain wall separating striped patterns rotated by $\pi/2$. The pinning range shrinks drastically to the vicinity of μ_M due to the more efficient 2D depinning mechanism. A zigzag instability was observed as well in simulations of the quintic model (Clerc et al, 2008, 2022):

$$\partial_t u = \mu u + \nu u^3 - (q^2 + \nabla^2)^2 u - u^5, \tag{4.4.17}$$

but it developed in a different way, as seen in Fig. 4.19b.

4.4.4 Pinning of Defects

The short-scale structure of the pattern may cause likewise pinning of defects – dislocations or domain walls – which appear to be mobile on the level of amplitude equations. Close to a primary bifurcation, this phenomenon can be understood as a "nonadiabatic" effect dependent on oscillatory corrections to the amplitude equations (Pomeau, 1984; Bensimon et al, 1988; Malomed et al, 1990). We recall that the common procedure of derivation of amplitude equations (Sects. 1.3.3, 1.5.2) based on computation of solvability conditions selects resonant terms only, while all terms oscillating with a period differing from the period of the basic structure, e.g., the term proportional to the cube of the amplitude, vanish upon integration. When the amplitudes change on an extended scale, the integrals including nonresonant terms do not, however, vanish precisely, but leave a remainder, which depends *exponentially* on the small parameter ϵ measuring the ratio of the basic scale of the pattern to the characteristic scale of the amplitude change. The amplitude equations corrected to include nonadiabatic effects *retain* these terms.

This introduces a short-scale perturbation which acts as a shallow periodic potential on the motion of defects described in the leading order by common amplitude equations. For example, a striped pattern with the wavenumber k_0 and the wave vector directed along the z-axis introduces through a cubic nonlinearity in underlying equations a potential proportional to $\cos\left(2\epsilon^{-1}k_0 x\right)$, where the large factor ϵ^{-1} appears when the coordinate x is measured in the long units of the amplitude equations. The

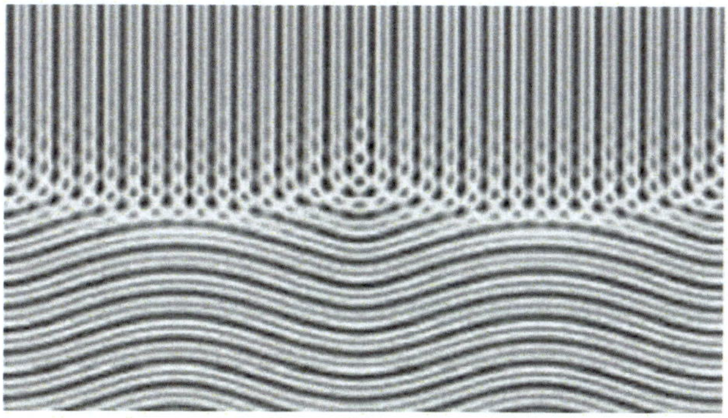

Fig. 4.20. A domain wall configuration with a curved set of stripes obtained by numerical solution of the SH equation (Boyer and Viñals, 2001, reproduced by permission, ©The American Physical Society)

corrected equation of motion of a defect describing its position $X(t)$ is obtained then in the form

$$X(t) = c_0 + c_1 \cos\left(2\epsilon^{-1}k_0 x\right), \qquad (4.4.18)$$

where c_0 is the unperturbed speed determined on the amplitude equation level, and the parameter c_1 is proportional to $\exp(-q/\epsilon)$ with $q = O(1)$. If $c_0 < c_1$, (4.4.18) has alternating stable and unstable fixed points, and the defect is pinned at one of stable positions. Otherwise, the propagation speed oscillates on a short time scale, and, while c_0 exceeds c_1 only slightly, the motion is of a "stick-and-slip" kind similar to that shown in Fig. 4.18a.

Bensimon et al (1988) applied (4.4.18) to propagation and pinning of a front between a patterned and homogeneous states in the SH equation modified by adding a quintic term, which is, in essence, the same problem that was treated more precisely in Sect. 4.4.3. Malomed et al (1990) applied a similar equation with a more complicated periodic term to a domain wall between a hexagonal and a striped pattern. Boyer and Viñals (2001, 2002) considered in the same way a domain wall parallel to the direction of stripes on the one side and normal to the direction of stripes on the other side described by the stationary equations (4.2.22), (4.2.23). This configuration turned dynamic through long-scale perturbations of the parallel set of stripes, as shown in Fig. 4.20, and the resulting motion was analyzed with the help of nonadiabatic equations applicable to a weakly distorted structure.

The same approach can be applied to motion of dislocations. The theory based on envelope equations does not distinguish between *climbing* of dislocations along the direction of stripes and their *gliding* in the direction of the wavenumber vector, so that interaction of dislocation pairs seen in Figs. 4.4d, e is indistinguishable in this approximation. However, motion across the stripes is affected by the discrete short-scale structure of the pattern, which may cause either motion with a variable speed or pinning. The action of a periodic potential on interaction of dislocation pairs is illustrated schematically in Fig. 4.21. The method used in Sect. 4.3.5 is not applicable in this situation, since the perturbation is short-scale, but one can expect (4.4.18) to be applicable, at least qualitatively, also in this situation.

Generally, nonadiabatic corrections are not based on the firm ground of a rational perturbation theory. Due to the exponentially weak dependence on the scale ratio, they can be substantial only when this ratio is not too small. Strictly speaking, all higher order algebraic terms, which are neglected in a usual amplitude expansion, decay at $\epsilon \to 0$ slower than exponentially, and retaining an exponentially small term

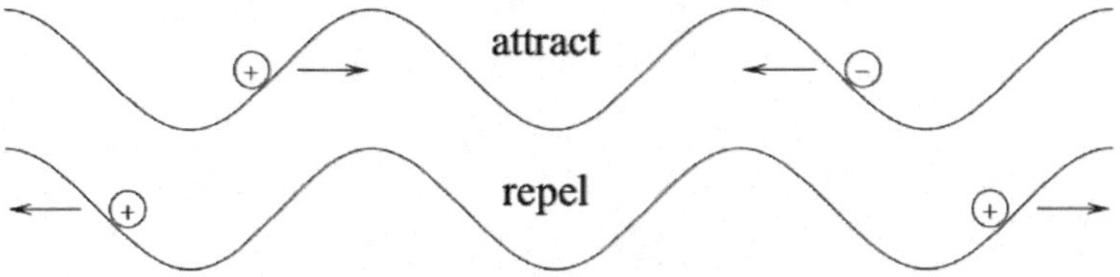

Fig. 4.21. Attracting and repelling dislocation pairs under the action of a periodic potential

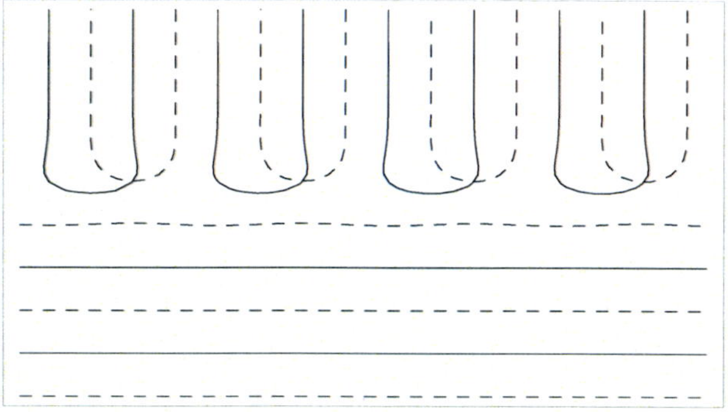

Fig. 4.22. Loci of Re $u = 0$ (*solid lines*) and Im $u = 0$ (*dashed lines*) intersecting at dislocations for the normal/parallel domain wall

while neglecting algebraically small ones cannot be formally justified. Nevertheless, adiabatic corrections give a faithful, at least in a qualitative sense, model of defect pinning conforming to direct numerical computations.

Beneath the level of envelope equations, there is a direct connection between dislocations and domain walls, since the latter can be viewed as *dislocation chains*. This short-scale structure revealed in both experiments and simulations, as e.g. in Figs. 4.3 and 4.20, but, of course, it remains hidden when the amplitude equations are solved as in Sect. 4.2.6. It is seen most clearly in Fig. 4.22 where the loci of Re $u = 0$ and Im $u = 0$ are drawn for the normal/parallel domain wall with $\rho_\pm$ solving (4.2.22), (4.2.23).

La Porta and Surko (2000) used the representation of a domain wall as a dislocation chain for reconstruction of convection patterns observed experimentally. The pattern in a circular cell seen in Fig. 4.23a contains a number of dislocation chains forming domain walls, as well as isolated dislocations. All these defects, supplemented by compensating defects with the opposite topological charge placed on the cell boundaries, are extracted in Fig. 4.23b. It is remarkable that the entire pattern can be approximated by the superposition of the "amplitude fields" (4.2.4) of these defects. The reconstruction becomes most faithful (Fig. 4.23c) if a pitch is added to the phase field of the dislocations, i.e., the symmetric structure $\theta_i = \phi_i$ in the polar coordinate system r_i, ϕ_i centered on the ith defect is replaced by a spiral phase field $\theta_i = \phi_i + \psi r_i$. In a later work (Aegerter and Surko, 2002) additional compression of the phase field with empirically adjusted parameters was introduced to further improve the fit.

This representation was motivated by the possibility of describing evolution of a distorted pattern in "particle-field" approximation, which reduces the continuous "field" problem to motion of a finite number of "particles" – defects in the manner of Sect. 4.3. This prospect is, however, illusory, since only widely separated defects rather than strongly coupled arrays are amenable to this technique.

Pinning of both dislocations and domain walls may arrest evolution to an ordered state even when the system has a gradient structure. Computations of Boyer and Viñals (2002) using the SH model suggested that the characteristic scale of the structure (or of a typical size of ordered domains), which increases with time as

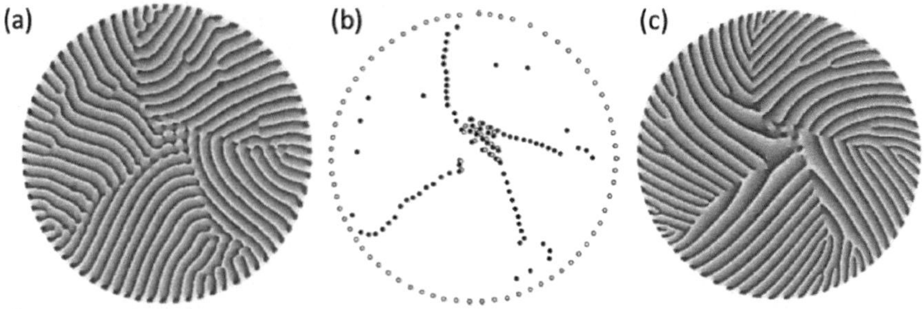

Fig. 4.23. A pattern in a circular cell (a) and its reconstruction (c) as a superposition of the phase field of the array of defects shown in (b). The filled and open circles mark the defects with the positive and negative charge, respectively (La Porta and Surko, 2000; reproduced by permission, ©Elsevier)

$t^{1/3}$ at intermediate stages of a computation starting from random initial conditions, saturates at long times, leaving a "glassy" state containing dislocations, disclinations, and domain walls, similar to that shown in Fig. 4.3.

4.5 Hexagonal Patterns

4.5.1 Triplet Amplitude Equations

As we have already noticed in Sect. 1.5.1, the hexagonal pattern in 2D is supported by resonant interaction among a triplet of modes with wave vectors forming an equilateral triangle. The spatial extension of the amplitude equations (1.5.14) can be written in a rescaled form

$$\partial_t u_j = (\mathbf{n}_j \cdot \nabla)^2 u_j + u_j \left[\mu - |u_j|^2 - \nu\left(|u_{j-1}|^2 + |u_{j+1}|^2\right)\right] + \overline{u}_{j-1}\overline{u}_{j+1}. \quad (4.5.1)$$

Here $\mathbf{n}_j$ is the unit vector along the wave vector corresponding to the jth mode (indexed modulo 3); otherwise, the notation is the same as in (1.5.14). These equations are derivable from the Lagrangian

$$\mathcal{L} = \sum_{j=0}^{2} (\mathbf{n}_j \cdot \nabla)\overline{u}_j \, (\mathbf{n}_j \cdot \nabla) u_j + V, \quad (4.5.2)$$

where the potential V is defined by (1.5.15). The differential term is written in the form first used by Pomeau (1986), including the derivative in the direction of the wave vector only. This is sufficient here since, unlike a striped pattern, a rigid structure of a hexagonal pattern is supported by resonant quadratic interactions, which makes operators of the NWS type (including higher order derivatives in transverse directions) superfluous.

If the amplitudes and the spatial derivatives are scaled as $u = O(\epsilon)$, $\nabla = O(\epsilon)$, the orders of magnitudes in (1.5.14) formally match only when parameters of an underlying problem are chosen in such a way that the lowest order (quadratic) interactions are weak. However, omitting the cubic terms is impossible, since the pattern cannot be otherwise stabilized at finite amplitudes (see Sect. 1.5.1). With this scaling, (1.5.14) can be supplemented by *nonlinear* differential terms of the same formal order of magnitude. The two additional terms in (4.5.1) admissible by symmetry are

$$\begin{aligned} &\mathrm{i}\beta_s(\overline{u}_{j+1}\,\mathbf{n}_{j-1}\cdot\nabla\overline{u}_{j-1}+\overline{u}_{j-1}\,\mathbf{n}_{j+1}\cdot\nabla\overline{u}_{j+1}) \\ &+\mathrm{i}\beta_a(\overline{u}_{j+1}\,\mathbf{t}_{j-1}\cdot\nabla\overline{u}_{j-1}-\overline{u}_{j-1}\,\mathbf{t}_{j+1}\cdot\nabla\overline{u}_{j+1}), \end{aligned} \tag{4.5.3}$$

where $\mathbf{t}_j$ is the vector normal to $\mathbf{n}_j$ and β_s, β_a are coefficients, which can be obtained by bifurcation expansion of a particular underlying system. The first, symmetric term, introduced by Brand (1989), describes nonlinear action of a small change of the wavelengths of the constituent modes, while the second, antisymmetric term (Kuznetsov et al, 1995) describes a more subtle effect of a change in their orientation. Both terms are incompatible with the inversion symmetry and are formally smaller than the algebraic quadratic term. Moreover, they cannot appear in a bifurcation expansion of a reaction–diffusion system with constant diffusivities, since they are dependent on the presence of nonlinear differential terms in underlying equations.

The only known system where the second (but not the first) term in (4.5.3) appears generically in the lowest order of the bifurcation expansion is convection in a fluid layer heated from below (Kuznetsov et al., 1995). In the commonly used Boussinesq approximation, both the algebraic quadratic term and the symmetric term are forbidden due to the momentum conservation law in the underlying hydrodynamic equations (Gershuni and Zhukhovitsky, 1976), and the only surviving quadratic term is the antisymmetric differential term, which appears whenever the boundary conditions on the top and at the bottom of the fluid layer are not symmetric. Computations of Kuznetsov et al., (1995) gave, however, a quite small value of β_a for a realistic case of rigid bottom and free top boundary conditions.

Generically, the orders of magnitude of all terms can match near a manifold of a higher codimension where the algebraic quadratic interaction term vanishes. Then both symmetric and antisymmetric terms can appear in the third order of the bifurcation expansion. For example, starting from the modified SH equation with two symmetry-breaking quadratic terms,

$$u_t = -(1+\nabla^2)^2 u + u(\mu - u^2) + \nu u^2 + \beta|\nabla u|^2, \tag{4.5.4}$$

one can annul the quadratic term in the second order of the bifurcation expansion near the critical point $\mu = 0$ by setting $\nu = \lambda/2$. Then the nonlinear differential term added to (4.5.1) is

$$-\mathrm{i}\beta\,(\overline{u}_{j+1}\,\mathbf{n}_{j+1}\cdot\nabla\overline{u}_{j-1}+\overline{u}_{j-1}\,\mathbf{n}_{j-1}\cdot\nabla\overline{u}_{j+1}), \tag{4.5.5}$$

which can be split into symmetric and antisymmetric terms as in (4.5.3) with $\beta_s = \beta/2$, $\beta_a = \sqrt{3}\beta/2$.

The polar form of (4.5.1) with added nonlinear differential terms is

$$\dot{\rho}_j = (\mathbf{n}_j \cdot \nabla)^2 \rho_j + \rho_j \left[\mu - (\mathbf{n}_j \cdot \nabla \theta_j)^2 - \rho_j^2 - \nu \left(\rho_{j-1}^2 + \rho_{j+1}^2\right)\right] \\ + f(\theta_{j-1}, \theta_{j+1}) \rho_{j+1} \rho_{j-1} \cos \Theta, \tag{4.5.6}$$

$$\dot{\theta}_j = (\mathbf{n}_j \cdot \nabla)^2 \theta + \frac{2\mathbf{n}_j \cdot \nabla \theta_j}{\rho_j} \mathbf{n}_j \cdot \nabla \rho_j - f(\theta_{j-1}, \theta_{j+1}) \frac{\rho_{j+1}\rho_{j-1}}{\rho_j} \sin \Theta, \tag{4.5.7}$$

where $\Theta = \sum \theta_j$ is the total phase, and the function

$$f(\theta_{j-1}, \theta_{j+1}) = 1 + (\beta_s \mathbf{n}_{j-1} + \beta_a \mathbf{t}_{j-1}) \cdot \nabla \theta_{j-1} + (\beta_s \mathbf{n}_{j+1} - \beta_a \mathbf{t}_{j+1}) \cdot \nabla \theta_{j+1} \tag{4.5.8}$$

lumps contributions of the resonant quadratic terms.

Out of the six real variables defining the complex amplitudes of the resonantly interacting triplet, four (the three real amplitudes ρ_j and the total phase Θ) affect the potential energy defined by (1.5.15), while the remaining two correspond to the global translational symmetries. On a scale large compared to the characteristic scale of (4.5.1), it is sufficient to consider the two translational modes. Equations of phase dynamics describing their evolution on an extended scale can be obtained by adiabatic elimination of ρ_j and Θ. The general form of the phase equation is quite complicated and cannot be reduced to a simple form $\dot{\phi} = D_{\parallel} \nabla^2 \phi + D_{\perp} \nabla(\nabla \cdot \phi)$ postulated by Lauzeral et al, (1993). In the absence of nonlinear differential terms, the phase equation can be derived from the Lagrangian (4.5.2) using the same method as in Sect. 4.1.4 (Hoyle, 2000). A simplified version of the phase equation will appear in the following subsections.

The nonlinear differential terms break the gradient structure of the amplitude equations and thereby make nontrivial dynamic behavior possible. This can be proved by trying to construct a suitable first-order Lagrangian, which should have the form $u_1 u_2 \, \mathbf{m}_3 \cdot \nabla u_3 +$ (circular transpositions), where $\mathbf{m}_j$ is a unit vector in some direction, so far unknown. Variation with respect to u_1 gives $u_2(\mathbf{m}_3 - \mathbf{m}_1) \cdot \nabla u_3 + u_3(\mathbf{m}_2 - \mathbf{m}_1) \cdot \nabla u_2$, and to obtain the required term, we need $\mathbf{m}_3 - \mathbf{m}_1 = \mathbf{n}_3$, etc., altogether six incompatible conditions on three vectors $\mathbf{m}_j$.

4.5.2 Skewed Triplets

Antisymmetric quadratic terms encourage formation of distorted "skewed" hexagonal patterns with angles between the constituent modes differing from $\pi/3$. This peculiar effect can be understood in its purest form by considering the amplitude equation retaining the antisymmetric quadratic terms only:

$$\dot{u}_j = (\mathbf{n}_j \cdot \nabla)^2 u_j + u_j \left[\mu - |u_j|^2 - \nu \left(|u_{j-1}|^2 + |u_{j+1}|^2\right)\right] \\ + \mathrm{i}\beta_a (\overline{u}_{j+1} \mathbf{t}_{j-1} \cdot \nabla \overline{u}_{j-1} - \overline{u}_{j-1} \mathbf{t}_{j+1} \cdot \nabla \overline{u}_{j+1}). \tag{4.5.9}$$

We restrict to patterns with a nearly optimal wavelength. Distorted patterns satisfying this condition can be constructed by setting $u_j = \rho_j \mathrm{e}^{\mathrm{i}\theta_j}$ with a phase distortion wave vector $\mathbf{q}_j = \nabla \theta_j$ normal to the basic wave vector of the respective mode, $\mathbf{n}_j \cdot \mathbf{q}_j = 0$. Under this assumption, the polar form of (4.5.9) becomes

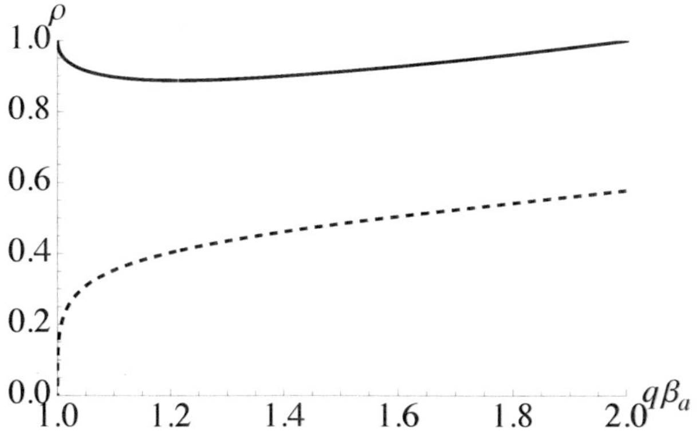

Fig. 4.24. Dependence of the amplitudes of the undistorted (ρ_1, solid line) and skewed ($\rho_2 = \rho_3$, dashed line) modes on the wave vector rotation q for $\nu = 2$

$$\dot{\theta}_j = -\beta_a \frac{\rho_{j+1}\rho_{j-1}}{\rho_j}(q_{j-1} - q_{j+1}) \sin\Theta, \tag{4.5.10}$$

$$\dot{\rho}_j = (\mathbf{n}_j \cdot \nabla)^2 \rho_j + \rho_j \left[\mu - \rho_j^2 - \nu\left(\rho_{j-1}^2 + \rho_{j+1}^2\right)\right] + \beta_a \rho_{j+1}\rho_{j-1}(q_{j-1} - q_{j+1}) \cos\Theta, \tag{4.5.11}$$

where $q_j = |\mathbf{q}_j|$, $\Theta = \sum \theta_j$.

Consider stability of a striped pattern with the optimal wavelength $u_1 = \sqrt{\mu}$, $u_2 = u_3 = 0$ to excitation of the side modes u_2, u_3. Assuming the amplitudes of the side modes to be equal ($\rho_2 = \rho_3 = \widetilde{\rho} \ll 1$), the phase equations for the combinations $\theta_\pm = \theta_2 \pm \theta_3$ derived from (4.5.10) become

$$\dot{\theta}_+ = \beta_a \sqrt{\mu}(q_3 - q_2) \sin\theta_+, \quad \dot{\theta}_- = \beta_a \sqrt{\mu}(q_3 + q_2) \sin\theta_+. \tag{4.5.12}$$

One can set here $q_2 = -q_3 = q$, $\theta_- = 0$. Then, provided q and β_a have the same sign, $\theta_+ = \Theta$ relaxes to zero (modulo 2π), and the equation for the perturbation $\widetilde{\rho}$ following from (4.5.11) is

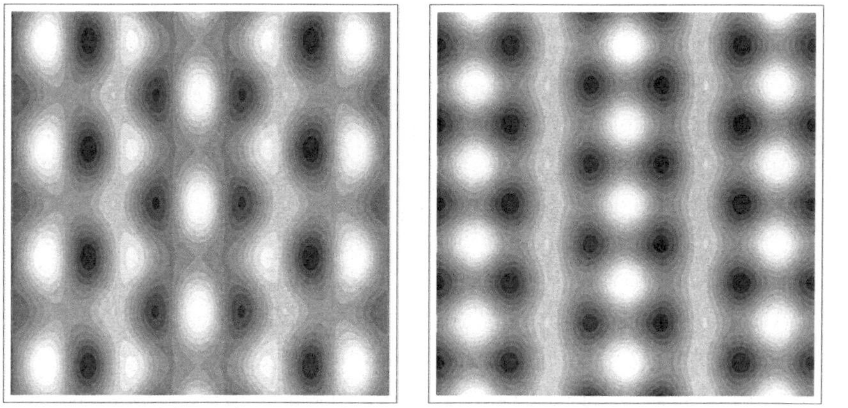

Fig. 4.25. Reconstructed patterns with $q = 1/12$ (left) and $q = -1/12$ (right)

$$\dot{\tilde{\rho}} = \tilde{\rho}\left[\mu(1-\nu) + q\beta_a\sqrt{\mu}\right], \tag{4.5.13}$$

indicating the sideband instability at $q\beta_a > \sqrt{\mu}(\nu - 1)$. The result does not change if either q or β_a flip the sign, since then θ_+ in (4.5.12) relaxes to π (modulo 2π), while (4.5.13) remains unchanged. Above this instability limit, there is a stable regular pattern formed by three modes with $\rho_1 \neq \rho_2 = \rho_3$. The dependence of both amplitudes on q, which can be obtained analytically in an implicit form[7], is shown in Fig. 4.24. The reconstructed patterns for $q = \pm 1/12,\ \beta_a = 24$ are shown in Fig. 4.25.

More general cases including all quadratic terms and patterns with a nonoptimal wavelength were considered by Echebarria and Pérez-García (1998), Nuz et al (1998, 2000), and Peña and Pérez-García (2000).

4.5.3 Penta–Hepta Defects

The resonant character of interactions among the modes forming a hexagonal pattern strongly affects the interaction of topological defects – dislocations in these modes. The experiments by Ciliberto et al. (1990) and Bodenschatz et al. (1992), as well as computations by Rabinovich and Tsimring (1993) showed that dislocations in two modes constituting the hexagonal pattern, created originally at arbitrary locations, are always *attracted* to each other, eventually forming an immobile bound pair corresponding to a penta–hepta defect in the hexagonal pattern (see Fig. 4.26). This behavior can be explained by the following qualitative argument (Ciliberto et al, 1990; Newell et al, 1993). If one of interacting modes has a defect, so that the respective phase has a quantum of circulation around a certain point (e.g., $\theta_2 = \phi$), the resonance condition $\Theta = \sum \theta_j = 0$ following from the polar form of the amplitude equations (1.5.9) will still be satisfied if another mode has a reverse circulation around the same point (e.g., $\theta_3 = -\phi$). The defects in resonantly interacting modes thus become *bound* to one another and, consequently, *confined*. However, pure phase rotation ($\theta_j = \pm\phi$) does not satisfy the amplitude equations (4.5.1) describing spatial

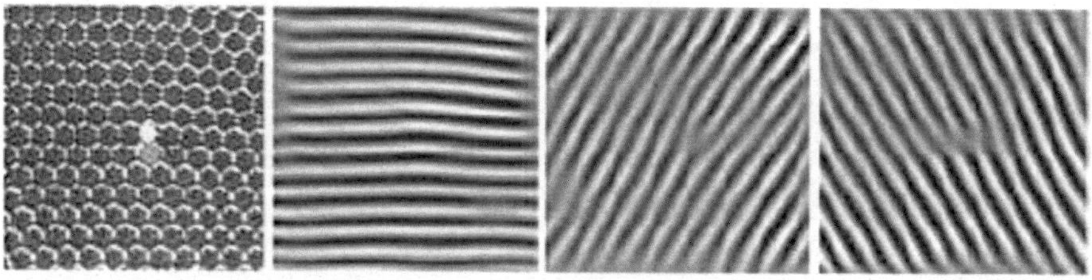

Fig. 4.26. Hexagonal pattern containing a penta–hepta defect (left) and its three constituent modes obtained by the Fourier filtering of the initial image (Abou et al, 2000; reproduced by permission, ©Cambridge University Press)

[7] All computations in this section are carried out assuming the value of the interaction parameter $\nu = 2$, which corresponds to a nonspecific (angle-independent) cubic interaction

modulation of amplitudes on an extended scale, so that the phase field of confined dislocations must have a more complicated structure, which can be obtained by solving these equations directly.

Pismen and Nepomnyashchy (1993) computed the *far field* of a dislocation by solving the equations of phase dynamics derived from a simplified form of (4.5.7), omitting the nonlinear differential terms and restricting to the band center. Under these conditions, the gradients of the real amplitudes can be neglected, and rescaling the coordinates by a small factor δ reduces (4.5.7) to

$$\partial_t \theta_j = \delta^2 (\mathbf{n}_j \cdot \nabla)^2 \theta_j - \sigma_j \sin\theta, \tag{4.5.14}$$

where $\sigma_j = \rho_{j+1}\rho_{j-1}/\rho_j > 0$.

On the $O(1)$ time scale, the total phase Θ relaxes to zero according to (1.5.9). The evolution equations for translationally invariant phase combinations, valid on an extended $O(\delta^{-2})$ time scale T, can be written as

$$\sigma_j^{-1} \left[\partial_T \theta_j - (\mathbf{n}_j \cdot \nabla)^2 \theta_j \right] = \text{idem}. \tag{4.5.15}$$

Setting

$$\mathbf{n}_1 = (1,\ 0), \quad \mathbf{n}_2 = (-1/2,\ \sqrt{3}/2), \quad \mathbf{n}_3 = (-1/2,\ -\sqrt{3}/2) \tag{4.5.16}$$

and presuming the real amplitudes ρ_j in the far field to be equal yields coupled equations for the phase combinations $\theta^{\pm} = \theta_2 \pm \theta_3$:

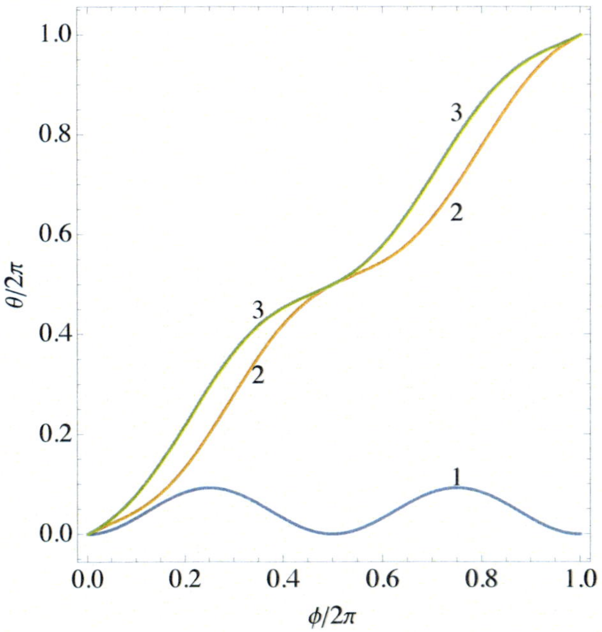

Fig. 4.27. The phases θ_1, θ_2, and $-\theta_3$ of the triplet of modes in the far field of the penta–hepta defect. The curves are marked by the indices of the respective modes

$$\frac{9}{4}\partial_x^2\theta^+ + \frac{3}{4}\partial_y^2\theta^+ = \frac{\sqrt{3}}{2}\partial_x\partial_y\theta^-, \quad \frac{1}{4}\partial_x^2\theta^- + \frac{3}{4}\partial_y^2\theta^- = \frac{\sqrt{3}}{2}\partial_x\partial_y\theta^+. \tag{4.5.17}$$

These equations are subject to the circulation conditions $\theta^-(2\pi) - \theta^-(0) = 4\pi$, $\theta^+(2\pi) = \theta^+(0)$.

Quite miraculously, eliminating θ^+ from this system of anisotropic equations yields an *isotropic* biharmonic equation $(\nabla^2)^2\theta^- = 0$. Transforming to the polar coordinates r, ϕ and assuming that the phases in the far field do not depend on the radial coordinate, we obtain

$$4\mathrm{d}_\phi^2\theta^- + \mathrm{d}_\phi^4\theta^- = 0. \tag{4.5.18}$$

The general solution (up to an arbitrary constant phase shift) of this equation, satisfying the circulation condition for θ^- and containing two integration constants a, b, is

$$\theta^- = 2\phi + a\sin 2\phi + b\cos 2\phi. \tag{4.5.19}$$

Using this in (4.5.17) one can see that both equations can be verified simultaneously if $a = -1/2$. The constant b is determined from the condition of zero circulation for θ^+. Adding also constant phase shifts to fix $\theta_j(0) = 0$ we cast the solution in the final form

$$\begin{aligned} \theta_1 &= -\theta^+ = \frac{1}{2\sqrt{3}}(1 - \cos 2\phi), \\ \theta_2 &= \frac{1}{2}(\theta^+ + \theta^-) = \phi - \frac{1}{2\sqrt{3}}\left[\frac{1}{2} + \cos\left(2\phi - \frac{2\pi}{3}\right)\right], \\ \theta_3 &= \frac{1}{2}(\theta^+ - \theta^-) = -\phi - \frac{1}{2\sqrt{3}}\left[\frac{1}{2} + \cos\left(2\phi + \frac{2\pi}{3}\right)\right]. \end{aligned} \tag{4.5.20}$$

The solution is plotted in Fig. 4.27. The phase field of each constituent mode contains, in addition to the appropriate circulation, a quadrupole component. The solution (4.5.20) plays, essentially, the same role as the basic solution $\theta = \phi$ for a vortex in

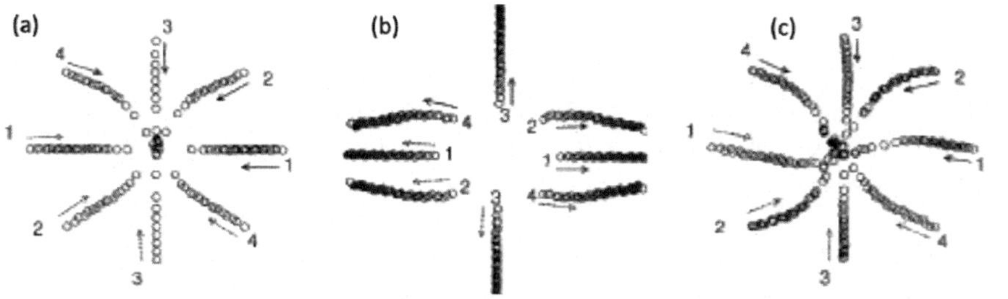

Fig. 4.28. Trajectories of interacting penta–hepta defects (Tsimring, 1996; reproduced by permission, ©Elsevier). Open circles indicate positions of the defect cores at constant time intervals. Arrows point toward directions of motion. The charges of dislocations in the individual modes (1,2,3) are $(0, 1, -1)$ and $(0, -1, 1)$ (a); $(0, 1, -1)$ and $(0, 1, -1)$ (b); $(0, 1, -1)$ and $(-1, 0, 1)$ (c). For each case, four sets of initial conditions are taken (labeled 1–4)

the isotropic RGL equation, or the vortex solution in Sect. 4.2.3 for a dislocation in the roll pattern described by the NWS equation.

The motion of the defect under the action of other defects or lateral boundaries must depend, as in the case of a striped pattern (Sect. 4.3), on the structure of the defect core where the real amplitudes and the total phase Θ change as well. It is clear that the real amplitudes of the modes containing dislocations (ρ_2 and ρ_3) should vanish at the origin, while the third amplitude (ρ_1) remains finite, though is likely to decrease due to suppression of the resonant term.

Mobility of interacting penta–hepta defects has been computed by Tsimring (1996). Interaction of defects is determined by the signs of dislocations in the individual modes: for example, the defects in Figs. 4.28a, c attract, while those in Fig. 4.28b repel. Trajectories of interacting defects are, generally, curvilinear, which can be attributed to the dipole component in the phase field computed above.

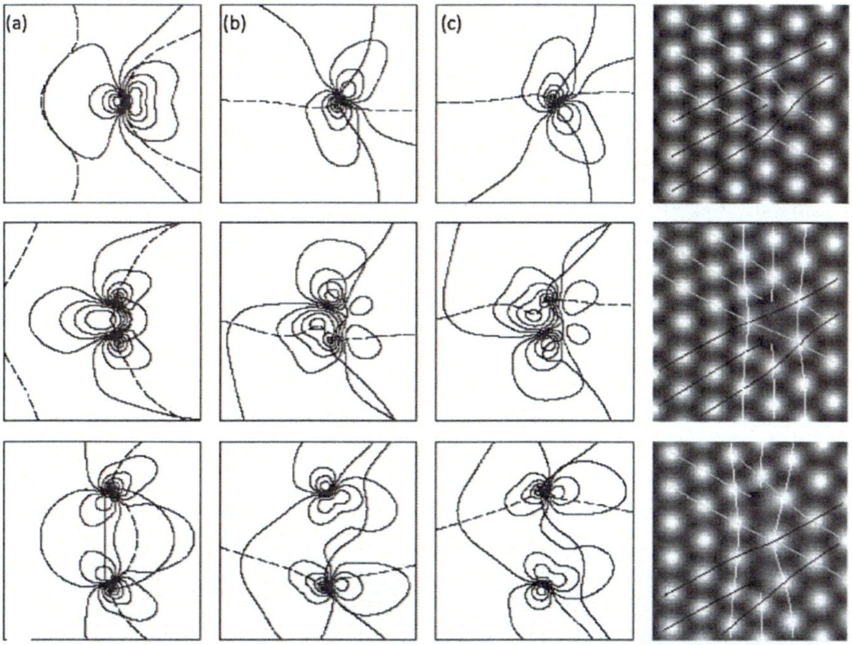

Fig. 4.29. Multiplication of a penta–hepta defect (Colinet et al, 2002, reproduced by permission, ©IOPscience). Plots (a)–(c) show contour lines of the three real amplitudes ρ_i (solid), together with the lines where the real and imaginary parts of the respective complex amplitudes u_i vanish (dashed). Defects are located at the intersection of these lines. The plots on the right show the reconstructed patterns. The lines drawn there by hand show that penta–hepta defects correspond to additional rows of hexagonal cells inserted in two of the three basic modes. Time increases from the top to the bottom row

Colinet et al. (2002) included in their study the influence of nonlinear differential terms (4.5.3). A curious phenomenon detected in the latter work is multiplication of defects at higher values of the supercriticality parameter μ. A new pair of dislocations is generated in the dislocation-free mode. Following this, the old penta–hepta defect separates into two dislocations, which recombine with the new pair of dislocations forming a pair of penta–hepta defects of the second generation (see Fig. 4.29). Such a multiplication can be repeated several times in a sufficiently large region.

4.5.4 Domain Boundaries

Both striped and hexagonal patterns are stable solutions of the amplitude equation (1.5.16) or (4.5.1) in the parametric region $\nu > 1$, $(\nu-1)^{-2} < \mu < (\nu+2)/(\nu-1)^2$ (see Sect. 1.5.2 and Fig. 1.11). In the subcritical region the hexagonal pattern coexists with a stable homogeneous state.

The structure of boundaries between alternative stable stationary patterns (stripes and hexagons), as well as between hexagons and a homogeneous state, was studied by Malomed et al (1990). In view of the resonant phase condition $\Theta = \sum \theta_j = 0$, one can restrict to real amplitudes when the wavelengths are optimal. The velocity c of a front separating the hexagonal pattern from either a striped pattern, where only a single mode of the resonant triplet persists, or a homogeneous state is obtained by solving the amplitude equations rewritten in the comoving frame:

$$\cos^2 \alpha_j\, \rho_j''(z) + c\,\rho_j'(z) + \rho_j\left[\mu - \rho_j^2 - \nu\left(\rho_{j-1}^2 + \rho_{j+1}^2\right)\right] + \rho_{j-1}\rho_{j+1} = 0, \quad (4.5.21)$$

where α_j is the angle between the axis z normal to the front and the wave vector of the jth mode.

In the amplitude equation formulation, the boundary is stationary when the respective potentials (1.5.14) are equal. Straightforward computation using the solutions obtained in Sect. 1.5.2 places this hexagon-stripe Maxwell construction μ_{13} and the hexagon-homogeneous Maxwell construction μ_{03} at

$$\mu_{13} = \frac{1 + 3\nu + \sqrt{2}(\nu+1)^{3/2}}{2(2\nu+1)(\nu-1)^2}, \quad \mu_{03} = -\frac{2}{9(1+2\nu)}. \quad (4.5.22)$$

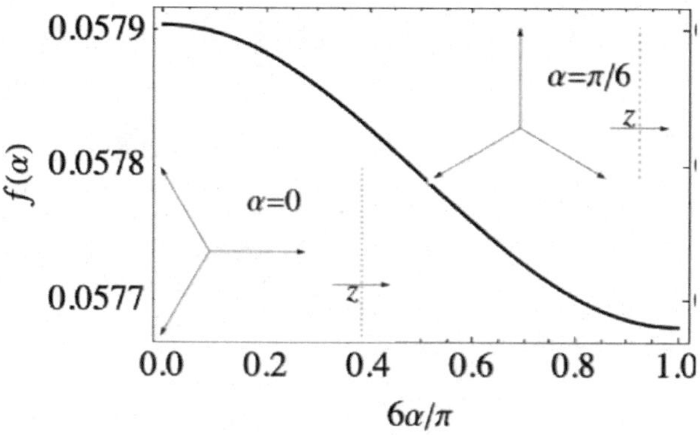

Fig. 4.30. The dependence of the interfacial energy of the stationary front between the hexagonal pattern and the homogeneous state on the orientation angle α for $\nu = 2$. The extreme configurations of the wave vectors of the constituent modes are shown in the insets

The front is *anisotropic*, and its energy depends on its orientation relative to the constituent modes. However, anisotropy is very weak, as seen from the orientation dependence of the interfacial energy of the front between hexagons and the homogeneous state shown in Fig. 4.30. The energy is maximal when the front is normal to the wave vector of one of the constituent modes ($\alpha = 0$) and minimal when the front is parallel to the wave vector of one of the constituent modes ($\alpha = \pi/6$). The angular dependence is well approximated by a simple harmonic function $f(\alpha) = 0.0578 + 1.12 \times 10^{-4} \cos(6\alpha)$. The same weak anisotropy persists in either stationary or propagating fronts between hexagons and stripes, where both the energy of a stationary front and the propagation speed depend, in addition, on the orientation of the stripes (Malomed et al, 1990).

It is notable that, due to the angular dependence of the effective amplitude diffusivities in (4.5.21), the more diffusive mode spreads out further into the region occupied by the homogeneous state (see Fig. 4.31). As a result, a band of a striped pattern parallel to the front may appear as an intermediate stage of the decay of the hexagonal pattern into homogeneous state. Domain walls may separate as well hexagonal patterns with different orientations. The short-scale structure of such a domain wall is resolved as a chain of penta–hepta defects (see Fig. 4.32a).

As in the case of striped patterns (Sect. 4.4.2), pinning of boundaries between the hexagonal pattern and homogeneous state may occur within a finite parametric interval near the Maxwell construction. Moreover, either isolated spots or finite clusters, which can be viewed as splinters of a hexagonal pattern, can be stable in a finite parametric interval. The "crystallization" and "melting" thresholds for such clusters can be depicted schematically as in Fig. 4.16. Some pictures obtained in 2D simulations (Coullet et al, 2000) are shown in Fig. 4.32. In Fig. 4.32b, corresponding to the interval between points C and 1-C in Fig. 4.16 where the patterned state is preferred, a single cell remains stationary, while larger clusters grow. As the balance shifts in favor of the uniform state, as in Fig. 4.32c, corresponding to the interval to the left of 1-M in Fig. 4.16, a single cell disappears, while larger clusters remain stationary. A great number of stable stationary states in the interval between the

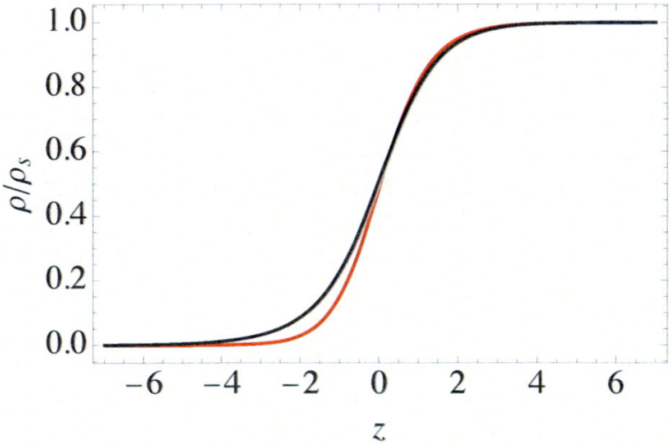

Fig. 4.31. The profile of a stationary front between the hexagonal pattern and the homogeneous state at $\nu = 2$, $\alpha = 0$. The black line denotes the amplitude of the mode parallel to the front, and the red line, the amplitudes of the two oblique modes

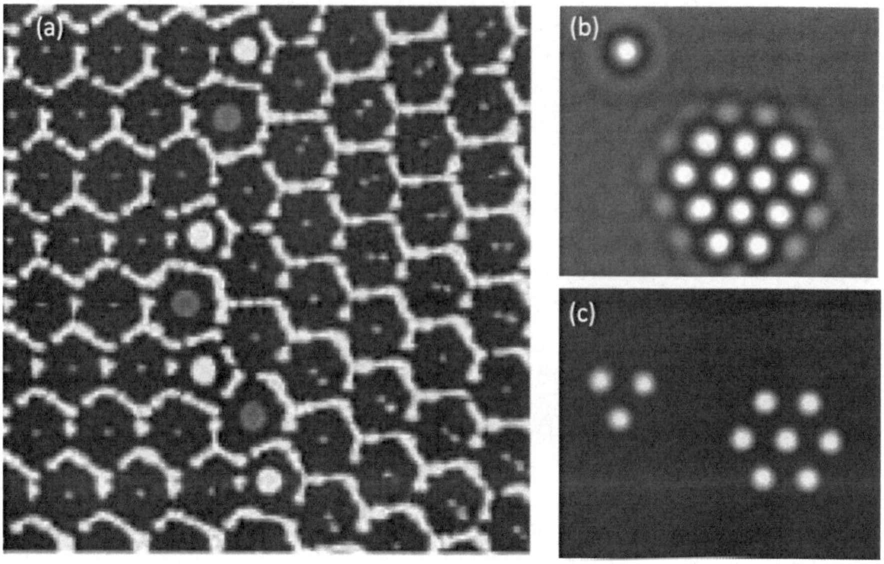

Fig. 4.32. (a) Domain boundary between hexagonal patterns with different orientations (Abou et al, 2000, reproduced by permission, ©Cambridge University Press). Gray and white circles mark heptagonal and pentagonal cells, respectively. (b,c) Localized structures obtained in two-dimensional simulations. In (a), the single cell is stable, but the larger structure is growing. In (b), one observes three-cell and seven-cell stationary clusters while a single cell cannot be stabilized (Coullet et al, 2000; reproduced by permission, ©The American Physical Society)

melting and crystallization transitions for finite clusters suggested a possible usage of such "incomplete" patterns for information storage (Coullet et al, 2000). Isolated cells or clusters can be "written" and "erased" at will by localized perturbations, and remain stationary otherwise within a certain parametric range, e.g., between points 1-C and 1-M in Fig. 4.16 for single cells or between points C and 1-M for clusters of arbitrary size.

4.5.5 Propagation of the Hexagonal Pattern

The angular dependence of the amplitude diffusivities also strongly influences the propagation of the hexagonal pattern into *unstable* equilibrium (Pismen and Nepomnyashchy, 1994). We shall again restrict to patterns with the optimal wavelength and use the real amplitude equations (4.5.21). For a pattern propagating along the direction of the wave vector of one of the modes, say, $\mathbf{k}_1$, taken as the x-axis, the amplitudes of the other two modes can be presumed equal due to the symmetry. Thus, (4.5.1) is rewritten as

$$\partial_t \rho_1 = \partial_x^2 \rho_1 + \rho_1 \left(\mu - \rho_1^2 - 2\nu\rho_2^2\right) + \rho_2^2, \tag{4.5.23}$$

$$\dot{\rho}_2 = \tfrac{1}{4}\partial_x^2 \rho_2 + \rho_2 \left[\mu - \nu\rho_1^2 - (1+\nu)\rho_2^2 + \rho_1\right]. \tag{4.5.24}$$

In the parametric interval $(\nu-1)^{-2} < \mu < (\nu+2)/(\nu-1)^2$, the hexagonal pattern is the only stable stationary state. It seems natural to apply under these conditions the linear theory of propagation speed selection at the leading edge of the propagating front (Sect. 2.1.4). The linear theory gives, however, no advantage whatsoever in the case of the hexagonal pattern, which is supported by *quadratic* interactions. Moreover, since the amplitude diffusivity of any particular mode is maximal in the direction of its wave vector, the mode with the wave vector parallel to the x-axis would have the largest propagation speed in this direction. According to the linear growth theory for a single reaction–diffusion equation in Sect. 2.1.6, the propagation speed of the front between the striped pattern and homogeneous state, denoted by the index (01), equals $c_{01} = 2\sqrt{\mu}$. The striped pattern emerging behind the front is, however, unstable. The question is whether the front of the stripes–hexagons (13) transition moving in the wake is able to catch up with the advancing front of the stripes.

The linear theory is also applicable to the leading edge of the hexagonal pattern propagating into the unstable striped pattern in the parametric range where hexagons is the only stable stationary state. The dispersion relation is obtained again by linearizing (4.5.24) in the vicinity of the single-mode solution $\rho_1 = \sqrt{\mu}$, $\rho_2 = 0$. Applying the general formula from (2.1.45) expressed as $c = 2\sqrt{\alpha D}$, where α is the linear growth coefficient of the side mode ρ_2 and $D = 1/4$, yields

$$c_{13} = \sqrt{\sqrt{\mu} - \mu(\nu - 1)}. \tag{4.5.25}$$

This speed exceeds c_{01} only at $\mu < (\nu + 3)^{-2}$, i.e., within a very narrow interval close to the critical point $\mu = 0$. Under these conditions, the hexagonal pattern advances directly into the homogeneous state with the speed $c_{03} = c_{01} = 2\sqrt{\mu}$. Faster propagation is, of course, impossible before a striped domain is created; thus, the (01) and (13) fronts *merge* into a direct (03) transition. At larger values of μ, the (01) front runs ahead, and, even though the striped pattern is unstable, a belt of stripes emerges as an intermediate state in the transition from the unstable homogeneous state

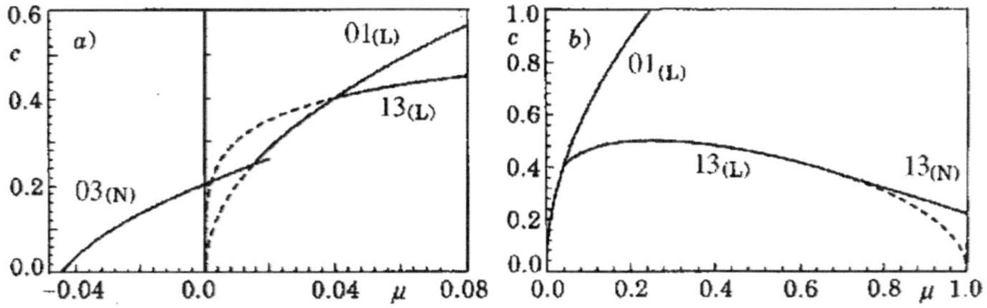

Fig. 4.33. The speed of propagating linear (L) and nonlinear (N) equilibrium–stripes (01), equilibrium–hexagons (03), and stripes– hexagons (13) fronts as a function of the supercriticality parameter μ for $\nu = 2$. (a) Blow-up near the critical point $\mu = 0$; (b) the supercritical region at large. The segments actually determining the propagation speed are shown by solid lines (Pismen and Nepomnyashchy, 1994)

to the hexagonal pattern. Moreover, this belt keeps widening with time as hexagons lag behind.

This picture has to be corrected, as in Sect. 2.1.6, by taking into account the competition between generic (linear, or "pulled") and nongeneric (nonlinear, or "pushed") propagating fronts. The nongeneric branch is obtained as a continuation of the branch of propagating fronts separating two stable stationary states into the parametric region where one of these states becomes unstable. A pushed front prevails when it is steeper and propagates with a higher speed than a pulled one. This necessarily happens in the vicinity of a marginal stability point where the speed of a pulled front vanishes, while that of a pushed one remains finite.

The nongeneric propagating front connecting the hexagonal pattern with the homogeneous state is obtained by solving (4.5.23), (4.5.24) with the boundary conditions $\rho_1 = \rho_2 = \rho_h$ at $x \to -\infty$, $\rho_1 = \rho_2 = 0$ at $x \to \infty$, where ρ_h is the stable stationary amplitude satisfying (1.5.17). This solution determines propagation of the hexagonal pattern near the lower limit of its existence, just above the Maxwell construction μ_{03}. For the pushed front connecting hexagonal and striped patterns, the appropriate boundary conditions are $\rho_1 = \rho_2 = \rho_h$ at $x \to -\infty$, $\rho_1 = \sqrt{\mu}$, $\rho_2 = 0$ at $x \to \infty$. This front is relevant at higher μ, close to the lower stability limit of the stripes. Both analytical and computational results are compiled in Fig. 4.33.

At low supercriticality, the pushed (03) front moves faster than the pulled (01) front at $\mu < \mu_{ln} \approx 0.016$. In this parametric region, the transitional striped belt does not widen with time. An narrow striped belt should exist, however, also in this region. The reason is that the mode with the wave vector parallel to the propagation direction is stronger than two others near the leading edge due to a higher amplitude diffusivity. This is clearly seen in Fig. 4.34 showing the computed profiles of both amplitudes at $\mu = 0.01$. One can also notice that, due to the same difference in amplitude diffusivities, the inclined modes slightly overshoot the parallel mode further in the transitional region.

At $\mu > \mu_{ln}$, the propagation speed selected linearly at the leading edge of the growing striped pattern is larger than the speed of the pushed (01) front. The striped pattern, however, cannot develop fully, as long as the speed of the (13) transition is higher, $c_{13} > c_{01}$. Apparently, the two side modes begin to grow at $\mu > \mu_{ln}$

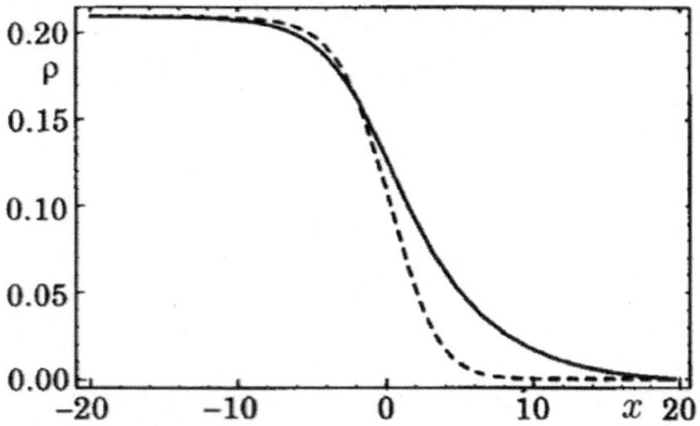

Fig. 4.34. The profiles of the amplitudes of the parallel (solid line) and inclined (dashed line) mode at $\mu = 0.01$, $\nu = 2$ (Pismen and Nepomnyashchy, 1994)

before the amplitude or the leading mode approaches its stationary value $\rho_1 = \sqrt{\mu}$. Computationally, one observes in this region propagation speeds corresponding to a smooth continuation of the nongeneric branch when the runs are not too long; the approach to higher values corresponding to the linear theory is very slow.

At $\mu \approx 0.04$, the speeds of the (13) and (01) fronts equalize. Beyond this point, the (03) front splits into two fronts: the fast (01) front and the (13) front lagging behind. The disparity between the two speeds grows further with increasing supercriticality, especially after the speed of the (13) front passes a maximum at $\mu = 1/4$. At $\mu >\approx 0.7$, the pushed (13) front overtakes the pulled one, and the front keeps slowing down. Thus, one should observe a transitional striped pattern even under conditions when the stationary striped pattern is unstable.

Csahók and Misbah (1999) carried out the same computation using the extended amplitude equation (4.5.3) and obtained qualitatively similar results, though the parametric region where the intermediate belt of unstable stripes is formed has shifted to higher values of μ. The front velocities obtained by direct numerical solution of the amplitude equations well approximated those obtained analytically. However, strong discrepancies were revealed when the solution of the amplitude equations was compared to the numerical solution of the simplest basic model – the modified SH equation (4.4.9) close to the symmetry-breaking threshold. It turned out that in the latter case the intermediate belt *does not* grow with time (Fig. 4.35a). The overall front velocity is fixed by the linear marginal velocity for the (01) front, while the (13) front goes faster than predicted by the linear theory, moving in a concerted fashion with the leading (01) front.

This reflects limitations of the amplitude equations approach similar to those discussed in Sect. 4.4. The discrepancy is, apparently, dependent on the short-scale structure of the pattern. As shown in Sect. 4.4, the wavelength of the pattern at the leading edge of a pulled front does not coincide with the optimal wavelength;

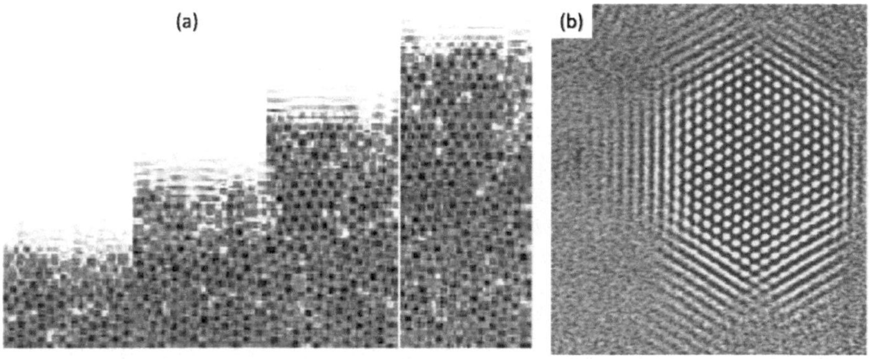

Fig. 4.35. (a) Temporal evolution (from left to right) of the propagating hexagonal front in the modified SH equation for $\mu = 0.05$ with stripes forming ahead of hexagons (Csahók and Misbah, 1999; reproduced by permission, ©IOPscience). (b) A transient between stripes and hexagons in thermal convection experiments by Bodenschatz et al (1992); reproduced by permission, ©Elsevier

moreover, it changes in the wake of the propagating front until it reaches an asymptotic wavelength, which also differs from the optimal one. Thus, the hexagonal pattern advances into the striped pattern with a variable wavenumber differing from the optimal one (unity) implied by the above theory, which takes the real amplitude as the only variable characterizing the pattern. Deviations from the optimal wavelength of the striped pattern increasing its local energy density are likely to accelerate its displacement by the advancing hexagonal pattern.

Experimentally, the peculiar structure of the boundary between a hexagonal pattern and homogeneous state was first detected by Bodenschatz et al (1992), both in the case when the hexagonal pattern spreads into an unstable homogeneous state, as shown in Fig. 4.35b, and in the case when the border between the hexagons and equilibrium is kept stationary due to a parametric ramp. The formation of an extended transitional zone should be a general phenomenon observable under certain conditions in various systems possessing three or more stationary states (Glasner and Almgren, 2000).

5 Irregular Wave Patterns

5.1 Plane Waves

5.1.1 Complex Ginzburg–Landau Equation

The complex Ginzburg–Landau (CGL) equation (1.1.6) has been derived in Sect. 1.3.3 as the amplitude equation (1.3.28) at the Hopf bifurcation that describes the evolution of the oscillation amplitude on a slow time scale far exceeding the oscillation period, as well as its modulation on an extended spatial scale. Its coefficients can be related to parameters of an underlying non-equilibrium system through a technically cumbersome but fully algorithmic procedure of multiscale expansion in the vicinity of a Hopf bifurcation, as described for a general reaction–diffusion system in Sect. 1.3.3. The appellation, which is in common use among physicists, is not literally justified, since neither Landau nor Ginzburg had anything to do with the complex form which makes its behavior so interesting. Its role is akin to that of the RGL equation in description of a second-order phase transition, bur the transition in question, leading to a persistent non-stationary state, is possible only far from equilibrium. The complex amplitude u of a periodic solution emerging as the result of a supercritical Hopf bifurcation plays the role of an order parameter.

As in the RGL equation, the real parts of all three coefficients can be reduced to unity by rescaling the amplitude, spatial coordinates, and time. Stable nontrivial solutions exist, provided the real parts of all three coefficients are positive, as they should be at a *supercritical* Hopf bifurcation. Thus, we set in (1.1.6) $\mu \to 1+\mathrm{i}\mu,\ \eta \to 1+\mathrm{i}\eta,\ \nu \to 1+\mathrm{i}\nu$. The imaginary part of the linear growth coefficient is irrelevant, since it can always be removed by a transformation $u \to u\,\mathrm{e}^{\mathrm{i}\mu t}$, so that the added frequency is absorbed into the frequency of the underlying oscillations. Thus, we adopt the standard form

$$\partial_t u = (1+\mathrm{i}\eta)\nabla^2 u + u - (1+\mathrm{i}\nu)|u|^2 u. \tag{5.1.1}$$

The polar form of (5.1.1) is obtained, as in Sect. 4.1.1, by setting $u = \rho(r)\mathrm{e}^{\mathrm{i}\theta}$, removing the phase factor, and separating the real and imaginary parts:

L. Pismen, *Patterns and Interfaces in Dissipative Dynamics*, Springer Series in Synergetics,
https://doi.org/10.1007/978-3-031-29579-9_6

$$\partial_t \rho = \nabla^2 \rho + (1 - |\nabla\theta|^2 - \rho^2)\rho - \eta(\rho\nabla^2\theta + 2\nabla\rho\cdot\nabla\theta), \tag{5.1.2}$$

$$\partial_t \theta = \nabla^2\theta + 2\rho^{-1}\nabla\rho\cdot\nabla\theta + \eta(\rho^{-1}\nabla^2\rho - |\nabla\theta|^2) - \nu\rho^2. \tag{5.1.3}$$

Another form of the CGL equation, particularly convenient for finding stationary solutions and their perturbations, is obtained by transforming $u \to u\mathrm{e}^{-\mathrm{i}\omega t}$ and rescaling the amplitude, spatial coordinates and time:

$$u \to u\sqrt{\frac{1+\omega\eta}{1+\nu\eta}}, \quad \mathbf{x} \to \mathbf{x}\sqrt{\frac{1+\eta^2}{1+\omega\eta}}, \quad t \to \frac{t}{1+\omega\eta}. \tag{5.1.4}$$

The resulting equation is

$$(1-\mathrm{i}\eta)\partial_t u = \nabla^2 u + (1+\mathrm{i}\Omega)u - (1-\mathrm{i}q)|u|^2 u, \tag{5.1.5}$$

where

$$q = \frac{\eta-\nu}{1+\eta\nu}, \quad \Omega = \frac{\omega-\eta}{1+\omega\nu}. \tag{5.1.6}$$

Equation (5.1.1) can be defined through a "complex energy" $\mathcal{F}$ as

$$\partial_t u = -\delta\mathcal{F}/\delta\overline{u}, \qquad \overline{u}_t = -\delta\overline{\mathcal{F}}/\delta u, \tag{5.1.7}$$

so that it can obtained as the Euler–Lagrange equation corresponding to a *complex* Lagrangian

$$\mathcal{L}(u) = (1+\mathrm{i}\eta)\,\nabla u\cdot\nabla\overline{u} - |u|^2 + \tfrac{1}{2}(1+\mathrm{i}\nu)\,|u|^4. \tag{5.1.8}$$

The complex energy $\mathcal{F} = \int \mathcal{L}d\mathbf{x}$ is neither evolving in a definite direction nor conserved. The CGL equation inherits therefore to a large extent sophisticated dynamic behavior of underlying non-equilibrium systems.

Generalizations of the CGL equation have also received much attention. Those are the *quintic* CGL equation (QCGL) (1.3.30) with an added term $|u|^4u$ (van Saarloos and Hohenberg, 1992), derived in the vicinity of the *codimension two* bifurcation points at a border between a supercritical and subcritical Hopf bifurcation, and the complex Swift–Hohenberg equation (CSH), with an added term $\nabla^4 u$, i.e., (1.1.3) with complex coefficients at all terms (Lega et al, 1994). We shall concentrate here on the basic generic form (5.1.1), which by itself generates rich spatiotemporal dynamics, as is emphasized in the title of a comprehensive review by Aranson and Kramer (2002): "The *world* of the complex Ginzburg–Landau equation".

5.1.2 Perturbations of Plane Waves

Traveling plane wave solutions of (5.1.1) or (5.1.2), (5.1.3) with the wave vector $\mathbf{k}$ is

$$u = \rho_0 \mathrm{e}^{\mathrm{i}(\mathbf{k}\cdot\mathbf{x}-\omega t)}, \quad \rho_0 = \sqrt{1-k^2}, \quad \omega = \nu + (\eta-\nu)k^2. \tag{5.1.9}$$

The waves are dispersive, and the group velocity is

$$\mathbf{v} = \mathrm{d}\omega/\mathrm{d}\mathbf{k} = 2\mathbf{k}(\eta-\nu). \tag{5.1.10}$$

Stability of plane waves in the CGL equation was analyzed by Malomed (1984) and in a more clear-cut form by Aranson et al (1992). Following the standard procedure, we define a perturbed solution in the form

$$\rho = \rho_0 + \varrho e^{\lambda t + i\mathbf{q}\cdot\mathbf{x}}, \qquad \theta = \mathbf{k}\cdot\mathbf{x} - \omega t + \Theta e^{\lambda t + i\mathbf{q}\cdot\mathbf{x}}, \tag{5.1.11}$$

where $\rho_0(k)$, ω are defined by (5.1.9) and ϱ, Θ are infinitesimally small. Linearizing (5.1.2), (5.1.3) leads to the eigenvalue problem $\lambda\mathbf{I} = \mathbf{L}$ where $\mathbf{I}$ is the 2D unity matrix and $\mathbf{L} = \mathbf{L}_0 - 2i(\mathbf{k}\cdot\mathbf{q})\mathbf{L}_1$ with

$$\mathbf{L}_0 = \begin{pmatrix} -(2\rho_0^2 + q^2) & \rho_0\eta q^2 \\ -(2\nu\rho_0 + \eta q^2\rho_0^{-1}) & -q^2 \end{pmatrix}, \qquad \mathbf{L}_1 = \begin{pmatrix} \eta & \rho_0 \\ -\rho_0^{-1} & \eta \end{pmatrix}. \tag{5.1.12}$$

Transverse and longitudinal perturbations can be considered separately. The former are defined exclusively by the matrix $\mathbf{L}_0$. Its eigenvalues are

$$\lambda_\pm = -(\rho_0^2 + q^2) \pm \sqrt{\rho_0^4 - 2\eta\nu\rho_0^2 q^2 - \eta^2 q^4}\,. \tag{5.1.13}$$

Clearly, $\lambda_- < 0$, and only λ_+ may change its sign. We expect the most dangerous perturbation modes to have a vanishing wavenumber, since they can be viewed as long-scale distortions of neutrally stable translational modes. Expanding λ_+ in $q \ll 1$ yields $\lambda_+ = -(1+\eta\nu)q^2$. This defines the transverse phase diffusivity, which must be positive for a stable wave :

$$D_\perp = 1 + \eta\nu\,. \tag{5.1.14}$$

The limit $D_\perp = 0$ defines the Benjamin–Feir (BF), or self-focusing, instability. One can check that λ_+ has no more zeroes at finite q. Setting $q = \rho_0\sqrt{Q}$ leads to

$$\lambda_+ = \rho_0^2\left[-(Q+1) + \sqrt{1 - 2Q\eta\nu - (Q\eta)^2}\right]. \tag{5.1.15}$$

This expression vanishes, besides $Q = 0$, at $Q = -2(1+\eta\nu)/(1+\eta^2)$, which is negative whenever $D_\perp > 0$.

The leading eigenvalue of the matrix $\mathbf{L}$ with $\mathbf{q}$ parallel to $\mathbf{k}$ is

$$\lambda_+ = 2i\eta kq - q^2(1+\rho_0^2) + \sqrt{\rho_0^4 - 2q^2[2 + \rho_0^2(\eta\nu - 2)] - \eta^2 q^4 + 4ikq(\nu\rho_0^2 + q^2\eta)}\,. \tag{5.1.16}$$

The long-scale limit of the Eckhaus instability is obtained by expanding (5.1.16) in $q \ll 1$. The first terms in the expansion are

$$\lambda_+ = -2ikq(\eta - \nu) - q^2\left[1 + \eta\nu + 2k^2\frac{1+\nu^2}{1-k^2}\right]. \tag{5.1.17}$$

The first term is purely advective and cancels exactly by transforming to the coordinate frame moving with the group velocity (5.1.10). The coefficient at the second term is the longitudinal phase diffusivity:

$$D_{\|} = 1 + \eta\nu + 2k^2 \frac{1+\nu^2}{1-k^2} \ . \tag{5.1.18}$$

It becomes negative at

$$k^2 > \frac{1+\eta\nu}{3+\eta\nu+2\nu^2} \ . \tag{5.1.19}$$

A *stationary* localized perturbation of a stable pattern is screened upstream from the disturbance (i.e., against the direction of the group velocity). The attenuation rate p can be obtained by setting in the dispersion relation (5.1.16) $\lambda = 0,\ q = -ip$; the direction of $\mathbf{q}$ is chosen parallel to the wave vector $\mathbf{k}$ and the group velocity $\mathbf{v}$. This leads to the equation for p:

$$(1+\eta^2)p^4 + 2\left[2k^2(1+\eta^2) - \rho_0^2(1+\nu\eta)\right]p^2 + 2\rho_0^2 v p = 0\ , \tag{5.1.20}$$

where $v = \pm|\mathbf{v}|$. Besides the trivial root $p = 0$ (reflecting the translational symmetry), there are three roots, which should have real parts of different sign, since their sum vanishes. Assuming $v > 0$, there is one root with a negative and two with a positive real part. The latter define the decay of the disturbance in the upstream direction. This pair of roots may collapse with growing k; at larger k, there will be then a pair of complex conjugate eigenvalues, and the decay will be oscillatory.

The instability arising beyond the threshold may be either absolute if perturbations amplify with time at a fixed location, or *convective* if they are washed away downstream with the phase velocity of the wave. Evolution of perturbations within a linear range is defined as in Sect. 2.1.4 (but in the fixed rather than moving frame) by Fourier integrals in the complex plane, separately for the components of $\mathbf{q}$ parallel and perpendicular to $\mathbf{k}$:

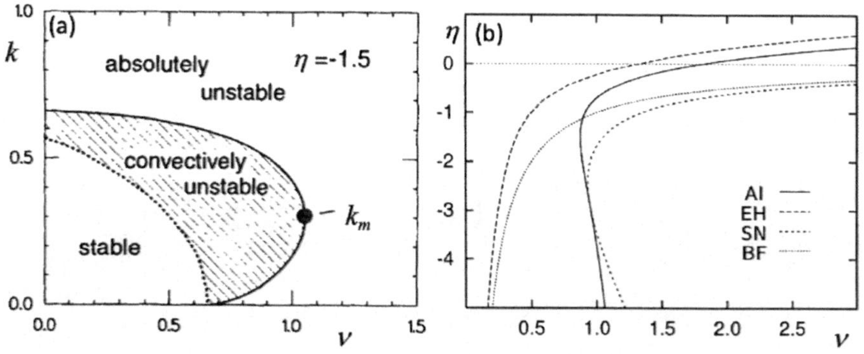

Fig. 5.1. (a) Limits of convective and absolute instabilities in the plane $(\nu,\, k)$ for $\eta = -3/2$. The *dot* marks the limit $k = k_m$ where the lower and higher limits of convectively unstable waves merge (Aranson et al, 1992, reproduced by permission, ©The American Physical Society. (b) Limits of convective (EH) and absolute (AI) Eckhaus instability and self-focusing instability (BF) in the plane (ν, η). The curve SN marks the location of the limiting points $k = k_m$ (Ipsen et al, 2000, reproduced by permission, ©Elsevier)

$$\widetilde{u}(\mathbf{x},t) = \frac{1}{2\pi}\int_{-\infty}^{\infty} \widehat{u}(q,t)e^{i[qx-\lambda(q)t]}dq \;, \tag{5.1.21}$$

Provided perturbations are sufficiently localized, evaluating the integral using the method of steepest descent in the complex plane as in Sect. 2.1.4 implies that the integral is dominated by the largest saddle point $q = \kappa$ of $\omega(q)$.

The saddle point is defined by requiring $\lambda'(q)$ to vanish at $q = \kappa$. The instability is absolute when the real part of $\lambda_r(\kappa) > 0$. These one complex and one real equations define both the location of the saddle point and the value of q at the locus of absolute instability. However, this problem is technically cumbersome: since κ does not need to be small, long-scale expansions are not applicable here, and the full expressions (5.1.13) or (5.1.16) should be used. Aranson et al (1992) carried out this procedure numerically by continuation method. The Eckhaus instability is proven more dangerous. The instability limits are shown in Fig. 5.1. Numerical simulations by Aranson et al (1992) showed that transition to turbulence indeed occurred only when the absolute stability condition was violated, but the system is very sensitive to noise in the convectively unstable region.

5.1.3 Phase Equations

Nonlinear phase equations describing phase dynamics on the background of either plane waves or homogeneous oscillations can be obtained by long-scale expansion of (5.1.2), (5.1.3) in the frame propagating with the group velocity (5.1.10). Phase perturbations, now denoted as $\Theta(\mathbf{X},T)$, are not assumed anymore to be small, but, due to rescaling $\nabla \to \epsilon\nabla$, perturbations of both $\mathbf{k}$ and ρ do not exceed $O(\epsilon)$. Plugging $\theta = \mathbf{k}\cdot\mathbf{x} + \Theta,\ \rho = \rho_0 + \epsilon\rho_1 + \ldots$ in (5.1.2) transformed to the comoving frame and expanding in ϵ yields

$$\begin{aligned}
\rho_1 &= -\rho_0^{-1}\mathbf{k}\cdot\nabla\Theta,\\
\rho_2 &= -\rho_0^{-1}\left(\tfrac{1}{2}\eta\nabla^2\Theta + \tfrac{1}{2}|\nabla\Theta|^2 + \nu\rho_0^{-1}\mathbf{k}\cdot\nabla\rho_1 + \tfrac{3}{2}\rho_1^2\right)\\
&= -\frac{1}{\rho_0}\left[\tfrac{1}{2}\eta\nabla^2\Theta + \tfrac{1}{2}|\nabla\Theta|^2 - \frac{\nu}{\rho_0^2}(\mathbf{k}\cdot\nabla)^2\Theta + \frac{1}{\rho_0^2}(\mathbf{k}\cdot\nabla\Theta)^2\right].
\end{aligned} \tag{5.1.22}$$

Using this in (5.1.2) transformed to the comoving frame yields in the leading second order of the expansion the anisotropic Burgers equation describing relaxation of phase inhomogeneities on a slow $O(\epsilon^{-2})$ time scale T:

$$\partial_T\Theta = D_{\|}\partial_X^2\Theta_{XX} + D_{\perp}\partial_Y^2\Theta_{YY} + (\nu-\eta)|\nabla\Theta|^2 - \frac{\nu k^2}{1-k^2}(\partial_X\Theta)^2, \tag{5.1.23}$$

where X, Y are extended coordinates directed, respectively, along and across $\mathbf{k}$, and the transverse and parallel diffusivities are defined by (5.1.14), (5.1.18). The isotropic form of this equation at $k = 0$ is

$$\partial_T\theta = (1+\nu\eta)\nabla^2\theta + (\nu-\eta)|\nabla\theta|^2, \tag{5.1.24}$$

or, after rescaling $u \to u\sqrt{(1+\omega\eta)/(1+\nu\eta)}$, $\mathbf{x} \to \mathbf{x}\sqrt{(1+\eta^2)/(1+\omega\eta}$, $t \to t/(1+\omega\eta)$,

$$(1+q\eta)\theta_T = \nabla^2\theta - q|\nabla\theta|^2. \tag{5.1.25}$$

At $k = 0$, the expansion can be continued near the self-focusing instability limit $D_\perp = 1+\nu\eta = 0$. The long spatial scale should be shortened now from $O(\epsilon^{-1})$ to $O(\epsilon^{-1/2})$. To make the expansion consistent, the nonlinear term also has to be reduced to $O(\epsilon)$, which requires the phase deviations to be small. The right-hand side of (5.1.23) vanishes after setting $k = 0$, $\eta = -\nu^{-1}(1-\widetilde{\eta})$ and rescaling $\Theta \to \epsilon\Theta$. The real amplitude has to be computed now by continuing the expansion of (5.1.2) to the third order:

$$\rho = 1 - \tfrac{1}{2}\epsilon^2\eta\,\nabla^2\Theta - \tfrac{1}{2}\epsilon^3\left(|\nabla\Theta|^2 + \tfrac{1}{2}\eta\nabla^2\Theta\right). \tag{5.1.26}$$

Expanding likewise (5.1.3) yields in the leading third order the Kuramoto–Sivashinsky (KS) equation describing phase dynamics on the $O(\epsilon^{-2})$ time scale:

$$\partial_T\Theta = \widetilde{\eta}\,\nabla^2\Theta + (\nu+\nu^{-1})|\nabla\Theta|^2 - \tfrac{1}{2}(1+\nu^{-2})\nabla^4\Theta. \tag{5.1.27}$$

A similar expansion near the limit of Eckhaus instability (Janiaud et al, 1992) is somewhat more problematic. In this case, only the longitudinal scale (along the X-axis) should be shortened, while the transverse axis retains the same scaling $\partial_Y = O(\epsilon)$ as in (5.1.23). The first derivative with respect to X does not appear when the equation is written in the comoving frame, but the third derivative cannot be excluded, so that the resulting equation is not properly balanced.

The KS equation was first derived by Kuramoto and Tsuzuki (1976) by direct long-scale small-amplitude expansion of a reaction–diffusion system in the vicinity of a spatially homogeneous periodic solution. The coefficients of the KS equation derived in this way are related to the parameters of the underlying system through averaging over the oscillation period. The same equation describes weakly nonlinear dynamics of transversely unstable reaction fronts (Sect. 3.2.6). The parametric range where the KS equation can be obtained by rational expansion of underlying equations is very limited; nevertheless, it is important theoretically as a basic equation generating spatiotemporal chaos. Its appeal is in its universality: it can be obtained out of scaling and symmetry considerations alone, in the way it was introduced in Sect. 1.1.1.

The KS equation has stationary periodic solutions, which are stable in a narrow wavenumber range (Frisch et al, 1986). These solutions, even when they do exist, have, apparently, small attraction basins, and spatiotemporal chaos is actually the relevant attractor. Chaotic solutions appear as well in the damped KS equation (with an added stabilizing linear algebraic term breaking translational invariance), which makes it possible to follow the parametric evolution from ordered cellular patterns immediately above the primary symmetry-breaking transition through intermittency to chaos with decreasing damping (Chaté and Manneville, 1987). Statistics of chaotic solutions both in the standard KS equation and its damped version has been studied in a number of publications (Egolf and Greenside, 1995; Elder et al, 1997; Bohr et al, 1998). Patterns generated by the KS equation seem to be similar to those appearing in simulations of the forced Burgers (KPZ) equation (Sect. 2.2.3), although fine distinctions can be found in statistical characteristics of chaotic solutions (Boghosian et al, 1999). The KS equation cannot imitate the complexity of behavior of the CGL equation, which is not reduced to mere phase chaos, but is typically dominated by the formation and motion of phase singularities, as we shall further discuss (Sects. 5.2.3, 5.2.4, and 5.4.5).

5.1.4 Propagation of Wave Pattern

If a plane wave propagates into the unstable trivial state $u = 0$, the wavelength and propagation speed selected according to the marginal stability criterion of Sect. 2.1.4 are determined by (2.1.38), where the dispersion relation $\omega(k) = \mathrm{i}\left[1 + (1 + \mathrm{i}\eta)k^2\right]$ is obtained by linearizing (5.1.1) in the vicinity of the trivial state. Accordingly, the group velocity is complex:

$$\widehat{c} \equiv \widehat{c}_r + \mathrm{i}\widehat{c}_i = \omega'(k) = 2\left[k_i + \eta k_r + \mathrm{i}(\eta k_i - k_r)\right]. \tag{5.1.28}$$

The value of the wavenumber verifying the stability condition $\widehat{c}_i = 0$ is $k_r = \eta k_i$. The condition of equality of the phase and group velocities, $\omega_i/k_i = \widehat{c}_r(k)$ is satisfied at $k_i = 1/\sqrt{1+\eta^2}$, which defines the slope of amplitude profile in accordance to the principle of marginal stability. The corresponding wavenumber and propagation speed are

$$k_r = \eta(1+\eta^2)^{-1/2}, \quad c = \widehat{c}_r = 2\sqrt{1+\eta^2}. \tag{5.1.29}$$

The wavenumber k^* of the pattern formed behind the propagating front does not coincide with k_r but is computed with the help of the wave crests conservation condition as in Sect. 4.4.1. The entire pattern should be stationary in the comoving

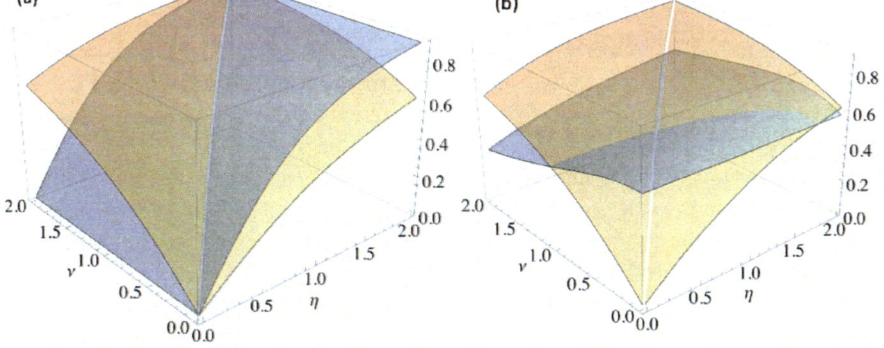

Fig. 5.2. The function $k^*(\eta, \nu)$ (orange-shaded surface) shown together with the blue-shaded surfaces representing the selected wavenumber at the leading edge $k_r(\eta)$ (a) and the upper stability limit for a plane wave (b). Semitransparent shading allows the lower-lying segments to be visible

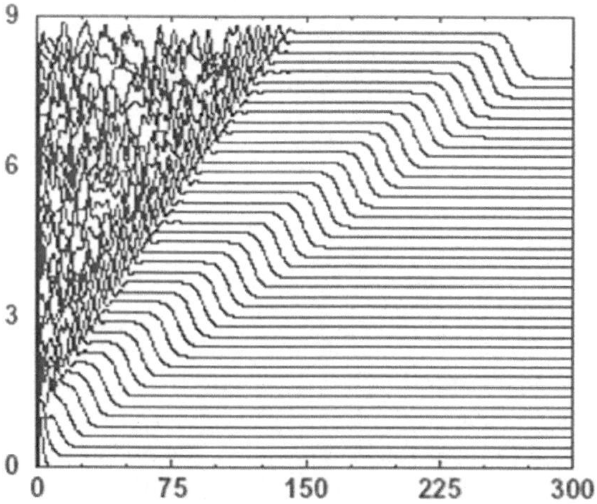

Fig. 5.3. Space–time plot of simulations results for pulled fronts in the CGL equation with $\eta = 1, \nu = -3$ (van Saarloos, 2003, reproduced by permission ©Elsevier). The lines denote the profiles of ρ at successive time intervals, and are shifted upward relative to each other.

frame, except for a common factor $e^{i\Omega t}$. Far behind the front, the relation between the frequency and the wavenumber is just the Doppler-shifted dispersion relation (5.1.9):

$$\Omega = \omega - ck = \nu + (\eta - \nu)k^2 - ck. \tag{5.1.30}$$

On the other hand, near the leading edge the same frequency is computed as

$$\Omega = \omega_r - ck_r = -\eta. \tag{5.1.31}$$

Equating the two expressions and solving for the root satisfying the inequality $|k| < 1$ yields the wavenumber k^* selected by the pulled front:

$$k^* = \left| \frac{\sqrt{1+\eta^2} - \sqrt{1+\nu^2}}{\eta - \nu} \right| . \tag{5.1.32}$$

The selected wavenumber may fall into the unstable range. In this case, the wave pattern may become chaotic in the wake of the propagating front, as seen in Fig. 5.3. This can be clearly seen in Fig. 5.2b: it happens when the yellow-shaded surface representing the function $k^*(\eta, \nu)$ lies above the blue-shaded surface showing the upper stability limit for a plane wave (5.1.19). In this case, the wave pattern may become chaotic in the wake of the propagating front, as seen in Fig. 5.3. An unstable wave pattern forming behind a pulled front is invaded by a second front establishing a chaotic state in its wake, and, due to a slower propagation speed of the trailing front, the segment occupied by the unstable state grows with time. Alternatively, the pattern can be stabilized further downstream through wavelength change facilitated by nonlinear interactions.

As in Sect. 2.1.6, there are also non-generic nonlinear front solutions, which can be obtained in a way similar to that described in Sect. 5.2.2 (van Saarloos and Hohenberg, 1992). However, these fronts are not selected by evolution starting from localized initial conditions.

5.2 One-Dimensional Structures

5.2.1 Converting CGL Equation into an ODE

There is a wide variety of nonuniform 1D wave patterns. Typical structures with a spatially varying modulus and wavenumber are stationary in a frame propagating with a certain speed c and depend on the comoving coordinate $z = x - ct$ only. The version of the CGL equation particularly convenient for finding nonuniform solutions in 1D and their perturbations is obtained by transforming the variable to the polar form, which is stationary in the comoving frame $z = x - ct$, except for oscillations with a constant frequency Ω:

$$u = \rho(z)\mathrm{e}^{\mathrm{i}[\theta(z)-\Omega t)]}. \tag{5.2.1}$$

The complex wavenumber is defined as

$$k \equiv k_r + \mathrm{i}k_i = -\mathrm{i}\,D(\ln u)/\mathrm{d}z = \theta'(z) - \mathrm{i}\,\mathrm{d}(\ln\rho)/\mathrm{d}z. \tag{5.2.2}$$

The derivatives entering the CGL equation (5.1.1) are expressed then as

$$\begin{aligned} u^{-1}\partial_t u &= -ck_i - \mathrm{i}(\Omega + ck_r) = -\mathrm{i}(\Omega + ck), \\ u^{-1}\partial_z^2 &= \mathrm{i}k_r'(z) - k_i'(z) + k_i^2 - 2\mathrm{i}k_r k_i - k_r^2 = \mathrm{i}k'(z) - k^2. \end{aligned} \tag{5.2.3}$$

The CGL equation is transformed using these expressions to the system of one real and one complex first-order dynamic equations with the coordinate z playing the role of time:

$$\rho'(z) = -k_i\rho, \tag{5.2.4}$$

$$\mathrm{i}k'(z) = -\frac{\mathrm{i}(\Omega + ck) - 1}{1 + \mathrm{i}\eta} + k^2 - \frac{1 + \mathrm{i}\nu}{1 + \mathrm{i}\eta}\rho^2. \tag{5.2.5}$$

The last equation turns after separating the real and imaginary parts to

$$k_r'(z) = -\frac{\Omega + c(k_r + \eta k_i) - \eta}{1 + \eta^2} + 2k_r k_i + \frac{\nu - \eta}{1 + \eta^2}\rho^2, \tag{5.2.6}$$

$$k_i'(z) = \frac{\eta\Omega + c(\eta k_r - k_i) + 1}{1 + \eta^2} + k_i^2 - k_r^2 - \frac{1 + \nu\eta}{1 + \eta^2}\rho^2. \tag{5.2.7}$$

This way of converting a PDE to ODE differs from that in Sect. 5.1.4, which is applicable in the vicinity of the trivial solution, since the nonlinear term cannot be rationally expressed in terms of a phase variable, even a complex one.

Plane waves are *nonlinear* (N) fixed points of the system (5.2.4) with $k_i = 0$, $k_r = \text{const} = k_0$, and $\rho = \rho_0 = \sqrt{1 - k_0^2}$. The relation between the frequency and the wavenumber is given by the Doppler-shifted dispersion relation (5.1.30). *Linear* (L) fixed points that correspond to the trivial state $\rho = 0$ with a nonzero k can be obtained by resolving the right-hand side of (5.2.4) to obtain

$$k = \frac{-c \pm \sqrt{c^2 - 4(\eta - \mathrm{i})(\Omega + \mathrm{i})}}{4(\eta - \mathrm{i})}. \tag{5.2.8}$$

Separating here the real and imaginary parts to find c and Ω as functions of k_r, k_i leads to awkward expressions, but implicit solutions can be obtained by resolving (5.2.6), (5.2.7), with respect to Ω, c:

$$\Omega = k_r k_i + \eta(k_r^2 + k_i^2) - \frac{k_r(1 - k_r^2)}{k_i}, \qquad c = k_i + 2\eta k_r + \frac{1 - k_r^2}{k_i}. \tag{5.2.9}$$

The system (5.2.4) can be viewed as a dynamical system with z playing the role of time, and its coherent structures, as homoclinic or heteroclinic trajectories connecting the fixed points. Homoclinic trajectories connecting linear fixed points are called pulses, or solitons. Trajectories connecting nonlinear fixed points are called *holes*, or dark solitons when they involve depressing the modulus ρ due to strong phase rotation. Heteroclinic trajectories connecting N-domains with distinct wavenumbers are also called *wavenumber kinks*. Heteroclinic solutions connecting N and L fixed points are pushed fronts describing the trivial state invaded by a wave train and competing with pulled fronts in Sect. 5.1.4 in a manner of Sect. 2.1.6.

Existence and robustness of these solutions can be estimated by the "counting argument" that appeared in its simplest form in Sect. 2.1.1. After linearizing (5.2.6) near a fixed point, one can count the number $n_\pm$ of roots with positive and negative real parts defining the dimensions of the stable and unstable manifolds. A family of trajectories starting from a fixed point with n_--dimensional unstable manifold has $n_- - 1$ free parameters, which can be adjusted to choose a trajectory orthogonal to the n_+-dimensional unstable manifold of the target fixed point. In addition, there are two free parameters, the frequency Ω and the propagation speed c. Thus, there is a n-parametric family of trajectories connecting the two points with $n = n_- - n_+ + 1$. Generally, if the equation contains n_p free parameters

$$n = n_- - n_+ + n_p - 1. \tag{5.2.10}$$

With $n < 0$, no trajectories exist; $n = 0$ indicates existence of an isolated trajectory or a discrete set, while a continuous n-parametric family exists when n is positive.

The counting argument is heuristic rather than strict. On the one hand, being based on *necessary* conditions, it gives the upper limit of multiplicity of solutions. On the other hand, symmetries of the dynamical system may increase the robustness of solutions, since fewer parameters may be needed in order to satisfy the constraints. A symmetry can be either obvious, e.g., related to a conservation law, or *hidden*. Van Saarloos and Hohenberg (1992) conjectured that some hidden symmetry of the CGL equation causes the solutions to be more robust than suggested by the counting argument.

The eigenvalues p fixing the dimensions of the stable and unstable manifolds are determined at $c = 0$ by (5.1.20). A generalized cubic equation for p at $c \neq 0$ (excluding the trivial root) can be obtained by linearizing (5.2.6) near a plane wave solution:

$$(1+\eta^2)p^3 + 2cp^2 - 2\left[2k_0^2(1+\eta^2) - \rho_0^2(1+\nu\eta) + c(c-4k_0\eta)\right]p + 2\rho_0^2 v = 0. \tag{5.2.11}$$

Domain boundaries can be either sources or sinks of waves, depending on the direction of the group velocity. Dynamically, sources are most important, since they create their own environment, while sinks play a passive role; perturbations travel away from sources and into sinks. The sign of the last term in (5.2.11) is opposite to the sign of the product of the three roots and coincides with the sign of the group velocity v, which, in the case of a source, is positive on the right and negative on the left. For a stationary source ($c = 0$), the sum of the real parts of all three roots is zero. At $\zeta \to -\infty$, the product of the roots is positive, so that there should be one positive real root and two roots (either real or complex conjugate) with a negative real part. Thus, the dimension of the unstable manifold is $n_- = 1$. In a similar way, we establish that $n_+ = 2$ for the fixed point at $\zeta \to \infty$ where the product of the roots is negative; this can be also inferred by symmetry to inversion of the direction of the ζ axis:

$$\zeta \to -\zeta, \quad z \to -z, \quad \rho \to \rho. \tag{5.2.12}$$

Since the only adjustable parameter on hand is ω, it appears that no trajectories can generically exist, as $n = n_- - n_+ < 0$. However, this conclusion is alleviated due to the symmetry relations (5.2.12). Any trajectory starting from a point $\rho_0 \neq 0,\ \kappa = 0,\ k_0 \neq 0$ should hit the plane $k = 0$ at some point $(\widehat{\rho},\ 0,\ \widehat{k})$, and the latter value can be brought to zero by adjusting $\omega(k_0)$. The remainder of the trajectory is continued symmetrically. This suggests that an isolated trajectory (or, possibly, a discrete set) can exist.

The dimensions $n_\pm$ may change with growing propagation velocity $|c|$. Without loss of generality, we can assume $c > 0$ (placing the advancing domain on the left). Since the sum of the roots is now negative, while the product retains the same sign on both sides, real parts of all roots may become negative beyond some critical propagation speed, so that n_+ reduces to zero, and a two-parametric family of trajectories may appear. For sinks at $c = 0$, the signs of the roots are inverted, so that $n_- = 2,\ n_+ = 1$, and a two-parametric family of trajectories is predicted. This conforms to the above-mentioned passive role of sinks, which do not reject any approaching trajectory.

5.2.2 Holes and Wavenumber Kinks

With good luck, non-uniform solutions can be obtained if a proper *ansatz* is found. The simplest particular case is a static solution with oppositely directed but otherwise identical waves on both sides. Nozaki and Bekki (1984) constructed this solution, which they called a *hole*, using the *ansatz*

$$k_i(\rho) = k_i^0\left(\frac{\rho_0}{\rho} - \frac{\rho}{\rho_0}\right), \qquad k_r(\rho) = k_r^0\frac{\rho}{\rho_0}. \tag{5.2.13}$$

With this *ansatz*, (5.2.4) readily integrates to

$$\rho(\zeta) = \rho_0 \tanh \kappa_0 \zeta, \quad k = k_0 \tanh \kappa_0 \zeta. \tag{5.2.14}$$

Inserting (5.2.13) in (5.2.4) yields a complex expression linear in ρ^2, and the four unknowns κ_0, ρ_0, k_0, and $\Omega = \omega$ can be chosen to cancel the four real coefficients. Of course, ρ_0, k_0, and ω are related by (5.1.9); these relations should be satisfied automatically and can be used to reduce the four equations to just two, defining κ_0 and ω. Thus, we obtain

$$k_0^2 = \frac{\omega - \nu}{\eta - \nu}, \quad \rho_0^2 = \frac{\eta - \omega}{\eta - \nu}, \quad \kappa_0^2 = \frac{1 + \eta\omega}{2(1 + \eta^2)}, \tag{5.2.15}$$

where ω satisfies the quadratic equation

$$9(1 + \eta^2)(\omega - \nu)(1 + \eta\omega) - 2(\eta - \nu)(\eta - \omega)^2 = 0. \tag{5.2.16}$$

Two alternative signs of k_0 and κ_0 should be chosen on the left and right semi-axes in such a way that the group velocity (5.1.30) be directed away from the origin and the hole be a *source* which selects the wavelength and frequency of the waves propagating to both sides. The hole solutions are stable in a narrow interval bounded, on the one side, by the core instability limit and, on the other side, by the Eckhaus instability limit of the emitted waves (see Fig. 5.4). In a bounded region or in a system encompassing a number of holes separated by shocks (serving as sinks of emitted waves), the absolute instability limit is likely to be more relevant than convective Eckhaus instability.

A heteroclinic trajectory connecting fixed points corresponding to two *distinct* wavenumbers $k_\pm$ describes a wavenumber kink propagating with some speed c, which has to be chosen to equalize the corresponding Doppler-shifted frequencies $\Omega(c, k_\pm)$. Because of (5.1.30), this speed is expressed as

$$c = \frac{\omega(k_+) - \omega(k_-)}{2(k_+ - k_-)} = (\eta - \nu)(k_+ + k_-), \tag{5.2.17}$$

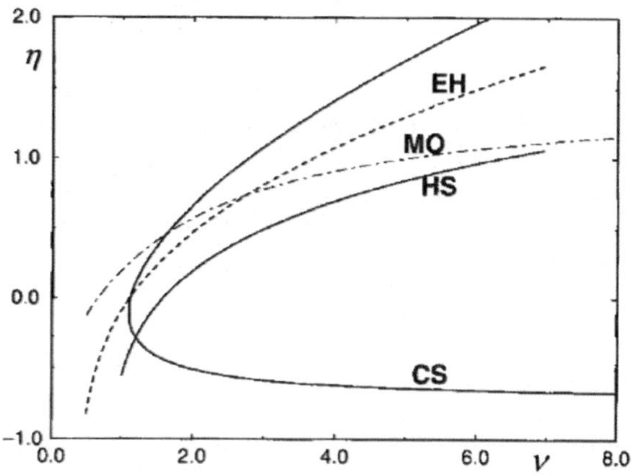

Fig. 5.4. Phase diagram of Nozaki–Bekki holes in the parametric plane (η, ν) (Popp et al, 1995, reproduced by permission, ©Elsevier). The core is unstable outside the region bounded by the curve CS. Emitted waves suffer convective and absolute Eckhaus instability below the lines EH and HS, respectively. The line MO is the boundary between monotonic (above) and oscillatory (below) interaction between standing holes.

where $\omega(k)$ is the dispersion relation for plane waves (5.1.9). This formula can be interpreted as a phase conservation relation. Comparing the last expression with (5.1.10), we see that the kink propagates with the average group velocity of the two plane waves which are approached asymptotically at $\zeta \to \pm\infty$.

The heteroclinic hole solution, also found by Bekki and Nozaki (1986), is not simply expressed in terms of polynomials $\kappa(\rho)$ and $k(\rho)$, but is written explicitly as

$$\begin{aligned} u(\zeta) &= u_0 \frac{1 + b\,\mathrm{e}^{-2\kappa_0\zeta}}{1 + \mathrm{e}^{-2\kappa_0\zeta}}\, \mathrm{e}^{\mathrm{i}\int k(\zeta)\mathrm{d}\zeta - \mathrm{i}\Omega t}, \\ k(\zeta) &= \tfrac{1}{2}\left[k_+ + k_- + (k_+ - k_-)\tanh \kappa_0\zeta\right], \end{aligned} \qquad (5.2.18)$$

where $k_\pm$ are the two asymptotic wavenumbers, u_0 and κ_0 are real, and b is a complex parameter. Inserting (5.2.18) in (5.1.1) yields a cubic polynomial in $\mathrm{e}^{-2\kappa_0\zeta}$, four complex coefficients of which should be set to zero. This can be done by adjusting $k_\pm$, c, Ω, u_0, κ_0, and b, altogether eight real constants. This suggests the existence of a discrete set of solutions. However, it turns out that the solution procedure leads eventually to a relation which is satisfied identically, so that a *continuous* family of solutions parametrized by the speed c exists. These solutions have a dip of the modulus, though not reaching zero as in the case of a standing hole, and are also *sources*. The constant κ_0 decreases with c and vanishes at the maximum velocity $c_{\max}$ where the solution merges with a plane wave having the group velocity $v = c_{\max}$.

Here again, solutions turn out to be more robust than suggested by the counting argument, which points out to a hidden symmetry of the CGL equation. It was found later that the family of moving holes is structurally unstable. If the CGL equation is slightly perturbed, e.g., by introducing a small quintic term, the solution family is destroyed, as the holes either accelerate or decelerate (Popp et al, 1995). On the other hand, Doelman (1995) has shown that even small quintic perturbations create large families of traveling localized structures that do not exist in the cubic case.

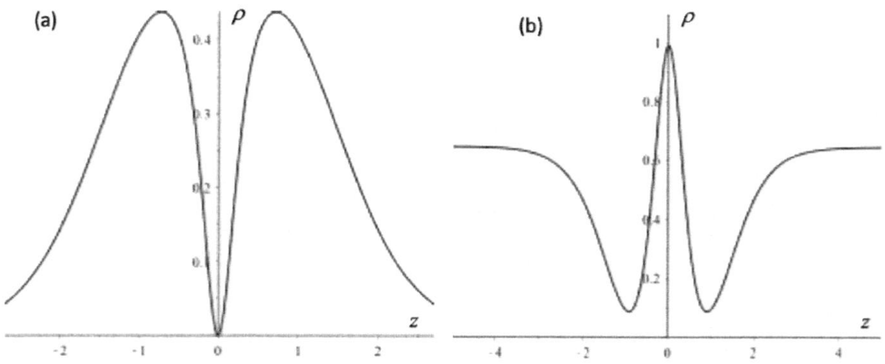

Fig. 5.5. (a) Coherent structures in the quintic CGL equation (Conte et al, 2022, reproduced by permission, ©The American Physical Society). (a) Homoclinic defect. (b) Homoclinic bound state of two dark solitons. The amplitude is shown against the rescaled coordinate

Beyond lucky *ad hoc* assumptions about the form of a solution, lies a deep mathematical structure of what Conte and Musette (1993) called "linearity inside nonlinearity": a set of linear equations that enables derivation of exact solutions to CGL and its analogs and extensions. A comprehensive study of 1D solutions of various generalizations of the CGL equation (with added fourth-derivative, quintic, or nonlinear derivative terms) was carried out by van Saarloos and Hohenberg (1992). Naturally, generalized equations are capable to sustain a still richer variety of behavior and generate different kinds of stable coherent structures. For example, the quintic CGL equation (QCGL) allows for a non-monotonic dispersion relation, which makes possible a stationary boundary between two wave domains with different wavelengths but identical frequencies. It makes possible as well coexistence of stable trivial and non-trivial solutions, which naturally translates into existence of stable pulses. Some solutions can be constructed analytically based on the forms suggested by hidden systems of linear equations. Two examples of coherent structures in the QCGL equation expressed through a combinations of hyperbolic functions (Conte et al, 2022) are shown in Fig. 5.5.

5.2.3 Modulated Waves and Phase Turbulence

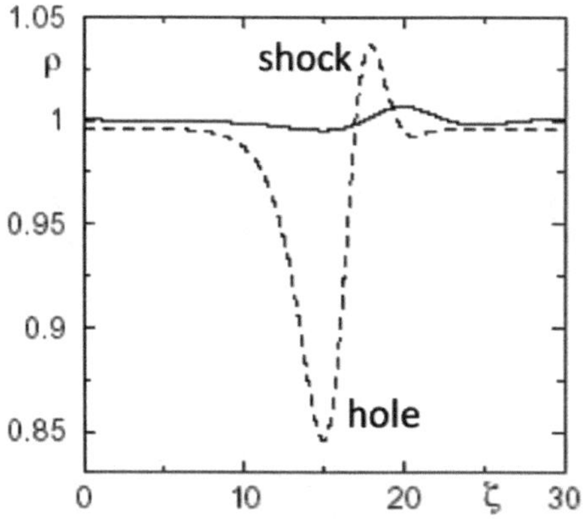

Fig. 5.6. Shallow (solid curve) and deep (dashed curve) MAW amplitude profiles (Brusch et al, 2001, reproduced by permission ©Elsevier

Other 1D solutions of the CGL equation, which have been constructed numerically using continuation methods (Brusch et al, 2000, 2001) are called modulated amplitude waves (MAW). Those are P-periodic solutions of (5.2.6), which correspond to *quasiperiodic* states of an underlying physical system. The limiting form at $P \to \infty$ is a "homoclon" (van Hecke, 1998) – a homoclinic solution approaching a plane wave at both extremes. Unlike a static Nozaki–Bekki hole, the direction of the asymptotic plane waves is now identical; a homoclon can be viewed therefore as a local phase twist that glides through a background plane wave. The amplitude has a peak as well as a dip, similarly to a moving hole-shock pair, as shown in Fig. 5.6. Likewise, the wavenumber has a minimum and a maximum, while the total change of the phase relative to the background wave is zero.

A homoclon with a moderate propagation speed connects the 1D unstable manifold of the background plane wave with its 2D stable manifold. Since there are two adjustable parameters, c and Ω, the counting argument suggests, according to (5.2.10), that a one-parametric family of homoclons should exist, and, adding the period as an additional parameter, a two-parametric family of MAW. The multiplicity of MAW also can be estimated by considering the instability of the plane wave solutions from

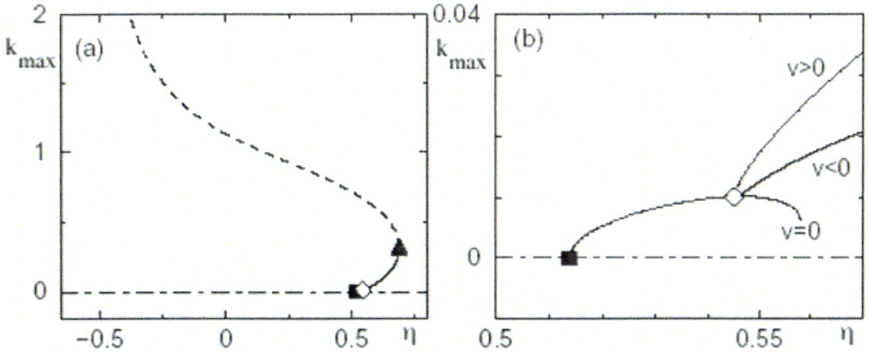

Fig. 5.7. Bifurcation diagrams for fixed $\nu = -2$ and $P = 30$, showing the Hopf (square), traveling (diamond) and saddle-node (triangle) bifurcations. The dot-dashed line represents the homogeneously oscillating solution of the CGL equation, while the lower and upper branches of MAW are represented by solid and dashed curves, respectively. (a) Overview of the maximum phase gradient of the MAW $k_{\max}$ as a function of η. (b) Close-up near $\eta = 1/2$ (Brusch et al, 2001, reproduced by permission ©Elsevier)

which the MAW emerge (Brusch et al, 2001). The plane waves form a one-parametric family, and in the Eckhaus-unstable regime, a plane wave with a certain wavenumber is unstable to a whole band of perturbations parametrized by their wavenumbers, leading, in agreement with the former argument, to a two-parametric family of MAW.

A typical bifurcation diagram of MAW constructed by Brusch et al (2001) is shown in Fig. 5.7. The lower branch of *shallow* MAW bifurcates from unstable plane waves and terminates at a saddle-node bifurcation point where it joins the upper branch of *deep* MAW. When, as in Fig. 5.7, the branch bifurcates off a homogeneous oscillating solution, a non-equilibrium traveling bifurcation (see Sect. 3.1.3) leading from a static to a pair of counter-propagating MAW takes place early on the lower branch; Brusch et al (2000, 2001) call it a drift pitchfork bifurcation. The shown bifurcation diagram is specific for the chosen period P. Generally, the saddle-node bifurcation shifts to smaller values of η and ν as the period grows.

Shallow MAW with mild amplitude modulation are typical elements of phase turbulence, which can be alternatively described by the KS equation (Sect. 5.1.3). MAW may remain stable as long as the system size small, while for larger systems periodic sequences of MAW are unstable with respect to splitting or interactions. A splitting instability leads to the growth of a new peak in the homogeneous part of MAW. As a result, two shorter-period MAW will appear. On the other hand, interaction between adjacent MAW shifts adjacent peaks in opposite directions, thereby creating occasional larger-period MAW, as well as causing peak collisions and merger of MAW. As a result of these instabilities, a perturbed unstable homogeneous state in a spatially extended system typically does not converge to a train of coherent MAW, but evolves instead to a state of phase chaos. Both splitting and interaction mechanisms are similar to instabilities observed in the KS equation (Chang et al, 1993).

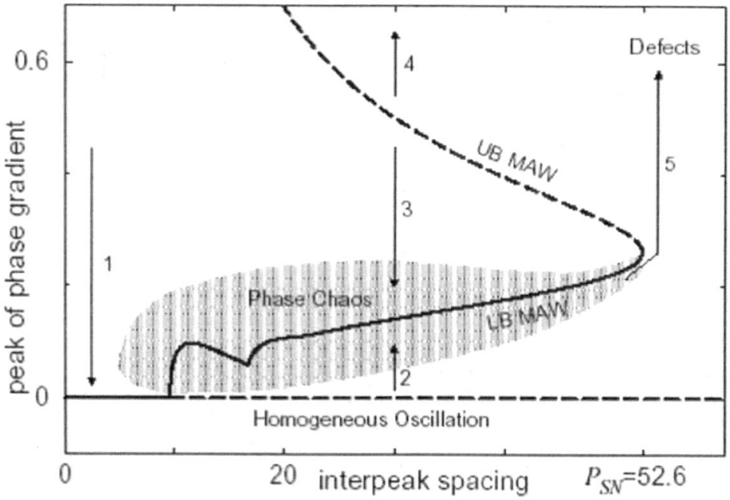

Fig. 5.8. Illustration of the relation between MAW structures and phase chaos for $\eta = 3,\ \nu = -0.65$. Solid (dashed) curves denote stable (unstable) solutions for the system size equal to the MAW period P. The shaded area indicates the typical values for near-MAW structures that occur in phase chaos. The arrows show the typical evolution directions (Brusch et al, 2001; reproduced by permission ©Elsevier)

Perturbations of shallow (lower branch) MAW lead, as a rule, to a mildly chaotic state, which remains close to the lower MAW branch (within the shaded area in Fig. 5.8). The exception is a perturbation pushing a MAW beyond the saddle-node bifurcation (arrow 5 in Fig. 5.8), which leads to defect chaos. Other arrows in this figure indicate typical ways of nonlinear evolution. Evolution starting from the unstable homogeneous state to the right of the bifurcation point (arrow 2) may lead to either stationary or moving MAW when the system size is small or to a state of phase chaos. Deep (upper branch) MAW are always unstable due to the positive eigenvalue associated with the saddle-node bifurcation. Depending on a perturbation, they either evolve into shallow MAW (arrow 3) or form defects as a result of further growth of local amplitude and phase gradients (arrow 4).

5.2.4 Transition to Defect Chaos

The CGL equation exhibits two qualitatively different kinds of spatiotemporal chaos known as phase chaos, when the modulus of the ρ is bounded away from zero, and defect chaos containing phase singularities with vanishing modulus. Several studies (e.g., Shraiman et al, 1992) tried to resolve the question whether phase turbulence is indeed a persistent state, and to draw a border of a parametric region where phase slips can be avoided altogether in a long-time evolution of an extended system. A combination of interacting shallow modulated waves and homoclons holds, apparently, a key to a mechanism that prevents phase slips in a well-defined range of parameters.

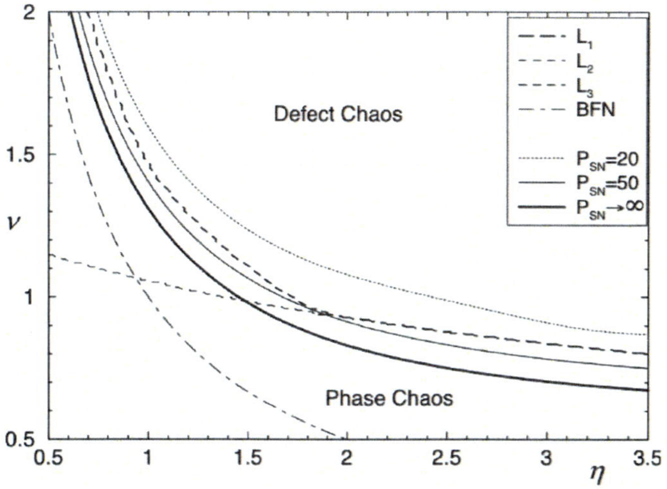

Fig. 5.9. Phase diagram of the CGL equation, showing the Benjamin–Feir limit curve BFN and the lines of transition from phase to defect chaos that correspond to different periods $P_{\rm SN}$ at the saddle-node bifurcation point, going to infinity at the solid line. The curves L_1, L_2, L_3 separate different chaotic states according to Shraiman et al (1992) and Chaté (1994). Between the L_2 and L_3 curves, there is a hysteretic regime where either phase or defect chaos can occur. (Brusch et al, 2000, reproduced by permission, ©The American Physical Society)

This points out to the infinite-period saddle-node bifurcation of MAW shown by the bold line[1] in Fig. 5.9 as the key feature that provides a strict lower boundary for the transition from the phase to defect chaos (Brusch et al, 2000). It is supported by direct simulations (Brusch et al, 2001) and the nonlinear bifurcation analysis near the Eckhaus instability of plane waves (Brusch et al, 2003).

The inter-peak distance distribution is established as a result of competition between the two kinds of instabilities of MAW mentioned in the preceding subsection. Splitting instability tends to reduce peak-to-peak distances (or the period P) and prevents MAW from crossing the saddle-node boundary; in the phase-chaotic regime, this tends to inhibit defect generation. On the contrary, interaction of MAW leads to an increase of the spacing between some peaks, thus enhancing the generation of defects. The "balance" point where the two instabilities are equally strong near the saddle-node bifurcation gives a rather good estimate of the apparent transition point from phase to defect chaos (Brusch et al, 2001).

Unlike phase turbulence, defect chaos is dominated by the formation, motion and interaction of holes. The intermittent regime where phase and defect chaos coexist was studied in detail by Chaté (1994) and Ipsen and van Hecke (2001). Since holes are not topological singularities, they can appear spontaneously, as well as merge and disappear. Holes have to be separated by sinks (shocks). Although the latter pay only passive role in the dynamics, interaction between neighboring

[1] Take note of the different orientation in Fig. 5.4 and Fig. 5.9

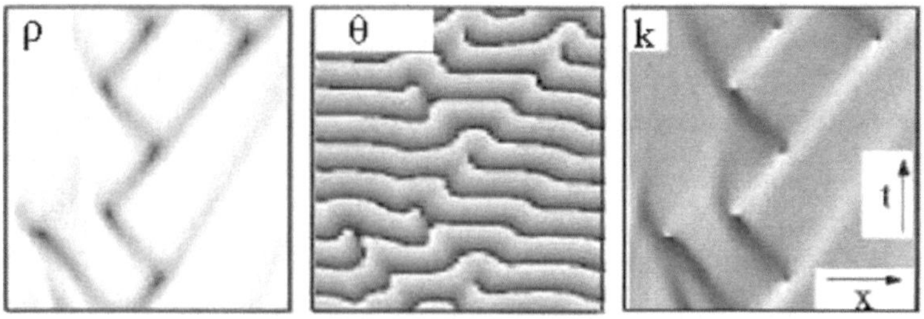

Fig. 5.10. Grayscale space–time plots of the amplitude ρ, phase θ and wavenumber k showing chaotic states in the intermittent spatiotemporal regime (van Hecke, 1998; reproduced by permission, ©The American Physical Society)

holes is sometimes described in terms of interaction between a hole and a shock. Hole-shock pairs are actually special cases of deep homoclons, as seen in Fig. 5.6. The hole-shock interaction is always attractive in the large separation limit when the decay rates defined by (5.2.11) are real. For static holes, this happens below the line MO in Fig. 5.4, but generally this limit is, of course, velocity-dependent. In the oscillatory range, and away from the core instability line, uniformly moving hole-shock arrangements were found to coexist with solutions having oscillatory hole velocities (Popp et al, 1995). Oscillating zigzagging holes, such as shown in Fig. 5.10, can be attractors in the intermittent regime, as shown in long-time simulations by van Hecke (1998).

5.2.5 Coupled CGL Equations

It was argued in Sect. 1.5.1 that degenerate modes with alternative directions described by coupled CGL equations, in particular, counter-propagating waves in 1D, tend to suppress each other. However, mutual suppression can be alleviated in the case of modulated waves. As follows from averaging (1.5.6), the linear growth coefficient of each mode is renormalized by adding the averaged squared amplitude of another mode times the mutual interaction coefficient. The globally coupled stationary equations for modulated waves with wavenumbers $k_\pm$ and frequencies $\omega_\pm$ are written then, each in its own moving frame, as

$$\mathrm{i}\omega_\pm \rho_\pm = [1 - (\gamma + \mathrm{i}\nu_-)\rho_\mp^2)\rho_\pm - (1 + \mathrm{i}\eta)k_\pm^2 \rho_\pm - (1 + \mathrm{i}\nu_+)\rho_\pm^3, \tag{5.2.19}$$

which leads after separating the real and imaginary parts to

$$\begin{aligned} \rho_\pm^2 &= \frac{\gamma(1-k_\mp^2) - (1-k_\pm^2)}{\gamma^2 - 1}, \\ \omega_\pm &= \frac{\nu[\gamma(1-k_\mp^2) - (1-k_\pm^2)] + \nu_-[\gamma(1-k_\pm^2) - (1-k_\mp^2)]}{\gamma^2 - 1} - \eta k_\pm^2 . \end{aligned} \tag{5.2.20}$$

At $\gamma > 1$, solutions exist, as for a single wave, whenever both $k_\pm$ are less then one.

The principal question of interest here is whether a modulated standing wave may exist at $\gamma > 1$, i.e., under conditions when excitation of a counterpropagating wave is suppressed. Restricting to amplitude perturbations only, we introduce in the standard way infinitesimal perturbations $\varrho_\pm e^{\lambda t}$ to arrive at the eigenvalue problem for the matrix

$$\mathbf{L} = \begin{pmatrix} A_+ & B \\ B & A_- \end{pmatrix}, \qquad A_\pm = -2\frac{\gamma(1-k_\mp^2)-(1-k_\pm^2)}{\gamma^2-1},$$
$$B = -\frac{2\gamma}{\gamma^2-1}\sqrt{[\gamma(1-k_+^2)-(1-k_-^2)][\gamma(1-k_-^2)-(1-k_+^2)]}. \tag{5.2.21}$$

The stability condition is $A_+A_- > B^2$, reduced to

$$(\gamma^2+2\gamma^3+\gamma^4-1)[\gamma(1-k_+^2)-(1-k_-^2)][\gamma(1-k_-^2)-(1-k_+^2)] > 0. \tag{5.2.22}$$

The first factor is positive at $\gamma > 1$, and the last two are positive whenever modulated solutions exist, so that modulated standing waves exist in the parametric region extending up to the stability limit of their modulation wavenumber. It should be noted that the coexistence region is narrow in physical units, since k is a modulation wavenumber, which is of $O(\epsilon)$ on the scale of an underlying system

The limits of Benjamin–Feir and Eckhaus instabilities for standing waves are determined in the long-wave limit in the same way as in Sect. 5.1.2. For $k = 0$, the stability condition is (Coullet et al, 1987)

$$1 + \frac{\eta(\nu_+ - g^2\nu_-)}{1-g^2} > 0. \tag{5.2.23}$$

The distinction between convective and absolute instabilities disappears in this case, since perturbation of one of the modes is transferred through the coupling term to its counterpart and propagates in the opposite direction; this exchange prevents it from being advected out of the system. Full stability analysis of modulated waves in coupled CGL equations is cumbersome; some limiting cases were resolved by Riecke and Kramer (2000).

Sources and sinks of counter-propagating waves arise in a natural way when one mode suppresses the other at $g > 1$. Then the system tends to form domains of either left-moving or right-moving waves separated by domain walls or shocks, which may be classified as sources or sinks according to whether the *total* group velocity $\widehat{v} = v \pm c_\pm$ of the asymptotic plane waves, where the modulation part $v = 2k(\eta - \nu_+)$ is defined as in (5.1.10), points outwards or inwards. Generically, the intrincic group velocity should dominate, so that the two wave trains meeting at a domain wall should belong to different families, as in Fig. 5.11. In this case, the motion of a domain wall cannot be obtained from integral conditions as in (5.2.17). Single-mode sources or sinks identical (in the comoving frame) to those constructed in Sect. 5.2.2 can appear when c_0 is small.

Various coherent structures can be obtained by solving the system of coupled equations similar to (5.2.6):

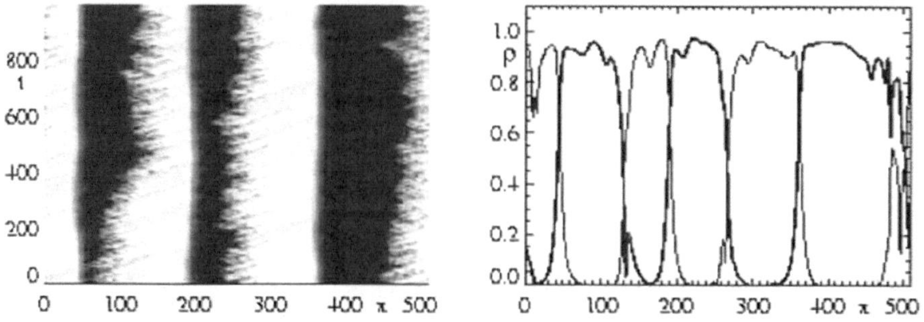

Fig. 5.11. Left: space–time plot showing the evolution of the amplitudes $\rho_\pm$ in the coupled CGL equations starting from random initial conditions in the parametric domain where bimodal patterns are weakly suppressed ($\gamma = 1.1$, $\eta = \nu_+ = 1$, $\nu_- = -1$, $c_0 = 0.5$). Gray shading is such that patches of ρ_+ mode are light and those of the ρ_- mode are dark. Right: amplitude profiles of the final state showing that a single wave propagating in either direction prevails on alternating segments (van Hecke et al, 1999; reproduced by permission ©Elsevier)

$$\partial_\varsigma \rho_\pm = \kappa \rho_\pm, \tag{5.2.24a}$$

$$\begin{aligned}(1 + \mathrm{i}\eta)(\partial_\varsigma z_\pm + z_\pm^2) + \mu + \mathrm{i}\Omega_\pm + (c \pm c_0)z_\pm \\ -(1 + \mathrm{i}\nu_+)\rho_\pm^2 - g(1 + \mathrm{i}\nu_-)\rho_\mp^2 = 0.\end{aligned} \tag{5.2.24b}$$

The coefficient at the linear growth term μ is *not* rescaled here to unity. The reason is that, due to unbalanced scaling discussed in Sect. 5.2.5, the small parameter and hence, dependence on the deviation from the bifurcation point, is not eliminated from the problem[2].

It was shown both numerically (Coullet et al, 1993) and by the counting argument (van Hecke et al, 1999) that sources form a discrete set, and therefore select the wavenumber of emitted waves; however, this may be not true under "abnormal" conditions when $|v| > |c|$. Sinks, being passive, form a two-parametric family, as in the case of a single CGL equation. Due to the advection direction, sources are, generally, wider than sinks. At small μ, or large $|c|$, there is a large interval where both amplitudes are close to zero. This can be easily understood, since at $c_0 \neq 0$ the trivial state is only *convectively* unstable.

Coullet et al (1993) found that there is a finite threshold value $\mu = \mu_c$, below which no coherent sources exist. This can be understood (van Hecke et al, 1999) by viewing dynamics in the vicinity of a sufficiently wide source as an "upstream" propagation of a wave pattern into the unstable trivial state against the linear group velocity. The motion of the front is governed by the single CGL equation in a frame moving with velocity c, and its propagation speed is defined by (5.1.29). Taking into account the rescaling and the change of the coordinate frame, the front propagation speed is $2\sqrt{\mu(1+\eta^2)} - c$. This speed vanishes at $\mu_c = \frac{1}{4}c^2/(1+\eta^2)$, and no stationary source can exist below this value. The width of coherent sources diverges

[2] Upon rescaling, this dependence is transferred to the dependence on the linear group velocity c_0, which is more difficult to monitor experimentally

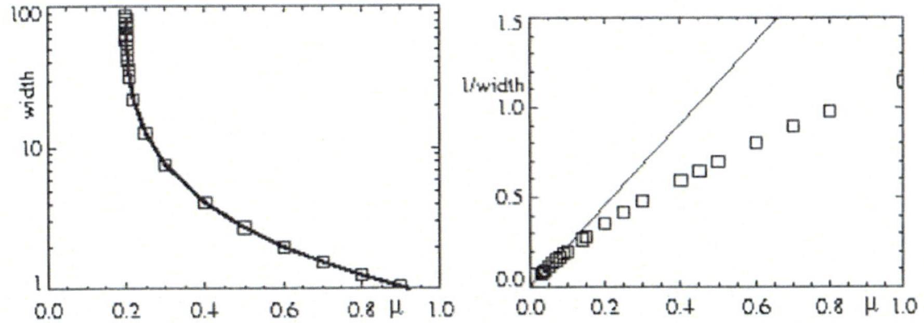

Fig. 5.12. Width of sources (left) and inverse width of sinks (right) as functions of the linear growth parameter μ. Squares indicate numerical results obtained by solving (5.2.24) with the parameter values $g = 2$, $\eta = 0.5$, $\nu_+ = -0.5$, $\nu_- = 0, c = 1$ for sources and $g = 2$, $\eta = 0.6$, $\nu_+ = -0.4$, $\nu_- = 0, c = 0.4$ for sinks. The straight line shows the asymptotic inverse width of sinks at $\mu \to 0$ following from (5.2.26) (van Hecke et al, 1999; reproduced by permission, ©Elsevier)

when μ_c is approached, as seen in Fig. 5.12. *Nonstationary* sources which cannot be obtained as solutions of (5.2.24) remain, however, possible also below μ_c.

Sinks become wider in the limit $\mu \to 0$, and their width can be estimated by solving (5.2.24) with the diffusional term neglected (Coullet et al, 1993). Then (5.2.24b) turn into algebraic equations. Assuming that the sink is stationary and therefore symmetric, we set $c = 0$, $\zeta = x$. After separating $z_\pm = \kappa_\pm + ik_\pm$, the real part of (5.2.24b) is easily solved to obtain

$$\kappa_\pm = \pm c_0^{-1} \left(\rho_\pm^2 + g\rho_\mp^2 - \mu\right) . \tag{5.2.25}$$

The two first-order equations obtained by using this in (5.2.24a) are integrated to yield

$$\rho_\pm^2 = \frac{\mu}{1 + \exp\left[\pm 2c^{-1}\mu(g-1)x\right]}, \tag{5.2.26}$$

which satisfies for $g > 1$ the asymptotic conditions $\rho_\pm \to 0$ at $x \to \pm\infty$. The asymptotic inverse width is linear in μ.

5.3 Waves in Two Dimensions

5.3.1 Symmetric Spiral

Formation of holes is a purely 1D phenomenon, since in 2D a locally initiated singularity of the wave front would develop into a *spiral wave*. The substantial difference lies in the topology of the phase field: in 2D the phase circulation must be conserved on an arbitrary contour surrounding a singularity.

A circularly symmetric vortex solution of (5.1.1) is a *rotating spiral wave*. The N-armed spiral wave solution (or a spiral vortex with the topological charge N) is obtained (Hagan, 1982) by assuming the *ansatz*

$$u = \rho(r)e^{i\theta}, \qquad \theta = N\phi + \psi(r) - \omega t. \tag{5.3.1}$$

This form differs from the vortex solution of the RGL equation (Sect. 4.2.4) by the presence of an explicit time dependence, as well as of a radial dependence of the phase $\psi(r)$. Moreover, the radial wavenumber $k = \psi'(r)$ generally approaches at $r \to \infty$ a constant non-vanishing value k_∞, so that the vortex appears to be radiating a wave with a certain asymptotic wavelength. The frequency ω should fit the asymptotic wavenumber, and is given by (5.1.9) with $k \to k_\infty$.

It is most convenient to use the stationary CGL equation in the form (5.1.5). The rescaling (5.1.4) is legitimate provided $1 + \eta\nu > 0$, since then, with ω defined by (5.1.9) and $k_\infty < 1$, we have also

$$1 + \omega\eta = 1 + \nu\eta + \eta(\eta - \nu)k_\infty^2 = (1 + \nu\eta)(1 - k_\infty^2) + (1 + \eta^2)k_\infty^2 > 0.$$

The rescaled asymptotic values of the real amplitude and wavenumber are related by (5.1.9), and the expression for Ω in (5.1.6) can be reduced thereby to

$$\Omega = -q(1 - k_\infty^2) = -q\rho_\infty, \tag{5.3.2}$$

The stationary polar form of (5.1.5) is

$$\nabla^2\rho + (1 - |\nabla\theta|^2 - \rho^2)\rho = 0, \tag{5.3.3}$$

$$\nabla^2\theta + 2\rho^{-1}\nabla\rho \cdot \nabla\theta = q(\rho_\infty^2 - \rho^2). \tag{5.3.4}$$

More explicitly, these equations are rewritten using the *ansatz* (5.3.1) in polar coordinates as

$$\rho''(r) + r^{-1}\rho'(r) + (1 - k^2 - N^2/r^2 - \rho^2)\rho = 0, \tag{5.3.5}$$

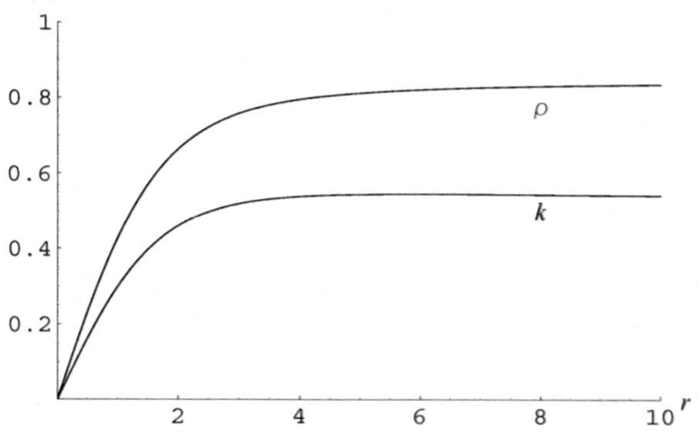

Fig. 5.13. Radial dependence of the wavenumber k and real amplitude ρ at $q = 2$ (after Hagan, 1982)

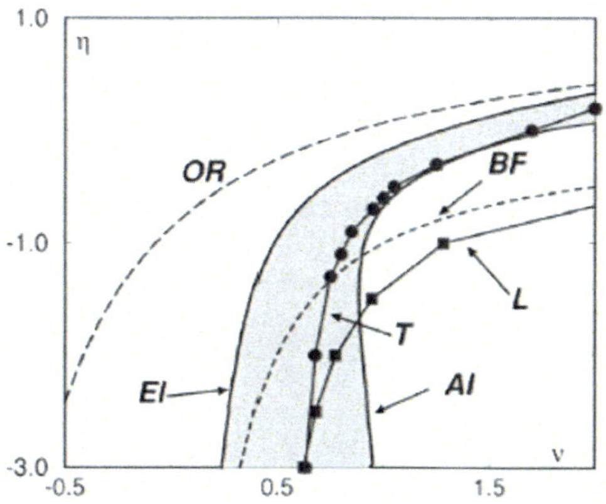

Fig. 5.14. Stability limits of a spiral wave solution in the parametric plane (η, ν). EI, AI – the limits of convective and absolute instability for the waves emitted by the spiral. Bound states exist to the right of the line OR – the boundary $q = 0.845$ of the oscillatory spatial decay for the emitted waves. BF – the Benjamin–Feir limit $\nu\eta = -1$. L – the limit of 2D phase turbulence. T – transition to defect turbulence for random initial conditions (Aranson and Kramer, 2002, reproduced by permission, ©The American Physical Society)

$$k'(r) + \frac{k}{r} + \frac{2}{\rho}k\rho'(r) \equiv \frac{1}{r\rho^2}\frac{\mathrm{d}}{\mathrm{d}r}(rk\rho^2) = q(\rho_\infty^2 - \rho^2). \tag{5.3.6}$$

where $k = \psi'(r)$ is the radial wavenumber.

Equations (5.3.5) and (5.3.6), subject to the boundary conditions

$$\rho(0) = k(0) = 0, \quad k(\infty) = k_\infty, \quad \rho(\infty) = \rho_\infty, \tag{5.3.7}$$

were solved numerically by Hagan (1982). As the problem is overdetermined, the numerical solution defines in a unique way the asymptotic wavenumber k_∞, dependent on the parameter q. A typical dependence of ρ and k on the radial distance is shown in Fig. 5.13. Take note that (5.3.6) is symmetric to the sign reversal of q and k; since both are of the same sign, the radial component of the group velocity (5.1.10) is always positive, and therefore all perturbations are swept from the core to the periphery of the spiral domain.This has nothing to do with an apparent direction of motion of spiral branches, which is determined by the *phase* velocity ω/k. The latter would also depend on an added basic oscillation frequency in an underlying system.

Standard stability analysis of plane waves in Sect. 5.1.2 should apply to far regions of spiral waves; one could expect therefore a transition to a turbulent state to occur under conditions when the selected asymptotic wavenumber k_∞ falls into the range where the corresponding plane wave solution of (5.1.1) is unstable. The respective limits of Benjamin–Feir, Eckhaus and absolute instability in the parametric plane (η, ν) are presented in Fig. 5.14.

5.3.2 Asymptotic Relations

At $q = 0$, the standard vortex structure obtained in Sect. 4.2.4 is recovered, so that $k_\infty(0) = 0$. One can attempt therefore to obtain the dependence $k_\infty(q)$ at $q \ll 1$ perturbatively (Hagan, 1982). Assuming $k = O(q)$, (5.3.5) gives just an $O(q^2)$

correction to the standard vortex solution $\rho_0(r)$, verifying (4.2.14), while (5.3.6) is readily integrated to yield, in the leading order

$$k(r) = \frac{q}{r\rho_0^2(r)} \int_0^r [1 - \rho_0^2(s)]\rho_0^2(s)s\,\mathrm{d}s. \tag{5.3.8}$$

The wavenumber is well behaved at $r \to 0$, as both $\rho_0(r)$ and $k(r)$ are linear in r in this limit. However, the integral (5.3.8) is logarithmically divergent at $r \to \infty$, so that k_∞ cannot be computed in such a straightforward manner. The asymptotic expression can be reduced by using (4.2.14) and integrating by parts to the familiar integral (4.3.10):

$$\begin{aligned} \lim_{r\to\infty} k(r) &= \lim_{r\to\infty} \frac{q}{r} \int_0^r \left[N^2 s^{-1} \rho_0^2(s) - \rho_0(s) \frac{\mathrm{d}}{\mathrm{d}s} \mathrm{d}s(s\rho_0'(s)) \right] \mathrm{d}s \\ &= \lim_{r\to\infty} \frac{q}{r} \int_0^r \left[N^2 s^{-1} \rho_0^2(s) + s(\rho_0'(s)]^2 \right) \mathrm{d}s = \frac{q}{r} \ln \frac{r}{a_0}, \end{aligned} \tag{5.3.9}$$

where $a_0 \approx 1.126$ for $|N| = 1$.

Matching with an outer solution, which is valid at distances $r = O(q^{-1})$, is required. Rescaling to the extended radial coordinate $R = qr$, we see that the real amplitude is slaved in the far field to the wavenumber, so that (5.3.3) or (5.3.5) reduces in the leading order to an algebraic relation $\rho^2 = 1 - |\nabla\theta|^2 = 1 - k^2 - N^2/r^2$. Using this in (5.3.4) yields the far field equation

$$\nabla^2\theta + q(k_\infty^2 - |\nabla\theta|^2) = 0. \tag{5.3.10}$$

The far field equation is now nonlinear; it is recognized as the stationary version of the Burgers equation. The nonlinearity is weak: since the radial component of the wave vector in the far field of a spiral wave is small only for $|q| \ll 1$, the terms containing q are formally of a higher order of magnitude under conditions when the radial wavenumber is small. The nonlinear terms, responsible for the screening effect, play, however, a crucial role in the theory, since their action accumulates at growing distances from the spiral core, and cannot be treated as a mere small perturbation.

Equation (5.3.10) can be linearized through the Hopf–Cole transformation $\theta = -q^{-1} \ln G$. The resulting equation is

$$\nabla^2 G - (k_\infty q)^2 G = 0. \tag{5.3.11}$$

In view of the circulation condition (4.2.1), G is not a univalued function, but the singularity can be removed by setting

$$G(r,\phi) = \mathrm{e}^{-Nq\phi}\chi(r). \tag{5.3.12}$$

Clearly, $\chi(r)$ is univalued, while the phase circulation equals $2\pi N$ as required. In polar coordinates, the equation of $\chi(r)$ reads

$$\chi''(r) + \frac{1}{r}\chi'(r) - \left(k_\infty^2 q^2 + \frac{N^2 q^2}{r^2}\right)\chi = 0. \tag{5.3.13}$$

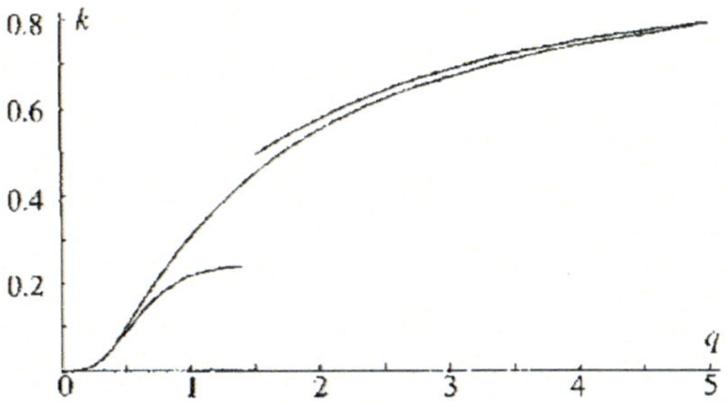

Fig. 5.15. Dependence of the asymptotic wavenumber k_∞ on the parameter q (after Hagan, 1982). The exact curve is flanked by the small q and large q asymptotics, expressed, respectively, by (5.3.16) and (5.3.17)

The symmetric solution corresponding to the far field of an isolated vortex is expressed by a modified Bessel function with an *imaginary* index:

$$\chi(r) = K_{\mathrm{i}|Nq|}(k_\infty qr). \tag{5.3.14}$$

In the following, we set $|N| = 1$. The outer limit of the inner solution (5.3.9) should be matched to the inner limit of the outer solution (Pismen and Nepomnyashchy, 1992):

$$\frac{1}{q}\lim_{r\to 0}\frac{\mathrm{d}}{\mathrm{d}r}\ln K_{\mathrm{i}|q|}(qk_\infty r) = \frac{1}{r}\,\frac{q\gamma\sin\left(q\ln\frac{1}{2}k_\infty qr\right) - \cos\left(q\ln\frac{1}{2}k_\infty qr\right)}{\sin\left(q\ln\frac{1}{2}k_\infty qr\right) + q\gamma\cos\left(q\ln\frac{1}{2}k_\infty qr\right)}, \tag{5.3.15}$$

where $\gamma \approx 0.5772$ is the Euler constant. The asymptotic expression (5.3.15) rapidly oscillates at $r \to 0$, and can match (5.3.9) only when the argument of the trigonometric functions is close to $-\frac{1}{2}\pi\sin q$. When (5.3.15) is expanded to $O(q)$, it reduces to $r^{-1}\left(\frac{1}{2}\pi + \gamma + q\ln\frac{1}{2}k_\infty qr\right)$. Matching with the inner solution (5.3.9) yields the asymptotic wavenumber

$$k_\infty = \frac{2}{a_0 q}\exp\left(-\frac{\pi}{2|q|} - \gamma\right). \tag{5.3.16}$$

This expression has been obtained in a somewhat more complicated way by Hagan (1982). It is plotted along with the "exact" numerical curve in Fig. 5.15.

Hagan also computed the asymptotics at $q \gg 1$:

$$\lim_{q\to\infty}\frac{q(1-k_\infty^2)}{k_\infty} \approx 2.34 \quad \text{at } |N| = 1. \tag{5.3.17}$$

However, this range is irrelevant, since (5.3.17) can be balanced only when k_∞ is close to unity, i.e., under conditions when the radiated waves are Eckhaus-unstable.

The asymptotic wavenumber at $|q| \ll 1$ turns out to be *exponentially* small. Such a quantity can be set to zero when a regular expansion in powers of a small parameter is carried out. Thus, a fair approximation to $k(r)$ can be obtained by solving (5.3.13) with k_∞ set to zero. The solution is

$$\chi(r) = \cos\left(q\ln\frac{r}{a}\right), \tag{5.3.18}$$

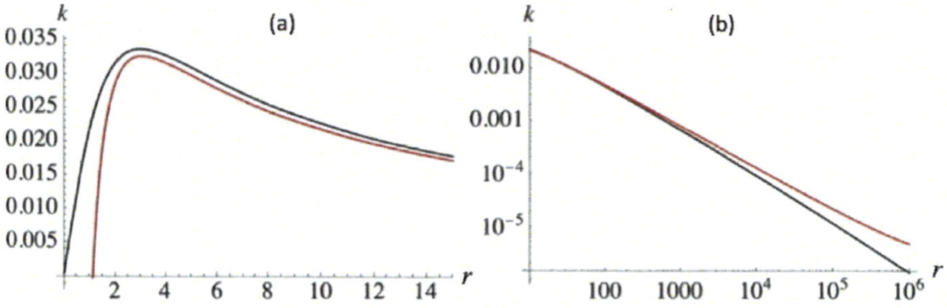

Fig. 5.16. Comparison of the approximations to $k(r)$ for $q = 0.1$ on moderate (a) and large (b) distances (after Pismen, 2003). Black line: the first-order solution obtained by integrating (5.3.8) numerically using $\rho_0(r)$ given by (4.2.15). Red lines: the near field asymptotics (5.3.9) (a) and the far field asymptotics (5.3.19) (b)

where the indefinite constant a fixes a phase shift. Choosing $a = a_0$ yields

$$\psi(r) = \frac{1}{q} \ln \cos\left(q \ln \frac{r}{a_0}\right), \quad k(r) = \frac{1}{r} \tan\left(q \ln \frac{r}{a_0}\right). \tag{5.3.19}$$

The last formula matches the asymptotics (5.3.9) when $r \gg 1$ but $\ln r = O(1)$. The approximation remains valid even at exponentially large distances $\ln r = O(q^{-1})$ but breaks down when the argument of the trigonometric function approaches $\pi/2$.

Matching (5.3.19) to (5.3.14) uses the asymptotics of the Bessel function at $\xi = k_\infty qr \ll 1$ expanded in powers of q:

$$\begin{aligned} K_{iq}(\xi) &\asymp -\frac{\pi}{\sinh(\pi q)} \mathrm{Im}\, \frac{(\xi/2)^{iq}}{\Gamma(1+iq)} = -q^{-1} \sin\left(q \ln \frac{\xi}{2}\right) - \gamma \cos\left(q \ln \frac{\xi}{2}\right) \\ &= -q^{-1} \cos\left(q \ln \frac{\xi}{2} + q\gamma + \frac{\pi}{2}\right). \end{aligned} \tag{5.3.20}$$

The cosine function in the last expression coincides with that in (5.3.19) after substituting (5.3.16). The radial wavenumber $k(r)$ computed using Eq. (5.3.14) is very well approximated by (5.3.19) up to exponentially large distances $r \approx a_0 \exp(\pi/2q)$. The radial wavenumber gradually decreases with r, but, although $k_\infty \to 0$ *beyond all orders* at $q \to 0$, this constant asymptotic value is approached at distances so huge that they are all but irrelevant.

The approximations to $k(r)$ on moderate, i.e., $O(q^{-1})$, and large, i.e., $O(\mathrm{e}^{1/q})$, distances for a moderately small value $q = 0.1$ are compared in Fig. 5.16. The diminutive asymptotic wavenumber $k_\infty \approx 1.5 \cdot 10^{-6}$ is not reached even at exponentially large distances shown in Fig. 5.16b, and the deviation from the asymptotic value reduces to a few percentage points only at $r > 10^8$.

5.3.3 Nondissipative Limit

The *non-dissipative* limit of the CGL equation,

$$-\mathrm{i}u_t = \nabla^2 u + (1 - |u|^2)u, \tag{5.3.21}$$

is known as the Gross–Pitaevskii (GP) equation in the theory of superfluids (Donnelly, 1991) and Bose–Einstein condensates (BEC), or as the defocusing nonlinear Schrödinger (NLS) equation in nonlinear optics (Newell and Moloney, 1992). It is derived from the Hamiltonian

$$u_t = -\mathrm{i}\frac{\delta \mathcal{H}}{\delta \overline{u}}, \quad \mathcal{H} = \int \left[\nabla\overline{u} \cdot \nabla u + \tfrac{1}{2}(1 - |u|^2)^2\right] \mathrm{d}\mathbf{x}, \tag{5.3.22}$$

which is conserved in the course of evolution. The linear term in (5.3.21) is, of course, inessential, since a transformation $u \to u\mathrm{e}^{-\mathrm{i}\omega t}$ adds the term ωu to the right-hand side. The choice of a coefficient at the linear term is actually dictated by desired properties of a "background" field, i.e., a spatially homogeneous solution of the field equations. When (5.3.21) is taken as the standard form, the background field has zero frequency and a non-vanishing amplitude, normalized to unity.

The NLS equation (5.3.21) is invariant to the Galilean transformation to a frame moving with an arbitrary velocity $v < 2$ (not exceeding the speed of sound), combined with a suitable gauge transformation of the field variable:

$$\mathbf{x} \to \mathbf{x} - \mathbf{v}t, \;\; \partial_t \to \partial_t - \mathbf{v} \cdot \nabla, \;\; u \to u\sqrt{1 - \tfrac{1}{4}v^2}\, \exp\left[(\mathrm{i}/2)\, \mathbf{v} \cdot \mathbf{x}\right]. \tag{5.3.23}$$

This hints at an intimate connection between (5.3.21) and hydrodynamic equations of ideal (inviscid) fluids. The NLS equation can be presented in a fluid-mechanical form with the help of a modified version of polar representation, often called the Madelung transform: $u = \sqrt{\varrho}\mathrm{e}^{\mathrm{i}\vartheta/2}$, where $\varrho = |u|^2$ is interpreted as the fluid density, and $\vartheta = 2\theta$ as the velocity potential defining the superfluid velocity $\mathbf{v} = \nabla\vartheta$. Then the non-dissipative counterparts of (5.1.2) and (5.1.3) are rewritten as the continuity equation

$$\varrho_t + \nabla \cdot (\varrho \mathbf{v}) = 0, \tag{5.3.24}$$

and the Bernoulli equation

$$\vartheta_t + \tfrac{1}{2}|\mathbf{v}|^2 + p = 0, \tag{5.3.25}$$

where p is pressure introduced via the "equation of state"

$$p = 2\left[(\varrho - 1) - \varrho^{-1/2}\nabla^2\varrho^{1/2}\right]. \tag{5.3.26}$$

Taking the gradient of (5.3.25) yields the Euler equation – the basic equation of ideal fluid mechanics:

$$\mathbf{v}_t + (\mathbf{v} \cdot \nabla)\mathbf{v} + \nabla p = 0. \tag{5.3.27}$$

Static vortex solutions of (5.3.21) are exactly the same as those of the RGL equation (1.1.7) – see Sect. 4.2.4. However, the dynamics is completely different; in this respect, both equations are polar opposites. In view of the Galilean invariance, the vortex remains symmetric when placed in a constant phase gradient and viewed in the respective Galilean frame, or, in other words, when advected with a constant velocity $\mathbf{v} = 2\nabla\theta$. The change in $\rho = |u|$ is a second-order effect when the velocity is small, and, for $v = O(\epsilon)$, the static solution remains invariant to $O(\epsilon^2)$. This can be contrasted to the $O(\epsilon)$ distortions of a dissipative vortex discussed in Sect. 4.3, which led there to a mobility relation containing a nonlinear logarithmic factor. This implies a remarkably simple law of vortex motion in a slowly varying field: the vortex is embedded in the ambient superfluid and moves with the local velocity; the latter can be considered constant as long as it does not change considerably on distances comparable with the size of the vortex core.

For $\mathbf{v} = O(\epsilon)$, the continuity equation (5.3.24) reduces in the leading order to the incompressibility condition $\nabla \cdot \mathbf{v} = 0$, so that the phase field obeys the Laplace equation $\nabla^2\theta = 0$. This equation can be solved exactly for any time-dependent configuration of vortices. The solution corresponding to a single vortex located at a point $\mathbf{x}_j(t)$ in the infinite plane is proportional to the polar angle counted relative to the vortex location: $\theta_j = N_j\phi(\mathbf{x}-\mathbf{x}_j)$; the respective velocity vector is $\mathbf{v} = 2N_j r_j^{-2}\mathcal{J}\mathbf{r}_j$, where $\mathbf{r}_j = \mathbf{x}-\mathbf{x}_j$, $r_j = |\mathbf{r}_j|$, and the operator $\mathcal{J}$ denotes rotation through $\pi/2$. The total phase field is obtained by superposition of the field induced by all extant vortices (and in a finite region, also of their images). This gives the equations of motion of vortices

$$\frac{d\mathbf{x}_j}{dt} = 2\sum_{k\neq j} N_k \mathcal{J}\,\frac{\mathbf{x}_j - \mathbf{x}_k}{|\mathbf{x}_j - \mathbf{x}_k|^2}. \tag{5.3.28}$$

The field problem has been reduced thereby to a dynamical problem with a finite number of degrees of freedom. This is the ultimate success of the particle–field approximation, which has not been achieved in any other problem of vortex motion. Solutions of the dynamical system (5.3.28) have been extensively studied in the context of classical ideal fluid mechanics (Batchelor, 1967); the motion can be chaotic when the number of vortices is larger than three.

The CGL equation approaches the NLS equation in the *weakly dissipative* limit $\eta \gg 1$, $\nu \gg 1$. The transition is most transparent when the form (5.1.5) is used. Requiring also $\omega \gg 1$, we see that both Ω and q, as defined by (5.1.6) are small in this limit, and the NLS equation is recovered in the limit $\eta \to \infty$. After rescaling $t \to \eta t$, this leads to the standard form

$$(\varepsilon - \mathrm{i})u_t = \nabla^2 u + (1 + \mathrm{i}\Omega)u - (1 - \mathrm{i}q)|u|^2 u = 0, \tag{5.3.29}$$

where $\varepsilon = 1/\eta$. Clearly, the basic vortex structure of the NLS and RGL equations with the vanishing radial wavenumber exists at $q = \Omega = 0$ for any η; this allows us to span the entire transition from conservative to dissipative vortex motion in a simple way.

5.3.4 Acceleration Instability

Aranson et al (1994) suggested that a spiral core may suffer an *acceleration instability*, which is analogous to the meandering instability of the spiral tip in reaction–diffusion models with separated scales (Sect. 3.5.4). The acceleration instability can be studied perturbatively when the real part of the amplitude diffusivity, rescaled to unity in (5.1.1), is small. When it is purely imaginary, as in the NLS equation (5.3.21), the system is Galilean invariant. The symmetry is broken due to both real correction to amplitude diffusivity and acceleration. A possible strategy is therefore to apply the transformation (5.3.23) and then to balance the extra terms that appear when the Galilean invariance is weakly broken.

Let $\mathbf{X}(\mathbf{x}, t)$ be a variable position of the phase singularity (i.e., of the zero of the complex order parameter). A linear relation between the velocity $\mathbf{v} = \partial_t \mathbf{X}$ and acceleration, presumed to be small and expressed as $\varepsilon \mathbf{a} = \partial_t \mathbf{v}$, must have the form $\mathbf{a} = \mathbf{F} \cdot \mathbf{v}$, where $\mathbf{F}$ is a "friction" matrix. In an isotropic system, the diagonal elements of the matrix $\mathbf{F}$ should be equal, and the off-diagonal ones should be antisymmetric. Acceleration instability is observed when $\mathbf{F}$ has an eigenvalue with a positive real part. At the instability threshold, the diagonal elements vanish. Since the off-diagonal elements, generally, differ from zero at this point, the matrix has a pair of imaginary eigenvalues. This means that acceleration is normal to velocity, as it should be in rotational motion. Meandering motion is possible in the supercritical region, provided the instability is saturated nonlinearly; otherwise, transition to turbulence is imminent.

The Galilean transformation (5.3.23) is applicable as well in the case when velocity is variable, and the coordinates are transformed as $\mathbf{x} \to \mathbf{x} - \mathbf{X}(\mathbf{x}, t)$. Starting with the weakly dissipative CGL equation (5.3.29) and applying the transformation (5.3.23), we obtain two non-vanishing $O(\varepsilon)$ terms lacking circular symmetry: one due to acceleration, and the other one due to the real part of the amplitude diffusivity. A solution of (5.3.29) corresponding to a slowly moving and weakly accelerating spiral vortex can be computed by expanding in ε in the same way as in Sect. 4.3.3. Assuming $\mathbf{v} = O(\varepsilon)$, $\mathbf{a} = O(\varepsilon^2)$, the dipole inhomogeneity is of $O(\varepsilon^2)$. The second-order equation obtained by linearizing the CGL equation has the general form (4.3.23) with the dipole inhomogeneity

$$\Psi(\mathbf{x}) = -\tfrac{1}{2}\mathbf{a} \cdot \mathbf{x}\, u_0(\mathbf{x}) + \mathbf{v} \cdot \nabla u_0(\mathbf{x}). \tag{5.3.30}$$

The first term is contributed by the time derivative of the transformed amplitude in (5.3.23). An additional circularly symmetric $O(\varepsilon^2)$ term stemming from this transformation modifies the frequency but does not affect the motion. The caveat is that the operator $\mathcal{H}$ is now *not self-adjoint*. The solvability condition requires (5.3.30) to be orthogonal to the eigenfunctions of the adjoint operator $\mathcal{H}^\dagger$ with zero eigenvalue. Unlike the translational eigenfunctions of $\mathcal{H}$, the respective eigenfunctions of the adjoint operator cannot be obtained simply by differentiating the circularly symmetric solution, and should be computed numerically.

One may find computing the eigenfunctions with the sole purpose of using them in the solvability condition impractical. The alternative is to solve the first-order perturbation equation directly. This was the approach adopted by Aranson et al (1994). Generically, the perturbation equation with the inhomogeneity (5.3.30) diverges exponentially at $r \to \infty$, but this behavior is suppressed, leaving a slower (power-like) growth, when the relation between the velocity and acceleration is chosen in a unique

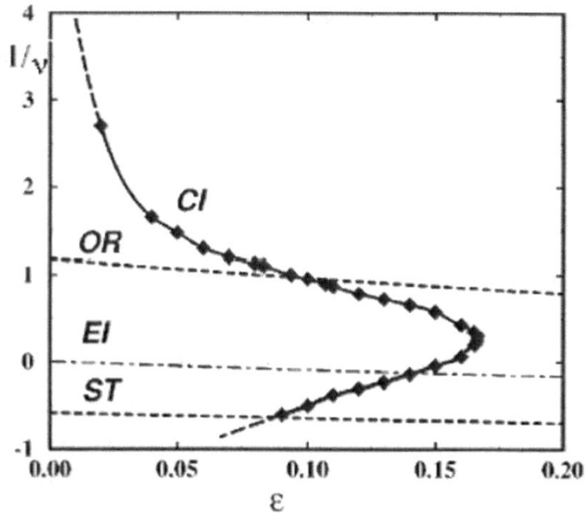

Fig. 5.17. Stability limits of a spiral wave in the large η limit. CI is the core instability limit (unstable to the left); the area below EI is the Eckhaus unstable region, ST denotes the transition line to strong turbulence, and the area below OR is the oscillatory range (Aranson et al, 1994; reproduced by permission, ©The American Physical Society)

way. The numerical results (Fig. 5.17) show that the spiral core is stable only when the real part of the amplitude diffusivity exceeds a certain finite positive value ε_c, going to zero in the limiting case $\nu \to \infty$ that corresponds to the defocusing NLS equation. The eigenvalues of the matrix $\mathbf{F}$ acquire an imaginary part when ν drops below a certain level, indicating a meandering instability similar to that observed in reaction–diffusion models (Sect. 3.5.4), where it typically stabilizes at a finite rotation amplitude. However, numerical simulations of the CGL equation have not produced such behavior. The state at $\varepsilon < \varepsilon_c$ was typically turbulent, but intermittent dynamics was observed near the instability threshold (Aranson et al, 1994).

Analytical computation is possible in the limit $q \to 0$, i.e., $\nu \geqslant \varepsilon^{-1}$. Then the translational eigenfunction $\nabla u(\mathbf{x})$ is used in the solvability condition. Comparing (5.3.30) with (4.3.42) we see that the problem is formally identical to that of motion in the supercriticality ramp, with the replacement $\mathbf{a} \to -2\mathbf{M}$. The area integrals of both terms in (5.3.30) diverge logarithmically, albeit with different constants. The divergence is eliminated by choosing $\mathbf{a} = 2\mathbf{v}$. The vortex turns to be always unstable under these conditions, and, as the off-diagonal elements of $\mathbf{F}$ vanish, it is accelerated along a straight line. Non-zero off-diagonal elements appear only at larger q.

5.3.5 Broken Phase Symmetry

The symmetry of the CGL equation to phase rotations is broken when the system is parametrically forced at a frequency ω_c *resonant* with the basic frequency at the Hopf bifurcation ω_0. For an integer ratio $\omega_c/\omega_0 = n$, the amplitude equation amending (5.1.1) can be written by adding the forcing term possessing the required symmetry:

$$u_t = (1 + \mathrm{i}\eta)\nabla^2 u + (\mu + \mathrm{i}\omega)u - (1 + \mathrm{i}\nu)|u|^2 u + \gamma \overline{u}^{n-1}, \qquad (5.3.31)$$

where γ is the forcing amplitude and $\epsilon^2\omega$ is a weak effective detuning, due to both parametric deviations from the Hopf bifurcation point and a weak mismatch between

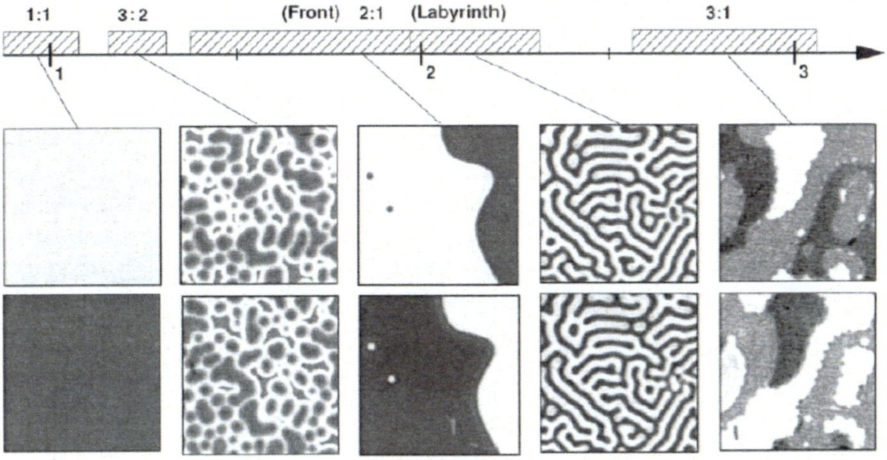

Fig. 5.18. Different frequency-locked regimes of the light-sensitive BZ reaction under periodic optical forcing (Petrov et al, 1997, ©Springer Nature). The axis above shows the ratio of the forcing frequency to the basic frequency of the system. The patterns are shown in pairs, one above the other, at times separated by the forcing period $2\pi/\omega_c$, except for the 1:1 resonance where the interval is π/ω_c

ω_c/n and ω_0. We retain here the linear growth coefficient μ, which can be normalized to ± 1 in the supercritical and subcritical region, respectively, since, in the presence of forcing, nontrivial solutions exist also for $\mu < 0$. Recalling the basic $u = O(\epsilon)$ scaling of (5.1.1), the orders of magnitude in (5.3.31) match in this case at a forcing amplitude $\gamma = O(\epsilon^{3-n})$. The polar form of (5.3.31) useful for computations is

$$\rho_t = \nabla^2\rho + (\mu - |\nabla\theta|^2 - \rho^2)\rho + \gamma\rho^{n-1}\cos n\theta - \eta(\rho\nabla^2\theta + 2\nabla\rho\cdot\nabla\theta), \quad (5.3.32)$$

$$\theta_t = \nabla^2\theta + 2\rho^{-1}\nabla\rho\cdot\nabla\theta + \eta(\rho^{-1}\nabla^2\rho - |\nabla\theta|^2) + \omega - \gamma\rho^{n-2}\sin n\theta - \nu\rho^2. \quad (5.3.33)$$

The forcing term in (5.3.31) reduces the invariance to arbitrary global phase shifts to a *discrete* symmetry $u \to e^{i\pi m}u,\ m = 1, \ldots, n-1$. For $n > 1$, this necessarily causes phase multiplicity of homogeneous solutions to (5.3.31) and changes the character of defects: instead of vortices, one can observed *fronts* separating alternative phase states. Various patterns at different forcing frequencies, which can be modeled by (5.3.31), were observed both in experiments and simulations (Petrov et al, 1997; Lin et al, 2004). Some typical patterns are shown in Fig. 5.18. One can see that behavior changes dramatically when the frequency ratio is changed.

Forcing at the same frequency ($n = 1$) is least interesting. At sufficiently large $O(\epsilon^2)$ forcing intensities, it just fixes a homogeneous state oscillating in phase with the forcing. However, a finite minimal intensity is required for synchronization under conditions when the unforced system is unstable, i.e., at $\nu\eta < -1$ (Chaté et al, 1999). For $n = 2$, two homogeneous phase states exist, separated by a front. Under certain conditions, the front destabilizes, giving rise to a chaotic labyrinthine pattern. For $n = 3$, there are three phase states and three kinds of fronts, which can rotate around their common point, such as one seen in Fig. 5.18. Resonances at fractional ratios require frequencies lower than the basic underlying frequency ω_0 of the CGL equation. In this case, no homogeneous states exist; thus, typical dynamics observed for 3:2 resonance is chaotic "bubbling" with the temporal spectrum at each point having peaks at $\omega_0/2$.

The strongest nontrivial resonance occurs at the frequency ratio 2:1 (Coullet and Emilsson, 1992; Yochelis et al, 2004). For $n = 2$, the residual symmetry is inversion, $u \to -u$, and the only possible singularity is a front separating the two alternative states. The formal expressions defining nontrivial homogeneous stationary solutions of (5.3.32), (5.3.33) are

$$\rho_\pm^2 = \frac{\mu + \omega\nu \pm \sqrt{(1+\nu^2)\gamma^2 - (\omega - \mu\nu)^2}}{1+\nu^2}, \quad \sin 2\theta_\pm = \frac{\omega - \nu\rho_\pm^2}{\gamma}. \tag{5.3.34}$$

To ensure existence, the expression for $\rho_\pm^2$ should be real and positive. This gives the bounds

$$\gamma > \gamma_c = \frac{|\omega - \mu\nu|}{\sqrt{1+\nu^2}}, \quad \gamma > \gamma_0 = \sqrt{\mu^2 + \omega^2}. \tag{5.3.35}$$

With increasing γ, a pair of solutions with moduli $\rho_\pm$ appear as a result of a saddle-node bifurcation at $\gamma = \gamma_c$, $\omega > \omega_c$. The upper solution $\rho = \rho_+$ is stable, and the lower one is unstable. At $\omega < \omega_c$, a single stable branch emerges at $\gamma_c = \gamma_0$.

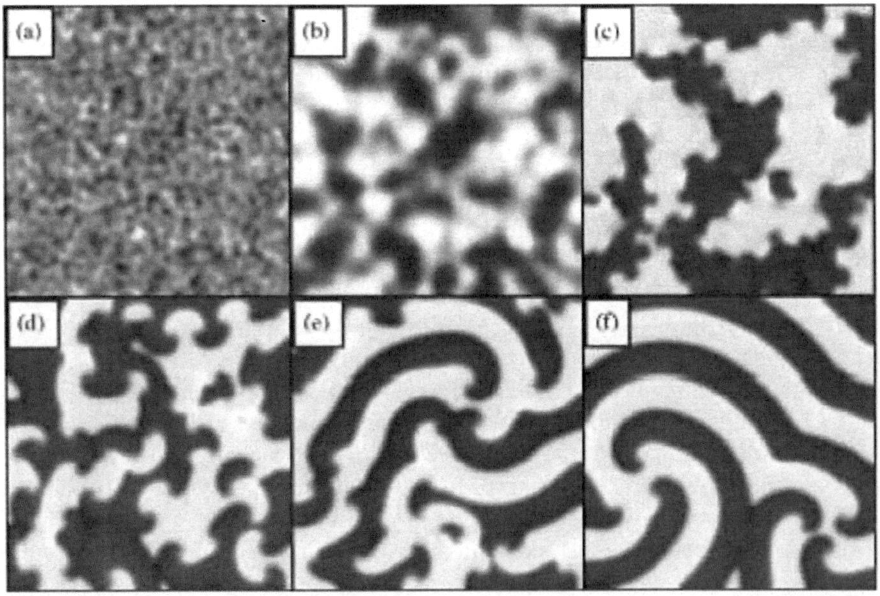

Fig. 5.19. (a–d) Development of a labyrinthine pattern through undulation instability of a stationary front. (e,f) Evolution starting from random initial conditions to a rotating spiral wave (Yochelis et al, 2004, reproduced by permission ©Elsevier Science)

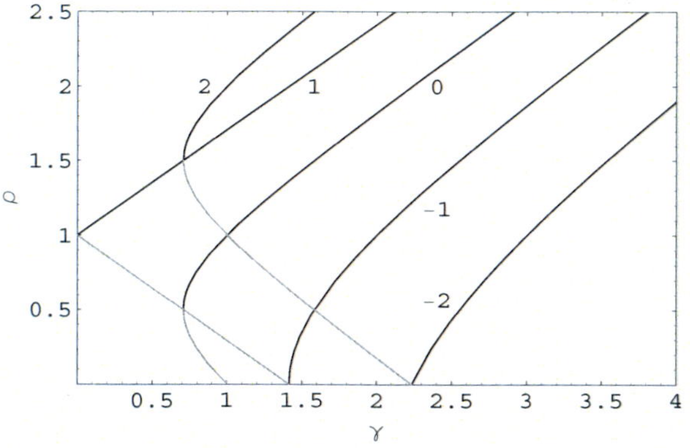

Fig. 5.20. Dependence of the modulus ρ on the forcing amplitude γ for $\mu = \nu = 1$. The stable and unstable branches are shown by the black and gray lines, respectively. The numbers at the curves show the values of ω. The line $\omega = 1$ is also the locus of the saddle-node bifurcation

Each solution exists in four phase states shifted by $\pi/2$. On the stable branch, these phase states are alternately stable and unstable. Some solution branches are shown in Fig. 5.20. Although alternative phase states are equivalent, fronts separating them are generally, nonstationary. Simulations of 2:1 resonance patterns by Yochelis et al (2004) showed undulations and oscillatory instabilities developing into either convoluted labyrinthine patterns or rotating spiral waves.

For the 3:1 resonance ($n = 3$) the three homogeneous solutions of (5.3.32), (5.3.33) are

$$\sin 3\theta = -\frac{\nu}{\gamma}\rho_{\pm}, \qquad \rho_{\pm}^2 = \frac{2\mu + \gamma \pm \sqrt{\gamma^4 + 4\mu\gamma^2 - 4\mu^2\nu^2}}{2(1+\nu^2)}, \tag{5.3.36}$$

where we set for simplicity $\omega = 0$. The solutions exist at

$$\gamma^2 > \gamma_c^2 = 2\mu\left(\sqrt{1+\nu^2} - 1\right). \tag{5.3.37}$$

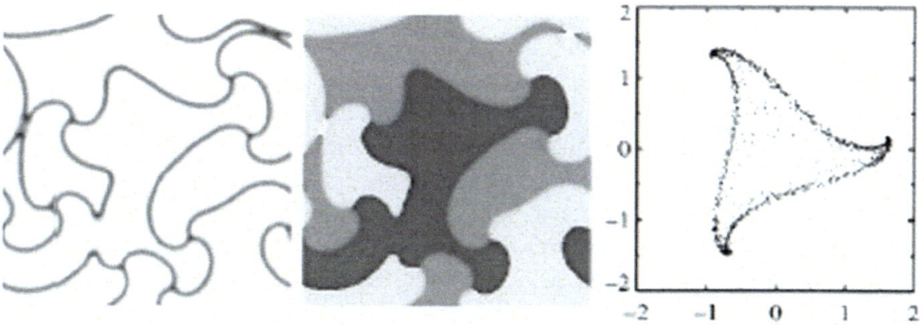

Fig. 5.21. The modulus (left) and phase (center) of the complex amplitude u, and the corresponding phase portrait in the complex plane taken at a fixed location (right) at the 3:1 resonance. Parameters: $\mu =, 1$ $\nu = -0.2$, $\eta = 2$, $\gamma = 0.25$. (Gallego et al, 2001, reproduced by permission, ©The American Physical Society)

For $\nu \neq 0$, the solutions appear at a finite forcing intensity. Of the three pairs of solutions that appear as a result of a saddle-node bifurcation at $\gamma = \gamma_c$, the triplet with $\rho = \rho_-$ is always unstable. The other triplet is stable in monotonous mode, but becomes oscillatory unstable for $\nu > \sqrt{3}$ in a certain range of γ (Gallego et al, 2001).

5.4 Interaction of Spiral Vortices

5.4.1 Nonradiative Limit

The structure of a symmetric spiral vortex can be perturbed by external phase and amplitude gradients, which would set the spiral core into motion. This problem appears at a first glance to be similar to that of perturbation of RGL vortices considered in Sect. 4.3, and, likewise, external gradients may be induced by interaction with neighboring vortices. Proceeding in the same way, one could try to impose on Hagan's solution (Sect. 5.3.1) a perturbation breaking the circular symmetry, and compute the drift velocity that would compensate this perturbation.

This approach fails for two reasons. The first obstacle is merely technical, and is connected to the fact that the linearized CGL equation is not self-adjoint, and the translational eigenfunction of the adjoint problem is not readily available, as already noted in Sect. 5.3.4. This obstacle can be overcome at the cost of numerical computation of the adjoint eigenfunction. Another impediment is more substantial, as it places the entire activity under a big question mark. Unlike the situation in Sect. 4.3, an external phase gradient, which may be generated, for example, by the phase field of another spiral vortex, generally, *does not penetrate* into the spiral wave domain, much less into the spiral core, due to effective screening caused by interference of nonlinear waves. This screening phenomenon is the principal reason for stability of spiral domain patterns with spiral domains separated by shocks, which are observed in experiments and simulations (more on this in Sect. 5.4.4). Moreover, all perturbations are washed out to the shocks, where they are absorbed, and never travel inwards. Clearly, a simple superposition of phase fields corresponding to isolated spiral wave solutions far from their cores cannot be a good choice of a basic solution in a pattern dominated by shocks, and the appropriate zero-order approximation should reflect the predominance of either vortex on the respective side of the shock.

A simple superposition of phase fields is possible only in the limit of small asymptotic wavenumbers, i.e., at $|\nu - \eta| \ll 1$, when shocks weaken and disappear. The non-radiative limit $\nu - \eta = 0$ encompasses the dissipative dynamics of Sect. 4.3 as well as the conservative dynamics briefly outlined in Sect. 5.3.3 (see Pismen, 1999 for a comprehensive review). In this limit, the CGL equation, written in the moving frame associated with a steadily propagating vortex, can be reduced by a gauge transformation to the RGL equation (Pismen and Rodriguez, 1990). This transformation does not affect the topological repulsion or attraction, but adds a velocity component normal to the straight line connecting a vortex pair, so that the overall direction of motion becomes oblique.

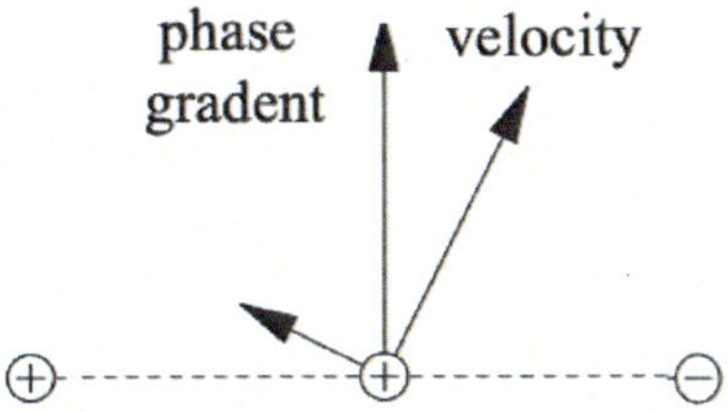

Fig. 5.22. Oblique motion of a positively charged vortex under the action of the phase gradient due to the presence of either a positive vortex on the left or a negative vortex on the right. According to (5.4.3), the external phase gradient is the sum of two vectors directed along and across the velocity vector. The motion is directed toward the negative vortex or away from the positive vortex

We start with the CGL equation in the form (5.1.5) rewritten in the comoving frame (with q and Ω set to zero):

$$(1 - i\eta)\mathbf{v} \cdot \nabla u + \nabla^2 u + u - |u|^2 u = 0. \tag{5.4.1}$$

Assuming the velocity to be small, we apply a combined Galilean and gauge transformation similar to (5.3.23):

$$\mathbf{x} \to \mathbf{x} - \mathbf{v}t, \quad \partial_t \to \partial_t - \mathbf{v} \cdot \nabla, \quad u \to u \exp\left[(i\eta/2)\, \mathbf{v} \cdot \mathbf{x}\right] + O(|\mathbf{v}|^2). \tag{5.4.2}$$

Unlike the NLS or GP equation (5.3.21) in Sect. 5.3.3 which possesses Galilean symmetry, the CGL equation retains after the gauge transformation the dissipative part of the advective term with dipole symmetry. Choosing the direction of motion as the y axis, we recover the stationary equation in the comoving frame (4.3.20). The perturbation expansion and matching to the far field proceed exactly as in Sect. 4.3.3, leading eventually to the relation (4.3.33) between the velocity and the phase gradient $\mathbf{k}$. Recalling the phase transformation (5.4.2), one has to extract from $\mathbf{k}$ the added phase gradient. Thus, (4.3.33) is modified to

$$\mathbf{k} - \frac{\eta}{2}\mathbf{v} = \frac{N}{2} \ln \frac{v_0}{|\mathbf{v}|} \mathcal{J}\mathbf{v}. \tag{5.4.3}$$

At $\eta \gg 1$, when the NLS limit is approached, the right-hand side can be neglected, so that the vortex is simply advected by the imposed phase gradient. If the latter is caused by another vortex removed at a distance large compared to the size of the vortex core, this causes translation of a pair of oppositely charged vortices and rotation of a pair of likely charged vortices. The dissipative interaction on the right-hand side causes *oblique* motion, with likely and oppositely charged vortices, respectively, repelling and attracting each other (see Fig. 5.22).

5.4.2 Weakly Radiative Vortices

Interaction of weakly radiative vortices at $q \ll 1$ can be considered as a perturbation of the non-radiative limit. Since, however, according to the results of Sect. 5.16, the limit $q \to 0$ is singular, the strategy, as well as chances for success, should depend on the range of separation between the interacting vortices. At "near" $O(q^{-1})$ distances seen in Fig. 5.16a, a weakly radiating spiral is similar to a non-radiating vortex. Since $\psi \approx O(q)$, this range of distances still falls within a single wavelength from the core, even though the separation is large compared to the core size.

One can expect that the radial component of the phase field of one vortex acting upon the core of the other one should cause a weak repulsion. An interesting question here is whether a bound vortex pair can be formed when this non-topological repulsion balances topological attraction of a pair of unlikely charged vortices. The relative strength of the two effects should be proportional to the ratio of the radial to the angular components of the wave vector. Since this ratio of $O(q)$, balancing them may be possible at $q \ll 1$ only when the system is close to the non-dissipative (NLS) limit, i.e., at $\eta \gg 1$. It is advantageous therefore to use the "weakly dissipative" scaling of the CGL equation in the form (5.3.29).

The phase gradient acting on the vortex core, which defines the velocity according to the mobility relation (5.4.3), should be obtained by solving the far field equation. The appropriate form is (5.1.25), modified in the current scaling to

$$(\epsilon + q)\theta_t = \nabla^2\theta - q|\nabla\theta|^2. \tag{5.4.4}$$

This form can be derived directly from the polar representation of (5.3.29) in a way similar to that described in Sect. 5.1.3. The asymptotic wavenumber k_∞ has been neglected here as an exponentially small quantity. On the other hand, although at $|q| = O(\epsilon) \ll 1,\ \nabla = O(\epsilon)$ the nonlinear term is formally of a higher order in the small parameter of the problem, it should be retained as the only source of non-topological interaction.

Consider a vortex pair with the positive and negative charges located at $x = \pm L/2,\ y = 0$. Assuming L^{-1}, as well as q, to be of $O(\epsilon) \ll 1$, rescale $q \to \epsilon q$ and $L \to \epsilon^{-1}L$. Up to $O(\epsilon^2)$ corrections, the vortex propagates, as in the NLS equation, along the y-axis with the speed $v = -2/L$. Introducing this book-keeping small parameter explicitly, we rewrite the far field equation (5.4.4) in the comoving frame as

$$\nabla^2\theta - \epsilon q|\nabla\theta|^2 + 2\epsilon^2 L^{-1}(1+q)\partial_y\theta = 0. \tag{5.4.5}$$

This equation is solved by expanding in ϵ as $\theta = \sum \epsilon^n \theta^{(n)}$ (Pismen, 2003).

We shall use the bipolar coordinate system with the foci at the vortex positions. The coordinates σ, τ (respectively, circular and unbounded) are related to the Cartesian coordinates x, y as

$$x = \frac{L}{2}\,\frac{\sinh\tau}{\cosh\tau + \cos\sigma}, \qquad y = \frac{L}{2}\,\frac{\sin\sigma}{\cosh\tau + \cos\sigma}. \tag{5.4.6}$$

The zero-order solution, satisfying the Laplace equation $\nabla^2\theta^{(0)} = 0$ and the topological conditions of $\pm 2\pi$ circulation around the foci at $x = \pm L/2$, is $\theta^{(0)} = \pi - \sigma$. The first-order radiative correction stems from the nonlinear term in (5.4.5). The inhomogeneity in the first order equation written in the bipolar coordinates turns out to be constant:

$$\theta^{(1)}_{\tau\tau} + \theta^{(1)}_{\sigma\sigma} - q = 0. \tag{5.4.7}$$

The boundary conditions should take into account radiation to infinity. A circle of an infinitely large radius is projected in bipolar coordinates onto a single point $\tau = 0$, $\sigma = \pm\pi$. If (5.4.7) is solved in the interval $-\pi \leqslant \sigma \leqslant \pi$, the appropriate boundary condition should contain two symmetric "point" sinks placed at $\tau = 0$ and $\sigma = \pm\pi$. The strength of the sinks will be further adjusted to verify the asymptotic matching condition (5.3.9) applied at $\tau \to \pm\infty$.

The obvious inhomogeneous solution independent of σ, $\theta^{(1)} = \frac{1}{2}q\tau^2$, does not match in the limit $\tau \to \infty$ the far field asymptotics of the core solution. It should be complemented therefore by a suitable solution of the Laplace equation with sinks at $\tau = 0$, $\sigma = \pm\pi$, which accounts for radiation by the spiral pair. It is easier to express the required solution as a symmetric combination of two sources at the foci, $c\,[\ln(r/L) + \ln(r'/L)]$, where r, r' are radial distances from the foci. The adjustable constant c can be chosen to satisfy the matching condition, which should be rewritten using the asymptotic relation between the bipolar and focus-centered polar coordinates. Near the positive focus, we have, in accordance with (5.4.6), up to the first order in r/L.

$$\tau = -\ln\frac{r}{L} + \frac{r}{L}\cos\phi, \quad \sigma = \pi - \phi + \frac{r}{L}\sin\phi. \tag{5.4.8}$$

The circularly symmetric component of $\theta^{(1)}$ matches in the limit $r \to 0$ (i.e., $\tau \to \infty$) the far field asymptotics of the core solution defined by (5.3.9) if one sets $c = -q\ln(a_0/L)$. Then $\theta^{(1)}$ is expressed in the vicinity of the focus as

$$\theta^{(1)} = \frac{q}{2}\ln^2\frac{r}{L} - q\ln\frac{r}{L}\ln\frac{a_0}{L} - q\frac{r}{L}\ln\frac{ra_0}{L^2}\cos\phi s. \tag{5.4.9}$$

5.4.3 Advective Correction

The inhomogeneity in the second-order equation is contributed by the advective term:

$$\partial_\tau^2\theta_2 + \partial_\sigma^2\theta_2 + (1+q)\frac{\cosh\tau\cos\sigma + 1}{(\cosh\tau + \cos\sigma)^2} = 0. \tag{5.4.10}$$

The inhomogeneity vanishes when averaged over σ; therefore there is no radiation to infinity that would necessitate a singular boundary condition, as in the first-order problem. In view of the symmetry $\sigma \to -\sigma$, the solution is sought for in the form of a cosine Fourier series:

$$\theta^{(2)}(\tau, \sigma) = \sum_{k=1}^{\infty}\psi_k(\tau)\cos k\sigma. \tag{5.4.11}$$

We only have to compute the dipole component ψ_1, which verifies the equation

$$\psi_{21}''(\tau) - \psi_{21} + 2(1+q)\mathrm{e}^{-|\tau|} = 0. \tag{5.4.12}$$

The unique solution is defined by the conditions of continuity and smoothness at $\tau = 0$ and absence of exponential growth at $\tau \to \pm\infty$:

$$\psi_{21} = (1+q)(1+|\tau|)\mathrm{e}^{-|\tau|}. \tag{5.4.13}$$

Using (5.4.8), this yields near the vortex location the dipolar term

$$\theta_2^{(1)} = \frac{r}{L}(1+q)\ln\frac{r}{\mathrm{e}L}\cos\phi. \tag{5.4.14}$$

The dipole contributions of both advective and radiative terms computed above are of the same $O(\epsilon^2)$. Collecting the dipole terms from Eqs. (5.4.9), (5.4.14), and adding the correction to the topological phase following from (5.4.8) yields the inner limit of the dipole component of the far field near the positive focus:

$$\theta_1 = -\frac{r}{L}\sin\phi + \frac{r}{L}\left[(1+q)\ln\frac{r}{\mathrm{e}L} - q\ln\frac{ra_0}{L^2}\right]\cos\phi. \tag{5.4.15}$$

The first term in the right-hand side can be removed by the gauge transformation (5.3.23) corresponding to the non-dissipative velocity component $v_y = 2/L$. The second term can be rearranged to the form (4.3.29):

$$\Theta_1 = r\chi(r)\cos\phi, \quad \chi(r) = \frac{1}{L}\left[\ln\frac{r}{a_0} + (1-q)\ln\frac{a_0}{L} + 1 + q\right]. \tag{5.4.16}$$

This is the inner limit of the outer solution, which can be used to evaluate the contour integral in the solvability condition (4.3.26) as in Sect. 4.3.3. The contribution of the logarithmic term in χ to the contour integral cancels with the area integral in (4.3.26). The remainder defines the dissipative velocity component v_x:

$$v_x = \frac{1}{L}\left[2(1-q)\ln\frac{a_0}{\epsilon L} + 1 + 2q\right], \tag{5.4.17}$$

or, returning to the original parameters,

$$v_x = \frac{1}{L}\left[2(1-\eta q)\ln\frac{a_0}{L} + 1 + 2\eta q\right]. \tag{5.4.18}$$

Since L must be large in the original units, the first term is dominating, unless at q close to $1/\eta$. At $q < 1/\eta$, v_x is negative, so that the interaction is attractive, as it is in the non-radiative case $q = 0$. At $q > 1/\eta$, repulsion prevails at large distances. The stationary separation,

$$L_s = a_0 \exp\left(\frac{1/2 + q\eta}{q\eta - 1}\right), \tag{5.4.19}$$

falls into the required range $L \gg 1$ only when $q\eta - 1$ is negative and small. Under these conditions, the interaction becomes repulsive at large distances, but the equilibrium is obviously unstable. Thus, oppositely charged spirals cannot form bound pairs in this limit, although topological attraction does switch to repulsion with growing separation in a certain parametric range.

The remaining possibility is that a stable equilibrium might be realized at exponentially large distances where screening becomes important, and the weak interaction limit is applicable. Aranson et al (1991) and Pismen and Nepomnyashchy (1991,

1992) computed a correction to the vortex mobility due to a finite but small asymptotic wavenumber. The perturbation theory was based on the usual core expansion in the comoving frame matched to a solution of the far field equation. The radial and angular components of the wavenumber equalize when $\ln r \sim O(|q|^{-1})$. Within this range, the far field equation (5.4.5) cannot be solved perturbatively as above. The phase equation can be linearized through the Hopf–Cole transformation as in Sect. 5.3.2. However, it is shown there that the transformed function becomes multi-valued when the phase field is topologically non-trivial. Pismen and Nepomnyashchy (1991, 1992) constructed an approximate solution by computing a correction to the circulation condition at small q. Remarkably, this topological correction resulted in a screening effect, so that vortex interaction turned out, indeed, to be exponentially decaying with separation.

Another way (Pismen, 2003) is to extract first the topological part by setting, say, $\theta = \Theta + \psi$ (e.g., $\Theta = \phi$ for the circularly symmetric case in Sect. 5.3.2), and to apply the Hopf–Cole transformation $\psi = -q^{-1} \ln \chi$ to the non-topological part only. Starting from (5.1.25) rewritten in the comoving frame, this transformation yields a linear equation with variable coefficients dependent on Θ:

$$(1 + q\eta)\chi_t = \nabla^2 \chi - 2q \nabla \chi \cdot \nabla \Theta + q\chi \left[q |\nabla \Theta|^2 + (1 + q\eta)\Theta_t \right]. \qquad (5.4.20)$$

Using as before the bipolar coordinate system, the extracted topological phase is $\Theta = \pi - \sigma$. Allowing for propagation along the y axis with the speed v, we write the far field equation in the comoving frame in the form

$$\chi_{\tau\tau} + \chi_{\sigma\sigma} + 2q\chi_\sigma + q^2 \chi + \tfrac{1}{2} L v (1 + q\eta) \mathcal{F}[\chi] = 0, \qquad (5.4.21)$$

where the coordinate dependence is contained only in the advective operator

$$\mathcal{F}[\chi] = \frac{(\cosh \tau \cos \sigma + 1)(\chi_\sigma + q\chi) - \chi_\tau \sinh \tau \sin \sigma}{(\cosh \tau + \cos \sigma)^2}. \qquad (5.4.22)$$

The order-of-magnitude estimates in (5.4.21) differ from those for the near field, since $1/\eta$ should not be presumed small, while velocity is expected to be exponentially small (though still far exceeding k_∞). The velocity of a hypothetical bound pair has to be determined from a solvability condition, since it is not approximated anymore by vortex dynamics in the NLS equation.

Using a power expansion in q to detect an *exponentially* small effect is precarious. This puts a question mark on the earlier analytical results. If the perturbation parameter is of $O(\mathrm{e}^{-|q|})$, all $O(q^n)$ quantities should be formally treated as being of $O(1)$, which introduces insurmountable difficulties, and even invalidates the approximation of the phase equation by an isotropic Burgers equation. Detecting an exponentially weak effect in this limit (at exponentially large separations which cannot be realized in practice) has little value, besides a purely intellectual challenge.

5.4.4 Strong Radiation and Shocks

The essential feature determining the interaction of *strongly* radiating spiral vortices at $q = O(1)$ is the formation of shocks – domain boundaries where waves with differently directed wave vectors emanating from different centers collide. The shocks are, generally, mobile. Their speed is determined by (5.2.17), where $k_\pm$ should be interpreted now as the normal components of the wave vectors on the two sides of the shock. For a shock separating two counter-propagating waves, the right-hand side of (5.2.17) is the difference of the normal components of their group velocities. According to this formula, the domain with a higher group velocity (i.e., a larger wavenumber k) advances at the expense of its rival.

A shock separating two extended symmetric spiral domains is expected to be stable, as long as the radial wave vector component decreases with the distance from the core at large distances. This is clearly seen in Fig. 5.16 drawn for $q \ll 1$; at larger q (Fig. 5.13), the decrease persists, though is not as pronounced. Under these conditions, a symmetry-breaking shift is counteracted by an increase of the group velocity on the receding, and a decrease on the advancing side, which returns the shock to its original symmetric position. If, however, the shock shifts very close to the core, into a region where the radial wavenumber still *increases*, stability is lost, and, as a result of symmetry breaking, only one extended spiral domain remains, whereas the other spiral, while being protected by conservation of the topological charge from disappearing altogether, is reduced to a "naked core".

Shocks effectively screen different spiral domains from radiation emitted by other spiral cores. Perturbations due to a neighboring spiral decay behind the shock with the attenuation rate defined by (5.1.20). An important threshold is the critical value $|q| \approx 0.845$ beyond which the decay becomes oscillatory. This limit, computed using in (5.1.20) the asymptotic wavenumber k_∞ selected by the spiral wave, is shown by the line OR in Fig. 5.14. At larger values of $|q|$ (to the right of this line), formation of bound spiral pairs is possible, as confirmed by computations of Aranson et al (1993). Both likely and unlikely charged spirals can form bound pairs, as seen in Fig. 5.23. The oscillatory decay at large $|q|$, creating effective traps for interacting spiral cores, is essential for this effect, and there is no wonder that bound pairs were not found in the analysis for $|q| \ll 1$ (Sect. 5.4.2).

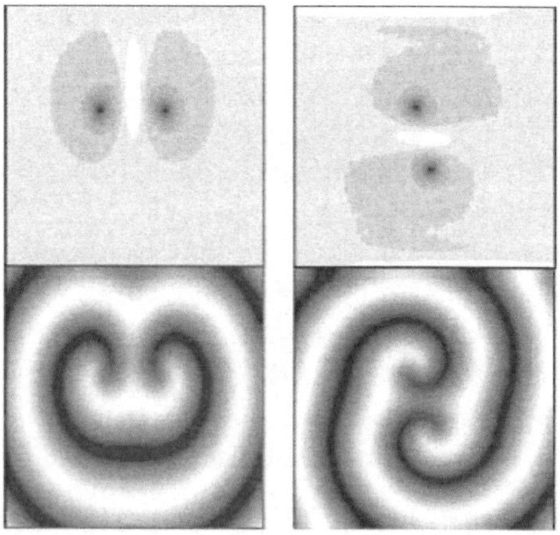

Fig. 5.23. The bound states of oppositely (left) and likely (right) charged spirals for $\eta = 0$, $\nu = 1.5$. The images show the modulus $\rho(x, y)$ (top) and Re(u) (bottom) (Aranson and Kramer, 2002, reproduced by permission, ©The American Physical Society)

Bound pairs are always mobile. Since velocities, being induced by interactions decaying exponentially with the distance, are slow, perturbation analysis is applicable here, but the difficulty lies in the non-Hermitian nature of the linearized problem at $q \neq 0$. Lacking a readily available translational eigenfunction, one has to solve the adjoint problem numerically to compute the adjoint eigenfunction with zero eigenvalue necessary for evaluation of the solvability condition. For $q \neq 0$, the adjoint eigenmode decays exponentially at large distances, thereby ensuring convergence of the area integral in the infinite plane. Therefore, it is no longer necessary to compute the solvability condition in a finite region as in Sect. 4.3.3. Although this seems to make the computations easier by eliminating the need in asymptotic matching of the near and far field solutions, it effectively disconnects the motion of the spiral core from its far field background. As a result, it becomes, for example, unclear whether the phase gradient added through the transformation (5.4.2) does indeed induce drift velocity as in Sect. 5.4.1 or is advected away from the core to be absorbed at a shock.

Aranson et al (1993) used another approach based on direct solution of the perturbation problem. The idea is that solutions exist only with "correct" velocities. Solutions are sought for in the form $u = u^{(0)}(\mathbf{x}) + \epsilon u^{(1)}(\mathbf{x})$, where ϵ is a formal small parameter scaling the velocity $\epsilon \mathbf{v}$, and $u^{(0)}(\mathbf{x})$ is the symmetric spiral solution within a limited domain, e.g., in a half-plane for a symmetric spiral pair. For a pair of spirals separated by a shock, this solution does not satisfy the boundary conditions at the shock taking into account the neighboring spiral, e.g., the symmetry condition $u_x(0, y) = 0$ when the symmetry axis is $x = 0$, as in the left panel of Fig. 5.23. Assuming the discrepancy to be of $O(\epsilon)$, the corrections are determined by the linearized first-order problem with the applicable boundary condition, e.g., $u_x^{(1)} = -\epsilon^{-1} u_x^{(0)}$ in the symmetric case. The advective term $\mathbf{v} \cdot \nabla u^{(0)}$ comes in as an inhomogeneity. The first-order problem is solved directly, and the velocity is adjusted in a unique way to avoid exponential divergence at large distances, which, in essence, is a numerical equivalent of computing a solvability condition.

Interaction of spirals with widely separated cores is only weakly dependent on their polarity, since the oscillatory asymptotic decay is the same for spirals of opposite charge. The only difference is in tangential motion defined by symmetry. For a pair of spirals separated by the shock at $x = 0$, the symmetry is $u(x, y) = u(-x, -y)$ for likely charged spirals and $u(x, y) = u(-x, y)$ for unlikely charged spirals. This leaves the symmetry condition $u_x(0, y) = 0$ used in the above computation identical in both cases, but the directions of motion of the two spirals in the bound pair should be identical when the charges are opposite (resulting in translation along the y axis) and opposite when the signs are identical (resulting in rotation). In aggregates of more than two bound spirals, each spiral core may move along a more complicated and possibly irregular trajectory.

5.4.5 Multispiral Patterns

The shock formation and its influence on spiral interactions are crucial for generic long-time evolution of large systems containing a number of spirals. A typical example of a spiral domain pattern in a stable parametric range obtained in a simulation run

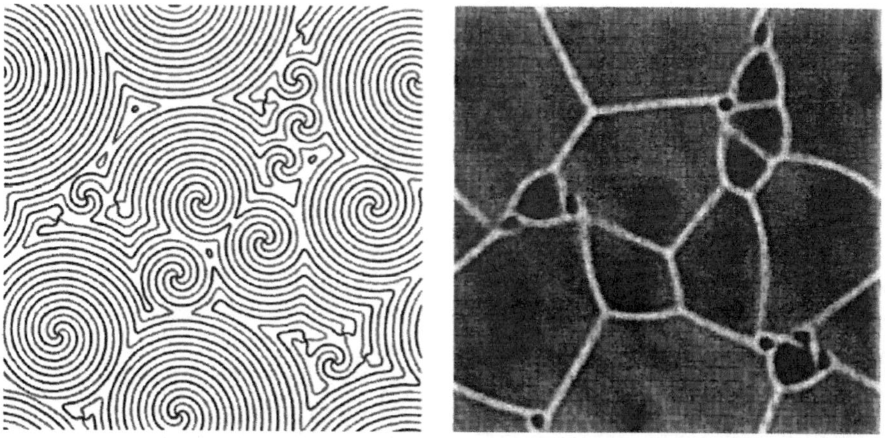

Fig. 5.24. Spiral domains. Left: levels of constant phase. Right: grayscale amplitude map showing enhanced amplitudes at the shocks (Chaté et al, 1996; reproduced by permission, ©Elsevier)

starting from random initial conditions is presented in Fig. 5.24. At the initial stage, the system tends to relax locally to the stable state with the unity real amplitude, but, as the phases are random, the relaxation is frustrated, and a large number of defects – vortices of unit charge – are formed. At the following coarsening stage, oppositely charged vortices annihilate, so that the density of defects decreases. However, the coarsening process stops halfway, leaving a certain number of single-charged spiral vortices.

The resulting stable spiral domain pattern is called *vortex glass* (Huber et al, 1992; Chaté et al, 1996). A rotating spiral wave of either sense emanating from each vortex fills a surrounding domain; the domain boundaries are shocks where waves with differently directed wave vectors emanating from different centers collide. Vortices that failed to conquer a sufficiently large domain and were reduced to a "naked core", as explained in the preceding subsection, may survive at shock junctions as phase singularities tightly bound by converging shocks. One cannot say for sure whether the "final" state seen in Fig. 5.24 is truly stationary.

In the final pattern, neighboring spiral domains may have any sense of rotation, i.e., carry a topological charge of any sign. Since the phase circulation around any domain should be equal to $\pm 2\pi$, a phase slip on a remnant vortex might be necessary to close the circulation balance. The waves always propagate *outwards* from the vortex cores; in fact, the entire domain structure is generated when local order spreads out from centers to the periphery as seen in Fig. 5.25. Perturbations, also traveling outwards with the prevailing group velocity, are absorbed at shocks, and therefore the pattern may survive beyond the convective instability threshold. The turbulent state takes over only when the waves become *absolutely* unstable, i.e., when some perturbations grow locally in the laboratory frame.

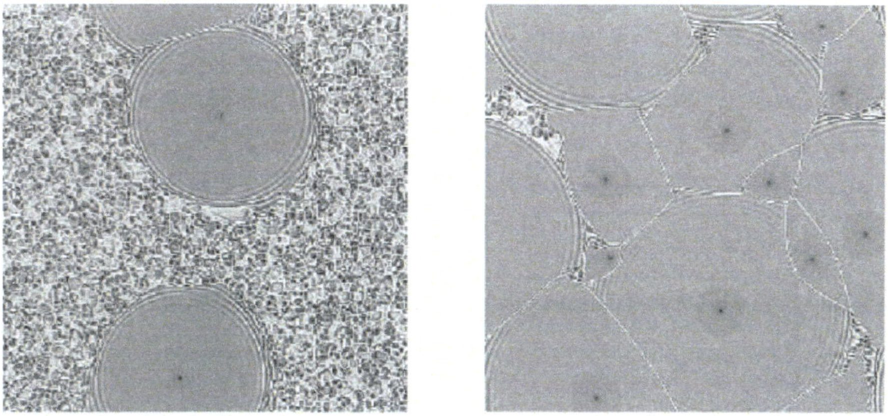

Fig. 5.25. Development of vortex glass in the convectively unstable range. Left: large spirals nucleating in turbulent sea, $\eta = -2,\ \nu = 0.75$. Right: developed vortex glass, $\eta = -2,\ \nu = 0.7$. Dark shades shows low amplitudes (Chaté et al, 1996; reproduced by permission, ©Elsevier)

The shapes of domain boundaries can be determined from the condition of phase continuity across the shocks (Bohr et al, 1997). Far away from the core of an unperturbed spiral, the phase is given by

$$\theta_i = k_\infty r_i \pm \phi_i + C_i, \tag{5.4.23}$$

where r_i, ϕ_i are the polar coordinates measured from the center of the ith spiral and C_i is a constant. At distances $r_i \gg 2\pi/k_\infty$, the angular term in (5.4.23) can be neglected; then the phase balance condition $\theta_i = \theta_j$ for two neighboring spirals yields

$$r_i - r_j = (C_j - C_i)/k_\infty = \text{const}, \tag{5.4.24}$$

which defines a hyperbola in the x, y plane. This simple formula was found to reproduce the structure of the spiral domains with high accuracy.

The overall structure of the pattern is different in the monotonic and oscillatory ranges separated by the line OR in Fig. 5.14. In the monotonic range above this curve, stable domains have to be of about the same size, as in Fig. 5.26a, since shocks are immobile when the normal components of the wave vector, which weakly decrease with distance, are equal. On the contrary, in the oscillatory range, domains may have various sizes, as in Fig. 5.26b, since the balance can be achieved at different separations.

Multispiral domain patterns were usually considered stationary, but shocks may still be moving very slowly due to a minute discrepancy of wavenumbers which have not quite attained their asymptotic values. Residual vortex interactions might cause creeping motion of vortex cores. Very long computation runs by Brito et al (2003) have shown that "frozen" patterns actually evolve on a very long time scale. The ultraslow evolution was characterized in this study by "activity" $\langle \rho_t \rangle$, where the average was taken over the entire computation domain. Since the amplitude ρ

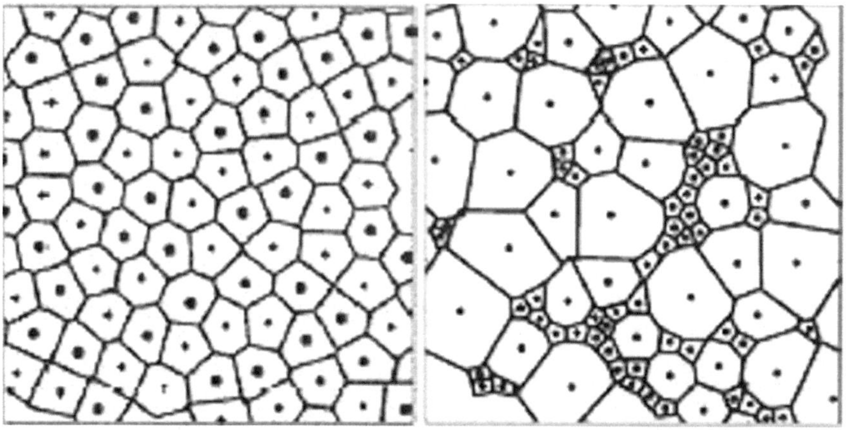

Fig. 5.26. Spiral domain pattern in the monotonic (left, $\eta = 0$, $\nu = 0.6$) and oscillatory (right, $\eta = 0$, $\nu = 1.2$) range. Oppositely charged vortices are distinguished in the left panel by the size of the dots indicating the core locations. The shocks are shown schematically by straight lines (Brito et al, 2003, reproduced by permission, ©The American Physical Society)

is almost constant everywhere except the core region, this quantity is related to the velocity of spiral cores.

In the monotonic range, spiral cores perform very slow diffusive motion, and $\langle \rho_t \rangle$ fluctuates around a certain value, while the pattern retains the general form of Fig. 5.26a; the apparent diffusivity increases with vortex density. In contrast, in the oscillatory range, the spiral population spontaneously segregates after a very long transient into two distinct phases seen in Fig. 5.26b: large and almost immobile spirals and clusters of trapped small vortices. When the "liquid fraction" is small, the resulting pattern exhibits slow intermittent dynamics: bursts of activity separated by long quiescent intervals. The system keeps evolving on an extremely slow scale, which is consistent with exponentially weak repulsion between well separated spiral cores.

Another possibility, realized in a different parametric region, is a dynamic chaotic state that shows no persistent features. This state is attained under conditions when either spiral waves or vortex cores, or both, are unstable. As in 1D, one can distinguish between mild phase turbulence, when no phase singularities occur, and the defect chaos characterized by persistent creation and annihilation of vortex pairs. In 2D, phase turbulence is always metastable with respect to transitions to either defect turbulence or vortex glass, but may persist in the parametric region between the Benjamin–Feir line and the line L in Fig. 5.14 (Chaté et al, 1996; Manneville and Chaté, 1996). Its range is somewhat smaller than in 1D. Beyond the line L, defects are created spontaneously, leading to the defect chaos.

The transition from vortex glass to defect turbulence in simulations starting from random initial conditions occurs at the numerically determined line T in Fig. 5.14 (Chaté et al, 1996). The transition occurs somewhat prior to the absolute instability limit determined by the linear stability analysis of plane waves emitted by spirals. However, the absolute instability limit can be approached by starting from carefully prepared initial conditions in the form of large spirals. Prior to the transition, one can observe transient defect turbulence which is unstable to spontaneous nucleation of spirals from the "turbulent sea", leading eventually to a vortex glass state as in Fig. 5.25. The nucleation time diverges at the line T.

5.4.6 Period Doubling in Spirals

A special kind of spiral wave patterns arises when the underlying dynamical system undergoes a period doubling transition. The period doubling causes the appearance of synchronization defect (SD) lines, which serve to reconcile the doubling oscillation period with the period of rotation of the spiral wave (see Fig. 5.27a). These lines are defined as the loci of those points in the medium where the two loops of the period two orbit exchange their positions in the local phase space. The period two oscillations on the opposite sides of an SD line are shifted relative to each other by 2π (i.e., a half of the full period), so that the dynamics projected on the rotation direction is effectively of period one, while it is of period two locally at any point in the medium (Fig. 5.27b). In a similar way, period four local dynamics generates two types of SD lines, with the phase jumping either by 2π or by 4π. SD lines of this kind were detected in experiments in a continuously fed gel reactor (Park and Lee, 1999). The local dynamics of the CGL equation is, of course, too simple to reproduce this phenomenon, but it has been successfully modeled (Goryachev et al, 1999) using the chaotic model due to Rössler (1976) complemented by diffusional terms with equal diffusivities:

$$\partial_t \mathbf{u} = \nabla^2 \mathbf{u} + \mathbf{f}(\mathbf{u}), \; f_1 = -(u_2 + u_3), \; f_2 = u_1 + a_1 u_2), \; f_3 = -u_1 u_3 - a_2 u_3 + a_3). \tag{5.4.25}$$

SD lines break the rotational symmetry and thus, on general grounds, spiral cores are expected to move. In the "laminar" regime at smaller values of the parameter a_2

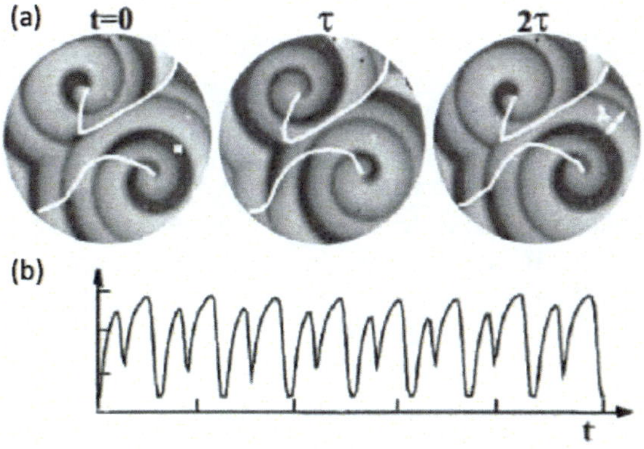

Fig. 5.27. (a) A pair of period two spiral waves with the fundamental period τ and the average wavelength λ. The white solid lines are synchronization defects (SD). (b) A period two time series measured at the point marked by the white square (Park and Lee, 2002; reproduced with permission, ©The American Physical Society)

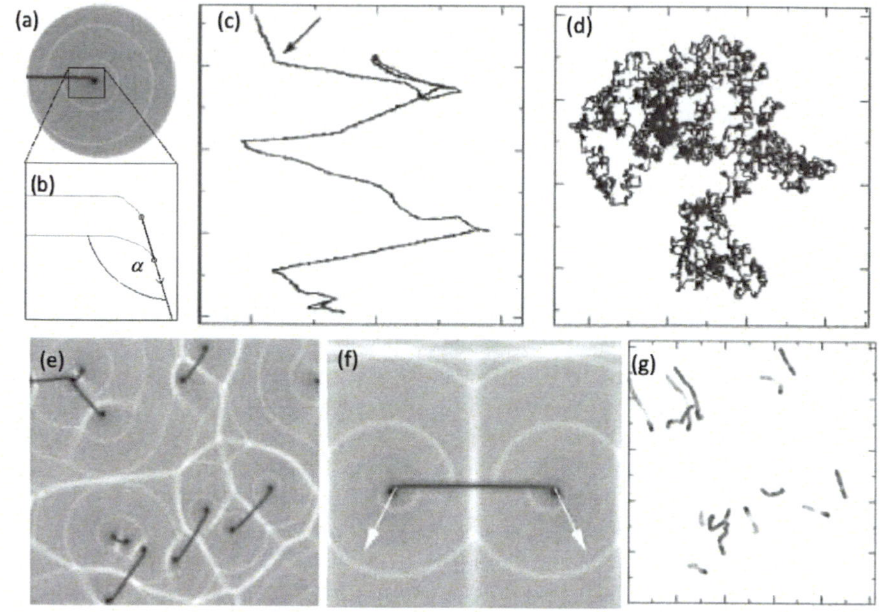

Fig. 5.28. Core dynamics of period two spirals in the extended Rössler model (Davidsen et al, 2004, reproduced by permission, ©The American Physical Society). (a) Snapshot of a rotating period two spiral in a laminar regime. The SD line originating from the spiral core is shown in black. (b) Magnification of the rectangular region in (a) showing the SD line at two moments of time; α is the angle between the spiral core trajectory (black line) and the attached SD line. (c,d) Spiral core trajectories in the turbulent region at $a_2 = 4.6$ (c) and $a_2 = 5.8$ (d). (e) Snapshot of a typical spiral patterns in the period two regime ($a_2 = 3.5$); SD lines are shown in black. The direction of the cores motion is highlighted by the white arrows in the blowup (f) showing a single pair of spirals. (g) Trajectories of spiral cores leading to the configuration shown in (e); changed shading indicates the elapsed time

in (5.4.25), the observed motion is very slow, and proceeds, after a transient, with a constant velocity directed at a certain angle to the SD line (Davidsen et al, 2004, see Fig. 5.28a,b). With growing a_2, the velocity increases, while the angle changes from almost π at the period doubling transition to close to $\pi/2$ at the transition to a turbulent regime. Close to the onset of turbulence, core trajectories take the form of "ballistic" flights with sporadically changing directions. Deeper in the turbulent region, the flights shorten, and the trajectory resembles Brownian motion (Fig. 5.28c,d).

Spiral domain patterns in extended systems undergo a drastic change following a period doubling transition. As discussed in Sect. 5.4.5, multispiral patterns in simply periodic systems typically relax to a frozen "vortex glass" state consisting of spiral domains with comparable areas separated by shock lines. This changes under the period two local dynamics (Fig. 5.28e–g): spiral pairs connected by SD lines now repeatedly rearrange, since the motion at an obtuse angle to the connecting line pushes them away one from the other.

Fig. 5.29. A spiral wave with a sinusoidal (left) and an irregular (center) SD line. The snapshot on the right shows the development of a chaotic domain with tangled SD lines (Park et al, 2004; reproduced by permission, ©The American Physical Society).

In multispiral disordered patterns, spiral cores move almost independently, stretching the SD lines to which they are attached, until they meet another core, leading to reconnections and creation or annihilation of SD lines. This process may be frustrated when spiral cores become attached to more than one SD line. The rearrangements occur mainly when two cores are close and may initiate a cascade of reconnection events. Repulsion of vortices of opposite sign connected by an SD line overcomes their topological attraction and prevents their annihilation at close approach. Thus, vortex glass gives way to a "vortex liquid" state where not only the total topological charge, but also the number of spiral cores is preserved due to complex dynamics of connecting SD lines (Davidsen et al, 2004). Transition to turbulence can be also initiated by bending SD lines, as seen in Fig. 5.29 (Park et al, 2004).

5.5 Line Vortices and Scroll Waves

5.5.1 Curvature-Driven Motion

CGL spirals, like spirals in the FN system (Sect. 3.5.6), can be extended in 3D into a scroll wave. We have seen in Sect. 3.5.6 that dynamics of a scroll wave is largely determined by the motion of its core filament. This motion can be analyzed most readily using the CGL model where the core is identified in a most natural way as a *line vortex* – the locus of the zero of the complex field. It should be noted that a line vortex in 3D, unlike a point vortex in 2D, is not protected by any topological conservation law, and may collapse.

Consider a curve $\mathcal{C}$ defined at any moment of time t in a parametric form $\mathbf{X}(s;t)$, where s is the arc length. As in 2D, the tangent to $\mathcal{C}$ at any point is the derivative $\partial_s \mathbf{X} = \mathbf{l}$. Since $\mathbf{l}$, by construction, it is a unit vector, the vector $\partial_s \mathbf{l}$ must be orthogonal to $\mathbf{l}$. By definition, the direction of $\partial_s \mathbf{l}$ is the normal $\mathbf{n}$, and its absolute value[3] is the *curvature* κ. The vector $\mathbf{n}_s$ measuring the rate of change of the normal must again

[3] Unlike Sect. 2.2.1 where the sign of the curvature was related to the direction of change of the order parameter, the curvature, as defined here, is always non-negative.

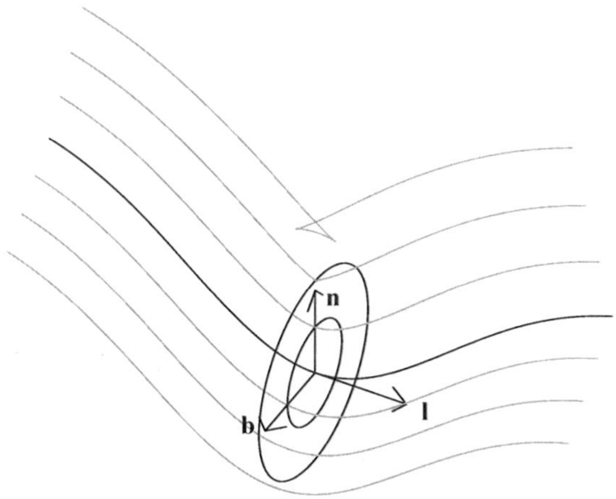

Fig. 5.30. Aligned coordinate frame in 3D. A sequence of coordinate lines (shown in gray), obtained by shifting points of the line vortex (shown in black) along the local normal, develops a singularity at an $O(\epsilon^{-1})$ distance when $\kappa = O(\epsilon^{-1})$, while displacement along the binormal is distortion-free. The Frenet trihedron and two circles in the local normal plane are shown in the center

be orthogonal to $\mathbf{n}$; its projection on the tangent must be equal to $-\kappa$ to insure the identity $\partial_s(\mathbf{l}\cdot\mathbf{n}) = 0$. The projection of $\mathbf{n}_s$ on the *binormal* $\mathbf{b}$, which is the unit vector orthogonal to both $\mathbf{l}$ and $\mathbf{n}$, is the *torsion* τ. Finally, the projections of the derivative $\mathbf{b}_s$ on $\mathbf{l}$ and $\mathbf{n}$ are fixed by the identities $\partial_s(\mathbf{l}\cdot\mathbf{b}) = \partial_s(\mathbf{n}\cdot\mathbf{b}) = 0$. This yields the Frenet–Serret equations

$$\mathbf{X}_s = \mathbf{l}, \quad \mathbf{l}_s = \kappa\mathbf{n}, \quad \mathbf{n}_s = -\kappa\mathbf{l} + \tau\mathbf{b}, \quad \mathbf{b}_s = -\tau\mathbf{n}. \tag{5.5.1}$$

The three orthogonal unit vectors $\mathbf{l}, \mathbf{n}, \mathbf{b}$ define at any point of the curve the *Frenet trihedron*. Curvature and torsion are the two intrinsic parameters that define locally the rotation of the Frenet trihedron, and therefore determine the shape of a smooth curve. If the functions $\kappa(s)$ and $\tau(s)$ are known, the curve can be reconstructed, in principle, by integrating (5.5.1), although this task is technically difficult due to the implicit form of these equations. For practical computations, the Frenet–Serret equations can be transformed to a more convenient set of intrinsic equations of motion in a way similar to Sect. 3.3.1, provided the normal and binormal velocities are known (see e.g., Pismen, 1999). The latter should depend in 3D, in addition to other factors, on local geometry of a line vortex.

When the curvature radius is much larger than the core size, it can be treated perturbatively by setting up an aligned coordinate frame in a way similar to Sect. 2.2.1. The plane normal to the curve at any point is spanned by the local normal and the binormal vectors. The aligned frame is built up, as in 2D, by shifting points of the line vortex along the local normal (Fig. 5.30). It is convenient to work with the CGL equation in the form (5.1.5), where we can set $\Omega = 0$ (absorbing it into the background frequency) Assuming both the curvature κ and the local normal velocity $\mathbf{v}$ to be small, we set $\kappa = \epsilon K$, $\mathbf{v} = \epsilon\mathbf{V}$ and rewrite (5.1.5), to the first order in a book-keeping small parameter ϵ, as a stationary equation in the aligned frame

$$\nabla_\perp^2 u + u - (1 - \mathrm{i}q)|u|^2 u + \epsilon\left[(1 - \mathrm{i}\eta)\mathbf{V} - K\mathbf{n}\right]\cdot\nabla_\perp u = 0, \tag{5.5.2}$$

where $\nabla_\perp$ is the 2D gradient operator in the local normal plane.

The perturbation term is independent of q, which creates an illusion that both radiative and nonradiative vortices can be treated in the same manner. If $\eta = 0$ (which at $q = 0$ corresponds to the gradient RGL equation), the first-order correction cancels exactly by setting $\mathbf{v} = \kappa\mathbf{n}$. Thus, the local law of dissipative motion is utterly simple: each point on the line vortex is translated along the normal with the speed proportional to the local curvature, while the vortex core remains undeformed in the first order. Clearly, the curve shrinks as a result of motion along the normal, resulting (in the gradient case) in decreasing energy integral, which is roughly proportional to the length of the line vortex. For example, a ring vortex of radius L should shrink with the speed $\mathrm{d}L/\mathrm{d}t = L^{-1}$, yielding upon integration the total lifetime $t_0 = L_0^2/2$ for a ring with the initial radius L_0.

Generally, the final result of dissipative evolution is collapse of a closed curve or straightening of a curve with fixed ends, but intermediate configurations of a curve evolving from smooth initial conditions may involve singularities formed as a result of local collapse of strongly curved sections. The local law of motion based on the expansion in small curvature becomes, of course, invalid shortly before collapse when the curvature radius becomes comparable to the core size.

With $\eta \neq 0$, the imaginary part of the advective term can be removed, as in 2D, by applying the transformation (5.4.2). Indeed, it is easy to see that the velocity and the added phase gradient can be chosen in such a way that the inhomogeneity vanishes exactly:

$$\mathbf{k} = \tfrac{1}{2}\kappa\eta\mathbf{n}, \quad \mathbf{v} = (1+\eta^2)\kappa\mathbf{n}. \tag{5.5.3}$$

Gabbay et al (1997) interpreted this as an indication that binormal motion is suppressed by an applied phase gradient. The phase gradient cannot be, however, chosen at will, but has to be obtained by solving applicable far field equations, e.g., the Burgers equation (5.1.25) or the diffusion equation in the nonradiative case $q = 0$. In the latter case, nonlocal induction by removed segments of the line vortex should be significant, as discussed below. At $q \neq 0$, nonlinear interference on shocks may serve to suppress distant action. Simulations by Frisch and Rica (1992) and Gabbay et al (1998) indicated no binormal velocity component, which may be taken as a sign that these poorly understood interactions conspire in a miraculous way to create a compensating phase gradient.

5.5.2 Nondissipative Motion

Lacking a freely adjustable phase gradient, there is no exact compensation in the non-gradient case, and a solvability condition has to be applied. As in Sect. 4.3.3, integration is carried out over a large circle of radius $1 \ll r_0 \ll \epsilon^{-1}$, and the solvability condition involves the contour integral dependent on the gradient of the phase field $\mathbf{k}_\perp = \nabla_\perp\theta$ induced by the entire line vortex. As in 2D, the relevant constituent of the phase field, which is also presumed to be of $O(\epsilon)$, should be *non-singular*, i.e., exclude the local phase circulation. The notion of "local" is not as obvious in 3D, where we deal with a continuous line vortex, as it was in the case of isolated point vortices in 2D. Separating local and nonlocal contributions necessitates, in fact, an additional asymptotic matching procedure, which can be carried out consistently in the non-dissipative case $\eta \to \infty$ when the nonlinear Schrödinger (NLS) equation is recovered upon rescaling (Pismen, 1999).

Starting from the NLS equation (5.3.21) rewritten in the aligned frame, we remove the term containing the drift velocity $\mathbf{v}$ with the help of the Galilean transformation (5.3.23). The resulting equation,

$$\nabla^2 u + (1 - |u|^2)u = \epsilon K \, \mathbf{n} \cdot \nabla_\perp u, \tag{5.5.4}$$

coincides with (4.3.20) if the y-axis is directed *against* $\mathbf{n}$, and $\kappa = \epsilon K$ is replaced by v. Thus, the expansion and asymptotic matching procedure from Sect. 4.3.3 leading to the solvability condition (4.3.26) is again applicable, with v replaced by κ as a small parameter.

Next, the phase field on the bounding circle entering the solvability condition has to be evaluated by solving the far field equation. The phase in the far field satisfies the Laplace equation $\nabla^2\theta = 0$ (Sect. 5.3.3), which can be solved exactly for any given configuration of the line vortex $\mathbf{X}(s;t)$. The solution should satisfy the circulation condition $\oint \mathrm{d}\theta = 2\pi$ on any contour surrounding the line vortex[4]. The phase gradient, which, in view of the Galilean symmetry, is equivalent to the velocity $\mathbf{v} = 2\nabla\theta$, is expressed by the Biot–Savart integral

$$\mathbf{v}(\mathbf{x}, t) = -\oint_C \frac{\mathbf{R}(s;t) \times \mathbf{l}(s)}{R^3} \, \mathrm{d}s, \tag{5.5.5}$$

where $\mathbf{R}(s,t) = \mathbf{x} - \mathbf{X}(s;t)$, $R = |\mathbf{R}|$. Small curvature presumes large length of the line vortex; thus, both $\mathbf{R}$ and the arc length s should be measurable on an outer scale extended by ϵ^{-1}. If the inner (core) scale is used, the estimate $\mathbf{v} = O(\epsilon)$ follows from $\partial_s \mathbf{R} = O(\epsilon^{-1})$.

The Biot–Savart formula is applicable far from the vortex core but diverges on the line vortex itself. One cannot compute therefore the vortex velocity in a straightforward way by applying the Galilean transformation (5.3.23) as in 2D. Nevertheless, the matching procedure is not impaired, since the integral is well defined on a surrounding circle lying well outside the vortex core.

We shall use now the outer (extended) scale to measure distances. Picking an arbitrary point $\mathcal{P}$ on the curve C, we take it as the origin of both the arc length s and the 3D coordinates $\mathbf{x}$. A point $\mathbf{X}$ on C in the vicinity of $\mathcal{P}$ has the representation valid as long as the arc length is small when measured on the outer scale:

$$\mathbf{X} = s\mathbf{l} + \tfrac{1}{2}Ks^2\mathbf{n} + O(s^3). \tag{5.5.6}$$

A point lying on a circle of radius r (small when measured on the outer scale) centered at $\mathcal{P}$ in the plane perpendicular to C which passes through $\mathcal{P}$ (Fig. 5.30) is presented as

[4] Throughout this section, we presume the charge $N = 1$; negative charges are superfluous, since the sign of circulation changes with the reversal of $\mathbf{l}$

$$\mathbf{x} = r(-\sin\phi\,\mathbf{n} + \cos\phi\,\mathbf{b}), \tag{5.5.7}$$

where ϕ is the angular coordinate in the $(\mathbf{n}, \mathbf{b})$ plane (counted from the direction of $\mathbf{b}$). In order to evaluate the line integral in (5.5.5), we divide it into two parts, $\mathbf{V}_1+\mathbf{V}_2$, where $\mathbf{V}_1$ is computed on the segment $|s| < \delta$, where $r \ll \delta \ll 1$, and $\mathbf{V}_2$, on the rest of the line vortex. The near integral $\mathbf{V}_1$ expresses the local contribution of the adjacent segment of the curve, diverging at $r \to 0$. On the near part of the contour, we rescale $s = r\sigma$, and compute the normal components of $(\mathbf{x}-\mathbf{X}) \times \mathbf{l}$ using the representation (5.5.6):

$$\begin{aligned}(\mathbf{x}-\mathbf{X}) \times \mathbf{X}_s &= r\left[-\sigma\mathbf{l} - (\sin\phi + \tfrac{1}{2}Kr\sigma^2)\mathbf{n} + \cos\phi\,\mathbf{b}\right] \times (\mathbf{l} + rK\sigma\mathbf{n}) \\ &\Rightarrow r\left[\cos\phi\,\mathbf{n} + (\sin\phi - \tfrac{1}{2}Kr\sigma^2)\mathbf{b} + O(r^2)\right]. \end{aligned} \tag{5.5.8}$$

In the last expression, the projection on $\mathbf{l}$ has been discarded. Using also the expansion

$$\begin{aligned}(\mathbf{x}-\mathbf{X})^3 &= \left[s^2 + r^2 + rKs^2\sin\phi + O(s^3)\right]^{3/2} \\ &= r^3(1+\sigma^2)^{3/2}\left[1 + \frac{3rK\sigma^2\sin\phi}{2(1+\sigma^2)} + O(r^2)\right], \end{aligned} \tag{5.5.9}$$

we compute

$$\begin{aligned}\mathbf{V}_1 = \lim_{\delta\to\infty} &\left\{ -\frac{1}{r}(\sin\phi\,\mathbf{b} + \cos\phi\,\mathbf{n}) \int_{-\delta/r}^{\delta/r} \left[\frac{1}{(1+\sigma^2)^{3/2}} - \frac{3}{2}\frac{rK\sigma^2\sin\phi}{(1+\sigma^2)^{5/2}}\right] \mathrm{d}\sigma \right. \\ &\left. + \frac{1}{2}K\mathbf{b}\int_{-\delta/r}^{\delta/r} \frac{\sigma^2\mathrm{d}\sigma}{(1+\sigma^2)^{3/2}} \right\} = \left(-\frac{2}{r} + K\sin\phi\right)\mathbf{n}\cos\phi \\ &+ \left(-\frac{2}{r}\sin\phi - K\cos^2\phi + K\ln\frac{2\delta}{r}\right)\mathbf{b} + O\left(\frac{r}{\delta}\right). \end{aligned} \tag{5.5.10}$$

When the induction $\mathbf{V}_2$ by the far part of the line $|s| > \delta$ is evaluated, one can neglect the core size and set $r = 0$. The logarithmic singularity in (5.5.5) at $s \to 0$ is removed by subtracting from the integrand the term $\frac{1}{2}K|s|^{-1}f(|s|)\mathbf{b}$, where $f(s)$ is any suitable function with correct asymptotic behavior, $f(0) = 1$. The remaining convergent integral expressing the contribution of the far part of C is denoted as $-\mathbf{V}_s$. This yields

$$\mathbf{V}_2 = \mathbf{V}_s + K\mathbf{b}\int_{|s|>\delta} |s|^{-1}f(|s|)\,\mathrm{d}s. \tag{5.5.11}$$

The limit of the last integral integral at $\delta \to 0$ can be expressed as

$$K\mathbf{b}\lim_{\delta\to 0}\int_{|s|>\delta} |s|^{-1}f(|s|)\,\mathrm{d}s = K\mathbf{b}\ln\frac{\lambda}{\delta}, \tag{5.5.12}$$

where λ is a constant dependent on the choice of the compensating function and the geometry of the vortex.

A simple example is a ring vortex with a radius L. The normal vector is directed against the radial coordinate r. If the direction of the tangent $\mathbf{l}$ (defining the sign of the circulation) is chosen as anticlockwise, the binormal is directed upwards along the axial coordinate z. The arc length is proportional to the angular coordinate, $ds = L d\phi$, and the curvature is L^{-1}. The nonlocal integral (5.5.11) yields in this case a purely binormal induction; it is evaluated most easily using the Cartesian components $\mathbf{X} = L\{1-\cos\phi,\ -\sin\phi,\ 0\}$, $\mathbf{l} = \{-\sin\phi,\ \cos\phi,\ 0\}$. Plugging this in (5.5.5) yields

$$\mathbf{V}_2 = \frac{1}{2L}\,\mathbf{b}\lim_{\delta\to 0}\int_{\delta/L}^{\pi}\frac{d\phi}{\sin\frac{\phi}{2}} = \frac{1}{L}\ln\frac{4L}{\delta}\,\mathbf{b}, \qquad (5.5.13)$$

which fits the form (5.5.11), (5.5.12) with $\lambda = 4L$, $\mathbf{V}_s = 0$.

After the near and far integrals (5.5.10) and (5.5.11) are combined, the auxiliary parameter δ falls out:

$$\mathbf{V} = \mathbf{V}_s + \left(-\frac{2}{r}\cos\phi + K\sin\phi\,\cos\phi\right)\mathbf{n} + \left[-\frac{2}{r}\sin\phi + K\left(\ln\frac{2\lambda}{r} - \cos^2\phi\right)\right]\mathbf{b}. \qquad (5.5.14)$$

Using this expression, the phase field in the vicinity of the line vortex can be reconstituted in a form similar to (4.3.29). One should also extract from the resulting expression the phase gradient $\frac{1}{2}\mathbf{V}\cdot\mathbf{x}$ removed by the Galilean transformation (5.3.23). The result is

$$\theta = \phi + \frac{1}{2}\left[Kr\cos\phi\,\ln\frac{2\lambda}{r} + (\mathbf{V}_s - \mathbf{V})\cdot\mathbf{x}\right]. \qquad (5.5.15)$$

The coordinates should be rescaled to the inner units before the solvability condition is evaluated. The entire expression is scale-invariant, except the argument of the logarithm, where one should set $r = \epsilon r_0$ to obtain the expression valid on the matching circle:

$$\theta = \phi + \epsilon\theta_1, \quad \theta_1 = \frac{1}{2}\left[Kr_0\cos\phi\,\ln\frac{2\lambda}{\epsilon r_0} + (\mathbf{V}_s - \mathbf{V})\cdot\mathbf{x}\right]. \qquad (5.5.16)$$

Using $u_1 = i\theta_1$ together with the asymptotic expression (4.3.30), the solvability condition (4.3.26) is evaluated as in Sect. 4.3.3. Replacing $r\cos\phi$ by $\mathbf{b}\cdot\mathbf{x}$, the resulting mobility relation is expressed in the vector form

$$\mathbf{v} = \mathbf{v}_s + \kappa\,\mathbf{b}\ln\frac{2\lambda}{\epsilon a_0\sqrt{e}}. \qquad (5.5.17)$$

The small parameter is retained in the argument of the logarithm. In the particular case of a ring vortex, the propagation velocity is evaluated as

$$\mathbf{v} = \frac{1}{L}\,\mathbf{b}\ln\frac{\lambda_0 L}{\epsilon}, \qquad \lambda_0 = \frac{8}{a_0\sqrt{e}} \approx 4.31. \qquad (5.5.18)$$

This is an exceptional case when no normal motion is induced nonlocally, and the ring steadily propagates along its axis. Take note that the dependence of velocity on curvature $\kappa = L^{-1}$ turns out to be weakly nonlinear. A somewhat unsettling presence of a small parameter has a simple explanation: the ring radius should be measured in short-scale (core) units to get a correct answer.

Generally, the non-locally induced velocity contained in the vector $\mathbf{v}_s$ would have both normal and binormal components. The binormal component is formally larger, since it is multiplied by the logarithm of a large number. Neglecting $\mathbf{v}_s$ altogether leaves only *locally* induced binormal motion. This localized induction approximation (LIA) has been extensively studied due to its simplicity and mathematical elegance (for review see Ricca, 1996). However, the small parameter enters only logarithmically, and therefore neglecting nonlocal terms makes a very poor approximation. Even more serious objection to LIA is its structural instability. Binormal motion *conserves* the length, while normal motion due to accumulated action of nonlocal induction, even weak, causes the vortex filament to shrink or stretch at different locations, and typically leads to highly convoluted or even singular shapes.

5.5.3 Dissipative Nonradiative Line Vortex

In a dissipative nonradiative case $\nu = \eta \neq 0$ when the asymptotic wavenumber vanishes, the motion of a line vortex is expected to interpolate, as in 2D (Sect. 5.4.1), between dissipative motion described by the RGL equation and conservative motion described by the NLS equation. The 3D and 2D equations in the comoving frame, (5.5.2) and (5.4.1), have identical structure, differing only by the added curvature-dependent term, which can be absorbed by replacing the real part of the advective term $\mathbf{V} \to \mathbf{V} - K\mathbf{n}$. One might be tempted therefore to use the results of Sect. 5.4.1 straightforwardly, and to adopt (5.4.3), with the matching replacement in right-hand side $\mathbf{v} \to \mathbf{v} - \kappa\mathbf{n}$, as the local mobility relation. Taking note that the planar rotation operator $\mathcal{J}$ is replaced in 3D by the vector product $\mathbf{l}\times$, we rewrite then (5.4.3) as

$$\mathbf{k} - \frac{\eta}{2}\mathbf{v} = \frac{1}{2}\mathbf{l} \times (\mathbf{v} - \kappa\mathbf{n}) \ln \frac{v_0}{|\mathbf{v} - \kappa\mathbf{n}|}. \tag{5.5.19}$$

An impediment is that (5.4.3) has been obtained by matching with the solution of the far field equation (4.3.1) obtained under the assumption of stationary motion. The nonlinear logarithmic correction in (5.4.3) is the result of this matching. Generally, the argument of the logarithm obtained by matching with a non-stationary solution of the dissipative far field equation would be history-dependent, in the same way as the argument of the logarithm in (5.5.17) depends on the global shape of the line vortex. The formal analogy between the terms containing the velocity and curvature ends at this point, since, on the one hand, motion with a constant curvature is not a reasonable assumption, and, on the other hand, making the local mobility relation dependent on the history of motion is not practical. The remaining alternative is to treat the constant in the argument of the logarithm as an adjustable parameter, keeping in mind that it must be an $O(\epsilon^{-1})$ quantity. An additional *coup de force* is to allow for different logarithmic factors in the velocity and curvature terms, and identifying the latter with the respective factor in (5.5.17). Then (5.5.19) is modified to a *linear* mobility relation describing dissipative motion under combined action of the local phase gradient $\mathbf{k}$ and the vortex line curvature:

$$2\mathbf{k} = \eta \mathbf{v} + \mathbf{l} \times \mathbf{v} \ln \frac{\lambda_1}{\epsilon} - \kappa \mathbf{b} \ln \frac{\lambda_0}{\epsilon}. \tag{5.5.20}$$

One can separate here the non-dissipative part

$$\mathbf{v}_0 = 2\mathbf{k} + \kappa \Lambda_0 \mathbf{b}, \qquad \Lambda_0 = \ln \frac{\lambda_0}{\epsilon}, \tag{5.5.21}$$

which has the same structure as (5.5.17) and describes the combined action of local curvature and "advection" by the phase gradient induced by nonlocal action of the line vortex. The parameter λ_0 remains indefinite, but so is the parameter λ in (5.5.17), as long as the global shape of the vortex is not defined. The mobility relation (5.5.20) can be further resolved with respect to the dissipative part of the vortex drift velocity, $\mathbf{v} - \mathbf{v}_0$:

$$\mathbf{v} = \frac{\eta \mathbf{v}_0 - \Lambda \mathbf{l} \times \mathbf{v}_0}{\eta^2 + \Lambda^2}, \qquad \Lambda = \ln \frac{\lambda_1}{\epsilon} > 0. \tag{5.5.22}$$

An equivalent mobility relation was postulated by Schwarz (1985) for the purpose of numerical computation of vortex motion in superfluids taking into account dissipation due to friction with a normal fluid component. The formula gives correct limiting behavior (up to logarithmic corrections) in both weakly and strongly dissipative limits ($\eta \gg 1$ or $\eta \ll 1$). For example, applying (5.5.22), to a ring vortex with $\mathbf{v}_0$ directed along the binormal $\mathbf{b}$, as defined by (5.5.18), we see that dissipation adds a velocity component directed along the normal $\mathbf{n} = -\mathbf{l} \times \mathbf{b}$. As the ring velocity acquires a normal component, the ring shrinks as it propagates, and eventually collapses. In the gradient case, the simple result from Sect. 5.5.1 remains valid, since the total phase gradient $\mathbf{k}$ induced by an undistorted ring is zero.

5.5.4 Instability of Line Vortices

Line vortices in the NLS equation are prone to instabilities and tend to develop into convoluted mazes, which repeatedly rearrange through reconnection events. Stability analysis taking a proper account of nonlocal induction can be carried out for the simplest shapes, such as a straight line, a ring, and a helix (for review, see Pismen, 1999). The former two turn out to be unstable to perturbations at short wavelengths only, i.e., under conditions when the applicability of the multiscale analysis lying at the basis of the analytical theory becomes questionable. On the other hand, a helix, which is always unstable even in the LIA limit, can be viewed as a finite-amplitude wave on a straight-line vortex. This suggests that perturbations with small but finite amplitudes may cause destabilization also on longer wavelengths. The nonlinear development of these instabilities seen in Fig. 5.31, leading to vortex stretching and hairpin formation, can be followed, short of direct simulation, with the help of the nonlocal filament equation derived by Klein and Majda (1991).

This instability persists in the weakly dissipative case, where it can be viewed as a 3D extension of the core acceleration instability of a point vortex (Sect. 5.3.4).

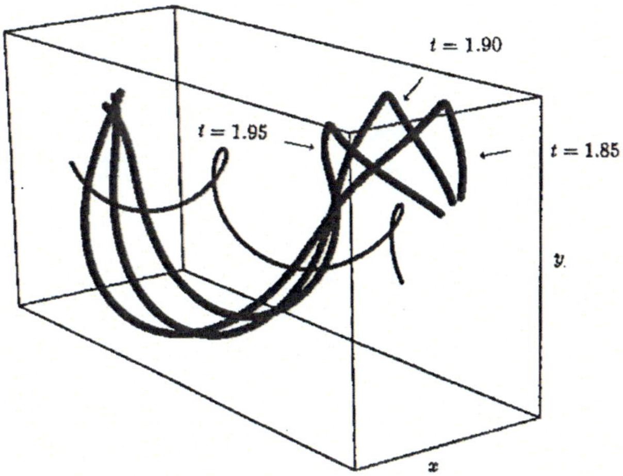

Fig. 5.31. Snapshots of a line vortex showing the initial helix and hairpin structures near the peak of the maximum stretching rate (Klein and Majda, 1991, reproduced by permission ©Elsevier)

Simulations by Aranson and Bishop (1997) and Aranson et al (1998) in the unstable regime revealed a spectacular effect of this instability on a scroll vortex. A series of snapshots in Fig. 5.32 shows how a slightly perturbed scroll vortex breaks down into a turbulent maze. Scroll vortices, as topological singularities of the phase field, play an important role in organizing the overall structure of the turbulent field, and may be called, in the fluid mechanic's idiom, "sinews of turbulence". Nevertheless, the local structure of individual vortices crowded in a turbulent maze becomes so distorted that vortex dynamics ceases to be a useful computational concept under these conditions, and simulation of the full CGL equation remains the only way to reproduce structures of this kind.

Dissipation stabilizes a straight-line vortex, which is the only stationary shape unaffected by curvature-driven normal motion. This can be checked with the help of

Fig. 5.32. Instability of a straight-line vortex filament ($\eta = 50$, $\nu = -0.03$). Isosurfaces of the modulus $\rho = 0.1$ are shown. (Aranson and Bishop, 1997, reproduced by permission, ©The American Physical Society)

the mobility relation (5.5.22) after presenting a perturbed vortex filament as

$$\mathbf{X} = \{e^{\lambda t + ikz}\boldsymbol{\xi}, z\}, \tag{5.5.23}$$

where the last component indicates the unperturbed position along the z axis, and $\boldsymbol{\xi}$ is an infinitesimally small 2D vector in the normal plane. Computing the vectors forming the Frenet trihedron and curvature according to Sect. 5.5.1 yields, to the leading order in ξ,

$$\begin{aligned} \mathbf{l} &= \{ike^{\lambda t + ikz}\boldsymbol{\xi},\ 1\}, & \mathbf{n} &= \{e^{i(\lambda_i t + kz)}\boldsymbol{\xi}/|\xi|,\ 0\}, \\ \kappa &= -k^2 e^{\lambda_r t}|\boldsymbol{\xi}|, & \mathbf{b} &= \{e^{i(\lambda_i t + kz)}\mathcal{J}\boldsymbol{\xi}/|\boldsymbol{\xi}|,\ 0\}, \end{aligned} \tag{5.5.24}$$

where $\mathcal{J}$ is the operator of rotation by $\pi/2$ in the plane normal to the z axis. Separating the real and imaginary parts of the eigenvalue $\lambda = \lambda_r + i\lambda_i$ is necessary for computation of absolute values. The action of the phase gradient $\mathbf{k}$ in (5.5.22) is negligible for infinitesimally small perturbations, and the stability analysis reduces after plugging (5.5.24) and (5.5.21) in (5.5.22) to computation of the eigenvalues of the matrix

$$\mathbf{M} = -\frac{k^2 \Lambda_0}{\eta^2 + \Lambda^2} \begin{Bmatrix} \Lambda & -\eta \\ \eta & \Lambda \end{Bmatrix}. \tag{5.5.25}$$

Evidently, the vortex is stable, as both Λ and Λ_0 are positive.

Stability can be, however, impaired in the presence of a *phase twist* when the phase field of the unperturbed filament changes linearly along the z-axis. In the nonradiative case, the unperturbed phase field is $\theta = \phi - \omega t + Qz$. Including the phase gradient $\mathbf{k} = \{0,\ 0,\ Q\}$ in (5.5.21) modifies (5.5.25) to

$$\mathbf{M} = -\frac{k^2 \Lambda_0}{\eta^2 + \Lambda^2} \begin{Bmatrix} \Lambda & -H \\ H & \Lambda \end{Bmatrix}, \quad H = \eta - \frac{2iQ\Lambda}{k\Lambda_0}. \tag{5.5.26}$$

The eigenvalues of this matrix are

$$\lambda_{\pm} = -\frac{k^2 \Lambda_0}{\eta^2 + \Lambda^2} \left(\Lambda \pm i\eta \pm \frac{2Q\Lambda}{k\Lambda_0} \right). \tag{5.5.27}$$

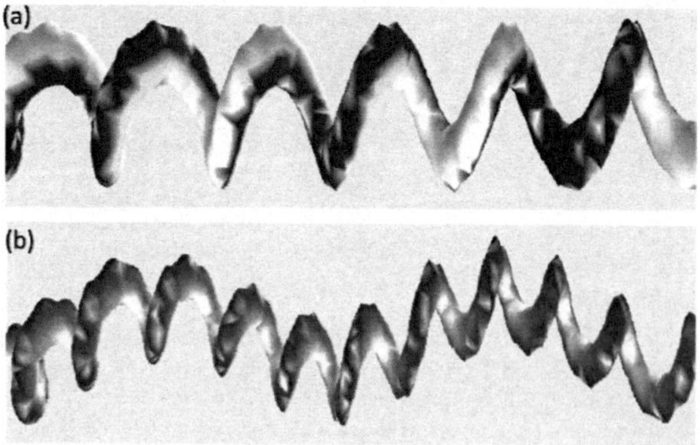

Fig. 5.33. (a) A restabilized helical vortex ($\eta = 1, \nu = 0, \gamma = 0.098$). (b) A doubly periodic "superhelix" ($\eta = 1.7, \nu = -0.5, \gamma = 0.27$). Isosurfaces of the modulus $\rho = 0.6$ shaded by phase field are shown (Rousseau et al, 1998; reproduced by permission, ©The American Physical Society)

The straight-line vortex turns out to be always unstable to perturbations with a sufficiently long wavelength (which, of course, can be limited by the total length of the filament anchored at bounding walls in a finite system). The growth rate of perturbations is maximal at $k = Q/\Lambda_0$. Numerical simulations (Rousseau et al, 1998) show that the filament can be restabilized as a finite-amplitude helical structure seen in Fig. 5.33a. A secondary bifurcation leading to a doubly periodic "superhelix" (Fig. 5.33b) has been also detected in these simulations.

6 Patterns of Active Matter

6.1 Dry Active Matter

6.1.1 Scalar Active Particles

Active systems do not require external energy input in an explicit form, as sketched in Fig. 1.1, but are capable of independent motion. Of course, a hidden energy input is always present, in the form of chemical reactions driving colloidal particles or, on a far more complicated level, sustaining living organisms, but this lies outside phenomenological description. Common nonequilibrium systems are at equilibrium locally, and obey all thermodynamic laws at this level, but in an active system *every* particle is out of equilibrium, as it has its own source of energy and a cause of motion.

There is a rather artificial distinction between "dry" and "wet" active matter, which originated more from the kind of modeling than from the nature of the "particles" themselves. Interactions in "wet" matter are mediated by the medium where they are immersed, while in the case of "dry" matter, the medium is ignored, sometimes rightly, when its influence is a minor factor, and sometimes just to make the problem tractable, e.g., ignoring fluid motion that carries interactions through the medium.

The simplest case is motion of *scalar* active particles, which have no internal degrees of freedom and are moving independently, without communicating with other particles of the same or different kind. A noninteracting scalar particle in a homogeneous isotropic medium does not have a preferred direction of motion, but changes it randomly. The way of doing it distinguishes between active Brownian and run-and-tumble particles. The former undergo orientational diffusion, similar to passive Brownian particles, but both motion and reorientation are driven by internal energy sources rather than by thermal noise. Run-and-tumble motion is specific to bacteria, which explore their surroundings by alternating straight-line motion with random changes of direction by a finite angle.

Most commonly, random motion is influenced by interactions with particles of the same or different kind. Interactions may have a solid physical basis, like hydrodynamic and chemotactic forces in active colloids or tendency to mutual alignment of non-spherical particles (Sect. 6.1.3). Chemical interactions may extend to quorum sensing

L. Pismen, *Patterns and Interfaces in Dissipative Dynamics*, Springer Series in Synergetics,
https://doi.org/10.1007/978-3-031-29579-9_7

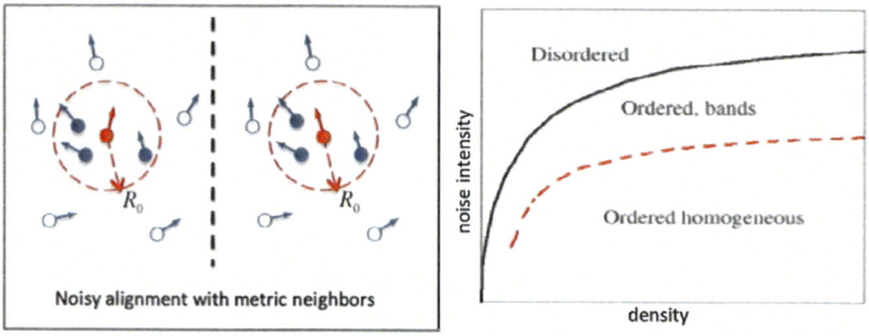

Fig. 6.1. (a) Scheme of the Vicsek model. A test particle (red) aligns imperfectly with neighbors (dark blue) within a range shown by the dashed circle, and then moves in this direction. (b) Qualitative phase diagram showing the dependence of ordering on the flock density and the strength of noise (Ginelli, 2016, CC 4.0)

in bacterial populations distinguishing between cooperating and competing species, and further to a "social force" between interacting higher animals. Helbing and Molnár (1995) introduced this notion to model pedestrian motion, defining its three principal components: acceleration to the desired velocity of motion, avoiding other pedestrians and borders, and attraction toward the target, but social interactions do not require high intelligence; for example, they were found to dominate speed control in bird flocks (Bialek et al, 2014).

The simplest and earliest approach to motion of interacting "dry" active particles is the Vicsek model (Vicsek et al, 1995) sketched in Fig. 6.1a. It lacks any specific physical mechanism, and just assumes that each particle adjusts its alignment angle θ to conform with the average alignment vector (shown by the arrows in the picture) of its neighbors within a certain range, and moves with a speed $\mathbf{v}$, $|\mathbf{v}| = \text{const}$ in the acquired direction. Thus, the rule of motion for any ith particle over a unit time step is

$$\mathbf{x}_i(t_{n+1}) = \mathbf{x}_i(t) + \mathbf{v}_i(t_n), \qquad \theta_i(t_{n+1}) = \langle \theta_j(t_n) \rangle + f, \tag{6.1.1}$$

where $\langle \dots \rangle$ denotes averaging over orientation angles θ_j of particles within a certain area and f is a random noisy bias. The adjustment keeps repeating, as motion brings the particle to another neighborhood. Since particles always move in the direction determined by their orientation with a constant speed, polarity is not distinguished in this model from velocity, and the particles can be still classified as scalars.

The Vicsek model is amenable to agent-based numerics rather than analysis, and therefore can be only mentioned in passing in this book. Depending on the average density and the level of noise, the particles may be perfectly aligned, or disordered, or ordered locally but disordered elsewhere (Fig. 6.1b); the ordering can also be intermittent when particles align for a while and then succumb to noise. Even when the ordered state persists, the interaction network is permanently rearranged due to the active motion of individuals, continuously changing their neighbors. The "particles" of this toy model have been early associated with "birds" (Toner and Tu, 1995), but alignment of real birds, besides being motivated by hydrodynamics, is based on a *topological* rather than metric distance. Field studies (Ballerini et al, 2008) confirmed that each individual interacts with a fixed number of neighbors, irrespective of their spacing. In this case, there is no segregation into an ordered "liquid" and a disordered "gas" phase, because neighbors in dilute regions are never disconnected, and therefore low density does not necessarily induce disorder.

There are many variations of the Vicsek model used in simulations, including, besides topological census-taking (Ginelli and Chaté, 2010), pairwise short-range attractive interactions (Grégoire et al, 2003), long-range cohesive interactions of hydrodynamic origin (Chaté et al, 2008), steric trapping in a model of soft repulsive disks with no alignment rule (Fily and Marchetti, 2012), local repulsive interactions (Strömbom et al, 2015). Both steric and repulsive interactions facilitate separation into dense and dilute phases. Simulations by Caprini et al (2020) have shown that both phase separation and velocity ordering occur in the absence of any microscopic force that explicitly aligns velocities, due to steric factors only. Most radically, Peruani (2017) demonstrated that active particles can exhibit various complex patterns of collective motion without explicit velocity alignment by restricting short-range position-based attractive interactions to their visibility cone, as shown in Fig. 6.2. This is in line with the primary role of visual contact in the behavior of animal groups (Rosenthal et al, 2015).

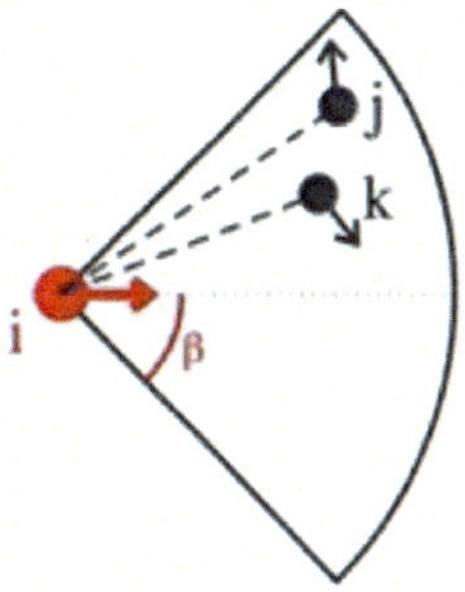

Fig. 6.2. Interactions within a visibility cone (Peruani, 2017, reproduced by permission ©The Physical Society of Japan)

Großmann et al (2015), besides quantifying the repulsion force, combined velocity alignment with nearest neighbors and *anti-alignment* with particles which are further away, which may be caused by hydrodynamic interactions. In particle-based

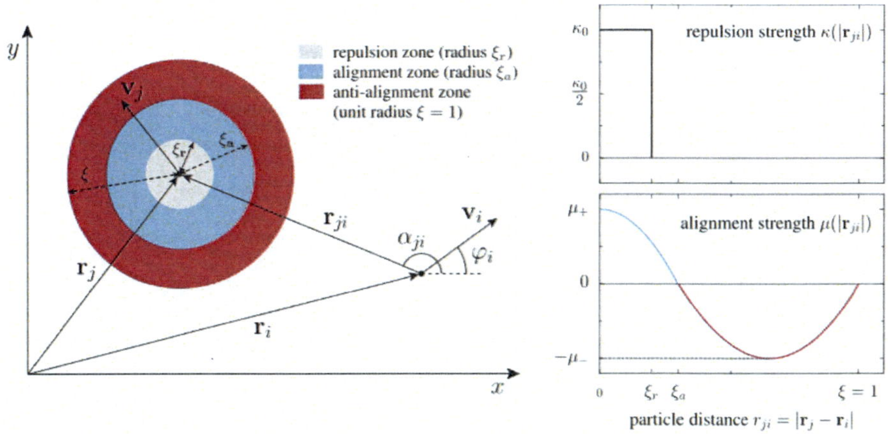

Fig. 6.3. The scheme of interactions in the model by Großmann et al (2015), ©Springer

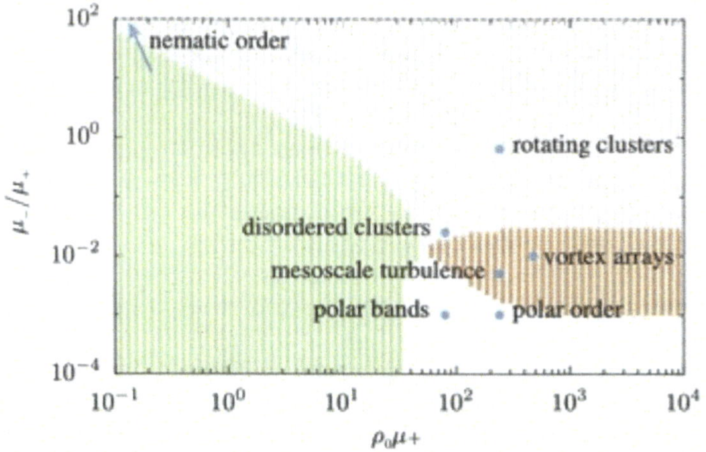

Fig. 6.4. The summary of results in computations by Großmann et al (2015), dependent on the average density ρ_0 and the strength of the alignment (μ_+) and antialignment (μ_-) interactions. The background coloring indicates the predictions of the kinetic theory: green – spatially homogeneous disordered isotropic state; white: emergence of polar order; gray and red – emergence of patterns with a characteristic length scale. The arrow indicates the shift of border lines with increased noise intensity D_φ. Representative simulation results are specified next to the blue points (©Springer)

computations, the interaction term is added to the second equation (6.1.1); their ranges are summarized in Fig. 6.3. Furthermore, they have framed the model into the nonlinear Fokker–Planck equation for the dynamics of the one-particle probability density $\mathcal{P}$ dependent on the spatial position $\mathbf{x}$, orientation angle φ and time:

$$\partial_t \mathcal{P}(\mathbf{x}, \phi, t) = -\mathbf{n} \cdot \nabla \mathcal{P} - \partial_\varphi (F[\mathcal{P}]\mathcal{P} + D_\varphi \partial_\varphi)\mathcal{P}), \tag{6.1.2}$$

where $\mathbf{n} = \{cos\varphi, \sin\varphi\}$, D_φ parametrizes the intensity of angular fluctuations, and $F[\mathcal{P}]$ is the nonlinear force defined as an integral over (anti)alignment and repulsion forces acting upon $\mathcal{P}$. Both the theory and simulations produced rich phenomenology summarized in Fig. 6.4. The averaged hydrodynamic theory, aimed at describing the pattern formation on macroscopic scales, turned out to be consistent with statistical theory and simulations only at a high level of noise.

6.1.2 Continuum and Statistical Description

The principal fault of the Vicsek model and its modifications, from the physicist's viewpoint, is the absence of the momentum conservation and of an equation of state. They do not even possess the Galilean invariance inherent in classical mechanics, since particles move with a constant velocity defined in a specific ("laboratory") frame of reference. This severely hinders the attempts of translating it into a continuous description., which, nevertheless, has been early undertaken by Toner and Tu (1995).

They proposed a model merging the Navier–Stokes equation, complete with the pressure gradient ∇p and shear and extensional viscosities, with an equation of state of a vectorial fluid with added vectorial noise $\mathbf{f}$:

$$\partial_t \mathbf{v} + \mathbf{v}\nabla \cdot \mathbf{v} = -\nabla p + \left[\nu \nabla^2 + \widetilde{\nu}\nabla(\nabla \cdot \mathbf{v}) + \gamma(\mathbf{v}\cdot\nabla)^2\right]\mathbf{v} + \alpha\mathbf{v} - \beta|\mathbf{v}|^2\mathbf{v} + \mathbf{f}. \tag{6.1.3}$$

It is complemented by the continuity equation $\partial_t \rho = \nabla\cdot(\rho\mathbf{v})$ and an equation of state relating pressure with density ρ. Later, more nonlinear terms were added, $(\nabla\cdot\mathbf{v})\mathbf{v}$ and $\nabla|\mathbf{v}|^2$, which, unlike the standard hydrodynamic inertial term, violate the Galilean invariance (Toner et al, 2005). However, the principal non-Galilean terms are not those, but the algebraic terms that select the preferred absolute value of velocity $|\mathbf{v}| = \sqrt{\alpha/\beta}$. This model has a far richer content than Vicsek's, which has been its original inspiration, and can be viewed as just another model, related in some respects but different in others. However, their common feature (and a common drawback) is identifying orientation and velocity, which has brought together terms of different nature in (6.1.3). The algebraic terms are a common device for phase separation; the viscous terms could be also derived from the splay distortion term in the elastic energy density of a fluid with the vectorial or nematic order, with the γ term viewed as a kind of nonvariational nonlinear elasticity; more "elastic" terms of this kind could have been added.

It is possible to apply to (6.1.3) methods of linear analysis and bifurcation expansion introduced in Sect. 1. The only simple solution that can be used as a starting point is what Toner et al (2005) call a "ferromagnetic state": the fluid moving with uniform velocity. Unlike the Vicsek model, (6.1.3) does not formally forbid perturbations of the absolute value of velocity, which would be retained up to $O(\alpha^{-1})$ only at $\alpha \gg 1$. In this case, the only relevant perturbation is that of the velocity component $\mathbf{w}$ normal to the unperturbed flow. In the following, we reduce $D_1 + D_2$ to unity by rescaling length, set the unperturbed velocity to unity by choosing the time scale, and use the equation of state to eliminate the pressure perturbation by expressing it through the density perturbation $p = \sigma\rho$; the unperturbed density can be also normalized to unity. The wave vector $\mathbf{k}$ is separated into the longitudinal $k_{\|}$ and normal $\mathbf{k}_{\perp}$ components, and, since the normal velocity components do not interact in the linearized equations, it is sufficient to consider only one of them and denote $k_{\|} = k\cos\theta$, $\mathbf{k}_{\perp} = k\sin\theta$. Then the spectral form of the linearized system is written as

$$\begin{aligned}[\lambda + ik\mathrm{Re}\cos\theta + k^2(1+\delta\sin^2\theta)]w - ik\sigma\rho\sin\theta &= 0,\\ -ikw\sin\theta + (\lambda - ik\cos\theta)\rho &= 0,\end{aligned} \tag{6.1.4}$$

where λ is the eigenvalue, $\delta = D_0/(D_1+D_2)$, and the random input is suspended. Following Toner et al (2005), we will consider long-scale perturbations with $k \ll 1$. Expanding the eigenvalues up to the second order in k yields

$$\begin{aligned}\lambda &= \frac{ik}{2}\left[(\mathrm{Re}-1)\cos\theta + \sqrt{S}\right] + \frac{k^2}{4}\left[-2 + \sqrt{2}\delta(1-\cos 2\theta)\right]\left[1 \pm \frac{\mathrm{Re}+1}{\sqrt{S}}\right],\\ S &= (1+\mathrm{Re})^2 + 4\sigma + [(1+\mathrm{Re})^2 - 4\sigma]\cos 2\theta > 0\end{aligned} \tag{6.1.5}$$

The real part of at least one eigenvalue is positive, indicating instability, for perturbations with a wave vector directed at such an angle that $\cos 2\theta < 1 + \sqrt{2}\delta$, which is possible at $\delta > \sqrt{2}$ and $|\theta|$ sufficiently close to $\pi/2$. This is a *wave* instability with the frequency of $O(k)$. Toner et al (2005) refer to the perturbations as the "sound" waves, but they are damped only when δ is sufficiently small and otherwise amplify to destroy the order.

The model by Toner and Tu has no gradient structure and a variety of its solutions should be enormous. Solon, Cassin, et al (2015), working with an equation lacking non-Galilean and nonlinear differential terms, obtained what one of the coauthors (Chaté, 2020) called an *embarrassingly* large family of linearly stable solutions: periodic patterns, solitary bands, phase-separated domains. All this variety disappeared when a noise term, missing in the simplified system, is reintroduced, causing a unique solution to be selected. This is consistent with the role of noise in equilibrium systems where noise facilitates transition from metastable states to a state with the minimal energy, but energy is neither well defined nor conserved in active matter. Chaté (2020) lists this among current riddles, asking: "How do we understand the selection of a unique solution observed at microscopic and fluctuating hydrodynamic levels but which is not present at the deterministic hydrodynamic level?"

The most serious objection to this kind of merger (or mix-up) of hydro- and thermodynamics is inapplicability of the notions of pressure and equation of state to active matter. One can compress a fluid, increasing its average density. In a common fluid, the required work is unequivocally determined by pressure, but in an active medium it depends on the way particles interact with the confining walls. Different forces and hence different amounts of work are needed to reach the same final density when compressing by a hard wall or by a soft enclosure, into which particles bump gently. Solon, Fily, et al (2015) proved this analytically for active Brownian and run-and-tumble particles and demonstrated it by simulating particles in a container separated in two parts by a mobile wall with asymmetric interaction potentials on its two sides. The partition moves to equalize the two wall-dependent pressures, resulting in a steady state with unequal densities in the two chambers (Fig. 6.5). In equilibrium fluids, the normal force per unit area on any part of the boundary is independent of its orientation. This is not so in active media, as long as the propulsion speed is anisotropic, even if the particles are oriented isotropically. The lack of an equation

Fig. 6.5. Simulated spontaneous compression/expansion of an active fluid due to an anisotropic wall potential (Solon, Fily, et al, 2015), ©Springer Nature

of state is a direct consequence of a violation of the momentum conservation (Fily et al, 2018).

Activity violates such basic principles as equipartition of energy among various degrees of freedom and detailed balance, and a structured noise in assemblies of active particles is unlike random thermal noise. The absence of meaningful definitions of such basic thermodynamic variables as energy and pressure makes it even more challenging to derive macroscopic dynamics from underlying microscopic interaction rules. Common hydrodynamic equations are rooted in statistical mechanics, and the ongoing work is aimed at building up a similar basis for dynamics of active matter. Bertin et al (2006, 2009) started from the Boltzmann equation

$$\partial_t \mathcal{P}(\mathbf{x}, \phi, t) + \mathbf{v} \cdot \nabla \mathcal{P}(\mathbf{x}, \phi, t) = \mathcal{I}_{\mathrm{dif}}[\mathcal{P}] + \mathcal{I}_{\mathrm{col}}[\mathcal{P}, \mathcal{P}]. \tag{6.1.6}$$

The second term on the left–hand side corresponds to the ballistic motion of particles. The first term in the right-hand side accounts for the self-diffusion and the second one, for the effects of collisions; both are defined through integrals over the Boltzmann distribution $\mathcal{P}(\mathbf{x}, \phi, t)$ dependent on the particle position, direction of velocity (still presumed to have a constant absolute value v_0), and time.

In the following, several approximations are made: it is assumed that the distribution is almost isotropic, so that the average hydrodynamic velocity is much smaller than velocities of individual particles, and varies over time and length scales far exceeding the microscopic scales defined by a velocity and free path of individual particles. This leads to a generalization of the Navier–Stokes equation to dynamics breaking both the Galilean invariance and the global momentum conservation of the assembly of particles. The equation for the momentum field $\mathbf{P} = \rho \mathbf{v}$ takes the form

$$\begin{aligned}\partial_t \mathbf{P} = -\gamma(\mathbf{P} \cdot \nabla)\mathbf{P} = -\frac{v_0^2}{2}\nabla\rho + \frac{\kappa}{2}|\nabla \mathbf{P}|^2 + \mu(1 - \xi|\mathbf{P}|^2)\mathbf{P} + \nu\nabla^2\mathbf{P} \\ - \kappa(\nabla \cdot \mathbf{P})\mathbf{P} + \nu'[\nabla\rho \cdot \nabla(\nabla\mathbf{P} + \nabla\mathbf{P}^\dagger) - \nabla\rho\nabla \cdot \mathbf{P}],\end{aligned} \tag{6.1.7}$$

where $\dagger$ denotes the transpose of the tensor $\nabla\mathbf{P}$. All coefficients are expressed through the parameters of the problem: v_0, the threshold distance d_0 causing a change of direction, and the Gaussian distribution variances of noise and self-diffusion. It should be complemented by the continuity equation $\partial_t \rho = \nabla \cdot \mathbf{P}$, which is valid at a coarse-grained level due to the mass conservation, but there is no need for pressure and an equation of state. Omitting spatial dependence leads to a simple equation

$$\partial_t \rho = \rho_0 + \epsilon \varrho e^{\mathrm{i}k \cdot x + \lambda t} = \mu(1 - \xi|\mathbf{P}|^2)\mathbf{P}, \tag{6.1.8}$$

which makes it clear that the stagnant solution is unstable at $\mu > 0$, giving way to uniform propagation with the momentum $P_0 = \sqrt{\mu/\xi}$ in an arbitrary direction. However, this solution is always unstable to transverse perturbations.

The assumptions that only binary interactions need to be considered and that particles decorrelate between collisions are questionable in the presence of dense clusters, but this applies to the classical theory of equilibrium liquids as well. Sophisticated attempts at a continuum statistical theory beyond the hydrodynamic level that would encompass phenomenology of Vicsek's model, simple as it is, remain unsettled, as reflected, in particular, by the two contending papers in a discussion issue on active matter (Peshkov et al, 2014; Ihle, 2014), which contain both justification of these efforts and mutual criticisms of the rival theories. This topic is beyond the scope of this book, concentrating upon macroscopic dynamics.

6.1.3 Rod-like Particles

A scalar particle with a spherically symmetric shape is a rare occasion in nature. The next step to reality is to recognize the influence of shape on motion and interactions. The notion of nematic and vectorial order originated on the molecular level (Sect. 6.2.1) can be naturally extended to macroscopic particles: elongated grains, oriented colloids, cells, intracellular filaments, microorganisms. The Vicsek model (Sect. 6.1.1) does not relate the direction of motion to particle's shape (even when referring to bird flocks). Its extension to rod-like particles (Peruani et al, 2006) recognized the significance of the distinction between the nematic and vectorial order. In their model, similar to Vicsek's, velocity is identified with alignment and the magnitude of the velocity is constant. However, the interaction rules had to be modified, since the connection between the velocity *vector*, and *nematic* alignment of rods is not as straightforward as between two vectors in a polar medium.

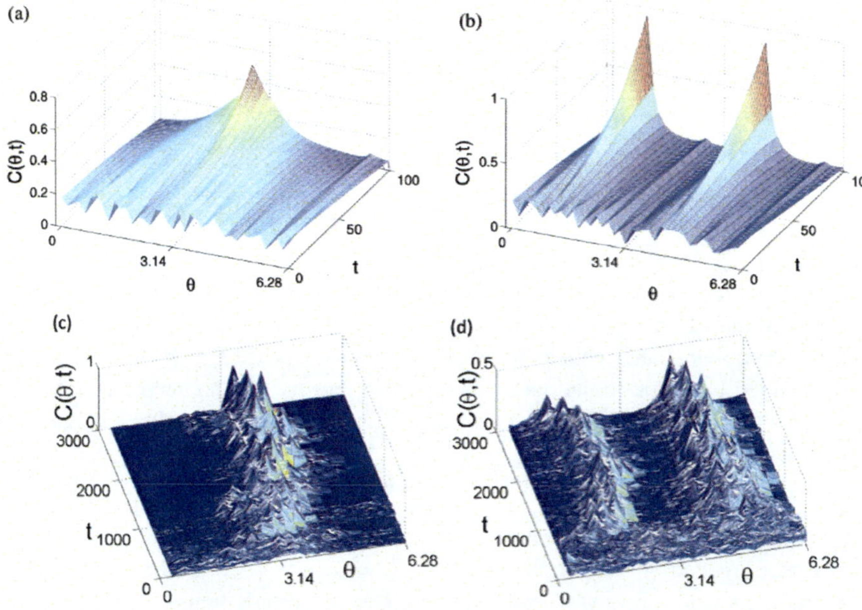

Fig. 6.6. Evolution of the distribution $C(\varphi, t)$ of the alignment angles φ with time t in a flock of particles with polar (a,c) and nematic (b,d) symmetry in continuous (a,b) and particle-based (c,d) computations (Peruani et al, 2008, CC-BY-NC)

Mobile rod-like particles, like scalar particles, align as a result of a *polar* collision at an acute angle, but anti-align following an *antipolar* collision at an obtuse angle. Rather than averaging the orientation angle of a particle at each interaction step over the propagation directions of the particles within a metric neighborhood, as in a polar flock, they inverted these directions when they were at an obtuse angle to the direction of motion of the test particle. This rule allows the particles to reverse their propagation direction, while still always moving with a constant speed. After a polar collision, rods tend to move in parallel, so that their positions and orientations are highly correlated. In contrast, particles are quickly separated far from another after an antipolar collision. The conservation of momentum, which does not hold for Vicsek's polar interaction rules, is violated still more severely in this case. Therefore Peruani et al (2008) did not attempt to build up hydrodynamic equations, but adopted a probabilistic approach based on a sole equation for the orientation distribution function similar to (6.1.2). Its solution evolved to two peaks of alignment angles, as can be seen in Fig. 6.6b, in contrast to the single peak for scalar particles in Fig. 6.6a. Particle-based simulations arrived at a qualitatively similar distribution, as shown in Fig. 6.6c,d, but with the smooth alignment peaks made rugged by noise.

Self-propelling rods reveal in particle-based simulations still richer phenomenology than scalar particle. The distinctions in dynamics are seen most clearly in simulations by Großmann et al (2020) mapping the influence of the aspect ratio $\phi = l_{\|}/l_{\perp}$ on transitions between different patterns. They operated with soft elliptic particles deforming upon collisions. In the limiting case of self-propelling discs $l_{\|}/l_{\perp} = 1$, motility-induced phase separation (MIPS) has been observed (more on this in Sect. 6.1.5), with the formation of dense immobile clusters displaying hexatic order, surrounded by a disordered gas of active particles, as shown in Fig. 6.7a. These

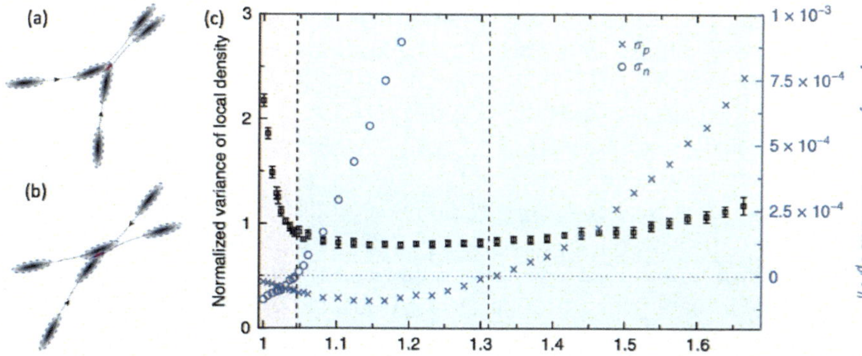

Fig. 6.7. (a,b) The overlap (highlighted in red) of soft rods upon a polar (a) and an antipolar (b) collision. (c) The influence of the aspect ratio $l_{\|}/l_{\perp}$ on patterns of soft self-propelled rods with nematic alignment interactions. Black symbols show the variance of the coarse-grained density field, with error bars indicating the confidence intervals. Blue circles and crosses show the growth parameters of the polar and nematic order σ_p and σ_n, respectively. The vertical lines show the onset local nematic and polar order. The background color corresponds to the regimes in Fig. 6.8 (Großmann et al, 2020, CC BY)

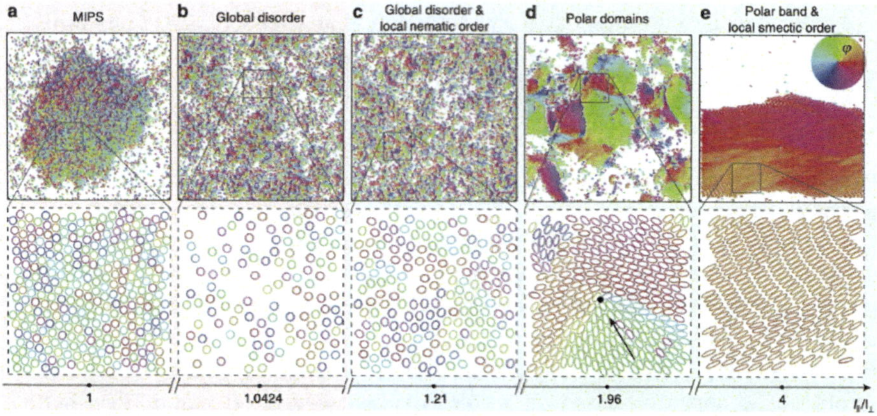

Fig. 6.8. Phenomenology displayed by self-propelled rods with nematic alignment interactions at different aspect ratios $l_\parallel/l_\perp$. The top row shows the entire simulation box; enlarged images are plotted in the corresponding panels below. The color coding represents the orientation of the direction of motion φ. (Großmann et al, 2020, CC BY)

clusters became unstable already for slightly anisotropic rods, giving way to global disorder, as seen in Fig. 6.8b. As the aspect ratio increases further, clusters with local nematic order emerge (Fig. 6.8c). The peculiar feature of self-propelled rods is a combination of nematic orientation alignment with polar alignment of velocity, which becomes prominent in Fig. 6.8d. The growth of polar and nematic order are characterized by separate parameters, σ_p and σ_n; their change with increasing aspect ratio is shown in Fig. 6.7c. Polar clusters are highly dynamic, as they form, merge and break. Local nematic orientation alignment is associated with the formation of half integer topological defects, mentioned already in Sect. 4.2, and to be further discussed in Sect. 6.2.5. A $+1/2$ defect is marked by the black dot in Fig. 6.8d; it is mobile, being propelled by its "comet tail", as shown by the arrow. At high isotropy, nematic order becomes global and, moreover, smectic order appears, as particles arrange in layers within large dense clusters (Fig. 6.8e).

Much efforts have been invested in the statistical theory of rod-like particles. Different approaches were compared by Bertin et al (2016). An elegant theory by Peshkov et al (2012) started, as the theory of scalar particles by Bertin et al (2006, 2009) mentioned in Sect. 6.1.2, from the Boltzmann equation applied to the fields with polar and nematic symmetries: the vector $\mathbf{P}$ and the tensor $\mathbf{Q}$ expressed in 2D as

$$\rho\mathbf{P} = \begin{pmatrix} \operatorname{Re} p \\ \operatorname{Im} p \end{pmatrix}, \qquad \rho\mathbf{Q} = \begin{pmatrix} \operatorname{Re} q & \operatorname{Im} q \\ \operatorname{Im} q & -\operatorname{Re} q \end{pmatrix}. \tag{6.1.9}$$

On the hydrodynamic level, the theory led to the equation system for the two complex fields p, q together with the continuity equation for the conserved density ρ, presented in the complex form

$$
\begin{aligned}
\partial_t p &= -\frac{1}{2}(\nabla \rho + \overline{\nabla} q) + \frac{\gamma}{2}\overline{q}\nabla q - (\alpha - \beta|q|^2)p + \zeta\overline{p}q, \\
\partial_t q &= -\frac{1}{2}\nabla p + \frac{\nu}{4}|\nabla|^2 q - \frac{\kappa}{2}\overline{p}\nabla q - \frac{\chi}{2}\overline{\nabla}(pq) + (\mu - \xi|q|^2)q + \omega|p|^2 + \tau|p|^2 q, \\
\partial_t \rho &= \mathrm{Re}(\overline{\nabla} p),
\end{aligned}
\tag{6.1.10}
$$

where $\nabla = \partial_x + \mathrm{i}\partial_y$, $\overline{\nabla} = \partial_x - \mathrm{i}\partial_y$ are complex differential operators, and all parameters are expressed through parameters of the microscopic model. The authors presented some numerical computations based on these equations, more difficult and less detailed than direct particle-based simulations. Even the linear stability analysis of states with uniform orientation based on this system (equivalent to five real fields) is technically difficult. Mahault et al (2018) extended the theory to 3D. The entire field of patterns of "dry" rod-like particles has been comprehensibly reviewed by Bär et al (2020) and Chaté (2020).

6.1.4 Broken Chiral Symmetry

Inanimate rod-like particles in a vertically vibrated granular layer acquire their activity from external forcing. There is no directed motion here, but elongated grains with nematic symmetry spontaneously align while forming patterns in the plane of the vibrating layer. However, unlike thin layers of nematic media, the alignment in a single dense layer may extend to the third dimension. When particles are allowed to deviate from a horizontal position, nematic 2D symmetry effectively turns into vectorial orientation. Blair et al (2003) observed, alongside a variety of 2D patterns, spontaneously formed moving ordered domains of slightly inclined nearly vertical rods on the background of immobile horizontal rods. The ordered domains coarsened with time, as shown in Fig. 6.9, to form large vortices slowly rotating in the direction of the tilt. Since no motion was observed in a horizontally vibrated layer, the experimentalists tentatively attributed this remarkable chiral symmetry breaking phenomenon to interactions of inclined rods with the bottom of the vertically vibrated container.

The starting point of the model by Aranson and Tsimring (2003) that fits these observations is the momentum conservation equation

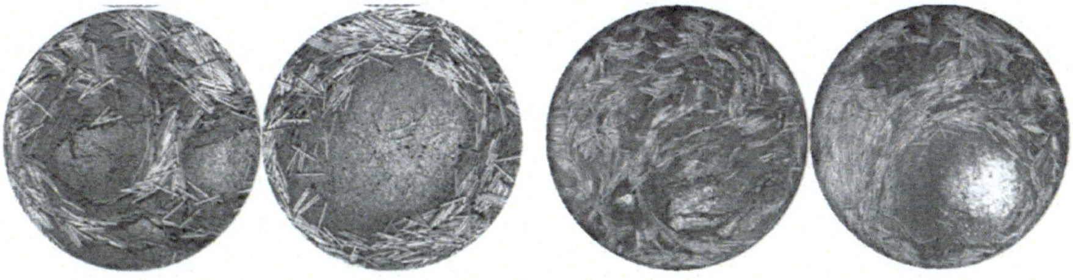

Fig. 6.9. Snapshots of the development of a vortex from an initial random state, showing the formation and coarsening of domains of vertical rods (Blair et al, 2003, reproduced by permission, ©The American Physical Society)

$$(\partial_t + \nu)\mathbf{v} = -\rho^{-1}\nabla p + \alpha \mathbf{n} f_n(n). \tag{6.1.11}$$

This equation does not include unconventional terms of the Toner–Tu model (6.1.3). The term $\nu\mathbf{v}$ describes the momentum dissipation due to bottom friction, which replaces dissipation by viscosity. The vector $\mathbf{n}$ is the projection of the rod director on the support plane. It is not a unit vector: the strength of the force αf_n, directed along the tilt, depends on its absolute value n. The force is proportional to the tilt of rods for small tilt values, but saturates and eventually decays to zero in the planar position at $n = 1$; the magnitude of the force is proportional to the vibration energy. The inertial term is missing, as velocities are not expected to be fast, and, as in the Stokes equation, the term $\partial_t \mathbf{v}$ can be neglected when motion is dominated by friction. Then the velocity is expressed as

$$\mathbf{v} = -\nu^{-1}[\nabla p - \alpha \mathbf{n} f_n \rho], \tag{6.1.12}$$

and the continuity equation takes the form

$$\partial_t \rho = \nu^{-1}[\nabla^2 p - \nabla \cdot (\rho f_n \mathbf{n})]. \tag{6.1.13}$$

The problematic part is the definition of pressure, which, as argued in Sect. 6.1.2, is not well defined in active matter. However, vibrated rods do not have their own energy source, being governed by gravity and vibration, and can be assigned free energy dependent on their orientation. Aranson and Tsimring have chosen the energy functional of the simple standard form (1.6.8):

$$\mathcal{F} = \frac{l^2}{2} \int \left[|\nabla \rho|^2 + V(\rho) \right] \mathrm{d}\mathbf{x}, \tag{6.1.14}$$

where $V(\rho)$ is a quartic potential, which, unlike the standard Landau energy, is chosen to be asymmetric. It is convenient to measure ρ with respect to a reference density ρ_0, identified with the minimal density for the onset of nematic order. Then the derivative of the potential is written as $V'(\rho) = \rho(c_0 - c_1\rho + \rho^2)$. The characteristic length l can be reduced to unity by rescaling the coordinates, and the friction coefficient ν eliminated by rescaling $t \to t/\nu l^2$. This leads to the mass conservation equation

$$\partial_t \rho = -\nabla^2 \left[\nabla^2 \rho - V'(\rho) \right] - \alpha \nabla \cdot \left[(\rho + \rho_0) f_n \mathbf{n} \right]. \tag{6.1.15}$$

We are left with *two* equations for the evolution of density. Combining them leads to the definition of effective pressure $p = -\delta\mathcal{F}/\delta\rho$, which would provide a direct route from (6.1.13) to (6.1.15).

The model is closed by adding the equation for the evolution of the tilt. Below the maximum packing density, the vertical orientation of rods corresponding to $n = 0$ is unstable, as rods spontaneously tilt. Assuming that the growth rate of n-instability depends on the density of rods, the local dynamics can be written in the form

$$\partial_t \mathbf{n} = f(\rho)(1 - n^2)\mathbf{n}, \qquad f(\rho) = a_0 - a_1\rho\,, \tag{6.1.16}$$

with $a_1 > 0$ taking into account stabilization of the vertical position in a dense layer. Since the tilt field is not divergence-free, its evolution can be characterized,

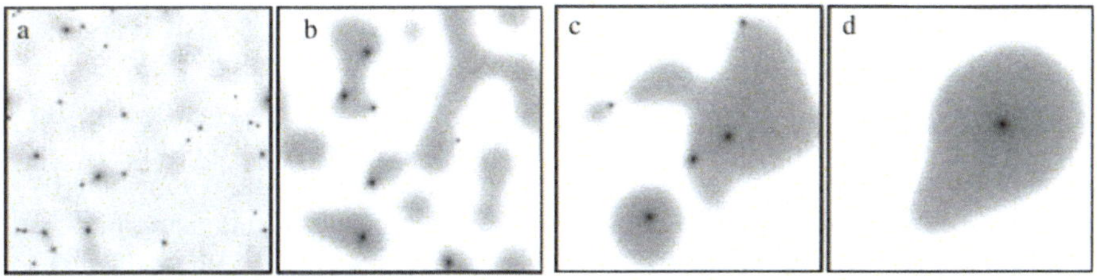

Fig. 6.10. Snapshots of the development of local vortices merging into the single global vortex in computations by Aranson and Tsimring (2003). Vortex cores are shown as black dots, white area corresponds to larger n. Reproduced by permission, ©The American Physical Society)

besides common shear viscosity $\nu(\rho)$, by the extensional viscosity $\widetilde{\nu}(\rho)$. The density dependence of both coefficients describes weakening of the spatial coupling strength with decreasing rods density. Including also the simplest term coupling the tilt and the density gradient, $\beta\nabla\rho$, leads to the evolution equation

$$\partial_t \mathbf{n} = f(\rho)(1-n^2)\mathbf{n} + \nu(\rho)\nabla^2\mathbf{n} + \widetilde{\nu}(\rho)\nabla(\nabla\cdot\mathbf{n}) + \beta\nabla\rho \,. \tag{6.1.17}$$

This sophisticated model reflects an unusual combination of 2D geometry with 3D orientations of the rods.

Omitting extensional viscosity and presuming density to be constant, this equation can be recast as the real Ginzburg–Landau equation in the polar coordinates r, ϕ for the complex variable $\psi = n\mathrm{e}^{\mathrm{i}\theta}$, where θ is the tilt angle:

$$\partial_t \psi = f(\rho)(1-|\psi|^2)\psi + \nu(\rho)\nabla^2\psi \,. \tag{6.1.18}$$

The remarkable property of this equation is the existence of a vortex solution $\psi = n(r)\mathrm{e}^{\mathrm{i}\phi}$ familiar from Sect. 4.2. It corresponds to the tilt rotating around the center of a circular enclosure with $n \to 0$ at the vortex core. Using this in (6.1.12) we can see that velocity and pressure gradient should also rotate in the same way; thus, it captures qualitatively the structure of the large vortex seen in the experiment. Of course, this oversimplifies the real situation. Aranson and Tsimring followed the evolution of the full system (6.1.15), (6.1.17) numerically, and observed development of local vortices merging into the single global vortex, as shown in Fig. 6.10, which faithfully reproduces experimental observations.

6.1.5 Active Phase Separation

It is an intuitive notion that crowding slows down motion. Only Vicsek's particles are able to marsh in ordered domains as soldiers on a parade. Crowding near a narrow exit slows down an emergency escape (Helbing et al, 2000). Jamming above a critical density, which is still far below the random close-packing density, has been observed in experiments with microscopic rollers (Geyer et al, 2019). A positive feedback between slowdown-induced accumulation and accumulation-induced slowdown leads to *motility-induced phase separation* (MIPS): particles stuck in jams create dense regions, while others move more or less freely in any remaining voids (Cates and Tailleur, 2015).

This may remind of equilibrium separation into liquid and gas phases, which are both disordered but differ by motility of molecules. But phase separation of active particles proceeds in a different way. In equilibrium processes, droplets or bubbles undergo Ostwald ripening (Sect. 2.3.4). A coarsening tendency remains in active systems, but this process does not lead to complete phase separation (Tjhung et al, 2018). Dense macroscopic "liquid" clusters formed during coarsening contain mesoscopic vapor bubbles, which are continuously created in its bulk, coarsen, and are ejected into the exterior "vapor" phase, as in the upper panels of Fig. 6.11. The lower panels of the same figure show a picture of arrested coarsening of dense droplets, coming to a steady state in the two right-hand panels.

Tjhung et al (2018) attributed the difference in behavior to violation of time reversal symmetry in active matter. In order to demonstrate this without specifying a particular mechanism, they added to the common definition of the flux (1.6.12) two nonequilibrium terms not derivable from an energy functional:

$$\partial_t u = -\nabla \cdot \mathbf{j}, \qquad \mathbf{j} = -\Gamma(u)\nabla \left[\frac{\delta \mathcal{F}}{\delta u} + \lambda |\nabla u|^2\right] + \zeta \nabla^2 u \nabla u, \tag{6.1.19}$$

where $\mathcal{F}$ is defined in the standard way by (1.6.8). The bracketed term in (6.1.19) is the nonequilibrium chemical potential μ. In a stationary state, the flux vanishes, and in 1D the ζ term can be incorporated in μ, so that its effect is indistinguishable from that of the λ term. Unlike equilibrium systems, where densities of coexisting phases maintained at constant temperature and pressure are fixed, they are variable in

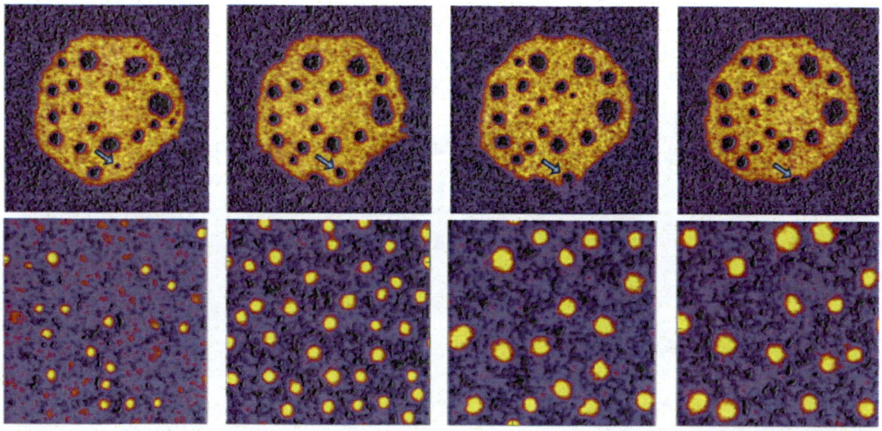

Fig. 6.11. Top: Bubbly phase coexistence of the "boiling liquid" (yellow) with the vapor phase (blue) obtained by simulating (6.1.19) with positive ζ, λ. The green arrow tracks a bubble created inside the liquid bulk, growing, moving toward the interface, and getting expelled. Bottom: Arrested phase separation in simulations of (6.1.19) with negative ζ, λ: the average cluster size saturates at a finite steady-state level (Tjhung et al, 2018, CC 4.0)

a collective of active particles. The static profile of the order parameter u, associated with density, is obtained then by solving the equation

$$\mu = -f(u) - Ku''(x) + \kappa u'(x)^2, \qquad f(u) = -\partial V/\partial u, \tag{6.1.20}$$

where $\kappa = \lambda + \zeta/2$. As noted in Sect. 2.1.1, a second-order ODE can be mapped onto the equation of motion of a Newtonian particle in the inverted double-well potential. In this case, the nonlinear differential term can be interpreted as a velocity-dependent force (Wittkowski et al, 2014). It breaks the symmetry of the system, and (6.1.20) has no stationary front solutions with $u'(x) = 0$ at $x \to \pm\infty$ if $\mu = 0$.

A stationary front solution of (6.1.20) with $u \to u_\pm$ at $x \to \pm\infty$ should satisfy, similar to its equilibrium version, the condition $\mu = f(u_\pm) = \text{idem}$. The second equilibrium condition, $V(u_\pm) = \text{idem}$, stems from multiplying the stationary equation by the translational Goldstone mode $f'(u)$ and integrating over the infinite line. Solon et al (2018) modified this condition by defining a "pseudopotential" $W(u)$ and "pseudodensity" $\psi(u)$, such that $\psi = u$ and $W = \mu$ at $\kappa = 0$, which satisfy

$$K\frac{\partial^2\psi}{\partial u^2} = -2\kappa\frac{\partial\psi}{\partial u}, \qquad \frac{\partial W}{\partial\psi} = \frac{\partial V}{\partial u}. \tag{6.1.21}$$

Multiplying (6.1.20) by $\psi'(x) = u'(x)\partial\psi/\partial u$ and integrating over the infinite line yields

$$\mu(\psi_+ - \psi_-) = \int \frac{\partial V}{\partial u}\psi'(x)\mathrm{d}x - K\int\frac{\partial\psi}{\partial u}u'(x)u''(x)\mathrm{d}x + \kappa\int\frac{\partial\psi}{\partial u}u'(x)^3\mathrm{d}x. \tag{6.1.22}$$

The first term in the right-hand side is converted using (6.1.21) to

$$\int\frac{\partial W}{\partial\psi}\psi'(x)\mathrm{d}x = W(\psi_+) - W(\psi_-). \tag{6.1.23}$$

The last two terms are transformed as

$$-K\int\left[\frac{\partial\psi}{\partial u}u'(x)u''(x) + \frac{1}{2}\frac{\partial^2\psi}{\partial u^2}u'(x)^3\right]\mathrm{d}x = -\frac{K}{2}\int\frac{\mathrm{d}}{\mathrm{d}x}\left[\frac{\partial\psi}{\partial u}u'(x)^2\right] = 0. \tag{6.1.24}$$

This leads to the condition

$$P(u_\pm) = \text{idem}, \qquad P(u) = \mu\psi(u) - W(u). \tag{6.1.25}$$

The values $u_\pm$ satisfying the above conditions are found by integrating (6.1.21). The solution of the first equation satisfying the condition $\psi \to u$ at $\kappa = 0$ is

$$\psi = \frac{K}{2\kappa}(\mathrm{e}^{-2u\kappa/K} - 1), \quad \text{or} \quad u = -\frac{K}{2\kappa}\ln\left(1 - \frac{2\kappa}{K}\psi\right). \tag{6.1.26}$$

Then the second equation (6.1.21) is integrated to

$$W = \frac{K}{8\kappa^4}\left[3K^3 - 2K\kappa^2 \right. \tag{6.1.27}$$
$$\left. - \,\mathrm{e}^{-2u\kappa/K}\left(3K^3 + 6K^2\kappa u - 2K\kappa^2(1-3u^2) - 4\kappa^3 u(1-u^2)\right)\right],$$

which reduces, as required, to $V = \frac{1}{4}u^2(u^2-2)$ in the limit $\kappa \to 0$. The dependence of μ and $u_\pm$ on κ at $K = 1$ for the cubic $f(u) = u - u^3$ is shown in Fig. 6.12. In order to draw it, we, first, make use of (2.1.14) with the replacement $v \to \mu$ to express $u_\pm$ through μ, whereupon the equation $P(u_+) = P(u_-)$ is solved numerically. The slopes of all three curves should pass en extremum and fall off to zero in the limit $\kappa \to \pm\infty$. Solon et al (2018) fitted densities of separated phases computed by this method to those observed in simulations of scalar self-propelled particles.

This solution still does not reveal the structure of the interface. As in Sect. 2.1.1, (6.1.20) can be recast into a a relation between what would be called in Newtonian mechanics the “momentum” $p = u'(x)$ and the "coordinate" u. Denoting $w = p^2$ leads to a linear equation

$$\frac{K}{2}w'(u) = \kappa w - f(u) - \mu, \tag{6.1.28}$$

which is readily resolved as

$$w(u) = \frac{1}{4\kappa^4}[(-3K^3+6K^2\kappa u-2K\kappa^2(1-3u^2)+4\kappa^3(\mu+u-u^3)]+C\,\mathrm{e}^{-2\kappa u/K}. \tag{6.1.29}$$

The integration constant C could be found, together with μ, to satisfy the boundary conditions $w(u) = w'(u) = 0$ at some $u = u_\pm$. This solution is, clearly, unacceptable, since it is singular at $\kappa \to 0$. The reason is that the profile $u(x)$ can be computed in this way only when it is *monotonic*, and the failure is a telltale sign that the transition between the asymptotic values $u = u_\pm$ should involve oscillations in the transitory region.

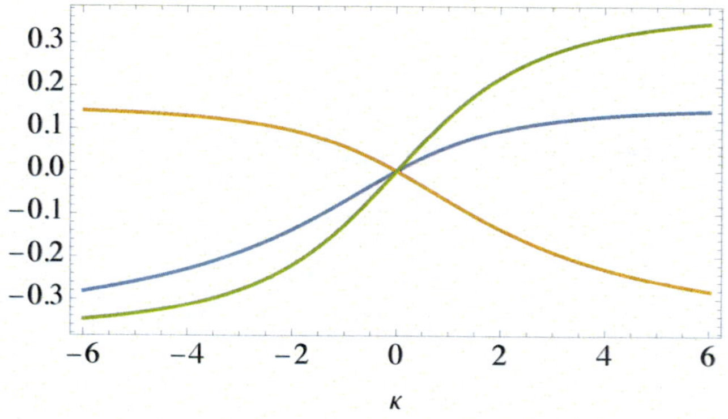

Fig. 6.12. The dependence of $u_+ - 1$ (blue curve), $u_- + 1$ (orange curve), and μ (green curve) on $\kappa = \lambda + \zeta/2$ for the cubic $f(u)$

6.1.6 Surface Tension and Pseudotension

Different roles of λ and ζ terms are revealed in the analysis of curved interfaces (Tjhung et al, 2018). The flux due to the ζ term in (6.1.19) can be presented using the radial symmetry as

$$\mathrm{j}_\zeta = \zeta \int \left[a \frac{u'(r)^2}{r} + u'(r)u''(r) \right] \mathrm{d}r = a\zeta \int \frac{u'(r)^2}{r} \mathrm{d}r + \frac{\zeta}{2} u'(r)^2, \qquad (6.1.30)$$

where $a = 1$ in 2D and $a = 2$ in 3D. If mobility $\Gamma(u)$ is assumed to be constant and reduced to unity by scaling time, this adds to (6.1.20) a *nonlocal* term:

$$\mu = -f(u) - Ku''(r) - aKr^{-1}u'(r) + \kappa u'(r)^2 + a\zeta \int_r^\infty \frac{u'(y)^2}{y} \mathrm{d}y. \qquad (6.1.31)$$

Different mobilities in the dilute and dense phases can be accommodated by allowing for different effective parameters $\zeta_\pm$; in this case, (6.1.31) remains applicable everywhere except a thin transitional layer.

The distinction from the flat interface with $\mu = -f(u_\pm)$ is due to the two curvature-dependent terms. This raises the problem of the role of surface tension in active matter. So far, we have encountered the notion of surface tension in two different contexts. It originates in thermodynamics as work needed to increase a surface area, and plays a crucial role in equilibria, transport processes and hydrodynamics of immiscible fluids. In passive nonequilibrium systems it is extended to fronts between alternative stationary states. In the simplest setting, it is related by (2.1.17) to diffusivity and reaction kinetics and determines the curvature dependence of the front propagation speed (Sect. 2.2.2). It drives Ostwald ripening under equilibrium (Sect. 2.3.4) and its nonequilibrium analog (Sect. 2.2.4), as well as the reverse phenomenon of interfacial instabilities discussed in Sect. 2.4 and Ch. 3. In all these cases, phenomenology at scales far extending the surface thickness depends on a single constant, determined by either molecular interactions or parameters of a nonequilibrium process, but in active systems various phenomena involving interfacial curvature may depend on effective surface tension defined in a different way.

The contrast between the macroscopic scale of droplets and bubbles and the "microscopic" scale of self-propelling particles is not as pronounced as the scale separation in either equilibrium or passive nonequilibrium systems, and therefore interfacial curvature may noticeably affect bulk densities. If in the stationary state $u(0) = u_+$, $u(\infty) = u_-$, in the limit when the radius R far exceeds the thickness of the interface we have $\mu(r) = -f(u_-) = -f(u_+) + \delta$, where

$$\delta = \frac{a}{R} \left[\zeta \int_0^\infty u'(r)^2 \mathrm{d}r - K(u_+ - u_-) \right] + O(R^{-2}). \qquad (6.1.32)$$

Computing σ in limit $R \gg 1$ in the sharp interface limit is fully justified: also in equilibrium systems, surface tension is well-defined only under this condition.

The second condition (6.1.25) is formally presented as

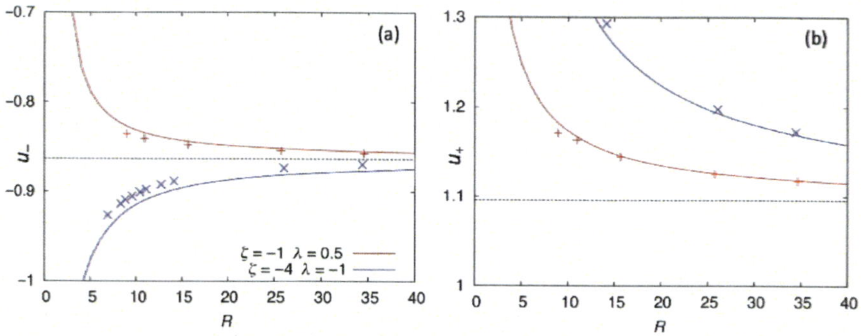

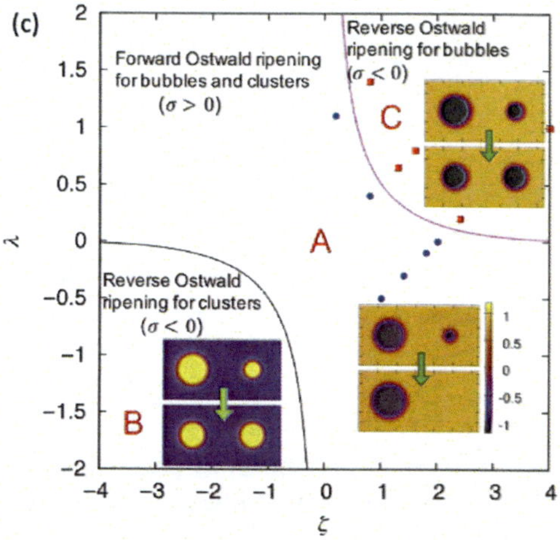

Fig. 6.13. The dependence of stationary values of $u = u_-$ (a) and $u = u_+$ (b) in a spherical dense ($u = u_+$) droplet in a dilute ($u = u_-$) environment at the indicated values of parameters. Red and blue lines show analytical results in the regions A and B in (c), respectively. Crosses indicate results from the numerical simulations. Dashed lines show the values $u_\pm$ for the flat interface. (c) Regions of normal Ostwald ripening (A) and reverse Ostwald ripening for bubbles (B) and droplets (C), dependent on the sign of λ and ζ (adapted from Tjhung et al, 2018, CC 4.0)

$$P(u_+) = P(u_-) + \frac{a\sigma}{R} + O(R^{-2}), \tag{6.1.33}$$

where σ is effective interfacial tension, called “pseudotension” by the authors. It is expressed, separating local and nonlocal effects, as

$$\sigma = \frac{K}{2\kappa}(\zeta S_0 + 2\lambda S_1), \tag{6.1.34}$$

$$S_0 = \mathrm{e}^{2\kappa u_+/K} \int_0^\infty \frac{u'(y)^2}{y} \mathrm{d}y, \qquad S_1 = \int_0^\infty u'(y)^2 \mathrm{e}^{2\kappa u(y)/K} \mathrm{d}y. \tag{6.1.35}$$

The computation is challenging due to nonlocality of the potential, which implies dependence on the entire shape of the interface. Tjhung et al (2018) adopted an iterative procedure. First, (6.1.31) is transformed to the variable $w = u'(r)^2$. The result, similar to (6.1.28) with added curvature-dependent terms, is written as

$$w'(r) = \frac{2\kappa}{K} w(u) + s[r, u_-, \mu(R)] + O(R^{-2}), \tag{6.1.36}$$

$$s[r, u_-, \mu(R)] = -\frac{2}{K}[f(u) + \mu(R)] - \frac{2a}{R}\sqrt{w(u)} + \frac{2a\zeta}{KR}\int_u^{u_-} \sqrt{w(u)}\,\mathrm{d}u. \tag{6.1.37}$$

The linear equation (6.1.36) is formally integrated, ignoring the dependence of s on w and assuming $\mu(R)$ to be known, to obtain

$$w(r) = \mathrm{e}^{-2\kappa r/K}\left[C + \int_1^r \mathrm{e}^{-2\kappa y/K} s[r, u_-, \mu(R)]\mathrm{d}y\right], \tag{6.1.38}$$

where the integration constant C can be tied to $u_\pm$, which define $\mu(R)$, and thus enable computing $w(r)$ from (6.1.38) and use it to obtain S_0, S_1 in (6.1.35). Once σ in (6.1.34) is known, (6.1.31) and (6.1.33) define $u_\pm$, which are fed back to compute $w(r)$ again. Precision is lost when iterations are repeated, but the results shown in Fig. 6.13 fit fairy well direct numerics.

The important result is that pseudotension may become *negative*, which changes the character of the dependence $u_\pm(R)$, as shown in Fig. 6.13. Although a difference in mobilities has not been taken into account here, it is unlikely to change this qualitatively. When $\sigma > 0$ (region A in Fig. 6.13c), both $u_\pm(R)$ decrease with R, while at $\sigma > 0$ (region B in Fig. 6.13c) u_+ is still decreasing but u_- increases with R. Negative pseudotension does not cause the interface to be unstable (hence the prefix "pseudo"), but reverses the direction of Ostwald ripening for either droplets or bubbles, i.e., dense or dilute clusters. Although Tjhung et al emphasize the reversal

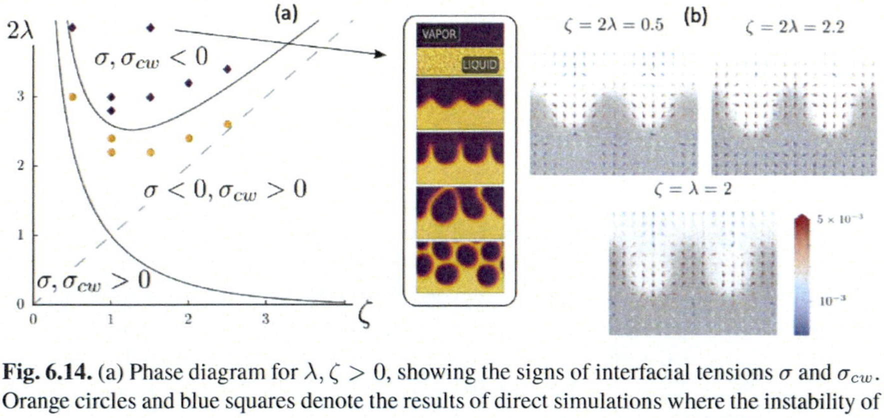

Fig. 6.14. (a) Phase diagram for $\lambda, \zeta > 0$, showing the signs of interfacial tensions σ and σ_{cw}. Orange circles and blue squares denote the results of direct simulations where the instability of the interface is or is not observed, respectively. The picture on the right shows the development of interfacial instability at $\lambda = 2$, $\zeta = 1.5$. (b) Simulated current $\mathbf{j}$ close to the stable interface with normal Ostwald ripening ($\lambda = 2\zeta = 0.5$, left), stable interface with reversed Ostwald ripening for bubbles ($\lambda = 2\zeta = 2.2$, right) and unstable interface ($\lambda = \zeta = 0.5$, below). The liquid is shown in dark gray and the magnitude of the current in colors (Fausti et al, 2021, reproduced by permission, ©The American Physical Society

of the dependence $u_-(R)$, the simple reason may be traced to (6.1.32), which shows that, when the nonlocal term is prevailing, chemical potential *decreases* with R at $\zeta < 0$, reversing the normal tendency. In both equilibrium and passive nonequilibrium systems, the curvature dependence of $u_\pm$ is negligible, and it is the shift of chemical potential that drives Ostwald ripening.

Fausti et al (2021) addressed the more common role of surface tension in sustaining stability of the interface by considering the spectrum of small amplitude capillary waves as in Sect. 3.1.2, but the nonlinearities and nonlocal interactions make the problem technically far more difficult. The main result is that the surface tension responsible for stability of the interface, denoted σ_{cw}, differs from pseudotension σ responsible for Ostwald ripening, The types of behavior in the parametric plane $\lambda, \zeta > 0$ and snapshots of developing interfacial instabilities at $\sigma_{cw} < 0$ are shown in Fig. 6.14a. Fig. 6.14b shows the flux near the interface, which is always stabilizing on the vapor side, while it is stabilizing on the liquid side only if $\sigma > 0$ However, this is not sufficient to drive the instability. For this, the destabilizing current on the liquid side needs to be stronger than the one on the vapor side. This happens only in the lower panel, which corresponds to $\sigma_{cw} < 0$.

6.2 Active Nematic Fluids

6.2.1 Dynamics of Nematic Director

The contrasting alternative to "dry" particle-based theory and computations is the continuous theory of active nematic fluids. Unlike self-propelling rod-like particles of Sect. 6.1.3, in nematic fluids there is no conflict between polar symmetry of propulsion and nematic alignment, since activity has nematic symmetry as well. The theory of active nematics is based on the well established nematodynamic theory (de Gennes and Prost, 1993; Kléman and Lavrentovich, 2003) including combination of hydrodynamic and elastic equations of passive nematics with added extensile or contractile activity along the nematic *director*.

Nematic order has been traditionally characterized by representing the director as the unit vector $\mathbf{n} = \{\cos\theta, \sin\theta\}$, where θ is the inclination angle. Such a description is flawed, since the director is not a well-defined geometric object, but makes all derivations transparent. When $\mathbf{n}$ is presented as a vector, its paired combinations have to appear in all expressions; otherwise they would not be invariant under rotations by π, which is the property of the nematic order. The starting point of the macroscopic continuum theory is the energy functional $\mathcal{F} = \int \mathcal{L}\,d\mathbf{x}$ with the scalar Landau–de Gennes Lagrangian $\mathcal{L}$. Recall that even in isotropic media (Sect. 1.6.1), this approach involves severe and, strictly speaking, unjustified simplifications. Nevertheless, it remains the only basis of rational theory.

The equations of nematodynamics taking into account viscous and elastic anisotropies but not changes of the modulus were developed by Ericksen (1961) and Leslie (1968), starting from the Landau–de Gennes Lagrangian and common hydrodynamic equations. Since $\mathbf{n}$ is a unit vector, the deviation of a static nematic

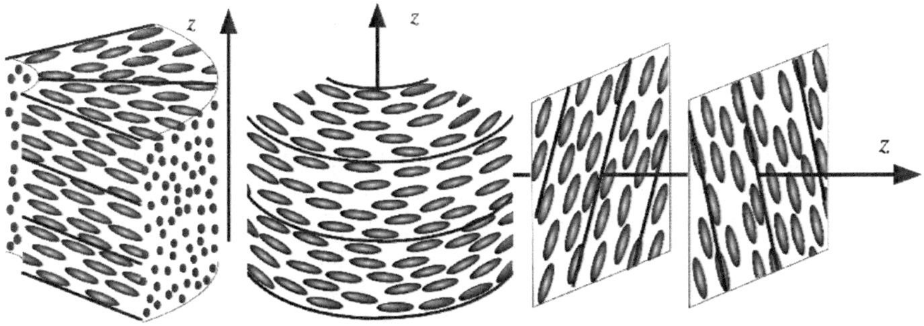

Fig. 6.15. *Left to right*: Splay, bend, and twist distortion of nematic alignment (Kléman and Lavrentovich, 2003, ©Springer)

medium from equilibrium is caused only by changes in orientation expressed by differential terms dependent on distortions of perfect alignment. The classical expression for the energy density of a uniaxial nematic in 3D is

$$\mathcal{L} = \frac{1}{2}\left[K_1(\operatorname{div}\mathbf{n})^2 + K_2(\mathbf{n}\cdot\operatorname{curl}\mathbf{n})^2 + K_3(\mathbf{n}\times\operatorname{curl}\mathbf{n})^2\right]. \tag{6.2.1}$$

The three terms in this expression correspond to the energies of splay, bend, and twist distortion of nematic alignment, sketched in Fig. 6.15. They include all possible quadratic combinations of the director with its divergence and curl invariant to rotations by π. The coefficients K_i, called Frank constants, are the respective elasticities. The *molecular field*[1] $\mathbf{h}$ resulting from virtual orientational distortions is defined by varying the elastic energy:

$$h_j = \partial\mathcal{L}/\partial n_j + \partial_i\pi_{ij}, \qquad \pi_{ij} = \partial\mathcal{L}/(\partial_i n_j), \tag{6.2.2}$$

where $\partial_i = \partial/\partial x_i$. In 2D, there is no twist, and, since $\operatorname{curl}\mathbf{n}$ is a pseudoscalar, the bend elasticity is expressed as $\frac{1}{2}K_2(\operatorname{curl}\mathbf{n})^2$. The resulting molecular field is

$$\begin{aligned} h_1 &= K_1(\partial_1^2 n_1 + \partial_1\partial_2 n_2) + K_2(\partial_2^2 n_1 - \partial_1\partial_2 n_2),\\ h_2 &= K_1(\partial_2^2 n_2 + \partial_1\partial_2 n_1) + K_2(\partial_1^2 n_2 - \partial_1\partial_2 n_1). \end{aligned} \tag{6.2.3}$$

The standard expression $\mathbf{h} = K\nabla^2\mathbf{n}$ is obtained in the commonly used "one constant approximation" $K_1 = K_2 = K$.

The principal idea of the Ericksen–Leslie nematodynamic theory is to couple changes of nematic orientation and fluid motion by tracing the change of entropy in the course of evolution and requiring the resulting equations to satisfy the Onsager reciprocity relations. In the absence of external fields, the energy density consists of the kinetic $\frac{1}{2}\rho v^2$ and distortion (6.2.1) energy densities. Its change, related at a constant temperature T to the change of entropy as $T\mathcal{D}_t S$, is expressed as

[1] The term is still applied to active nematics even though "microscopic" active elements far exceed a molecular size

$$T\mathcal{D}_t S = -\mathcal{D}_t \int \left(\tfrac{1}{2}\rho v^2 + \mathcal{L}\right) \mathrm{d}\mathbf{x}, \tag{6.2.4}$$

where ρ is density and $\mathcal{D}_t = \partial_t + \mathbf{v} \cdot \nabla$ denotes the substantial derivative accounting for advection with flow velocity $\mathbf{v}$. The acceleration of the fluid is determined by the generalized Navier–Stokes equation

$$\rho \mathcal{D}_t v_i = \partial_j \sigma_{ij}, \tag{6.2.5}$$

where $\boldsymbol{\sigma}$ is the total stress tensor, which includes the viscous stress $\boldsymbol{\sigma}^{\mathrm{s}}$ and the Ericksen stress $-\boldsymbol{\sigma}^{\mathrm{e}}$ stemming from the change of the elastic energy due to advection; summation over repeated indices is implied. Using (6.2.5) in (6.2.4) and integrating by parts transforms the first term as

$$-\mathcal{D}_t \int \tfrac{1}{2}\rho v^2 \mathrm{d}\mathbf{x} = \int \sigma_{ij}\partial_i v_j \mathrm{d}\mathbf{x}, \tag{6.2.6}$$

The incompressibility condition is accounted for by varying the enthalpy

$$\mathcal{H} = \mathcal{F} - \int p(\nabla \cdot \mathbf{v})\mathrm{d}\mathbf{x}. \tag{6.2.7}$$

In the case of an incompressible fluid, pressure p is not related to an equation of state but serves to enforce the continuity equation $\nabla \cdot \mathbf{v} = 0$. Varying $\mathcal{H}$ with respect to advective distortions defines the distortion stress $\sigma^{\mathrm{d}}_{ij} = \sigma^{\mathrm{e}}_{ij} - p\delta_{ij}$:

$$\delta\mathcal{H} = \int \sigma^{\mathrm{d}}_{ij}\partial_i v_j \mathrm{d}\mathbf{x}, \qquad \sigma^{\mathrm{d}}_{ij} = -\pi_{ik}\partial_j n_k. \tag{6.2.8}$$

The velocity gradient (strain rate) tensor is further separated into symmetric and antisymmetric parts:

$$s_{ij} = \tfrac{1}{2}(\partial_i v_j + \partial_j v_i), \qquad \Omega_{ij} = \tfrac{1}{2}(\partial_i v_j - \partial_j v_i). \tag{6.2.9}$$

Since $\boldsymbol{\sigma}$ is invariant to adding a tensor satisfying $\partial_i \sigma_{ij} = 0$, it can be complemented by an antisymmetric tensor $\boldsymbol{\sigma}'$, which represents the rate of change of the angular momentum due to the torque applied to the liquid. The energy does not change when both the nematic alignment and the fluid as a whole are rotated with the same angular velocity:

$$\partial_t n_i = \Omega_{ij} n_j, \quad \partial_t x_i = \Omega_{ij} x_j. \tag{6.2.10}$$

The antisymmetric tensor satisfying this relation is $\boldsymbol{\sigma}' = \frac{1}{2}\mathbf{n} \times \mathbf{h}$.

Combining (6.2.6) and (6.2.8), the total change of entropy is presented as

$$T\mathcal{D}_t S = \int \left(\sigma^{\mathrm{s}}_{ij} s_{ij} + \mathbf{h} \cdot \mathbf{N}\right) \mathrm{d}\mathbf{x}, \tag{6.2.11}$$

where $\boldsymbol{\sigma}^{\mathrm{s}} = \boldsymbol{\sigma} - \boldsymbol{\sigma}^{\mathrm{e}}$ is the viscous stress and $\mathbf{N} = \mathcal{D}_t \mathbf{n} - \boldsymbol{\Omega} \cdot \mathbf{n}$.

In the theory of irreversible processes (de Groot and Masur, 1984), contributions to the entropy source are presented as products of fluxes by conjugate forces with coefficients having the dimension of viscosity. Based on (6.2.11), σ^{s}_{ij} is identified as the force conjugate to s_{ij} and $\mathbf{h}$ as the force conjugate to $\mathbf{N}$ (de Gennes and Prost, 1993). The forces are tied to the fluxes by linear relations with coefficients having the dimension of viscosity arranged into tensors of an appropriate rank:

$$\sigma^{\mathrm{s}}_{ij} = S_{ijkl}s_{kl} + M_{ijk}N_k, \qquad h_k = M'_{ijk}s_{ij} + P_{ki}N_i. \tag{6.2.12}$$

Since the structure of these matrices must be compatible with the local symmetry of the nematic, they must contain only the director $\mathbf{n}$ and be symmetric to its reversal. Thus, the matrix $\mathbf{S}$ must contain an even number of $\mathbf{n}$, up to four. Since $\mathbf{h}$ and $\mathbf{N}$ are odd while s and $\boldsymbol{\sigma}^{\mathrm{s}}$ are even under reversal of $\mathbf{n}$, the matrices $\mathbf{M}$ and $\mathbf{M}'$ should contain a single $\mathbf{n}$. By the Onsager principle, these matrices should be identical.

Due to complexity of full tensor expressions and lack of data, simplified isotropic expressions are commonly used. We restrict to the relations containing tensors of the lowest rank, since only they are retained in a natural way when the modulus is variable as well (Sect. 6.2.3):

$$\sigma^{\mathrm{s}}_{ij} = \eta s_{ij} + \tfrac{1}{2}\gamma_2(n_iN_j + n_jN_i), \tag{6.2.13}$$

$$h_i = \gamma_2 n_j s_{ij} + \gamma_1 N_i. \tag{6.2.14}$$

The dissipative parameters are the common dynamic viscosity η, and the rotational viscosity γ_1, usually replaced by the mobility coefficient $\Gamma = 1/\gamma_1$.

The dynamic equation of the director follows from (6.2.14):

$$\mathcal{D}_t n_i = -\chi s_{ij}n_j + \Omega_{ij}n_j + \Gamma h_i. \tag{6.2.15}$$

where the alignment parameter $\chi = \gamma_2/\gamma_1$ determines the response of the director to local shear. Collecting all components of the total stress in (6.2.5), including also the antisymmetric stress $\boldsymbol{\sigma}'$ leads to the nematic flow equation

$$\rho\mathcal{D}_t v_i = \eta\nabla^2 v_i - \partial_i p + \partial_j\sigma^{\mathrm{p}}_{ij}, \tag{6.2.16}$$

where $\boldsymbol{\sigma}^{\mathrm{p}}$ is the part of the viscous stress $\boldsymbol{\sigma}^{\mathrm{s}}$ related to nematic alignment:

$$\sigma^{\mathrm{p}}_{ij} = \tfrac{1}{2}\chi(n_ih_j + h_in_j) - \tfrac{1}{2}(n_ih_j - h_in_j) + \pi_{ik}\partial_j n_k. \tag{6.2.17}$$

The last term is the distortion stress. In the theory of active nematics $\boldsymbol{\sigma}^{\mathrm{p}}$ is called passive stress and is supplemented by active stress, which can be defined as either $\sigma^{\mathrm{a}}_{ij} = -\zeta n_in_j$ or through the respective traceless tensor with the parameter ζ positive for extensile and negative for contractile activity (more on this below).

6.2.2 Instabilities in the Director-based Description

Activity disturbs monotonic relaxation to equilibrium. The dynamics of the director and velocity fields are tied by a feedback loop. Hydrodynamic instabilities give rise to distortions in the director field, which, in their turn, drive flow. The positive feedback makes the ordered state of an active nematic intrinsically unstable (Simha and Ramaswamy, 2002). Activity-induced instabilities in a confined geometry were investigated numerically by Edwards and Yeomans (2009) in the framework of the Ericksen–Leslie theory.

Onsager's reciprocity relations do not apply to active media, so that, strictly speaking, the entire nematodynamic theory based on these relations should be revised. Current theoretical studies do not go this far but only add the active stress σ^{a}, which can be expressed as $\sigma^{\mathrm{a}}_{ij} = -\zeta n_i n_j$ with the parameter ζ positive for extensile and negative for contractile activity. This expression, taken literally, is problematic, since entropy-changing contributions to the stress tensor have to be odd under time reversal, as emphasized by Brand et al (2014) who noted "that the dissipative driving into non-equilibrium does not come from the 'active term', but from the coupling to a chemical energy-consuming subsystem". The contradiction can be resolved by letting $\zeta = \partial_t c$, where c is the concentration of an active chemical, such as ATP in intracellular processes. Then the term becomes odd under time reversal, as required, and can shift entropy either way. If there is no feedback of alignment on the active chemical and its consumption rate is constant, chemistry may remain behind the scenes.

We first test a passive system, where a perfectly ordered state is expected to be stable. However, even in this case, there are two questions to decide: whether perturbations of a perfectly aligned state may cause a deviation of the modulus from unity and whether arbitrary values of the alignment parameter are admissible. Consider perturbations of an ordered state with $|\mathbf{n}| = 1$ in an unbounded domain. In view of the circular symmetry with respect to the unperturbed alignment, it is sufficient to view the problem in 2D. A constant background velocity is irrelevant, since it can be eliminated by transforming to a comoving frame. Accordingly, the unperturbed alignment direction is also irrelevant, and can be chosen along the x_1 coordinate, so that the 2D components of the director are $n_1^0 = 1$, $n_2^0 = 0$. In the leading order, only a small perturbation ϵn_2 does not perturb the modulus. We expand it in the Fourier series $n_2 = \epsilon \widehat{n}_2 \mathrm{e}^{\mathrm{i}\mathbf{k}\cdot\mathbf{x}}$ where $\mathbf{k} = k\{\cos\varphi, \sin\varphi\}$ is the wave vector. The molecular field (6.2.3) is transformed as

$$\widehat{h}_1 = -\tfrac{1}{2}k^2\widehat{n}_2(K_1 - K_2)\sin 2\varphi, \quad \widehat{h}_2 = -k^2\widehat{n}_2(K_1\sin^2\varphi + K_2\cos^2\varphi). \tag{6.2.18}$$

The distortion stress vanishes upon linearization, and the components of passive stress (6.2.17) are linearized and Fourier-transformed as

$$\widehat{\sigma}^{\mathrm{p}}_{11} = \chi\widehat{h}_1, \quad \widehat{\sigma}^{\mathrm{p}}_{12} = \tfrac{1}{2}(\chi - 1)\widehat{h}_2, \quad \widehat{\sigma}^{\mathrm{p}}_{21} = \tfrac{1}{2}(\chi + 1)\widehat{h}_2, \quad \widehat{\sigma}^{\mathrm{p}}_{22} = 0. \tag{6.2.19}$$

It is advantageous to express the perturbation velocity $\widetilde{\mathbf{v}}$ through the stream function, $\widetilde{v}_i = \varepsilon_{ij}\partial_j\Psi$, expand the latter as $\Psi = \epsilon\widehat{\Psi}\mathrm{e}^{\mathrm{i}\mathbf{k}\cdot\mathbf{x}}$, and take the curl of (6.2.16). The contribution of the passive stress to the hydrodynamic equation is expressed then as $\varepsilon_{ik}\partial_k\partial_j\widehat{\sigma}^{\mathrm{p}}_{ij}$, which leads to the Fourier-transformed hydrodynamic equation

$$\left(\rho\mathcal{D}_t + \eta k^2\right)\widehat{\Psi} = \tfrac{1}{2}k^4\widehat{n}_2[(\chi - 1)K_1\sin^2\varphi - (\chi + 1)K_2\cos^2\varphi)]. \tag{6.2.20}$$

The transformed linearized equation (6.2.15) also defines dynamics of $\widehat{n}_1$:

$$\mathcal{D}_t\widehat{n}_1 = \Gamma\widehat{h}_1 + \tfrac{1}{2}k^2\chi\widehat{\Psi}\sin 2\varphi, \quad \mathcal{D}_t\widehat{n}_2 = \Gamma\widehat{h}_2 - \tfrac{1}{2}k^2\widehat{\Psi}(1 + \chi\cos 2\varphi). \tag{6.2.21}$$

Any non-zero n_1 causes deviation of the modulus from unity. The analysis is still justified as long as the alignment is stable, but an instability excites n_1 and, hence, changes the modulus as well.

Stability requires the eigenvalues λ of the following Jacobi matrix to have negative real parts at any φ:

$$\mathbf{J} = \begin{pmatrix} -\eta & -\frac{1}{2}\left[K_1(\chi-1)\sin^2\varphi + K_2(\chi+1)\cos^2\varphi)\right] \\ \frac{1}{2}(\chi\cos 2\varphi - 1) & -\Gamma(K_1\sin^2\varphi + K_2\cos^2\varphi) \end{pmatrix}. \quad (6.2.22)$$

Common powers of k are cancelled here in each row, so that if instability arises, it happens simultaneously on all wavelengths. The trace is always negative, so that an oscillatory instability is excluded, and the determinant is

$$\det(\mathbf{J}) = \eta\Gamma(K_1\sin^2\varphi + K_2\cos^2\varphi) + \tfrac{1}{4}(\chi\cos 2\varphi - 1)\left[K_1(\chi-1)\sin^2\varphi + K_2(\chi+1)\cos^2\varphi)\right]. \quad (6.2.23)$$

It turns out that stability is not warranted in the entire parametric domain, since the second term in (6.2.23) is not positive definite. This can be seen most clearly at $K_1 = K_2 = K$ when the determinant simplifies to $Kf(\varphi)$ with

$$f(\varphi) = \eta\Gamma + \tfrac{1}{8}\left[(\chi^2(1+\cos 4\varphi) - 2\right]. \quad (6.2.24)$$

The part of (6.2.24) excluding $\eta\Gamma$ is negative in the dark-shaded areas in Fig. 6.16 symmetric to the reversal of χ and translation by $\pi/2$. The most dangerous perturbation angle is $\varphi = \pi/4$ modulo $\pi/2$ when the second term in (6.2.24) reaches the minimal value -1/4. Thus, stability is retained at $\eta\Gamma > 1/4$, but not in the entire parametric domain, as it should be in a passive system.

This computation confirms that the standard theory (de Gennes and Prost, 1993) does not ensure relaxation to equilibrium. The likely reason why the derivation based on the Onsager relations in Sect. 6.2.1 failed to ensure stability is, as noted by Brand et al (2018), that a combination of variables even $(\mathbf{n}, \mathbf{h})$ and odd $(\mathrm{s}, \mathbf{N})$ under time reversal in (6.2.12) is "physically unsound", and cannot ensure evolution to thermodynamic equilibrium. Another factor is inadequacy of the director-based description: we shall see in Sect. 6.2.4 that perturbations of the modulus serve as an efficient stabilizing factor.

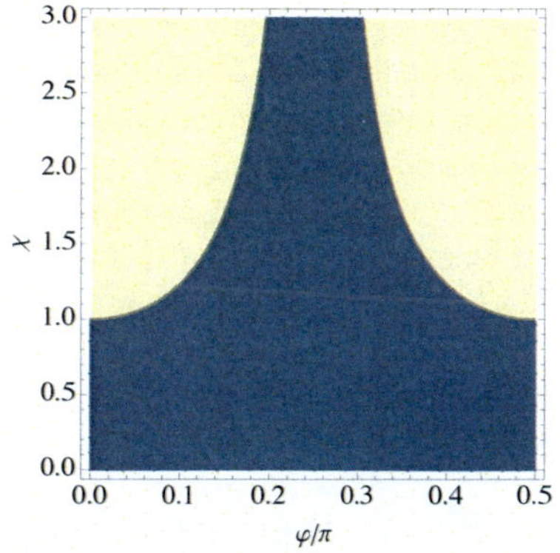

Fig. 6.16. Dark-shaded instability region in a passive system in the parametric plane φ, χ at $K_1 = K_2$

Active stress adds to the hydrodynamic equation the term $\frac{1}{2}k^2\zeta\widehat{n}_2\cos 2\varphi$ containing a lower power of k, so that it dominates the determinant at long scales, and an arbitrarily small activity causes instability at some perturbation angle φ. Instability is constrained in a domain bounded by solid walls with no-slip boundary conditions, as in computations by Edwards and Yeomans (2009).

6.2.3 Nematodynamics with Variable Modulus

Tensor representation becomes necessary when not only the orientation but the modulus of the director is variable. The tensor order parameter is expressed through twice the inclination angle, and therefore is invariant to rotation by π:

$$M_{ij} = \varrho n_i n_j, \qquad Q_{ij} = \varrho(n_i n_j - \delta_{ij}/d), \tag{6.2.25}$$

where ϱ is the modulus and d is the number of dimensions – this term serves to ensure that the tensor $\mathbf{Q}$ be traceless.

The tensor description has been pioneered by Doi (1981) and extended to include spatial inhomogeneities (Edwards and Beris, 1989) and interactions between the nematic alignment and flow (Beris and Edwards, 1994). The tensor-based nematodynamic theory has been further independently developed[2] by Qian and Sheng (1998) and Stark and Lubensky (2005). The tensor-based Landau–de Gennes Lagrangian is

$$\begin{aligned}\mathcal{L} = &\frac{\alpha_0}{2} Q_{ij}Q_{ij} + \frac{\alpha_1}{3} Q_{ij}Q_{jk}Q_{ki} + \frac{\alpha_2}{4} Q_{ij}Q_{jk}Q_{kl}Q_{li} \\ &+ \frac{K_1}{2}|\partial_k Q_{ij}|^2 + \frac{K_2}{2}|\partial_j Q_{ij}|^2 - \lambda Q_{ij},\end{aligned} \tag{6.2.26}$$

where λ is a Lagrange multiplier required to satisfy the zero trace condition $S_{ii} = 0$. Varying the energy functional $\mathcal{F} = \int \mathcal{L} d\mathbf{x}$ defines the molecular field tensor

$$h_{ij} = -\Lambda_{ij} + \partial_k \pi_{ijk}, \qquad \Lambda_{ij} = \frac{\partial \mathcal{L}}{\partial Q_{ij}}, \qquad \pi_{ijk} = \frac{\partial \mathcal{L}}{\partial_k Q_{ij}}. \tag{6.2.27}$$

Unlike (6.2.1), there are only two independent elastic constants. A third constant could be only obtained by adding a higher-order term $Q_{ij}\partial_i Q_{kl}\partial_j Q_{kl}$. It did not look contentious that different terms in (6.2.1) contain different powers of $\mathbf{n}$, but the tensor order parameter, unlike $\mathbf{n}$, has a physical reality, and we should be aware that elastic anisotropies cannot be fully captured in the framework of the Landau model restricted, as it is common in multiscale expansions (Sect. 1.1.5), to the fourth combined order in $\mathbf{Q}$ and spatial derivatives.

The tensor-based derivation by Qian and Sheng (1998) follows the same route to nematodynamic equations as the director-based method outlined in Sect. 6.2.1, leading to a relation between conjugate forces and fluxes analogous to (6.2.13), (6.2.14). The tensor analogue of (6.2.15) in the Beris–Edwards model is

$$\mathcal{D}_t\mathbf{Q} = \Gamma\mathbf{h} + (\chi\mathbf{s} + \boldsymbol{\omega})\cdot\mathbf{M} + \mathbf{M}\cdot(\chi\mathbf{s} + \boldsymbol{\omega}) - 2\chi\mathbf{M}\,\mathrm{Tr}(\mathbf{Q}\cdot\mathbf{h}). \tag{6.2.28}$$

The passive stress in (6.2.16) is expressed as

$$\boldsymbol{\sigma}^{\mathrm{p}} = \chi\left[\mathbf{M}\cdot\mathbf{h} - \mathbf{h}\cdot\mathbf{M} + 2\mathbf{M}\,\mathrm{Tr}(\mathbf{Q}\cdot\mathbf{h})\right] + \mathbf{h}\cdot\mathbf{Q} - \mathbf{Q}\cdot\mathbf{h} + \partial_k Q_{ij}\pi_{ijk}. \tag{6.2.29}$$

In the following, we restrict to 2D, which affords substantial simplifications. 2D patterns are commonly observed in thin layers with tangential alignment on containing

[2] None of these authors cited preceding work

walls, as well as in thin elastic sheets (Sect. 6.3.1) and cellular layers (Sect. 6.4.4). Computational models are commonly restricted to 2D, as 3D simulations are too cumbersome and hard to present on a 2D paper sheet. Often they are based on the Beris–Edwards 3D model, lacking particular parametrization and adjusted to 2D just by setting $d = 2$ in (6.2.25). In 2D, it is advantageous to parametrize the tensor $\mathbf{Q}$ by introducing the vector $\mathbf{q}$ with the components $q_1 = \varrho \cos 2\theta$, $q_2 = \varrho \sin 2\theta$, leading to $Q_{11} = -Q_{22} = \frac{1}{2}q_1$, $Q_{12} = Q_{21} = \frac{1}{2}q_2$.

The theory based on the Q-tensor, though well justified, has its specific problems. In 2D, the cubic term in (6.2.26) vanishes identically. The resulting Lagrangian is

$$\mathcal{L} = -\tfrac{1}{2}\alpha_0 \varrho^2 (1 - \tfrac{1}{2}\varrho^2) + \mathcal{L}_{\rm d}. \tag{6.2.30}$$

The coefficient at the fourth-order term is chosen to set the value of the modulus ϱ to unity in a perfectly aligned nematic. The parameter α_0 will be further chosen as the scale of energy per unit area and set to unity when equations are written in the dimensionless form. The two constituents of the distortion energy $\mathcal{L}_{\rm d}$ are

$$\begin{aligned}\mathcal{L}_1 &= K_1 |\partial_k Q_{ij}|^2 = \tfrac{1}{2} K_1 (|\nabla q_1|^2 + |\nabla q_2|^2),\\ \mathcal{L}_2 &= K_2 |\partial_j Q_{ij}|^2 = \tfrac{1}{4} K_2 [(\partial_1 q_1 + \partial_2 q_2)^2 + (\partial_2 q_1 - \partial_1 q_2)^2].\end{aligned} \tag{6.2.31}$$

Another possible construction, $|\varepsilon_{jk}\partial_k Q_{ij}|^2$, where ε is the antisymmetric tensor, leads to the expression identical to $\mathcal{L}_2$. Moreover, the extra terms in $\mathcal{L}_2$, $\partial_1 q_1 \partial_2 q_2 - \partial_2 q_1 \partial_1 q_2$ vanish upon variation, so that they do not contribute to the molecular field expressed as $h_{ij} = K\nabla^2 Q_{ij}$ or, equivalently, $h_i = \frac{1}{2} K \nabla^2 q_i$ with $K = K_1 + 2K_2$.

Most commonly, the "one-constant approximation" is used in simulations, assuming all elastic coefficients to be equal. Since the Landau–de Gennes Lagrangian is, in any case, a crude approximation allowing for qualitative insights, equality of bend and splay elasticities may appear to be not quite disturbing: the distinction would certainly appear in higher orders. However, this turns out to be an artifact of the tensor formalism in 2D. The tensor $\mathbf{Q}$ can be constructed by merging two vectors, $\mathbf{q}/\sqrt{2}$ and $\mathbf{q}'/\sqrt{2}$, where $q'_i = \varepsilon_{ij} q_j$. Hence, the vector $\mathbf{q}$, which is invariant under rotation by π, alone carries all necessary information (Pismen, 2022).

The distinction between bend and splay elasticities is restored in the vector-based description. In 2D, $\operatorname{curl} \mathbf{q}$ is a pseudoscalar, and the bend term is expressed without involving higher orders of $\mathbf{q}$, so that the Lagrangian based on the vector order parameter becomes

$$\mathcal{L} = -\tfrac{1}{2}\alpha_0 \varrho^2 (1 - \tfrac{1}{2}\varrho^2) + \tfrac{1}{2} K_1 (\operatorname{div} \mathbf{q})^2 + \tfrac{1}{2} K_2 (\operatorname{curl} \mathbf{q})^2, \tag{6.2.32}$$

where, for the sake of similarity to the director-based description, the elasticities K_i are multiplied by two. The explicit form of the tensor $\pi_{ij} = \partial\mathcal{L}/(\partial_i q_j)$ is

$$\boldsymbol{\pi} = \begin{pmatrix} K_1(\partial_1 q_1 + \partial_2 q_2), \; K_2(\partial_2 q_1 - \partial_1 q_2) \\ K_2(\partial_1 q_2 - \partial_2 q_1), \; K_1(\partial_2 q_2 + \partial_1 q_1) \end{pmatrix}. \tag{6.2.33}$$

This leads to the molecular field

$$
\begin{aligned}
\mathbf{h} &= \alpha_0 \mathbf{q}(1-\varrho^2) + \mathbf{h}^{\mathrm{d}}, \quad h_k^{\mathrm{d}} = \partial_i \pi_{ik}, \\
h_1^{\mathrm{d}} &= K_1(\partial_1^2 q_1 + \partial_1\partial_2 q_2) + K_2(\partial_2^2 q_1 - \partial_1\partial_2 q_2), \\
h_2^{\mathrm{d}} &= K_1(\partial_2^2 q_2 + \partial_1\partial_2 q_1) + K_2(\partial_1^2 q_2 - \partial_1\partial_2 q_1).
\end{aligned} \tag{6.2.34}
$$

Setting $K = K_1 + K_2$, $K' = K_1 - K_2$ redefines $\mathbf{h}^{\mathrm{d}}$ as

$$
\begin{aligned}
h_1^{\mathrm{d}} &= K\nabla^2 q_1 + K'[(\partial_1^2 - \partial_2^2)q_1 + 2\partial_1\partial_2 q_2], \\
h_2^{\mathrm{d}} &= K\nabla^2 q_2 + K'[(\partial_2^2 - \partial_1^2)q_2 + 2\partial_1\partial_2 q_1].
\end{aligned} \tag{6.2.35}
$$

The simple expression $\mathbf{h}^{\mathrm{d}} = K\nabla^2\mathbf{q}$ is obtained at $K' = 0$, but otherwise the distinction between the two constants is retained in the vector-based theory. This rank reduction is specific to 2D, since the bend and twist terms in 3D would require, similar to (6.2.1), a higher power of $\mathbf{q}$.

Constructing the Lagrangian based on the vector $\mathbf{q}$ in leu of the Q-tensor allows us to apply the Ericksen–Leslie formalism to systems with a variable modulus, so that all equations of Sect. 6.2.1 are retained under the replacement $\mathbf{n} \to \mathbf{q}$. The only slight modification concerns the action of rotation. Since $\mathbf{q}$ rotates with doubled director's speed, the energy-conserving relation (6.2.10) is modified to

$$
\partial_t q_i = 2\omega\varepsilon_{ij}q_j, \quad \partial_t x_i = \omega\varepsilon_{ij}x_j. \tag{6.2.36}
$$

Accordingly, the antisymmetric part of $\boldsymbol{\sigma}^{\mathrm{s}}$ is doubled to $2\Omega_{ij} = (\partial_i v_j - \partial_j v_i)$. We also take note that, since 2D vorticity is a pseudoscalar, the expression $\boldsymbol{\omega} \times \mathbf{q}$ makes no sense in 2D, and the change of $\mathbf{q}$ due to advection and rotation is presented by vector $\mathbf{N} = \mathcal{D}_t\mathbf{q} - 2\boldsymbol{\Omega}\cdot\mathbf{q}$.

Another qualification concerns higher-rank tensors accounting for viscous anisotropy, which are composed by adding higher powers of $\mathbf{n}$. A similar construction with higher powers of $\mathbf{q}$ would add higher powers of the modulus, which contradicts the restriction of the Landau–de Gennes Lagrangian to lower powers of the order parameter. Therefore it is appropriate to restrict to the lowest-order expressions. The resulting equations of vector nematodynamics modifying (6.2.15), (6.2.17) are

$$
\mathcal{D}_t q_i = -\chi s_{ij}q_j + 2\Omega_{ij}q_j + \Gamma h_i \, , \tag{6.2.37}
$$

$$
\sigma_{ij}^{\mathrm{p}} = \tfrac{1}{2}\chi(q_i h_j + h_i q_j) - (q_i h_j - h_i q_j) + \pi_{ik}\partial_j q_k \, . \tag{6.2.38}
$$

The dimensionless form of (6.2.37), (6.2.34), obtained by scaling length by the healing length $\xi = (K_1/\alpha_0)^{1/2}$, time by $\tau = (\alpha_0\Gamma)^{-1}$, and h by α_0, is

$$
\tau\mathcal{D}_t q_i = -\chi s_{ij}q_j + 2\Omega_{ij}q_j + h_i, \tag{6.2.39}
$$

$$
\begin{aligned}
h_1 &= q_1(1-\varrho^2) + (\partial_1^2 q_1 + \partial_1\partial_2 q_2) + \kappa(\partial_2^2 q_1 - \partial_1\partial_2 q_2)\, , \\
h_2 &= q_2(1-\varrho^2) + (\partial_2^2 q_2 + \partial_1\partial_2 q_1) + \kappa(\partial_1^2 q_2 - \partial_1\partial_2 q_1)\, .
\end{aligned} \tag{6.2.40}
$$

where $\kappa = K_2/K_1$. The dimensionless hydrodynamic equation (6.2.16) with stress and pressure scaled by α_0/ξ takes the form

$$\mathcal{D}_t v_i = \nu \nabla^2 v_i - \partial_i p + \partial_j \sigma^{\mathrm{p}}_{ij}, \tag{6.2.41}$$

where $\nu = \eta\alpha_0\Gamma/(\xi\rho)$ is the dimensionless kinematic viscosity and $\boldsymbol{\sigma}^{\mathrm{p}}$ is the part of the dimensionless viscous stress $\boldsymbol{\sigma}^{\mathrm{s}}$ related to nematic alignment defined by (6.2.38).

A quasi-2D formulation of a problem in the 3D world is possible in a thin layer $0 < z < l$ with tangential alignment on the upper and lower boundaries. The nematic alignment is constant across a thin layer, but the velocity is constant only in a nonrealistic case of free-slip boundary conditions. They might be realized in a suspended film, which is hard to sustain in any fluid, and is known to be unstable in nematics. In the most common case of solid confining walls, no-slip boundary conditions apply, leading to the parabolic velocity profile

$$\mathbf{v}(z) = \tfrac{1}{6}\mathbf{v}_0 l^{-2} z(l - z), \tag{6.2.42}$$

where $\mathbf{v}_0$ is the average in-plane velocity. The momentum flux to the walls $-2\eta\partial_z v_{|z=0} = \frac{1}{3}\eta v_0 l^{-1}$ is compensated by the force $\mathbf{F} = -\nabla \cdot (p + \boldsymbol{\sigma})$ driving fluid's motion. This defines the average velocity $\mathbf{v}_0 = 3\mathbf{F}l/\eta$, and the total flux is $\mathbf{j} = \mathbf{F}l^2/\eta$, which is equivalent to Darcy's law with the coefficient $\gamma = 3l^2/\eta$. Accordingly, the hydrodynamic equation (6.2.16) is replaced by the frictional equation

$$\rho\mathcal{D}_t v_i = -\eta' v_i - \partial_i p + \partial_j \sigma^{\mathrm{p}}_{ij} \tag{6.2.43}$$

containing the effective friction coefficient $\eta' = \frac{1}{3}\eta/l^2$. The dimensionless version (6.2.41) is

$$\mathcal{D}_t v_i = -\nu' v_i - \partial_i p + \partial_j \sigma^{\mathrm{p}}_{ij} \tag{6.2.44}$$

with $\nu' = \eta'\alpha_0\Gamma/(\xi\rho)$. The frictional dynamics with a friction coefficient dependent on properties of the particles and the substrate is also observed in an adsorbed nematic layer.

6.2.4 Instability of Flow and Nematic Alignment

Adjusting the analysis in Sect. 6.2.2 to the case of a variable modulus, we start again with testing a passive system and consider perturbations of an ordered state $q_1^0 = 1,\ q_2^0 = 0$ in an unbounded 2D domain. Since the circular symmetry allows, as in Sect. 6.2.2, for a 2D description even in the 3D world, it is possible to neglect friction for the time being in order to see most clearly the effect of a variable modulus. We will work here with the dimensionless variables and parameters defined above. As before, deviations from the base state $\epsilon\widetilde{q}_i$ are expanded in the Fourier series $\widetilde{\mathbf{q}} = \epsilon\widehat{\mathbf{q}}e^{i\mathbf{k}\cdot\mathbf{x}}$. Unlike Sect. 6.2.2, the molecular field has now an algebraic part $q_i(1 - \varrho^2)$, reduced to a single component $\widehat{\mathbf{h}}_1^0 = -2\epsilon\widehat{q}_1$. The distortion part (6.2.34) is transformed as $\widehat{\mathbf{h}}^{\mathrm{d}} = -\epsilon k^2 \mathbf{H} \cdot \widehat{\mathbf{q}}$ with

$$\mathbf{H} = \begin{pmatrix} \cos^2\varphi + \kappa\sin^2\varphi, & \frac{1}{2}(1-\kappa)\sin 2\varphi \\ \frac{1}{2}(1-\kappa)\sin 2\varphi, & \sin^2\varphi + \kappa\cos^2\varphi \end{pmatrix}. \tag{6.2.45}$$

The passive stress is linearized and Fourier-transformed as

$$\widehat{\sigma}^{\mathrm{P}}_{ij} = \tfrac{1}{2}\chi(q^0_i \widehat{h}_j + \widehat{h}_i q^0_j) - (q^0_i \widehat{h}_j - \widehat{h}_i q^0_j). \tag{6.2.46}$$

After taking the curl of (6.2.41), the contribution of the passive stress to the Fourier-transformed hydrodynamic equation is expressed as $\varepsilon_{ik}\partial_k\partial_j\widehat{\sigma}^{\mathrm{P}}_{ij} = -k^2\mathbf{f}\cdot\widehat{\mathbf{q}}$, where

$$\mathbf{f} = \begin{pmatrix} \chi \sin 2\varphi \\ 0 \end{pmatrix} + \frac{k^2}{2}\begin{pmatrix} \frac{1}{2}[\chi - 2 + \kappa(\chi+2)]\sin 2\varphi \\ (\chi-2)\sin^2\varphi - \kappa(\chi+2)\cos^2\varphi \end{pmatrix}. \tag{6.2.47}$$

The resulting concise form of the hydrodynamic equation is

$$(\partial_t + \nu k^2)\widehat{\Psi} = \mathbf{f}\cdot\widehat{q}. \tag{6.2.48}$$

Turning to the dynamic equation of $\widehat{q}$, we compute the Fourier transform of (6.2.9)

$$\widehat{\mathbf{s}} = \tfrac{1}{2}\epsilon k^2\widehat{\Psi}\begin{pmatrix} -\sin 2\varphi & \cos 2\varphi \\ \cos 2\varphi & \sin 2\varphi \end{pmatrix}, \qquad \widehat{\boldsymbol{\Omega}} = \tfrac{1}{2}\epsilon k^2\widehat{\Psi}\varepsilon, \tag{6.2.49}$$

to obtain the transform of the linearized equation (6.2.39)

$$\tau\partial_t\widehat{\mathbf{q}} = \widehat{\mathbf{h}} + \tfrac{1}{2}k^2\mathbf{g}\widehat{\Psi}, \qquad \mathbf{g} = \begin{pmatrix} \chi\sin 2\varphi \\ -(2+\chi\cos 2\varphi) \end{pmatrix}. \tag{6.2.50}$$

Stability requires the eigenvalues λ if the following matrix to have negative real parts:

$$\mathbf{J} = \begin{pmatrix} -k^2\nu/\tau, & -f_1/\tau, & -f_2/\tau \\ \frac{1}{2}k^2 g_1, & -2-k^2H_{11}, & -k^2H_{12} \\ \frac{1}{2}k^2 g_2, & -k^2H_{21}, & -k^2H_{22} \end{pmatrix}. \tag{6.2.51}$$

The Routh–Hurwitz stability criterion is $a_0 > 0$, $a_1a_2 - a_0 > 0$, where a_j are the coefficients of $\mathrm{Det}(\lambda\mathbf{I} - \mathbf{J}) = \sum_0^3 a_j\lambda^j$; $\mathbf{I}$ is the identity matrix. The full expressions are quite involved, but a transparent expression for a_0 is obtained in the one-constant approximations $\kappa = 1$:

$$\tau a_0 = k^4(1+\tfrac{1}{2}k^2)\left[2\nu + \tfrac{1}{4}\chi^2 + (1+\chi\cos 2\varphi)\right]. \tag{6.2.52}$$

The last two terms in the bracketed expression reach the zero minimum at $\varphi = \pi/2$, so that $a_0 > 0$ and the system is stable to monotonic perturbations under this assumption. The stabilizing factor is the lower degree of k in f_1 and in the matrix element J_{22} contributed by perturbations of the modulus. However, detailed analysis reveals narrow parametric areas where instability is not excluded at low friction, which reminds again of the problematic mixing of variables with different time-reversal symmetry. The oscillatory stability criterion is still more involved, but the leading term determining the long-scale stability is definitely positive:

$$a_1a_2 - a_0 = k^2[4(\nu + \sin^2\varphi + \kappa\cos^2\varphi) + \tfrac{1}{2}\tau^{-1}\chi^2\sin^2 2\varphi + O(k^4). \tag{6.2.53}$$

Dimensionless active stress $\sigma^{\rm a} = -\zeta Q_{ij}$ adds to the hydrodynamic equation (6.2.48) the term $\frac{1}{2}k^2\zeta(\widehat{q}_1 \sin 2\varphi - \widehat{q}_2 \cos 2\varphi)$, which modifies the entries in the first row of the matrix $\mathbf{J}$ to

$$J_{12} = (\tfrac{1}{2}\zeta \sin 2\varphi - f_1),$$
$$J_{13} = -(\tfrac{1}{2}\zeta \cos 2\varphi + f_2). \quad (6.2.54)$$

This contributes to $\det(\mathbf{J})$ the $O(k^2)$ leading term $\frac{1}{2}\zeta \cos 2\varphi(2 + \chi \cos 2\varphi)$, so that, as in Sect. 6.2.2, an arbitrary small activity always causes instability at some perturbation angle φ in an unconstrained system.

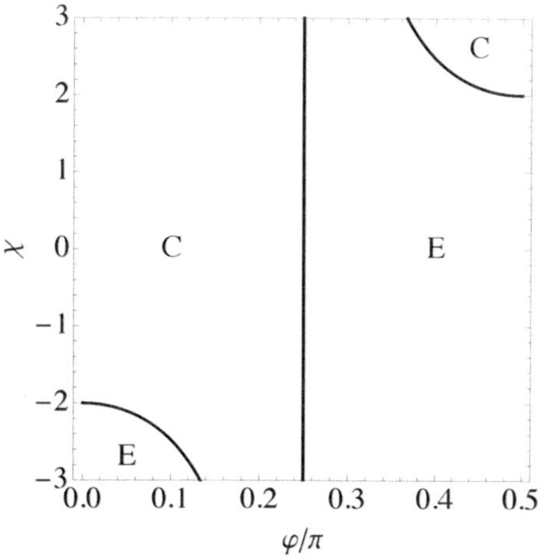

Fig. 6.17. Domains of instability due to contractile (C) and extensile (E) activity

This is corrected in a realistic case of a confined geometry. The setup most relevant for applications is a thin layer with frictional flow described by (6.2.44). The only, but most substantial, difference in the stability analysis is the loss of the Laplacian in the frictional term, so that the matrix element J_{11} in (6.2.51) switches to ν'. Then the leading term $O(k^2)$ term in the criterion of monotonic instability is

$$a_0 = k^2 \left[\tfrac{1}{4}\zeta\tau^{-1} \cos 2\varphi(2 + \chi \cos 2\varphi) + 2\nu'(\cos^2 \varphi + \kappa \sin^2 \varphi\right]. \quad (6.2.55)$$

In the one-constant approximation $\kappa = 1$, the determinant vanishes at

$$\frac{\zeta}{\nu'\tau} = -\frac{8}{\cos 2\varphi(2 + \chi \cos 2\varphi)} \quad (6.2.56)$$

The instability domain switches between the extensile ($\zeta > 0$) and contractile ($\zeta < 0$) activity at $\phi = \pi/4\,(\mathrm{mod}\pi/2)$ and $\chi = -2/\cos 2\varphi$ with activity required for instability diverging at these borders. The long-scale oscillatory instability is excluded, as the leading term $2\nu'(2 + \nu')$ is purely frictional.

6.2.5 Topological Defects

As instabilities develop, nematic alignment becomes disordered, but perfect order is hard to attain in the first place, even in passive nematics. When an isotropic liquid is "frozen" into the nematic state, different orientations may be chosen at different locations as a consequence of rotational symmetry, in the same way as when incipient striped pattern develops with different orientations at removed locations or when a wave pattern emanates from a number of active spots. When these domains come into contact, various kinds of defects and imperfections arise on their boundary. An equilibrium system possessing a gradient structure evolves to a minimum of the applicable energy functional, but even then the resulting stationary pattern may retain defects, if they are unavoidable for *topological* reasons.

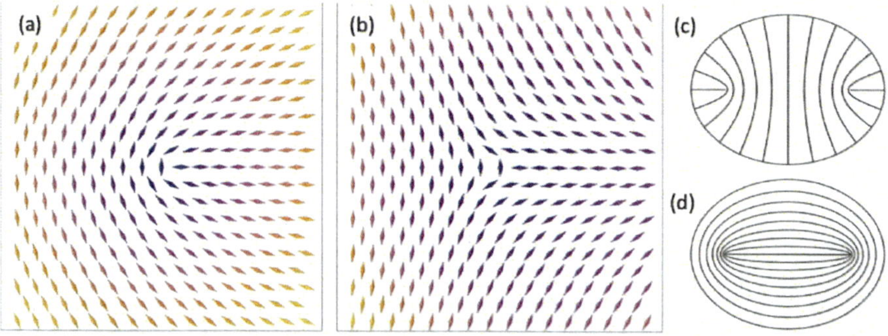

Fig. 6.18. The structure of a positive (a) and negative (b) half-charged defects. Nematic texture in an ellipse with normal (c) and parallel (d) anchoring at the boundary.

The minimal number of topological defects is determined by the nature of the order parameter and either boundary conditions or topology of a closed manifold. For example, if any 2D surface is enclosed by smooth line with the boundary conditions specifying a fixed alignment angle of the director or a vector order parameter, the total circulation around the boundary equals 2π. This must be compensated by defects with equal total circulation. The *charge* of a defect equals the total circulation of the order parameter around an enclosing contour, divided by 2π, and its minimal absolute value, determined by the symmetry of the order parameter, equals 1 for the vectorial and 1/2 for the nematic order. The energy of a defect usually grows with its charge (with some exceptions, which may be caused by anisotropies of the medium), therefore their charges should be minimal. In the case of nematic order, two positive half-charge defects are necessary. If a hole is cut within the enclosed area, the circulation on its boundary, viewed from the inside, is negative, and no defects are necessary. With two holes, two defects with the charge $-1/2$ are required to compensate the rotation.

The nematic texture containing defects can be visualized most easily within an ellipse. Lines with the circulation of the director by π anchored normal or parallel to the boundary coincide with lines v = const or u = const in the elliptic coordinate system, which are defined as $x = \cosh v \cos u$ and $y = \sinh v \sin u$, respectively. The two half-charged defects are located at the foci of the ellipse (Fig. 6.18c,d). The structure of positive and negative defects is completely different. The $+1/2$ defect, schematized in Fig. 6.18a, is asymmetric, with a "comet tail" radiating from its core. The "comet tails" of defects are directed toward the boundary in Fig. 6.18c and connect at the center in Fig. 6.18d. Contrariwise, the $-1/2$ defect (Fig. 6.18b) has the threefold symmetry.

On a surface without a boundary with the topology of a sphere, the total circulation is 4π, which should be compensated by four $+1/2$ defects. A pattern on a sphere is not easy to visualize on a flat page, but the lowest energy nematic pattern on a sphere can be expressed analytically when the bend and splay elastic constants are assumed to be equal (Vitelli and Nelson, 2006). The texture is shown in the the Mercator projection (grossly distorting distances close to the poles) in Fig. 6.19a. The defects

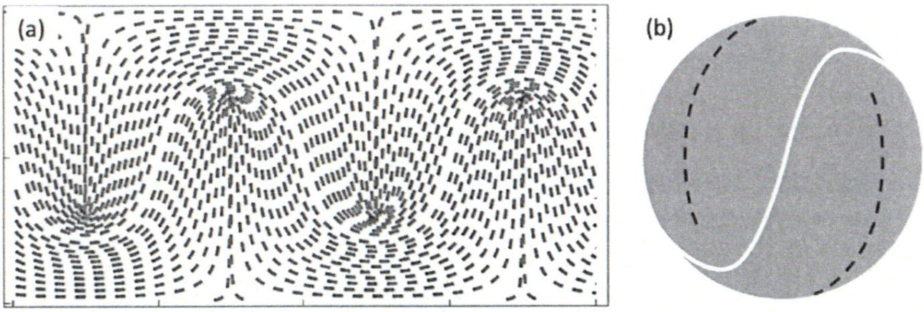

Fig. 6.19. Nematic texture on a sphere in the Mercator projection (a) and on a tennis ball (b). The dashed line shows the "comet tails" of the defects, and the white line, the seam on the ball

are situated at the vertices of a tetrahedron inscribed in the sphere, and the white line on both panels goes as a seam on a baseball (Fig. 6.19b). The symmetry can be distorted due to high elastic anisotropy (Shin et al, 2008), which may cause just two unit-charged defects to have the lowest energy. A wider physicist's perspective of topological defects is found in a review by Mermin, 1979 and at a more elementary level in my book (Pismen, 1999).

In active media, pairs of defects emerge spontaneously, which leads to turbulent dynamics. Thumpi et al (2014) simulated patterns in a 2D active nematic fluid using simplified Ericksen–Leslie equations, omitting elastic and viscous anisotropies. This does not affect in a qualitative way the principal features of active flow brought about by the simple input σ^{a}. It is the active stress that brings the entire system to life. High extensile activity brings about turbulent dynamics with spontaneously

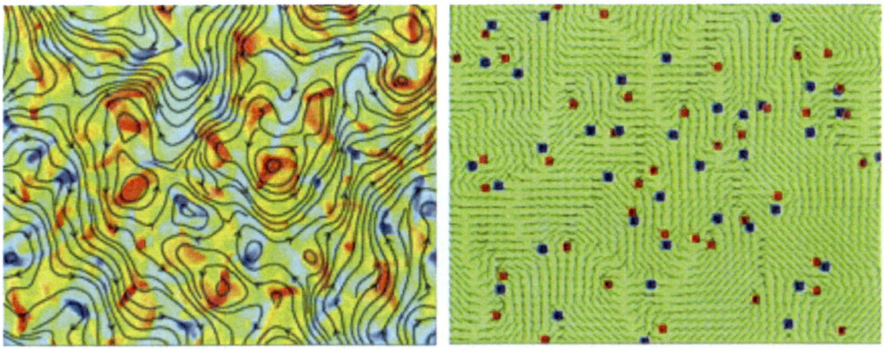

Fig. 6.20. Snapshot of the flow (left) and alignment (right) patterns in a nematic fluid with high extensile activity. In the left panel, lines with arrows indicate streamlines; the coloring is graded from red to blue, which correspond to a high positive or negative vorticity, respectively. Small bars in the right panel show the director field; positive half-charged defects are marked by red, and negative, by blue dots (Thumpi et al, 2014, reproduced by permission, courtesy Prof. Julia Yeomans)

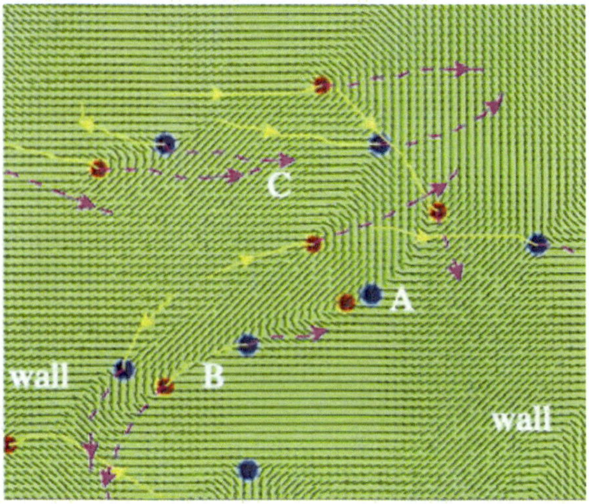

Fig. 6.21. Snapshots of defect dynamics in an extensile system. Positive half-charged defects are marked by red, and negative, by blue dots. The past and future trajectories of the defects are shown by continuous yellow and dashed magenta lines, respectively. The labels mark: A – a newly created defect pair; B – a pair of defects moving from their creation site along a wall; C – a future annihilation site (Thumpi et al, 2014, reproduced by permission, courtesy Prof. Julia Yeomans)

emerging, moving, annihilating and emerging again topological defects. A snapshot of the simulation of the alignment and flow field is shown in Fig. 6.20. The picture for a contractile activity is qualitatively similar, but it turns out to contain a lower density of defects at an equal level of activity. In the blowup picture of this simulation shown in Fig. 6.21, trajectories of defects are traced by yellow arrows from their creation sites, where defect pairs are born. Some future trajectories traced by magenta arrows terminate in annihilation of defect pairs. The picture also indicates "walls", which are elongated distortions in the director field.

6.2.6 Interactions of Half-Charged Defects

The dynamic equation for a half-charge defect in the absence of flow is defined by using $\mathbf{q} = \frac{1}{2}\varrho(r)\{\cos\phi, \pm\sin\phi\}$ in (6.2.34). The curl of $\mathbf{q}$ vanishes, and the equation of the modulus depends only on the splay elasticity K_1:

$$\Gamma \mathcal{D}_t \varrho = \tfrac{1}{2}\alpha_0 \varrho(1-\varrho^2) + \tfrac{1}{2}K_1[\varrho''(r) + r^{-1}\varrho'(r) - r^{-2}\varrho(r)]. \tag{6.2.57}$$

The dimensionless form obtained by scaling time by $\frac{1}{2}(\alpha_0\Gamma)^{-1}$ and length by the healing length $\xi = (K_1/\alpha_0)^{1/2}$ is

$$\mathcal{D}_t \varrho = \varrho(1-\varrho^2) + \varrho''(r) + r^{-1}\varrho'(r) - r^{-2}\varrho. \tag{6.2.58}$$

The dynamic equation for the complex order parameter $\psi = \varrho \mathrm{e}^{\pm 2i\theta}$ coincides with the Ginzburg–Landau equation (4.2.14) that describes a vortex of unit charge.

The motion and interactions of nematic defects with the core structure given by (6.2.58) has been studied analytically (Pismen, 2013) using the same method of matched asymptotic expansion of the inner (core) and outer (long-scale) solutions as interaction of disclinations in striped patterns in Sect. 4.3. Like there, the method is restricted to the case when defects are removed at a distance far exceeding the core size, but additional complications come from the necessity to take into account self-propulsion and fluid motion induced by active defects. Interaction of defects due to the phase field, which coincides in this case with the orientation field, is identical to interaction of vortices of unit charge, and the results of Sect. 4.3 apply here as well. However, the additional factors working in nematics is the flow induced by moving defects and self-propulsion due to activity. The phase interaction between two defects always induces motion along the line connecting their locations, repelling or attracting for defects of the same or opposite charge, respectively. Interactions turning the orientation of defects should be important even if they decay with distance faster than the phase interaction. Therefore we need to consider in detail hydrodynamic interactions due to induced flow.

As noted in Sect. 6.2.3, there are two versions of 2D hydrodynamics, either free, (6.2.16), or frictional, (6.2.43). In both cases, the inertial term $\mathcal{D}_t\mathbf{v}$ is negligible at small velocities, and it is advantageous to express velocity through the stream function Ψ. Taking the curl of the Stokes equation to eliminate pressure leads to the dimensionless equation

$$\nabla^4\Psi = \gamma\nabla\times\mathbf{F}(\mathbf{x}), \qquad \mathbf{F}(\mathbf{x}) = \nabla\cdot\left(\boldsymbol{\sigma}^{(\mathrm{p})} + \boldsymbol{\sigma}^{\mathrm{a}}\right), \tag{6.2.59}$$

where velocity is scaled by $\Gamma\alpha_0/\xi$, stress, by α_0, and $\gamma = (\eta\Gamma)^{-1}$ is the dimensionless hydrodynamic mobility. The dimensionless Darcy's equation of flow dominated by friction is similarly transformed to

$$\nabla^2\Psi = -\gamma'\nabla\times\mathbf{F}(\mathbf{x}), \qquad \gamma' = 3(l/\xi)^2\gamma = 3(l/\xi)^2(\eta\Gamma)^{-1}. \tag{6.2.60}$$

We start with computing the contribution of the active stress in its standard form described at length in a different way by Angheluta et al (2021) and Rønning et al (2022). In the case of frictional dynamics, the equation of the stream function is

$$\nabla^2\Psi = -\zeta\gamma'\varepsilon_{ik}\partial_k\partial_j Q_{ij} = \zeta\gamma'\left[\partial_1\partial_2 q_1 + \tfrac{1}{2}(\partial_1^2 - \partial_2^2)q_2\right]. \tag{6.2.61}$$

It is rewritten in polar coordinates for a positive and a negative defect as

$$\begin{aligned} &\nabla^2\Psi_+ = -\tfrac{1}{4}\zeta\gamma' g_+(r)\sin\phi, \quad \nabla^2\Psi_- = -\tfrac{1}{4}\zeta\gamma' g_-(r)\sin 3\phi, \\ &g_+(r) = \varrho''(r) + r^{-1}\varrho'(r) - r^{-2}\varrho, \quad g_-(r) = \varrho''(r) - 3r^{-1}\varrho'(r) + 3r^{-2}\varrho. \end{aligned} \tag{6.2.62}$$

The symmetry of $\Psi_\pm$ reflects the symmetry of the defect cores (Fig. 6.18): axial for the positive and threefold for the negative defect, and the solutions are expressed as

$$\begin{aligned} &\Psi_+ = G_+(r)\sin\phi, \quad \Psi_- = G_-(r)\sin 3\phi, \\ &G_+''(r) + r^{-1}G_+'(r) - r^{-2}G_+(r) = -\tfrac{1}{4}\zeta\gamma' g_+(r), \\ &G_-''(r) + r^{-1}G_-'(r) - 9r^{-2}G_-(r) = -\tfrac{1}{4}\zeta\gamma' g_-(r). \end{aligned} \tag{6.2.63}$$

The resulting local velocities for positive and negative defects are

$$
\begin{aligned}
v_x^+ &= \partial_y \Psi_+ = r^{-1} G_+(r) \cos^2\phi + G'_+(r) \sin^2\phi, \\
v_y^+ &= -\partial_x \Psi_+ = \tfrac{1}{2} \sin 2\phi \left[r^{-1} G_+(r) - G'_+(r) \right], \\
v_x^- &= \partial_y \Psi_- = \tfrac{3}{2} r^{-1} G_-(r)(\cos 2\phi + \cos 4\phi) + \tfrac{1}{2} G'_-(r)(\cos 2\phi - \cos 4\phi), \\
v_y^- &= -\partial_x \Psi_- = -\tfrac{1}{2} \sin 2\phi \left[3r^{-1} G_-(r)(1 - 2\cos 2\phi) + G'_-(r)(1 - 2\cos 2\phi) \right].
\end{aligned}
\tag{6.2.64}
$$

As expected by symmetry, all velocities except v_x^+ vanish upon angular averaging, while the positive defect is self-propelled along the x axis, as if pushed or pulled by its "comet tail" due to extensile or contractile activity. Serendipitously (or due to a deeper reason), the definition of $g_+(r)$ in (6.2.62) makes it clear that $G_+(r)$ is proportional to $\varrho(r)$. The propagation speed induced by the active stress, as measured at the origin using the numerical value $\varrho'(0) \approx 0.583$, is $-0.146\zeta\gamma'$. Another reasonable choice for an estimate of the propagation speed of the defect as a whole is the average velocity that gives a more than fivefold larger value:

$$
\langle v_x^+ \rangle = -\tfrac{1}{4}\pi\zeta\gamma' \int_0^\infty [rG'_+(r) + G_+(r)]\mathrm{d}r = -\tfrac{1}{4}\pi\zeta\gamma'[\varrho(\infty) - \varrho(0)] = -\tfrac{1}{4}\pi\zeta\gamma'.
\tag{6.2.65}
$$

The difference should be attributed to reshaping of the core structure by local flow. All velocities $\mathbf{v}^\pm$ decay in the far field as r^{-1}, at the same rate as the alignment field, so that their influence is substantial. Some typical trajectories of defects driven by the alignment gradient combined with self-propulsion of positive defects are shown in Fig. 6.22. In a large area, pairs do not necessarily annihilate but may be driven outside the interaction range because of a suitably directed self-propulsion.

Active flow with free-slip boundary conditions is described by (6.2.62) with $\nabla^2\Psi_\pm$ replaced by $\nabla^4\Psi_\pm$. Respectively, the Laplacian should be applied to the left-hand sides of (6.2.63). We can formally retain (6.2.64) but redefine $G_\pm(r)$, so that we have, replacing $g_\pm(r)$ in (6.2.63) by $G_\pm(r)$ as defined before, for a positive defect

$$
\begin{aligned}
&G''_+(r) + r^{-1}G'_+(r) - r^{-2}G_+(r) \\
&= \varrho''''(r) + 2r^{-1}\varrho'''(r) - 3r^{-2}\varrho''(r) + 3r^{-3}\varrho'(r) - 3r^{-4}\varrho(r).
\end{aligned}
\tag{6.2.66}
$$

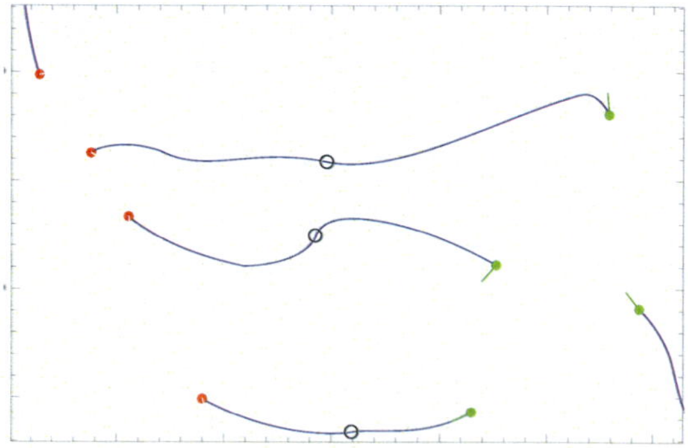

Fig. 6.22. Typical trajectories starting from 4 positive and 4 negative defects scattered randomly within two adjacent unit squares. The "comet tails" of $+\frac{1}{2}$ defects, marked near their initial positions, are directed at random angles distributed within the intervals $(-\pi/2, \pi/2)$. The annihilation sites (the earliest below and the latest above) are marked by circles (Pismen, 2013, ©The American Physical Society)

The right-hand side decays in the far field as r^{-4}, so that $G_+(r) \asymp r^{-2}$, and the velocity decays as r^{-1}, at the same rate as in friction-dominated flow. Flow due to passive stress dependent on the molecular field $\mathbf{h}$ defined in its dimensionless form by the right-hand side of (6.2.58) decays in the far field at a higher rate, since $\mathbf{h} \asymp r^{-2}$ at $r \to \infty$, as compared to $\mathbf{q} \asymp 1 - r^{-2}$, and therefore plays no substantial role in far-field interactions. Recall, however, that this case is not realistic.

6.3 Nematic Elastomers

6.3.1 Nematoelastic Interactions

Nematic elastomers, a soft elastic medium with nematic order, were first envisaged by de Gennes (1975) and soon synthesized (Finkelmann et al, 1978) by polymerizing nematic molecules (mesogen moieties). This material is attracting ever growing attention due to its ability to undergo deformations that can be triggered and controlled by temperature, applied electric fields, irradiation, or chemical transformations. Strong coupling between the director orientation and mechanical deformations leads to phase transitions and instabilities (Warner and Terentjev, 2003). These properties, approaching flexibility and mechano-responsiveness of biological tissues, can be exploited for the construction of actuators operating as artificial muscles (Hébert et al, 1997). Instabilities and restructuring of the kind observed in nematic elastomers, not driven by internal energy sources, can be characterized as *geometric* activity, employed also by plants, in particular, for seed dispersal (see e.g. Harrington et al, 2011).

An equilibrium state should minimize the energy functional dependent on the deformation tensor F_{ij} and the nematic tensor order parameter Q_{ij}. Since the mechanical and nematic elastic energies are strongly interconnected, separating them is a non-trivial problem resolved in alternative ways in different publications; here we follow Köpf and Pismen (2013a). Within this framework, it is necessary to keep track of both the unperturbed reference state $\mathbf{X}$ and the current deformed state $\mathbf{x}(t)$. The energy corresponding to a given elastic deformation, determined by comparing these two configurations, depends on the deformation tensor $F_{ij} = \partial x_i/\partial X_j \equiv \widehat{\partial}_j x_i$, where $\widehat{\partial}_j$ denotes the derivative with respect to the reference coordinates X_j, to distinguish it from the derivative ∂_j in the current (deformed) coordinate frame.

Following an intrinsic transformation, e.g., due to growth or a change of nematic order, the material will not return to its original configuration, which is then no longer stress-free. Deformations due to intrinsic transformations can be separated from elastic deformations with the help of the formalism developed in the context of elasticity combined with growth (Rodriguez et al, 1994), where the deformation tensor is decomposed into the product $F_{ij} = A_{ik}G_{kj}$, where A_{ij} denotes the elastic deformation tensor and G_{ij}, the growth tensor. The latter plays the same role as the tensor defining extension or contraction in the directions parallel and perpendicular to the director following a change of the order parameter.

We will consider a thin incompressible flat layer with tangential alignment on non-deformable upper and lower boundaries, as in Sect. 6.2.4. Under these conditions, transverse deformations are negligible compared to planar ones, which, in turn, can be assumed to be uniform in the vertical direction. The elastic energy of the layer is expressed through the 2D strain tensor $u_{ij} = \frac{1}{2}(A_{ij} + A_{ji}) - \delta_{ij}$:

$$\mathcal{L}_{\rm el} = \tfrac{1}{2}\nu|u_{ij}|^2 - p\det \mathbf{A}, \tag{6.3.1}$$

where ν is the shear modulus, and pressure p serves as the Lagrange multiplier enforcing the incompressibility condition $\det \mathbf{A} = 1$. The nematic energy per unit thickness is expressed by the Landau–de Gennes Lagrangian (6.2.30), (6.2.31). Mechanical deformations appear at a first sight to be absent in these formulas, but they are hidden in the gradient operator denoting the derivatives in the *deformed* frame x_i. This brings about an implicit dependence on the growth factor contingent on deviations of the order parameter $\widetilde{Q}_{ij}$ or $\widetilde{q}_i$ from a reference level $\mathbf{Q}_0$ or $\mathbf{q}_0$ in the stress-free state:

$$G_{ij} = \gamma^{-1/2}(\delta_{ij} + g\widetilde{Q}_{ij}). \tag{6.3.2}$$

The coefficient g describes the coupling between the nematic order parameter and the shape change. The elastomer expands parallel to the director and contracts perpendicular to it when $g > 0$, and vice versa for $g < 0$. Both situations can be realized, depending on the molecular links between mesogen moieties and the polymer (Ohm et al, 2010). The normalization factor γ is introduced to satisfy the incompressibility condition $\mathrm{Tr}(\mathbf{G}) = 2$. Since $\mathbf{Q}$ is traceless, the deviation of γ from unity is of the second order in $\widetilde{\mathbf{Q}}$:

$$\gamma = 1 + \tfrac{1}{2}g^2 Q_{ij}Q_{ij} = 1 + \tfrac{1}{4}g^2(\widetilde{q}_1^2 + \widetilde{q}_2^2). \tag{6.3.3}$$

The resulting expression for the tensor $\mathbf{A}$ is, up to the second order in $\widetilde{\mathbf{Q}}$, is

$$A_{kj} = F_{ki}(G^{-1})_{ij} = F_{kj}\left(1 + \tfrac{3}{2}g^2\widetilde{Q}_{il}\widetilde{Q}_{il}\right) - gF_{ki}\widetilde{Q}_{ij}. \tag{6.3.4}$$

Evolution to equilibrium follows the same kind of gradient dynamics as in nematic fluids (Sect. 6.2.3). In contrast to wave propagation common to elastic materials, this kind of dynamics is characterized by diffusive spreading of mechanical perturbations. The relaxation dynamics is overdamped, so that velocity rather than acceleration is proportional to the force, and the momentum is not conserved, being dissipated on the microscale by internal friction accompanying displacement and deformation of polymer chains. The mechanical driving force is the elastic stress $\boldsymbol{\sigma}^{\rm el}$. It is a two point tensor that relates forces acting on the current deformed state to the unperturbed state, and is generally not symmetric. Computations in the Lagrangian frame are facilitated by using, instead of the common Cauchy stress tensor (applicable at small deformations), the Piola–Kirchhoff stress $\boldsymbol{\sigma}$:

$$\sigma^{\rm el}_{ij} = \varsigma_{ik}(G^{-1})_{jk}, \quad \varsigma_{ij} = \partial\mathcal{L}_{\rm el}/\partial u_{ij} = \nu u_{ij}. \tag{6.3.5}$$

The complementing force comes from the nematic distortion stress $\sigma^{\rm d}_{ij} = -\pi_{ik}\partial_j q_k$, where the tensor $\boldsymbol{\pi}$ is defined by (6.2.33). This leads to the dynamic equation

$$\Gamma_{\rm el}^{-1}\partial_t x_i = \partial_i(\mathcal{L}_{\rm el} + \mathcal{L}_{\rm d}) = \widehat{\partial}_j \left[\varsigma_{ik}(G^{-1})_{jk} + \sigma^{\rm d}_{ij}\right] - \partial_i(p \det \mathbf{A}) \tag{6.3.6}$$

where $\Gamma_{\rm el}$ is the effective elastic mobility.

Although the molecular field (6.2.34) formally remains unchanged, its distortion part is tied to deformations, since all derivatives are computed in the current (deformed) coordinate frame. The direct effect on the dynamics of the nematic order parameter comes from replacing the term containing s_{ij} in (6.2.39) by the deformation-induced input, which brings it to the form

$$\partial_t q_i = \Gamma(h_i - \partial\mathcal{L}_{\rm el}/\partial q_i). \tag{6.3.7}$$

The last term, which is absent in a non-deformable material, is computed using (6.3.5) together with the relation following from (6.3.4):

$$\partial A_{kl}/\partial Q_{ij} = -g\delta_{jl}F_{ki} + 3g^2F_{kl}Q_{ij}. \tag{6.3.8}$$

Using this in (6.3.1) yields

$$\begin{aligned} \frac{\partial\mathcal{L}_{\rm el}}{\partial\widetilde{Q}_{ij}} &= \tfrac{1}{2}\nu\varsigma_{kl}\frac{\partial(A_{kl}+A_{lk})}{\partial\widetilde{Q}_{ij}} - p\frac{\partial \det \mathbf{A}}{\partial\widetilde{Q}_{ij}} \\ &= \varsigma_{kl}[-\tfrac{1}{2}(g\delta_{lj}F_{ki} + g\delta_{kj}F_{li}) + \tfrac{3}{2}g^2(F_{kl}+F_{lk}) - 2pg^2\widetilde{Q}_{ij}\det\mathbf{F}, \end{aligned} \tag{6.3.9}$$

which is expressed in terms of the vector $\widetilde{q}$ as

$$\frac{\partial\mathcal{L}_{\rm el}}{\partial\widetilde{q}_i} = \varsigma_{kl}\left[\tfrac{1}{2}gf_{ikl} + \tfrac{3}{8}g^2(F_{kl}+F_{ik})\widetilde{q}_i\right] - pg^2\widetilde{q}_i\det\mathbf{F}, \tag{6.3.10}$$

$$f_{1kl} = \begin{pmatrix} F_{11} & \frac{1}{2}(F_{21}-F_{12}) \\ \frac{1}{2}(F_{21}-F_{12}) & -F_{22} \end{pmatrix}, \; f_{2kl} = \begin{pmatrix} F_{12} & \frac{1}{2}(F_{11}+F_{22}) \\ \frac{1}{2}(F_{11}+F_{22}) & F_{21} \end{pmatrix}.$$

This expression is plugged in (6.3.7) together with h_i defined in the current frame by (6.2.34).

6.3.2 Doped Elastomers

The nematic-isotropic transition may be shifted by varying the concentration c of a nonmesogenic component in a doped liquid-crystalline medium (Rey, 2010). This may lead to the emergence of a biphasic region combining isotropic and nematic phases. Biphasic patterns would be easily disrupted by convection in liquid nematics, and are more likely to be observed in nematic elastomers. The concentration Lagrangian expressing the direct contribution of the composition to the free energy is

$$\mathcal{L}_{\rm c} = \tfrac{1}{2}\Lambda|\nabla c|^2 + \tau\left(c\ln c + (1-c)\ln(1-c)\right). \tag{6.3.11}$$

The total number density is taken here as the concentration scale; τ is the thermal energy per unit area scaled by the nematic energy scale α_0 introduced in (6.2.30). This is the simplest expression neglecting the mixing energy and retaining only the rigidity and entropic terms. The first term, containing the rigidity coefficient Λ acts within a narrow interphase boundary region (see Sect. 2.3.1).

The direct effect of a variable admixture concentration on the nematic order parameter is reducing the coefficient at the quadratic term, equal to unity in (6.2.30), to $1-\alpha_c c$. Another addition to the nematic Lagrangian is the interaction term coupling the gradients of the concentration and the nematic order parameter (Rey, 2004):

$$\mathcal{L}_{\rm i} = -\beta \partial_i c \partial_j Q_{ij} = \tfrac{1}{2}\beta[\partial_1 c(\partial_2 q_2 + \partial_1 q_1) + \partial_2 c(\partial_1 q_2 - \partial_2 q_1)]. \qquad (6.3.12)$$

The role of the gradient interaction term is made transparent when only the scalar nematic order parameter ϱ changes along the dopant concentration gradient, taken as the x_1-axis, while the nematic orientation angle θ remains constant. In this case, the interaction term reduces to

$$\mathcal{L}_{\rm i} = -\tfrac{1}{2}\beta \partial_1 c(\partial_1 \varrho \cos 2\theta + \partial_2 \varrho \sin 2\theta). \qquad (6.3.13)$$

Since ϱ decreases with c ($\alpha_c > 0$), the optimal angle, reducing the overall energy at $\beta < 0$, is $\theta = 0$, so that the director tends to orient along the concentration gradient. At $\beta > 0$, on the opposite, the lowest energy is attained at $\theta = \pi/2$, when the director is oriented across the gradient; the change of ϱ along the x_2 axis does not affect this argument. This term is therefore responsible for the spontaneous anchoring at a nematic-isotropic interface.

In the absence of external fluxes, the equilibrium state corresponds to a constant value of chemical potential $\mu = \delta\mathcal{F}/\delta c$, which is given, using (6.2.30), (6.3.11), by

$$\mu = -\Lambda \nabla^2 c + \tau \ln \frac{c}{1-c} - \tfrac{1}{2}\alpha_c |q|^2 + \beta\left[\partial_1\partial_2 q_2 + \tfrac{1}{2}(\partial_1^2 - \partial_2^2)q_1\right]. \qquad (6.3.14)$$

The dynamic equation of the dopant concentration in the absence of external fluxes is, as in (1.6.14),

$$\partial_t c = \Gamma_c \nabla^2 \mu, \qquad (6.3.15)$$

where the concentration mobility Γ_c is proportional to the effective diffusivity.

The molecular field in the dynamic equation of the nematic order parameter (6.3.7) is modified by adding the vector $\mathbf{h}^{\rm c}$ stemming from the dopant-dependent terms:

$$h_1^{\rm c} = -\alpha_c c q_1 - \tfrac{1}{2}\beta(\partial_1^2 - \partial_2^2)c, \quad h_2^{\rm c} = -\alpha_c c q_2 - \beta\partial_1\partial_2 c. \qquad (6.3.16)$$

Assuming that the dopant does not affect mechanical properties of the material, its influence on the deformation dynamics (6.3.6) reduces to adding the capillary stress and modifying the distortion stress. The stress dependent only on the concentration gradient stems from the first term in (6.3.11) and is effective only in a narrow interphase boundary region. Varying this term and adding the term expressing the virtual energy increment due to deformation of the contour through the respective stress tensor, yields

$$\delta\mathcal{F}_{\rm c} = -\Lambda \int \nabla^2 c\, \delta c\, {\rm d}^2\mathbf{x} + \oint \left[\frac{\partial \mathcal{L}_{\rm c}}{\partial(\partial_i c)}\delta c - u_j \sigma_{ij}^{\rm c}\right] n_i\, ds, \qquad (6.3.17)$$

where the first term in the contour integral results from integration by part, and $\mathbf{n}$ is the normal to a contour parameterized by the arc length s. Displacing the concentration distribution by a deformation field $\mathbf{u}$ leads to $\delta c = u_j \nabla_j c$. Due to the translational invariance of $\mathcal{F}$, the contour integral must cancel for an arbitrary displacement; this defines the capillary stress

$$\sigma^{\mathrm{c}}_{ij} = \Lambda \partial_i c \partial_j c. \tag{6.3.18}$$

Varying (6.3.12) is carried out in a similar way, displacing also the tensor $\mathbf{Q}$ to replace $\delta Q_{jk} = u_l \partial_l Q_{jk}$. We will write explicitly only the contour integral stemming from integration by parts:

$$\oint \left[\frac{\partial \mathcal{L}_{\mathrm{i}}}{\partial(\partial_i c)} \delta c + \frac{\partial \mathcal{L}_{\mathrm{i}}}{\partial(\partial_i Q_{jk})} \delta Q_{jk} \right] n_i\, ds = -\beta \oint (\partial_j Q_{ij} \delta c + \partial_k c\, \delta_{ij} \delta Q_{jk})\, n_i ds$$
$$= -\beta \oint (\partial_k c \partial_j Q_{ij} u_k + \partial_k c\, \delta_{ij} u_l \partial_l Q_{jk})\, n_i ds = \oint u_k \sigma^{\mathrm{i}}_{ki} n_i ds, \tag{6.3.19}$$

leading to

$$\sigma^{\mathrm{i}}_{ij} = -\beta\, (\partial_i c\, \partial_k Q_{jk} + \partial_k c\, \partial_i Q_{jk}) \tag{6.3.20}$$
$$= -\beta \begin{pmatrix} \partial_1 c \partial_1 q_1 + \frac{1}{2}(\partial_1 c \partial_2 q_2 + \partial_2 c \partial_1 q_2) & \partial_1 c \partial_1 q_2 - \frac{1}{2}(\partial_1 c \partial_2 q_1 + \partial_2 c \partial_1 q_1) \\ \partial_2 c \partial_2 q_2 + \frac{1}{2}(\partial_1 c \partial_2 q_1 + \partial_2 c \partial_1 q_1) & -\partial_2 c \partial_2 q_1 + \frac{1}{2}(\partial_1 c \partial_2 q_2 + \partial_2 c \partial_1 q_2) \end{pmatrix}.$$

6.3.3 Phase Separation and Patterns

We shall use the equations derived so far to investigate stability of a uniform state $|\mathbf{q}|^2 = 1 - \alpha_c c_0$. According to (6.3.14), the chemical potential of a uniform state is

$$\mu_0 = \tau \ln \frac{c_0}{1 - c_0} + \tfrac{1}{2}\alpha_c(1 - \alpha_c c_0), \tag{6.3.21}$$

Different concentrations may correspond to equal values of μ_0, which would lead to *spontaneous demixing*. The right-hand side of (6.3.21) passes a minimum at

$$\frac{\alpha_c}{\tau} = \sqrt{\frac{2}{c_0(1 - c_0)}}. \tag{6.3.22}$$

The minimal level $\alpha_c/\tau = 2\sqrt{2}$ is attained at $c_0 = \frac{1}{2}$. This condition defines the threshold of a biphasic region where separation into prevalently isotropic and nematic domains occurs. At larger values of α_c, (6.3.14) has three solutions, of which the upper and lower ones are expected to be stable. Since $\alpha_c < 1$, phase separation is possible only when the thermal energy sufficiently exceeds the nematic energy.

Typical snapshots of concentration patterns developing from an originally uniform state are shown in Fig. 6.23. Take note that the stripes are oriented relative to the nematic director depending on the sign of the gradient interaction parameter, as argued above. These patterns are transient: they coarsen with time to minimize the length of the boundary between the dopant-rich isotropic and dopant-poor nematic phases.

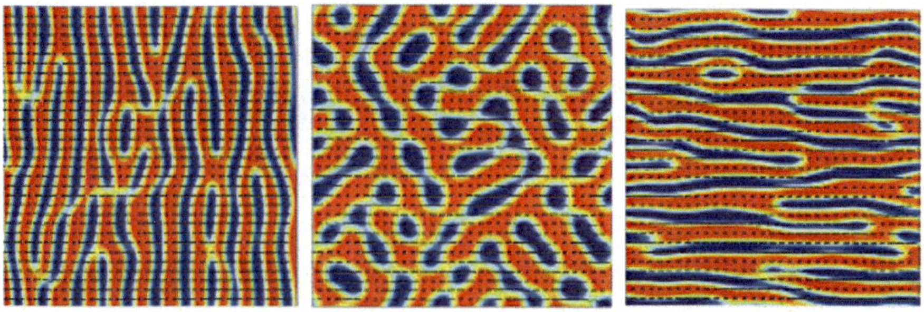

Fig. 6.23. Snapshots of an intermediate stage of the coarsening sequence at different values of the interaction parameter β: negative (left), zero (center), and positive (right). The dopant concentration distribution is color-coded, changing from blue to red in different phases. The nematic director orientation is shown by dashed lines (after Köpf and Pismen, 2013a)

The alternative domains coarsen on a long time scale to minimize the interfacial energy, leaving eventually two separated phases with a boundary of a minimal length and relative areas determined by the elastomer composition. The phase separation is accompanied by deformations, as shown in Fig. 6.24. While the total free energy decreases with time due to the gradient nature of the dynamical model, there is a strong exchange between the nematic and chemical energies $\mathcal{F}_{\mathbf{n}}$, $\mathcal{F}_c$. The change of the elastic energy $\mathcal{F}_{\rm el}$ is significantly less pronounced.

Symmetry-breaking bifurcations leading to persistent patterns can be located by linearizing the dynamic equations (6.3.15) together with (6.3.7) with the added term $\mathbf{h}^{\rm c}$ in (6.3.16) and (6.3.6). The latter is insensitive to the dopant-related stresses, since (6.3.13) and (6.3.20) vanish upon linearization, and so does $\sigma^{\rm d}_{ij} = -\pi_{ik}\partial_j q_k$. Therefore the nematic order parameter affects deformations in the linear approximation only through the growth tensor, and the dopant does it only indirectly. The distinction between the operators ∂ and $\widehat{\partial}$ disappears as well, so that $F_{ij} \approx \delta_{ij} + \partial_j u_i$ and the tensor $\mathbf{A}$ defined by (6.3.4) is linearized to $A_{ij} \approx \delta_{ij} + \partial_j u_i - g\widetilde{Q}_{ij}$. Then, for an incompressible sheet in the absence of an externally imposed finite strain, the elastic stress (6.3.5) is linearized as

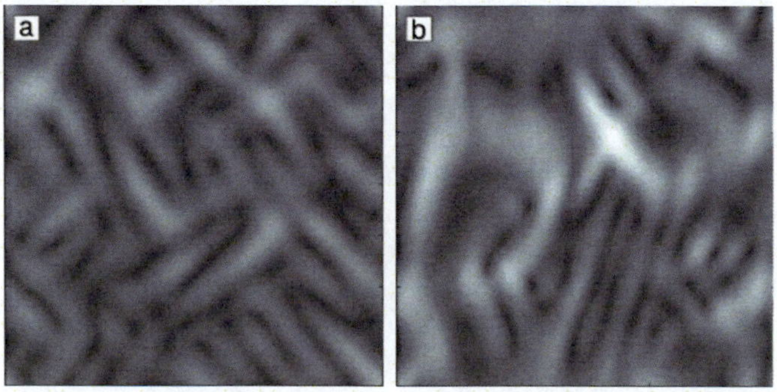

Fig. 6.24. Snapshots of the deformation field $|u|$ at $\beta = 0$ (a) and $\beta < 0$ (b) Lighter shades show higher deformations (Köpf and Pismen, 2013a, ©Springer)

$$\sigma_{ij}^{\text{el}} \approx \varsigma_{ij} \approx \nu(\partial_j u_i + \partial_i u_j - 2g\widetilde{Q}_{ij}). \tag{6.3.23}$$

In an incompressible material, the deformations can be conveniently expressed, similar to Sect. 6.2.4, as $u_i = \varepsilon_{ij}\partial_j \Psi$. Taking the curl of the deformation dynamic equation (6.3.6) eliminates pressure, leading to the dynamic equation of Ψ

$$\Gamma_{\text{el}}^{-1}\partial_t \Psi = \nabla^2 \Psi - 2g\partial_1\partial_2\widetilde{q}_1 + g\left(\partial_1^2 - \partial_2^2\right)\widetilde{q}_2. \tag{6.3.24}$$

The mechanical source (6.3.10) of deviations of $\widetilde{\mathbf{q}}$ is linearized as

$$\frac{\partial \mathcal{L}_{\text{el}}}{\partial \widetilde{q}_1} = \nu g(\partial_1 u_1 - \partial_2 u_2) - 2g\widetilde{q}_1, \quad \frac{\partial \mathcal{L}_{\text{el}}}{\partial \widetilde{q}_2} = \nu g(\partial_1 u_2 + \partial_2 u_1) - 2g\widetilde{q}_2. \tag{6.3.25}$$

Using also the linearized molecular field (6.2.34) supplemented by (6.3.16), the linearized dynamic equation (6.3.7) becomes, with Γ incorporated into the time scale,

$$\begin{aligned}
\partial_t \widetilde{q}_1 &= [f_1(c_0) + 2g]\widetilde{q}_1 + \tfrac{1}{2}K_1(\partial_1^2\widetilde{q}_1 + \partial_1\partial_2\widetilde{q}_2) + \tfrac{1}{2}K_2(\partial_2^2\widetilde{q}_1 - \partial_1\partial_2\widetilde{q}_2) \\
&\quad - \alpha_c c_0 q_1^0 \widetilde{q}_1 - \tfrac{1}{2}\beta(\partial_1^2 - \partial_2^2)\widetilde{c} - \nu g(\partial_1 u_1 - \partial_2 u_2), \\
\partial_t \widetilde{q}_2 &= [f_1(c_0) + 2g]\widetilde{q}_2 + \tfrac{1}{2}K_1(\partial_2^2 q_2 + \partial_1\partial_2 q_1) + \tfrac{1}{2}K_2(\partial_1^2 q_2 - \partial_1\partial_2 q_1) \\
&\quad - \alpha_c c_0 q_2^0 \widetilde{q}_2 - \beta\partial_1\partial_2\widetilde{c} - \nu g(\partial_1 u_2 + \partial_2 u_1),
\end{aligned} \tag{6.3.26}$$

where $f_i(c_0) = 1 - \alpha_c c_0 < 0$ in the isotropic state at $\alpha_c c_0 > 1$ and depends on the original orientation at $\alpha_c c_0 < 1$ The linearization of (6.3.15) is

$$\begin{aligned}
\Gamma_c^{-1}\partial_t \widetilde{c} &= -\Lambda\nabla^4\widetilde{c} + \frac{\tau}{c_0(1-c_0)}\nabla^2\widetilde{c} \\
&\quad - \alpha_c\nabla^2(q_1^0\widetilde{q}_1 + q_2^0\widetilde{q}_2) + \beta\nabla^2\left[\partial_1\partial_2\widetilde{q}_2 + \tfrac{1}{2}(\partial_1^2 - \partial_2^2)\widetilde{q}_1\right].
\end{aligned} \tag{6.3.27}$$

The linearized equations are Fourier-transformed as in Sect. 6.2.4 with the wave vector $\mathbf{k} = k\{\cos\varphi, \sin\varphi\}$. We consider first perturbations of the isotropic state $\mathbf{q}^0 = 0$, and assume for simplicity $K_1 = K_2 = K$. Then the Fourier-transformed system becomes

$$\begin{aligned}
\Gamma_{\text{el}}^{-1}\partial_t \Psi &= -k^2\Psi + gk^2(\widetilde{q}_1 \sin 2\varphi - \widetilde{q}_2 \cos 2\varphi), \\
\partial_t \widetilde{q}_1 &= (1 - \alpha_c c_0 + 2g)\widetilde{q}_1 - \tfrac{1}{2}k^2 K\widetilde{q}_1 + \tfrac{1}{2}\beta k^2\widetilde{c}\cos 2\varphi + \nu k^2 g\Psi \sin 2\varphi, \\
\partial_t \widetilde{q}_2 &= (1 - \alpha_c c_0 + 2g)\widetilde{q}_2 - \tfrac{1}{2}k^2 K\widetilde{q}_2 + \tfrac{1}{2}\beta k^2\widetilde{c}\sin 2\varphi - \nu k^2 g\Psi \cos 2\varphi, \\
\Gamma_c^{-1}\partial_t \widetilde{c} &= -\Lambda k^4\widetilde{c} - \frac{\tau}{c_0(1-c_0)}k^2\widetilde{c} + \tfrac{1}{2}\beta k^4(\widetilde{q}_1 \cos 2\varphi + \sin 2\varphi\widetilde{q}_2).
\end{aligned} \tag{6.3.28}$$

At this point, we need to recall that K and Λ act on small lengths, respectively, near nematic defects and interphase boundaries, and can be neglected in relatively smooth emerging domains, which we expect to observe near incipient symmetry breaking at moderate wavenumbers. Then the determinant of the above system reduces to

$$\begin{aligned}
&k^4\frac{(1-\alpha c_0 + 2g)^2}{c_0(1-c_0)} + k^6(1 - \alpha c_0 + 2g)\left[\frac{g^2\nu\tau}{c_0(1-c_0)} - \tfrac{1}{8}\beta^2(1 + \cos 4\varphi)\right] \\
&- \tfrac{1}{8}k^8\beta^2[(1 - \alpha c_0 + 2g)(1 + \cos 4\varphi) + g^2\nu(1 - \cos 4\varphi)] - \tfrac{1}{4}k^{10}\beta^2\nu\cos^2 2\varphi.
\end{aligned}$$

The instability leading to a stationary pattern takes place at $g < \frac{1}{2}(\alpha c_0 - 1)$. The parameter β plays the essential role of ensuring saturation of the instability at short lengths. An oscillatory instability is also possible. The lowest-order expression for the respective Hurwitz determinant,

$$-2k^2\Gamma_{\rm el}^2\Gamma_{\rm c}^2(1-\alpha c_0+2g)^5\left(\Gamma_{\rm c}+\frac{\Gamma_{\rm el}}{c_0(1-c_0)}\right), \tag{6.3.29}$$

indicates a long-scale Hopf instability at $g > \frac{1}{2}(\alpha c_0 - 1$, alternative to the necessary condition for bifurcation to a stationary pattern.

When the system is linearized near a perfectly oriented nematic state $q_1^0 = \sqrt{1-\alpha_c c_0}$, $q_2^0 = 0$, the only change takes place in the first term of the growth rate of $\widetilde{q}_1$, modified to $2(1-\alpha_c c_0 + g)\widetilde{q}_1$

$$\partial_t\widetilde{q}_1 = 2(1-\alpha_c c_0+g)\widetilde{q}_1+\ldots,\quad \partial_t\widetilde{q}_2 = (\alpha_c c_0+2g)\widetilde{q}_2+\ldots, \tag{6.3.30}$$

The determinant is now expressed, omitting higher-order terms, as

$$2k^4\tau\frac{2g^2+3g(1-\alpha_c c_0)+(1-\alpha c_0)^2}{c_0(1-c_0)}+\frac{k^6g^2\nu\tau}{c_0(1-c_0)}\left[2+3(g-\tfrac{1}{2}\alpha_c c_0)\right.$$
$$\left.+(1+g-\tfrac{1}{2}\alpha_c c_0)\cos 4\varphi\right]-\tfrac{1}{2}\beta^2k^6(1+g-\alpha_c c_0)\sin^2 2\varphi+\ldots$$

Now the leading term is negative, indicating a long-scale instability (phase separation) within the interval $-(1-\alpha c_0) < g < -\frac{1}{2}(1-\alpha c_0)$. Outside this interval, bifurcation to a stationary pattern takes place at $g < 0$; an oscillatory instability is possible as well. Activity further enhances instabilities, as analyzed by Pismen and Köpf (2014).

6.3.4 Bending Elastomer Sheets

Internal stresses arising in flat sheets due to changes of nematic alignment would relax when 3D deformations are allowed, since bending energy of a thin sheet is lower than energy of stretching or compression. Actuation by heating or illumination that induces the phase transition from the nematic to the isotropic state causes shortening by a factor a along the nematic director $\mathbf{n}$, compensated by extension in the normal directions by the factor a^σ, where σ is the Poisson ratio equal to 1/2 in an incompressible material. Assuming that the nematic alignment is constant over the thickness of the sheet, this modifies the flat metric to a curved metric (Aharoni et al, 2014)

$$\mathrm{g} = \mathcal{R}(\theta)\begin{pmatrix}1/a & 0\\ 0 & a^\sigma\end{pmatrix}\mathcal{R}^\dagger(\theta),\quad \mathcal{R}(\theta)=\begin{pmatrix}\cos\theta(\mathbf{x}) & -\sin\theta(\mathbf{x})\\ \sin\theta(\mathbf{x}) & \cos\theta(\mathbf{x})\end{pmatrix}, \tag{6.3.31}$$

where $\theta(\mathbf{x})$ is the local alignment angle θ (assumed to be constant over the thickness of the sheet) and $\mathcal{R}$ is the rotation matrix. This yields the metric tensor of the actuated surface

$$\mathbf{g} = \left(a^2 - a^{-2\sigma}\right) n_i n_j - a^{-2\sigma}\delta_{ij} \tag{6.3.32}$$
$$= \begin{pmatrix} a^{-2\sigma} + \left(a^2 - a^{-2\sigma}\right)\cos^2\theta(\mathbf{x}) & \frac{1}{2}r\left(a^2 - a^{-2\sigma}\right)\sin 2\theta(\mathbf{x}) \\ \frac{1}{2}r\left(a^2 - a^{-2\sigma}\right)\sin 2\theta(\mathbf{x}) & a^{-2\sigma} + \left(a^2 - a^{-2\sigma}\right)\cos^2\theta(\mathbf{x}) \end{pmatrix}.$$

The most important characteristic or the emerging surface is its Gaussian curvature K (the product of two principal curvatures) related to the metric tensor through Gauss's *Theorema Egregium*:

$$K = \frac{1}{g}\left[\partial_2\left(\frac{g}{g_{1,1}}\Gamma^2_{1,1}\right) - \partial_1\left(\frac{g}{g_{1,1}}\Gamma^2_{1,2}\right)\right], \tag{6.3.33}$$

where $g = \sqrt{\det(\mathbf{g})}$ and Γ^i_{jk} are Christoffel symbols[3]

$$\Gamma^k_{ij} = \tfrac{1}{2}g^{kl}(\partial_j g_{il} + \partial_i g_{jl} - \partial_l g_{ij}). \tag{6.3.34}$$

Generally, the resulting shape is determined by the balance between bending and stretching/compression energies, but in thin sheets the former is much larger, and the energy is directly determined by the metric of the actuated surface (Efrati et al, 2013):

$$\mathcal{F} = \int \frac{gE}{24(1+\sigma)}\left(\frac{\sigma}{1+\sigma}g^{ij}g^{kl} + g^{ik}g^{jl}\right)R_{ij}R_{kl}\mathrm{d}\mathbf{x}, \tag{6.3.35}$$

where E is the Young modulus and R_{ij} is the Ricci tensor (called also the second fundamental form):

$$R_{ij} = \partial_l\Gamma^l_{ij} - \partial_j\Gamma^l_{il} + \Gamma^l_{ij}\Gamma^k_{kl} - \Gamma^k_{il}\Gamma^l_{jk}. \tag{6.3.36}$$

Visualizing the shape is, as a rule, a nontrivial computational problem. The computation is facilitated in the "natural" coordinate system where the coordinate lines are tangent and normal to the director. This is the case for a planar nematic pattern is an ellipse where the nematic director is aligned along coordinate lines of the elliptic coordinate system: u = const (if it is anchored normal to the boundary) or v = const (if the anchoring is tangential), as sketched in Fig. 6.18c,d. Two defects with the charge +1/2 are located at the foci. The metric tensor remains diagonal upon actuation, and is readily obtained from the metric $g_{ij} = g\delta_{ij}$ in the elliptic coordinates u, v by extending in one and contracting in another direction:

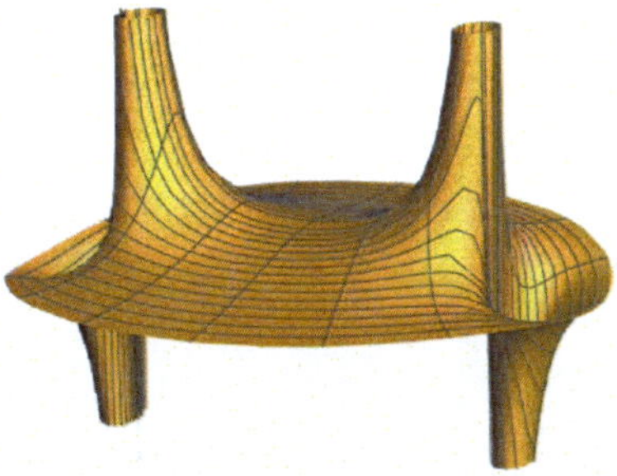

Fig. 6.25. The function $f(u, v)$ proportional to the Gaussian curvature of the actuated elliptical pattern

[3] The summation is carried out here over identical *upper and lower* indices. The indices in (6.3.34) can be interchanged; g^{kl} is the inverse metric

$$\mathbf{g} = g \begin{pmatrix} a & 0 \\ 0 & a^{-\sigma} \end{pmatrix}, \quad g = \tfrac{1}{2}(\cosh 2v - \cos 2u), \tag{6.3.37}$$

where the half-focal distance is normalized to unity. This expression is equally applicable to textures with the director along the coordinate lines $u = \text{const}$ or $v = \text{const}$; one needs only to interchange the metric factors corresponding to the two variables. The Gaussian curvature computed according to (6.3.33), (6.3.34) is

$$K = 4a^{-\sigma}\left[1 - a^{-(1+\sigma)}\right] f(u, v), \quad f(u, v) = \frac{1 - \cosh 2v \cos 2u}{(\cosh 2v - \cos 2u)^3}. \tag{6.3.38}$$

The function $f(u, v)$ is drawn as a function of original positions in Fig. 6.25.

The divergence of the curvature at the defect locations can be corrected by taking into account the attenuation of the modulus in the defect core. Following the procedure applied by Mostajeran et al (2016), modified by allowing for a dependence $a(r)$, we start with the Cartesian metric (6.3.32) and transform it to to the metric $\mathbf{g}_p$ in the polar coordinates r, ϕ:

$$\mathbf{g}_p = \mathbf{J}^\dagger \cdot \mathbf{g} \cdot \mathbf{J}, \quad \mathbf{J} = \begin{pmatrix} \cos\phi & -r\sin\phi \\ \sin\phi & r\cos\phi \end{pmatrix}. \tag{6.3.39}$$

This yields the elements of the metric tensors $\mathbf{g}_\pm$ for half-charged defects with $\theta = \pm\phi/2$

$$\begin{aligned}
g^+_{rr} &= \tfrac{1}{2}\left[a(r)^2(1+\cos\phi) + a(r)^{-2\sigma}(1-\cos\phi)\right], \\
g^+_{r\phi} &= g^+_{r\phi} = \tfrac{1}{2} r \sin\phi \left[-a(r)^2 + a(r)^{-2\sigma}\right], \\
g^+_{\phi\phi} &= \tfrac{1}{2} r^2 \left[a(r)^2(1-\cos\phi) + a(r)^{-2\sigma}(1+\cos\phi)\right], \\
g^-_{rr} &= \tfrac{1}{2}\left[a(r)^2(1+\cos 3\phi) + a(r)^{-2\sigma}(1-\cos 3\phi)\right], \\
g^-_{r\phi} &= g^-_{r\phi} = \tfrac{1}{2} r \sin 3\phi \left[a(r)^2 - a(r)^{-2\sigma}\right], \\
g^-_{\phi\phi} &= \tfrac{1}{2} r^2 \left[a(r)^2(1-\cos 3\phi) + a(r)^{-2\sigma}(1+\cos 3\phi)\right].
\end{aligned} \tag{6.3.40}$$

Computing as above the Gaussian curvature $K_\pm$ we obtain

$$\begin{aligned}
K_+ = {}& \tfrac{1}{2}\left[a^{-3}\sigma(1+\cos\phi) - a^{2\sigma-1}(1-\cos\phi)\right] a''(r) \\
& - \tfrac{1}{2}\sigma\left[a^{-4}(2+\sigma)(1+\cos\phi) + a^{2(\sigma-1)}(1-\cos\phi)\right][a'(r)]^2 \\
& + \tfrac{1}{4} r^{-1}\{a^{-3}[1+\cos\phi+\sigma(3+\cos\phi)] \\
& - a^{2\sigma-1}[3-\cos\phi+\sigma(1-\cos\phi)]\}a'(r) - \tfrac{1}{4} r^{-2}\left(a^{2\sigma} - a^{-2}\right)\cos\phi,
\end{aligned} \tag{6.3.41}$$

$$\begin{aligned}
K_- = {}& \tfrac{1}{2}\left[a^{-3}\sigma(1+\cos 3\phi) - a^{2\sigma-1}(1-\cos 3\phi)\right] a''(r) \\
& - \tfrac{1}{2}\sigma\left[a^{-4}(2+\sigma)(1+\cos 3\phi) + a^{2(\sigma-1)}(1-\cos\phi)\right][a'(r)]^2 \\
& - \tfrac{1}{4} r^{-1}\{a^{-3}[1+3\cos\phi+\sigma(5\cos 3\phi - 1)] \\
& - a^{2\sigma-1}[1+5\cos 3\phi-\sigma(1-\cos 3\phi)]\}a'(r) + \tfrac{3}{4} r^{-2}\left(a^{2\sigma} - a^{-2}\right)\cos 3\phi.
\end{aligned} \tag{6.3.42}$$

With $a = \text{const}$, $K_\pm$ diverge as r^{-2} at $r \to 0$, but the factor a should depend on the modulus as $a(r) = a[1 - \varrho(r)/\varrho(\infty)]$. Taking into account that $\varrho(r) \propto r$

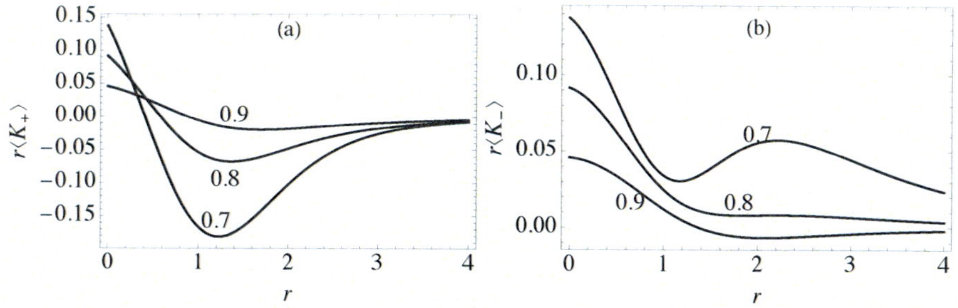

Fig. 6.26. The angular averages of $r\langle K_+\rangle$ (a) and $r\langle K_-\rangle$ (b) for $\sigma = 1/2$

at $r \to 0$ while expanding the above expressions, shows that $K_\pm \propto r^{-1}$ at $r \to 0$, so that $\int rK_\pm(r)\mathrm{d}r$ is finite. The angular dependence reflects the symmetry of the defects and can be integrated analytically to compute the radial dependence of their angular averages $r\langle K_\pm\rangle$, plotted in Fig. 6.26 for $\sigma = 1/2$ with $\varrho(r)$ given by (4.2.15).

Generally, the director field can be imposed by a substrate pretreatment, so that an unlimited variety of patterns can be prepared for actuation to desired shapes. Aharoni et al (2014) expressed the metric generated by a a continuously varying director field in terms of the scalar field $\tau = \tan\theta(\mathbf{x})$ as

$$\mathbf{g} = \frac{1}{1+\tau^2}\begin{pmatrix} a^2 + a{-}2\sigma\tau^2 & \left(a^2 - a{-}2\sigma\right)\tau \\ \left(a^2 - a{-}2\sigma\right)\tau & a^2\tau^2 + a{-}2\sigma \end{pmatrix}, \tag{6.3.43}$$

and computed the Gaussian curvature

$$K(\mathbf{x}) = \left(a^2 + a^{-2\sigma}\right)\left[\frac{(\tau^2-1)\partial_1\partial_2\tau - \tau(\partial_1^2 - \partial_2^2)\tau}{(\tau^2+1)^2} + \frac{(3\tau^2-1)(\partial_1\tau^2 - \partial_2\tau)^2 - 2\tau(\tau^2-3)\partial_1\tau\partial_2\tau}{(\tau^2+1)^3}\right]. \tag{6.3.44}$$

This expression can be viewed as a partial differential equation, which can be solved for $\tau(\mathbf{x})$ for any given $K(\mathbf{x})$, thereby "reverse engineering" the problem to find a director field that would induce a desired geometry. Aharoni et al (2014) give an example of a solution for a field dependent on a single variable. Griniasty et al (2019) consider more complicated cases, but warn that "an exact solution to the full inverse problem was not found in the general case, nor was it shown to exist". Reversely, multiple surface geometries can be obtained from the same flat sheet (Griniasty et al, 2021).

6.3.5 Inhomogeneous Actuation

Textures leading to various shapes upon actuation of films with tangential nematic orientation can be prepared in a more natural way by taking advantage of a gradual change of dopant concentration that could be attained without externally imposed surface patterning or anchoring at the edges. This mimics at a basic level biological signaling (Wolpert, 2007) that modifies the properties of heretofore undifferentiated cells in an embryo and thereby initiates complex reshaping necessary for morphogenetic development.

A scheme of a suitable experimental procedure is shown in Fig. 6.27a. The nematic alignment is established in the liquid state before polymerization. A nonmesogenic dopant would concentrate at the center to avoid formation of defects, and, by symmetry, the emerging pattern on a disk would have the circulation ± 1 but depend in a crucial way on the sign of the coupling parameter β in (6.3.12). As argued in Sect. 6.3.2, the energy is minimized when the director orients across or along the concentration gradient at $\beta > 0$ and $\beta < 0$, respectively. In the former case, it orients in a circular domain on lines $r =$ const. Upon actuation, the elastomer shortens along these lines and elongates in the radial direction. The resulting shape is a cone smoothed out at the apex where the material is isotropic (Fig. 6.27b). More variegated shapes are obtained at $\beta < 0$ when the nematic orients aster-like in the radial directions, shortening along them upon actuation while enlarging the circumference. The result is an *anticone* with $n \geqslant 2$ "petals" (Fig. 6.27c).

Assuming that diffusion is frozen in the polymer, the concentration and polarization fields are determined in the liquid state by minimizing the nematic and concentration energy including the interaction terms. However, this cumbersome computation is relevant only in a narrow (or the order healing length) transition belt between the isotropic and nematic domains and does not affect the overall bending energy accumulated in the actuated nematic. Therefore it is sufficient to consider the shape of the outer nematic region with $\varrho = 1$. The case of a cone is trivial: its apex angle Θ depends on the actuation factor as $\sin \frac{1}{2}\Theta = a^{1+\sigma}$ and no qualitative changes occur as the factor a increases. Contrariwise, anticone's optimal shape depends on bending energies of different configurations.

A point originally at $r, \varphi, z = 0$ in cylindrical coordinates is displaced upon actuation to a location $\rho, \phi, z(\phi)$ on a curve lying on the sphere with the radius

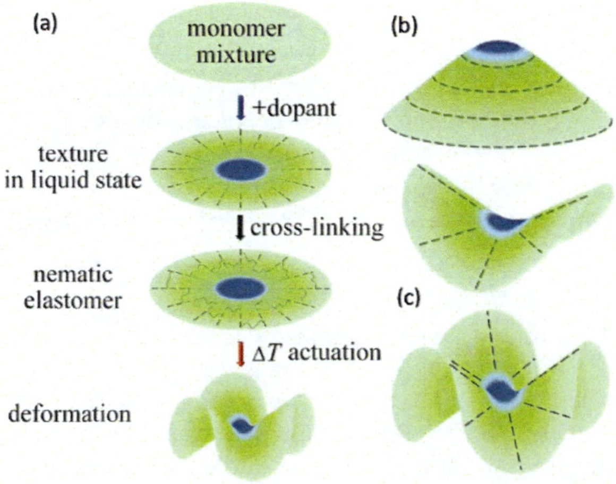

Fig. 6.27. (a) A scheme of the film preparation and actuation. (b) A graded cone at $\beta = 1$ (c) Shapes of anticones with $n = 2$ and $n = 3$, at $\beta = -1$. The blue shading marks the isotropic domain; dashes mark the director orientation before actuation (Zakharov and Pismen, 2017, ©The Royal Society of Chemistry)

$R = \sqrt{\rho^2 + z^2} = ar$, $a < 1$. The function $z(\phi)$ must be periodic, and can be chosen in the harmonic form $z = \rho A \sin n\phi$ with an integer n. The angular coordinate ϕ must be distinct from φ, since a harmonic dependence on φ would cause either nonuniform stretching of a deformed circle or formation of creases (Modes and Warner, 2015). The infinitesimal length element along the curve is

$$\mathrm{d}s = \mathrm{d}\phi\sqrt{\rho^2 + (\partial_\phi \rho)^2 + (\partial_\phi z)^2}. \tag{6.3.45}$$

It is necessary for uniform extension that $R = ar$, $\mathrm{d}s = a^{-\sigma} r \mathrm{d}\varphi$, so that

$$\rho(\phi) = ar(1 + A^2 \sin^2 n\phi)^{-1/2}, \quad z(\phi) = arA \sin n\phi(1 + A^2 \sin^2 n\phi)^{-1/2}. \tag{6.3.46}$$

The length element (6.3.45) is expressed using these relations as

$$\mathrm{d}s = ar\mathrm{d}\phi \frac{\sqrt{1 + A^2(\sin^2 n\phi + n^2 \cos^2 n\phi)}}{1 + A^2 \sin^2 n\phi} = a^{-\sigma} r \mathrm{d}\varphi. \tag{6.3.47}$$

Dividing this by $r\varphi$ yields the relation between ϕ and φ:

$$\frac{\partial \phi}{\partial \varphi} = \frac{a^{1+\sigma}(1 + A^2 \sin^2 n\phi)}{\sqrt{1 + A^2(\sin^2 n\phi + n^2 \cos^2 n\phi)}}. \tag{6.3.48}$$

The amplitude A is defined implicitly by the integral relation

$$\int_0^{2\pi} \mathrm{d}\phi \frac{\sqrt{1 + A^2(\sin^2 n\phi + n^2 \cos^2 n\phi)}}{1 + A^2 \sin^2 n\phi} = 2\pi a^{-(1+\sigma)}. \tag{6.3.49}$$

For $n = 1$, the integral equals to 2π, so that this relation is satisfied only at $a = 1$. Indeed, this case is trivial, as it is simply a uniform rotation leaving the disk flat. The dependence $A(a)$ at $n = 2, 3, 4$ for $\sigma = 1/2$ computed by integrating (6.3.49)

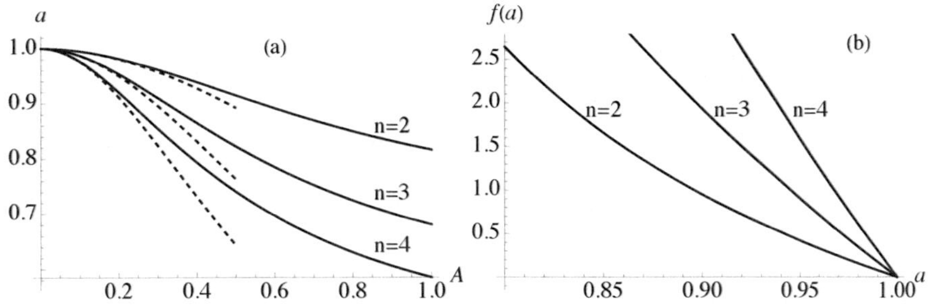

Fig. 6.28. (a) The dependence of the amplitude A on the actuation factor a at $n = 2, 3, 4$ for $\sigma = 1/2$. Dashed lines show the small-amplitude approximation. (b) The small amplitude approximation of the integral $f(a)$ in (6.3.52)

numerically is shown in Fig. 6.28a together with the small-amplitude approximation $A \approx 2\sqrt{(a^{-(1+\sigma)})/(n^2-1)}$.

It is clear from the picture that different structures are possible at the same value of the factor a, and it is important to compare their bending energy determined by the curvature integrated over the entire deformed surface. The easiest way to compute the curvature (Modes and Warner, 2015) is to consider a curve $bs\gamma(s)$ on the sphere of unit radius, which can be viewed in the Darboux frame spanned by the three unit vectors: tangent $\boldsymbol{t} = \gamma'(s)$, normal $\boldsymbol{n}(s)$, and binormal $\boldsymbol{b} = \boldsymbol{n}(s) \times \boldsymbol{t}(s)$. The rate of change in the tangent,

$$\boldsymbol{t}'(s) = \kappa_n \boldsymbol{n} + \kappa_b \boldsymbol{b}. \tag{6.3.50}$$

gives the normal and binormal curvatures, $\kappa_n = \boldsymbol{n} \cdot \boldsymbol{t}'(s)$, $\kappa_b = \boldsymbol{b} \cdot \boldsymbol{t}'(s)$. Since $\gamma \cdot \gamma'(s) = 0$, it coincides with $\boldsymbol{b}$, and it follows from the Darboux triad that $\boldsymbol{n} = \gamma \times \gamma'(s)$, so that taking the scalar product of (6.3.50) with $\boldsymbol{n}$ yields

$$\kappa_n = \boldsymbol{n} \cdot \boldsymbol{t}'(s) = [\gamma \times \gamma'(s)] \cdot \gamma''(s). \tag{6.3.51}$$

The entire surface is described as the merger of the scale-invariant curves $\mathbf{x}(s) = R\gamma(s)$, or rather bent rods of infinitesimal thickness $\mathrm{d}s$ with the height equal to the thickness l_0 of the disk, increased to $l = l_0 a^{-\sigma}$ upon actuation. The energy of such a bent rod is $\mathrm{d}\mathcal{F} = \frac{1}{24} E z^2 l^3 R\,\mathrm{d}R\,\mathrm{d}s$, where E is the Young modulus (Landau and Lifshitz, 1970). The curvature κ differs from κ_n just by the scaling factor R^{-1}, so

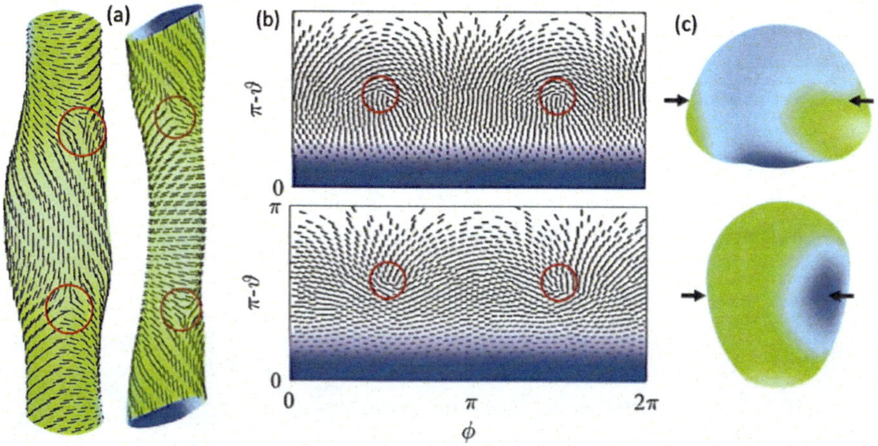

Fig. 6.29. (a) Deformed cylinders with two point concentration sources at the edges for $\beta = 1$ (left) and $\beta = -1$ (right). (b) The nematic texture on a sphere of unit radius (shown in the Mercator projection) for $\beta = -1$ (above) and $\beta = 1$ (below). The dopant source is at the south pole. The blue shading corresponds to a lower modulus and dashes mark the director orientation. (c) The respective shapes following the transition to the isotropic state with shading showing the local curvature. The defects locations (prior to the transition to the isotropic state) are shown by arrows. Red circles in (a,b) mark the locations of defects. (Zakharov and Pismen, 2017, ©The Royal Society of Chemistry)

that $\mathrm{d}\mathcal{F} = \frac{1}{24}E\kappa_{\boldsymbol{n}}^2 l^3 R^{-1}\mathrm{d}R\,\mathrm{d}s$. Therefore the total energy is computed as

$$\mathcal{F} = \frac{1}{24}El_0^3\ln(r_1/r_0)f(a), \quad f(a) = \int_0^{2\pi}[\gamma\times\gamma'(s)]^2\cdot\gamma''(s)]\mathrm{d}s, \tag{6.3.52}$$

where r_0, r_1 are the radii of the inner transition zone (eliminating the singularity that would arise otherwise) and of the disk edge, respectively. At small amplitudes, the integrand is expressed as $n^4(1+s^2)^{-3}A^2\sin^2 ns + O(A^4)$. The function $f(a)$ based on this expression with the above approximation $A(a)$ is plotted in Fig. 6.28b. The computation shows that energy increases with n. If this trend ever reverses, it would happen at unrealistically low values of the actuation factor a.

Whereas adding a dopant eliminates defects on a disk, it may induce them on a cylindrical surface where they are not demanded by topology. A point source of a dopant at the edges of a cylinder creates there an isotropic region (painted blue in Fig. 6.29a), and the circulation around interphase boundary should be compensated by two $+1/2$ defects. Reversing the sign of β just causes the entire distribution to rotate by $\pi/2$ but the shapes following actuation are very much different, with radial bulging and shrinking interchanged. The circulation due to a point source on a spherical surface reduces the required number of $+1/2$ defects by two. The resulting nematic textures at $\beta = \pm 1$ and the shapes after transition to the isotropic state are shown in Fig. 6.29b,c.

6.3.6 Janus Filaments

Another way to reshaping is using a material with different properties at different locations. The simplest example is bending of a stripe as a result of one-sided illumination or heating that causes shrinking due to the local phase transition. Similar phenomena of the intrinsic curvature arising due to nonuniform expansion go back to Timoshenko plates (Timoshenko, 1925) bending due to non-uniform thermal expansion, as well as to deformation of prestressed and an unstressed plates or rods glued together. Buckling of "birods" provides much more leeway than that of plates, resulting in variegated helical shapes (Lestringant and Audoly, 2017). It can be realized in *Janus filaments* containing nematic and isotropic elastomers, which can be fabricated by using two connected extruders (Ionov et al, 2017).

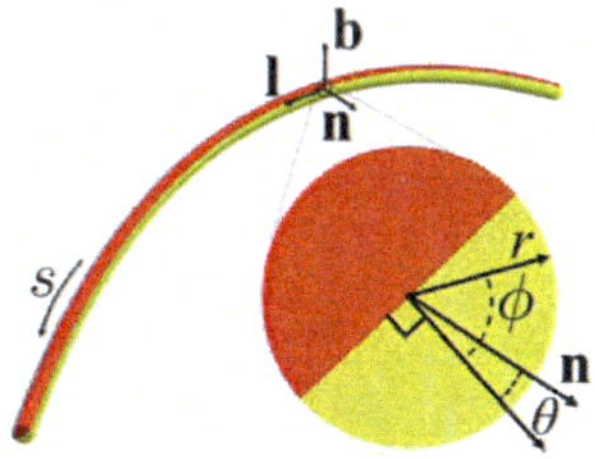

Fig. 6.30. A segment of Janus filament showing the Darboux frame and a cross-section geometry with a mismatched orientation of the Darboux normal vector **n** and the internal normal

Consider a Janus filament of a radius r that consists of a driven component occupying the sector $|\phi| > \pi/2$ of its circular cross-section and a passive component in the sector $|\phi| < \pi/2$. The driven component of the filament is assumed to be a nematic elastomer that polarizes upon phase transition along the filament axis thereby causing elongation, assumed to be slight, by a factor $a = 1 + \epsilon$, while the passive component remains unchanged. A longitudinal polarization is natural for thin filaments and can be ensured by anchoring conditions on the interface. When the nematic component elongates, the centerline of the filament develops an intrinsic curvature, which, when unforced and unconstrained, is directed along the normal to the dividing plane. The curvature radius R is proportional to r/ϵ, and therefore small extensions are sufficient to significantly bend a thin filament. The direction of bending may, however, change in the presence of constraints, in particular, when the filament is clamped or has a form of a closed loop. A spontaneous twist may arise under these conditions.

The elastic energy $\mathcal{F}$ of an inextensible filament can be written in the slender body (Kirchhoff rod) approximation (Landau and Lifshitz, 1970) as the integral of the sum of bend and twist energies (respectively, $\mathcal{F}^b$ and $\mathcal{F}^t$) over the arc length s of its neutral line:

$$\mathcal{F} = \frac{1}{2}AE \int \left(\mathcal{F}^b + \mathcal{F}^t\right) \mathrm{d}s, \quad \mathcal{F}^b = I(\kappa, \theta), \quad \mathcal{F}^t = J\widehat{\tau}^2, \tag{6.3.53}$$

where A is the cross-sectional area, E is the Young modulus; $\kappa = 1/R$ is the curvature of the centerline with the curvature radius R, and $I(\theta)$ is the flexural rigidity dependent on the angle θ between the normal to the dividing plane (to be called the internal normal) and the normal vector $\mathbf{n}$ in the Darboux frame (see Fig. 6.30); $J = r^2/2$ is the torsional rigidity. The twist $\widehat{\tau} = \tau + d\theta/ds$ coincides with the geometric torsion τ in the Darboux frame only when the internal normal does not deviate from the Darboux normal vector.

The flexural rigidity is computed by adding the contributions of the driven (I_a) and passive (I_p) sectors:

$$\begin{aligned} I_a &= \frac{1}{\pi r^2}\int_0^r r\,\mathrm{d}r \int_{\pi/2}^{3\pi/2} \left(\frac{r}{R}\cos(\phi-\theta) - \epsilon\right)^2 \mathrm{d}\phi \\ &= \frac{1}{8}\left(\frac{r}{R}\right)^2 - \frac{4\epsilon}{3\pi}\frac{r}{R}\cos\theta + \frac{\epsilon^2}{2}, \end{aligned} \tag{6.3.54}$$

$$I_p = \frac{1}{\pi r^2}\int_0^r \frac{r^3}{R^2}\,dr \int_{\pi/2}^{-\pi/2} \cos^2(\phi-\theta)\,\mathrm{d}\phi = \frac{1}{8}\left(\frac{r}{R}\right)^2, \tag{6.3.55}$$

$$I(\kappa,\theta) = I_a + I_p = \frac{1}{4}(\kappa r)^2 - \frac{4\,\epsilon}{3\pi}\kappa r\cos\theta + \frac{\epsilon^2}{2}. \tag{6.3.56}$$

Clearly, the lowest energy state is attained at $\theta = 0$ when the internal normal is directed along the normal vector n in the Darboux frame. The intrinsic curvature $\widehat{\kappa}$ is defined by the condition $dI/d\kappa = 0$, yielding $\widehat{\kappa}r = \frac{8}{3}\epsilon/\pi$. The energy of an inextensible filament is expressed after plugging this together with (6.3.56) in (6.3.53) and omitting the constant term as

$$\mathcal{F} = \tfrac{1}{2}AEr^2 \int \left[\tfrac{1}{4}\kappa(\kappa - 2\widehat{\kappa}\cos\theta) + \tfrac{1}{2}\tau^2\right] \mathrm{d}s. \tag{6.3.57}$$

This formula does not coincide with the standard expression proportional to $(\kappa-\widehat{\kappa})^2$, which vanishes at $\kappa = \widehat{\kappa}$, since a Janus filament remains under compression even when optimally bent.

Dynamics of a filament can be studied in the coordinate-independent form with the help of the Frenet–Serret equations (5.5.1). Using them to evolve the filament numerically is not practical, but they are advantageous for locating instabilities of an original shape by expressing with their help the infinitesimal energy change under virtual displacements:

$$\delta\mathcal{F} = \frac{1}{4}AEr^2 \int [(\kappa - \widehat{\kappa})\,\delta\kappa + \kappa\widehat{\kappa}\sin\theta\,\delta\theta \\ + 2(\tau\,\delta\tau + \tau\,\partial_s\delta\theta + \partial_s\theta\delta\tau + \partial_s\theta\partial_s\delta\theta)]\,\mathrm{d}s. \tag{6.3.58}$$

The change of curvature and torsion due to the tangential, normal and binormal velocities u, v, w is expressed by the equations

$$\delta\kappa = V'(s) - \tau W, \quad \delta\tau = \kappa W + U'(s), \tag{6.3.59}$$
$$U = \kappa^{-1}(W'(s) + \tau V).\ \ V = v'(s) + \kappa u - \tau w, \quad W = w'(s) + \tau v.$$

Omitting terms of higher order in small displacements, we obtain

$$\delta\kappa = v''(s) + \kappa^2 v + u\kappa'(s) + \kappa u'(s) - w\tau'(s) - 2\tau w'(s), \tag{6.3.60}$$
$$\delta\tau = \kappa w'(s) + \kappa^{-1}w'''(s). \tag{6.3.61}$$

Besides physical displacements v, w that affect curvature and torsion, these equations contain an axial displacement u, which is a mere reparametrization of the curve that can be adjusted with the help of the gauge equation $u'(s) = \kappa v$ to satisfy the incompressibility constraint.

We apply these equations to study linear stability of a circle with the curvature κ_0 to harmonic perturbations

$$v = \widehat{v}\mathrm{e}^{\mathrm{i}ks}, \quad w = \widehat{w}\mathrm{e}^{\mathrm{i}ks}, \quad u = \widehat{u}\mathrm{e}^{\mathrm{i}ks}, \quad \delta\theta = \widehat{\theta}\mathrm{e}^{\mathrm{i}ks}, \tag{6.3.62}$$

where the wavenumber $k = n\kappa_0$ should be an integer multiple of κ_0. The change of energy $\delta\mathcal{F}$ is quadratic in the perturbation amplitudes $\widehat{v}, \widehat{w}, \widehat{u}, \widehat{\theta}$, and the basic configuration is unstable when $\delta\mathcal{F} < 0$.

Stability of an untwisted ($\tau_0 = 0$) circular filament with the curvature κ_0 is tested by computing the lowest order (quadratic) resonant terms in Eq. (6.3.58) to determine whether the energy decreases upon virtual small displacements of the filament. In the lowest energy configuration, $\theta_0 = 0$, so that small deviations τ, θ can be presented in the harmonic form $\tau = \widehat{\tau}\mathrm{e}^{\mathrm{i}ks}$, $\theta = \widehat{\theta}\mathrm{e}^{\mathrm{i}ks}$. The curvature is expressed as

$$\kappa = \kappa_0 + \overline{\kappa} + \widehat{\kappa}\mathrm{e}^{\mathrm{i}ks}, \tag{6.3.63}$$

where $\overline{\kappa}$ denotes constant terms quadratic in displacements. The relations between the perturbation amplitudes follow from the linearized Eqs. (6.3.60), (6.3.61):

$$\widehat{\kappa} = -(k^2 - \kappa_0^2)\widehat{v}, \qquad \widehat{\tau} = -\frac{\mathrm{i}k}{\kappa_0}(k^2 - \kappa_0^2)\widehat{w}. \tag{6.3.64}$$

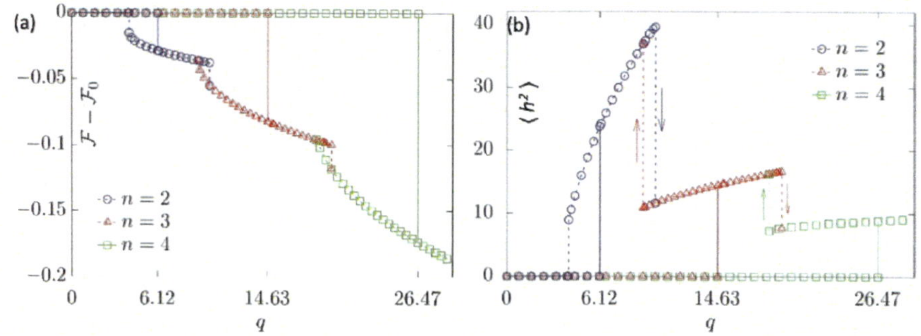

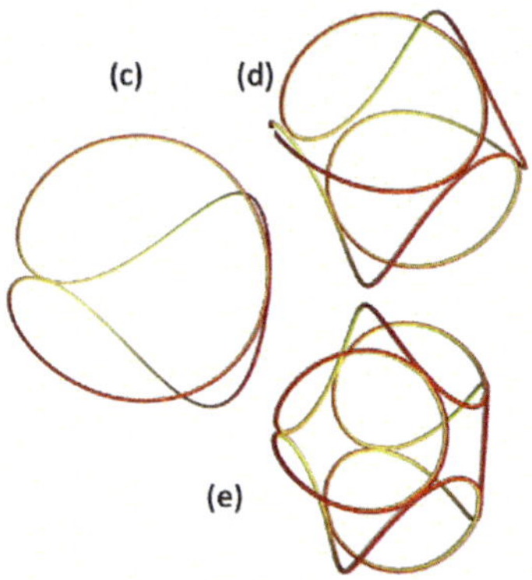

Fig. 6.31. The dependence of the energy decrements (a) and the mean square deviation $\langle h^2 \rangle$ from the original plane (b) at different looping numbers n. Configurations with the looping numbers $n = 2$ at $q = 9$ (c) , $n = 3$ at $q = 16$ (d), and $n = 4$ at $q = 27$ (e). The driven and passive components are colored red and yellow, respectively (Zakharov and Pismen, 2018, ©The American Physical Society)

In the lowest order, the tangential displacement amplitude is expressed using the gauge equation as $\widehat{u} = -\mathrm{i}(\kappa_0/k)\widehat{v}$. Substituting this together with (6.3.62), (6.3.63) in (6.3.60) yields the second-order resonant term

$$\overline{\kappa} = 2(k^2/\kappa_0)(k^2 - \kappa_0^2)|\widehat{w}|^2, \tag{6.3.65}$$

which, combined with (6.3.64), generates the second-order resonant terms in the integrand of (6.3.58), leading to the energy increment

$$\delta\overline{\mathcal{F}} = (\kappa_0 - \widehat{\kappa})\overline{\kappa} + 2|\widehat{\kappa}|^2 + 2(\kappa_0\widehat{\kappa} + 2k^2)|\widehat{\theta}|^2 + 4|\widehat{\tau}|^2 + 2k\,\mathrm{Im}(\widehat{\theta}\widehat{\tau}). \tag{6.3.66}$$

Normal displacements affect only the second term of this expression, which reduces to $\kappa_0^4(n^2-1)^2|v|^2 > 0$. Hence, the filament is always stable to normal displacements. Instability to binormal displacements may be caused either by the first term in (6.3.66), which is negative at $\widehat{\kappa} > \kappa_0$, $n > 1$, or by the last term dependent on the phase difference between the binormal displacement and rotation of the internal normal. The latter term can be presented using (6.3.64) as $-4\kappa_0^3 n^2(n^2-1)|\theta w|\cos\psi$, where ψ is the difference between the phases of $\widehat{w}$ and $\widehat{\theta}$. Thus, the instability is most likely when the binormal displacement and rotation of the internal normal are *in phase*. Setting $\psi = 0$, omitting the v-dependent term, and denoting $q = \widehat{\kappa}/\kappa_0$, $\widehat{\vartheta} = \widehat{\theta}/\kappa_0 = \zeta\widehat{w}$, (6.3.66) is rewritten as

$$\delta\overline{\mathcal{F}} = 2\kappa_0^4 \left[n^2(n^2-1)(2n^2-q-1) + (q+2n^2)\zeta^2 - 2n^2(n^2-1)\zeta\right] |\widehat{w}|^2. \tag{6.3.67}$$

This expression is at minimum with respect to the amplitude ratio ζ at $\zeta_c = n^2(n^2-1)/(q+2n^2)$. Using this in (6.3.67), it is easy to see that zeroes of $\delta\overline{\mathcal{F}}$ satisfy the quadratic equation $q(1-q) = n^2(3n^2-1)$, which yields the instability threshold

$$q_c = \frac{1}{2}\left(\sqrt{1-4n^2+12n^4} - 1\right). \tag{6.3.68}$$

The numerical values are $\zeta_c \approx 0.848$, $q_c \approx 6.152$ at $n = 2$, $\zeta_c \approx 2.195$, $q_c \approx 14.8$ at $n = 3$, $\zeta_c \approx 4.073$, $q_c \approx 26.9$ at $n = 4$. Computation of large-amplitude structures emerging due to this instability shows that the bifurcation is in fact *subcritical*, and three-dimensional configurations with lower energy exist already at $q < q_c$. The dependences of the energy decrements relative to the planar circle and of the mean square deviation $\langle h^2\rangle$ from the original plane on q are shown in Fig. 6.31a,b. Some shapes are shown in Fig. 6.31c–e.

While spontaneous reshaping of closed loops is compelling, open Janus filaments are more attractive from the practical point of view. A variable zonal actuation generating variable shapes can lead to artificial swimmers (Zakharov et al, 2016). Variegated shapes can be produced by weaving Janus filaments in *active textiles* (Zakharov and Pismen, 2019).

6.4 Active Polar Media

6.4.1 Dynamics of Polar Order Parameter

Studies of polar media have been traditionally motivated by ferroelectric structures, but a renewed interest is prompted by the polar character of cellular cytoskeleton and epithelial tissues. Rheology of biological materials is far more complex than in simple fluids and elastic solids. Pleiner et al (2016) developed a sophisticated theory of active polar materials, which systematically incorporates transient networks into hydrodynamics. A widely used approach is to present the medium as a two-phase continuum. The poroelastic theory, originated in the studies of fluid-saturated porous soils (Biot, 1941) but widely applied to biological materials (Mogilner and Manhart, 2018), treats the material as a liquid-filled elastic base.

On a more basic level, Kruse et al (2005) treated a polar gel as an intermediate between a fluid and a liquid described by the Maxwell model. This model relates the symmetric stress tensor to the symmetric part of the strain rate tensor s_{ij} as $(1+\tau\partial_t)\sigma_{ij} = 2\eta s_{ij}$, where τ is the viscoelastic relaxation time: at short times, the gel behaves as a solid with the Young modulus $E = \eta/\tau$, and at long times, as a viscous fluid. However, the distinction between the current (deformed) and reference frames (see Sect. 6.3.1) is missing in this study. An appropriate description of diverse rheology of gels is beyond the scope of this book, and we will consider fluid and elastic polar media along the same lines as nematic fluids and elastomers in Sects. 6.2 and 6.3.

The theory of passive polar fluids follows the same strategy as in Sect. 6.2.1, rooted in the Ericksen –Leslie theory starting with the rate of entropy production. In spite of different symmetry, the terms of the polar Lagrangian in 2D are identical, up to the change of notation, to those of the nematic Lagrangian (6.2.32):

$$\mathcal{L} = -\tfrac{1}{2}\alpha_0\varrho^2(1 - \tfrac{1}{2}\varrho^2) + \tfrac{1}{2}K_1(\operatorname{div}\mathbf{p})^2 + \tfrac{1}{2}K_2(\operatorname{curl}\mathbf{p})^2. \tag{6.4.1}$$

Giomi et al (2008), following the work related to ferroelectrics (Blankschtein and Hornreich, 1985) added the term $\frac{1}{2}|\mathbf{p}|^2\nabla\cdot\mathbf{p}$ compatible with polar but not nematic symmetry. The Lagrangian may also contain a linear term $\nabla\cdot\mathbf{p}$, allowing for a base state with a constant splay but vanishing upon variation; both extra terms introduce an additional scale. The added nonlinear term is problematic, since it causes the deviation of $|\mathbf{p}|$ from unity to decay in the far field as r^{-1}, in contrast to the r^{-2} decay of other terms. This would cause, for example, anomalously strong interaction between far removed defects, and therefore we will further avoid this term retaining the basic form (6.4.1). The dynamic equations (6.2.15)–(6.2.17) remain applicable with $\mathbf{n}$ replaced by $\mathbf{p}$. These assignments differ from (6.2.39), (6.2.38) because $\mathbf{p}$ rotates with the same speed as the nematic director, twice slower than $\mathbf{q}$, so that antisymmetric tensors are halved only in the case of a $\mathbf{q}$-vector.

Activity may cause polar fluids, similar to nematics, to spontaneously perturb a perfectly ordered immobile state. Kruse et al (2005) emphasized that the source of activity is the energy contributed by chemical transformations, incorporated in the extended version of (6.2.11):

$$T\mathcal{D}_tS = \int\left(\sigma^{\rm s}_{ij}s_{ij} + \mathbf{h}\cdot\mathbf{N} + \Delta\mu\partial_t c + \partial_i j_i^{(k)}\mu^{(k)}\right)\mathrm{d}\mathbf{x}, \tag{6.4.2}$$

where $\mathbf{N} = \mathcal{D}_t\mathbf{p} - \boldsymbol{\Omega}\cdot\mathbf{p}$ with the tensor $\boldsymbol{\Omega}$ defined by (6.2.9). The novelty is in the last two terms, which include the chemical force due to a shift of the chemical potential $\Delta\mu$ conjugate to the consumption rate $\partial_t c$ of an active chemical (e.g., ATP) and the exchange of other constituents of the gel due to fluxes $j_i^{(k)}$ conjugate to their chemical potentials $\mu^{(k)}$. However, chemical inputs do not serve the same purpose of deriving dynamic equations compatible with the Onsager reciprocity relations as in the standard theory, since they are capable to decrease entropy.

The active stress tensor stemming from the active terms in (6.4.2) should be added to the passive stress defined by (6.2.17) with $\mathbf{n}$ replaced by $\mathbf{p}$. Kruse et al constructed for this purpose all possible tensorial combinations with phenomenological coefficients $\zeta, \zeta', \zeta'', \nu, \nu'$:

$$\sigma^{\rm a}_{ij} = -(\zeta p_i p_j + \zeta'\delta_{ij} + \zeta''|p|^2\delta_{ij})\Delta\mu + \tfrac{1}{2}\nu(p_i h_j + p_j h_i) + \nu' p_k h_k\delta_{ij}. \tag{6.4.3}$$

The first term in parentheses, equivalent to that commonly applied to nematics, was used by Giomi et al (2008), among others, and criticized by Brand et al (2014) as being even under time reversal. However, here it is multiplied by a shift of the chemical potential, which should be a result of an irreversible chemical reaction. This effectively corrects the time reversal signature in same way as discussed in the beginning of Sect. 6.2.2. These terms make it possible to input essential experimental information, and eliminates the necessity of adding an *ad hoc* active tensor. It should be noted that the scalar terms containing δ_{ij} do not generate flow but only contribute to pressure in an incompressible fluid.

A problematic point is that active stress defined by (6.4.3) is quadratic in the order parameter and does not affect linear stability analysis. Voituriez et al (2005) avoided this problem by restricting the modulus to unity with the help of a Lagrange multiplier, so that the nonlinear stress reduces just to a trigonometric function. In this way, the problem maps onto that of perturbations of the nematic director and, indeed, it was a source of the work by Edwards and Yeomans (2009). Recall, however, that instability turns out there to be possible even in the absence of activity when stabilizing influence of perturbations of the modulus is not taken into account (Sect. 6.2.2).

Giomi et al (2008) suggested another form of active stress, $\sigma^{\mathrm{a}}_{ij} = \beta(\partial_i p_j + \partial_j p_i)$, which may arise from self-propulsion of active units and is exclusive to polar media[4]. This form affects linear stability, and incorporating it in the hydrodynamic equation (6.2.16) leads, following the Fourier transform, to the equation of the stream function equivalent to (6.2.48) but with $\widehat{\mathrm{q}}$ replaced by $\widehat{\mathrm{p}}$ and an imaginary right-hand side:

$$f_1 = \tfrac{1}{2}\mathrm{i}k^3 p_1 \sin\varphi, \quad f_2 = \tfrac{1}{2}\mathrm{i}k^3 p_2 \cos\varphi. \tag{6.4.4}$$

This introduces imaginary entries into the matrix (6.2.51), so that the Routh–Hurwitz stability criterion is no longer applicable. There are suitable algorithms, though quite complicated, in mathematical literature (Frank, 1946; Agashe, 1985), but one can see without applying them that the active term defined in this way, having an extra derivative, comes with a higher power of the wavenumber k and therefore the lowest-order expressions of the common Routh–Hurwitz stability criteria remain real, and the long-scale stability conditions are not affected. We will see in Sect. 6.4.3 that the ordered polar state is always unstable even without active stress when a variable concentration of active particles is included.

Topological defects in a polar medium have a unit charge, like dislocations in striped patterns described by the real Ginzburg–Landau equation (RGL) (1.1.7) and spiral waves in the complex Ginzburg–Landau equation (1.1.6). The phase θ of the polar vector order parameter $\mathbf{p} = \varrho\{\cos\theta, \sin\theta\}$ equals the alignment angle rather

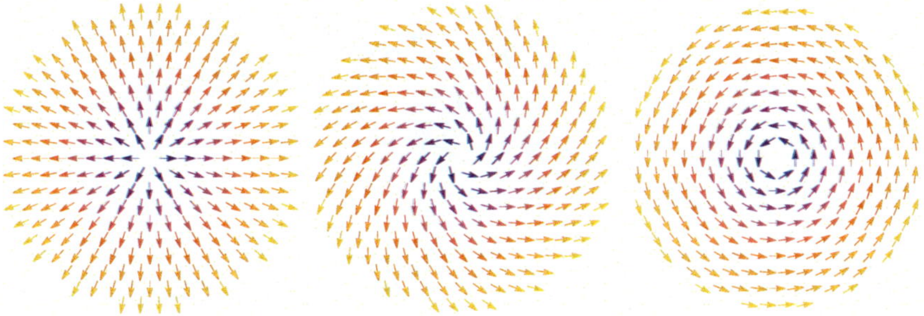

Fig. 6.32. Aster, spiral, and vortex alignment patterns in a unit-charge defect

[4] Other active terms in this work include concentration of an active agent as in Sect. 6.4.4

than its double, which comes to the same relation $\theta \propto \phi$. However, unlike half-charged defects, which have, up to rotation, a unique structure, the orientation of the polar order parameter $\mathbf{p}$ in unit-charge defects is variable, dependent on a constant phase shift ϕ_0. The generic shape is a spiral, interpolating between an aster at $\phi_0 = 0$ or $\phi_0 = \pi$ and a vortex at $\phi_0 = \pm\pi/2$, as shown in Fig. 6.32. These structures have the same energy in the RGL equation, but generally are not equivalent. The distinction can be clearly seen when the Lagrangian is expressed in the polar form

$$\mathcal{L} = -\tfrac{1}{2}\alpha_0\varrho^2(1 - \tfrac{1}{2}\varrho^2) + \tfrac{1}{2}r^{-2}(K_1 \cos^2\phi_0 + K_2 \sin^2\phi_0)\partial_r(r\varrho)^2. \qquad (6.4.5)$$

Since an aster has only splay and a vortex, only bend distortion energy, it is clear that the aster structure has the lowest energy at $K_2 > K_1$ and the vortex structure otherwise. This decides in a trivial way which structure is preferred in the absence of flow and chemical inputs.

In the polar representation, the molecular field has a single component:

$$h = \alpha_0\varrho(1 - \varrho^2) + (K_1 \cos^2\phi_0 + K_2 \sin^2\phi_0)[\varrho''(r) + r^{-1}\varrho'(r) - r^{-2}\varrho]. \quad (6.4.6)$$

The stationary structure defined by $h = 0$ reduces to the stationary solution of the RGL equation under rescaling $r \to r\alpha_0^{-1}(K_1 \cos^2\phi_0 + K_2 \sin^2\phi_0)^{1/2}$. Long-scale interactions between defects driven by this molecular field are independent of their structure, up to this scaling factor, and are identical to interactions of disclinations in striped patterns (Sect. 4.3) or of half-charged defects (Sect. 6.2.6), excluding the latter's self-propulsion.

6.4.2 Dynamic Selection of the Structure of Defects

A branched interconnected dynamic structure of the cytoskeleton sustained by a complex set of chemical transformations is beyond the scope of the basic model at our disposal, but it can address patterns characteristic to polar fluids observed in *in vitro* experiments with a mixture of microtubules and molecular motors in a constrained

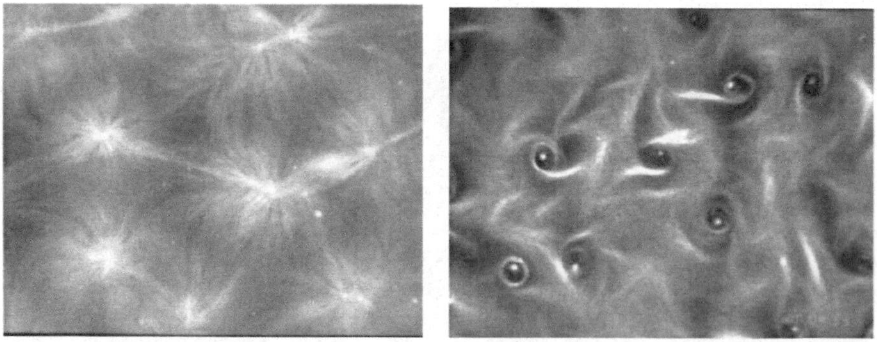

Fig. 6.33. Aster and vortex array patterns in a mixture of microtubules and molecular motors (Nédélec et al, 1997, ©Springer Nature)

2D geometry, such as shown in Fig. 6.33 (Nédélec et al, 1997). We concentrate therefore upon understanding these patterns, emphasized by Kruse et al (2005) as their essential result. The filaments in these experiments are not interconnected, and therefore the medium can be viewed as an active polar fluid and analyzed along the lines of Sects. 6.2.1 and 6.2.3.

While relations between the elastic constants affect only equilibrium energy levels, the effect of active stress should play a major role in the transition from static to dynamic structures, and thereby it may elucidate the difference between the experimental patterns in Fig. 6.33 where dynamic asters were observed at higher, and vortex arrays at lower motor concentrations. In the realistic case of frictional dynamics, the equation of the stream function Ψ is obtained by setting in (6.2.43) $v_i = \varepsilon_{ij}\partial_j \Psi$ and taking the curl to eliminate pressure:

$$(\rho\partial_t + \eta' v_i)\nabla^2\Psi = -\varepsilon_{ik}\partial_k\partial_j\partial_j(\sigma^{\rm p}_{ij} + \sigma^{\rm a}_{ij}). \tag{6.4.7}$$

We will concentrate upon the first term in (6.4.3). Its contribution is evaluated as

$$\Psi''(r) + r^{-1}\Psi'(r) = -\zeta\Delta\mu \sin 2\phi_0 r^{-1}[r\varrho\varrho''(r) + r\varrho'(r)^2 + 3\varrho\varrho'(r)]. \tag{6.4.8}$$

The terms containing δ_{ij} do not generate convection, and, since $\mathbf{h}$ is parallel to $\mathbf{p}$ in the stationary state, the contribution of the ν-term just modifies the radial dependence of the right-hand side of (6.4.8). The induced velocity is angular, but depends on r, decaying in the far field. Denoting the bracketed expression as $f(r)$, it is expressed in quadratures as

$$v_\phi(r) = -\frac{1}{r}\Psi'(r) = \frac{\zeta\Delta\mu\sin 2\phi_0}{r^2}\int_0^r f(r')r'{\rm d}r'. \tag{6.4.9}$$

The numerical solution using the approximant (4.2.15) of $\varrho(r)$ is shown in Fig. 6.34a. Passive stress, given by (6.2.17) with $\mathbf{n}$ replaced by $\mathbf{p}$ and $\boldsymbol{\pi}, \mathbf{h}$ defined by (6.2.33), (6.2.34) with $\mathbf{q}$ replaced by $\mathbf{p}$, cannot generate convection in a stable stationary structure, and one can check that it does not.

Velocity vanishes at ϕ_0 equal to any multiple of $\pi/2$, so that neither asters nor vortices rotate, and we need to investigate stability of these stationary states to advective perturbations. For an aster, setting $\phi_0 = \epsilon^{1/2}\varphi(r,\phi)$ and expanding (6.4.5) in $\epsilon \ll 1$, we obtain the $O(\epsilon)$ Lagrangian

$$\begin{aligned}\widetilde{\mathcal{L}} &= -\alpha_0\varrho(1-\varrho^2)\widetilde{\varrho} - \tfrac{1}{2}\alpha_0\varrho(1-\varrho^2)\varphi \\ &\quad + r^{-2}K_1\partial_r(r\varrho\widetilde{\varrho}) + \tfrac{1}{2}r^{-2}(K_2 - K_1)\partial_r(r\varrho)^2\varphi,\end{aligned} \tag{6.4.10}$$

where the function $\varrho(r)$ is determined by the stationary solution and $\widetilde{\varrho}(r,\phi)$ is an $O(\epsilon)$ perturbation. The tensor $\boldsymbol{\pi}$ has a single component $\pi_{r\widetilde{\varrho}} = \varrho/r$, and the $O(\epsilon)$ molecular field depends on ϱ only:

$$\begin{aligned} h_{\widetilde{\varrho}} &= \alpha_0\varrho(1-\varrho^2) - 2K_1 r^{-2}\varrho, \\ h_\varphi &= \tfrac{1}{2}\alpha_0\varrho(1-\varrho^2) + \tfrac{1}{2}(K_1 - K_2)[r^{-1}\partial_r(r\varrho)]^2. \end{aligned} \tag{6.4.11}$$

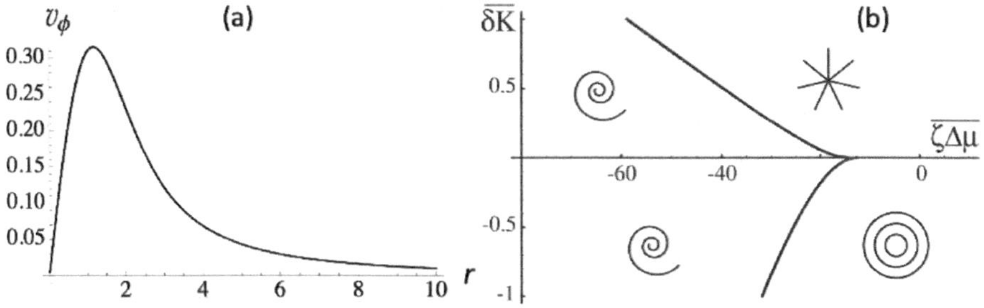

Fig. 6.34. (a) The angular velocity v_ϕ (scaled by $\zeta\Delta\mu \sin 2\phi_0$) induced by the active stress, plotted as a function of the rescaled radius. (b) Stability diagram of asters, vortices and spirals dependent on activity $\zeta\Delta\mu$ and the relation between splay and bend elasticities $\delta K = (K_2 - K_1)/(K_2 + K_1)$ (Kruse et al, 2005, ©Springer)

The perturbation problem can be solved by evaluating the $O(\epsilon)$ passive stress tensor (6.2.17) with $n_1 = \widetilde{\varrho}$, $n_2 = \varphi$, $h_1 = h_{\widetilde{\varrho}}$, $h_2 = h_\varphi$, computing the stream function generated by this stress, and substituting the symmetric and antisymmetric tensors defined thereby in the dynamic equation (6.2.15) with the variables replaced in the same way. This should result in rather complicated coupled linear differential equations for $\widetilde{\varrho}, \varphi$, burdened by the radial dependence through the function $\varrho(r)$. Similarly, for a vortex, this expansion is carried out in the vicinity of $\phi_0 = \pi/2$. This route to the solution of the problem is outlined here for the benefit of a reader who might undertake this task combining analytical transformations and numerics.

Kruse et al (2005) did not follow this painful route, but worked in an informal way. They replaced the algebraic part by a Lagrange multiplier ensuring $\varrho = 1$, so that the decay of the modulus in the defect core was ignored. Without explanations, they restricted the study to modes varying along a single direction, assuming the system to be homogeneous in the other directions. The reader can judge whether this is possible when studying stability of a circularly symmetric structure. Also, they used a different Lagrangian, where, in contradiction to the principle of Landau's expansion, the role of the common bend term plays a term of the sixth combined order in ∇ and $\mathbf{p}$, but this does not alleviate the computations.

Nevertheless, the result, illustrated by Fig. 6.34b, should be qualitatively correct to some extent, reflecting the expected general form of this diagram. Prevalence of either aster or vortex static structures, dependent on the difference of elasticities, follows from (6.4.5), as we have already noticed. These structures must become unstable at a certain level of activity, although asters and vortices are not likely to be destabilized at the same threshold of contractile activity as shown in the picture. Indeed, a systematic analysis by means of lattice Boltzmann simulations (Elgeti et al, 2011) has shown that extensile activity favors spirals and contractile activity favors asters. It is also possible that selection of different patterns in the experiments by Nédélec et al is not related to convective instabilities at all, but is caused by specific interactions between the filaments and molecular motors, according to the model by Aranson and Tsimring (2006).

6.4.3 Instabilities Due to Variable Concentration

Polar media often owe their properties to a variable concentration of polar particles, the effect of the opposite nature, but subject to the same kind of analysis as doped nematic media. Giomi et al (2008), addressing this problem, included concentration as the additional variable complementing the polar order parameter in the Lagrangian (6.4.1). We will write it in the dimensionless form, as defined in the end of Sect. 6.2.3, and scale the concentration by its critical value at the threshold of the polar order:

$$\mathcal{L} = -\tfrac{1}{2}\varrho^2(c - 1 - \tfrac{1}{2}\varrho^2) + \tfrac{1}{2}(\operatorname{div}\mathbf{p})^2 + \tfrac{1}{2}\kappa(\operatorname{curl}\mathbf{p})^2 + \beta c \operatorname{div}\mathbf{p}. \qquad (6.4.12)$$

The algebraic term describes enhancement of polarization by active particles; c_0 is the threshold level beyond which a polarized state exists. The interaction term is qualitatively different from that in nematic media, and it affects only the splay of the polar vector. The molecular field is

$$\begin{aligned} h_1 &= p_1(c - 1 - \varrho^2) + \partial_1^2 p_1 + \partial_1\partial_2 p_2 + \kappa(\partial_2^2 p_1 - \partial_1\partial_2 p_2) + \beta\partial_1 c, \\ h_2 &= p_2(c - 1 - \varrho^2) + \partial_2^2 p_2 + \partial_1\partial_2 p_1 + \kappa(\partial_1^2 p_2 - \partial_1\partial_2 p_1) + \beta\partial_2 c. \qquad (6.4.13) \end{aligned}$$

Since the particles are supposed to be macroscopic, and sharp interphase boundaries cannot form in a fluid, we do not introduce here the concentration Lagrangian (6.3.11), so that chemical potential $\mu = \delta\mathcal{F}/\delta c$ is derived exclusively from (6.4.12), but retain effective diffusivity D, which may be caused by random collisions between particles. Then the convective-diffusion equation defining dynamics of concentration subject to the conservation law takes the form

$$\mathcal{D}_t c = \nabla^2(\gamma\mu + Dc), \quad \mu = \tfrac{1}{2}\varrho^2 + \beta \operatorname{div}\mathbf{p}, \qquad (6.4.14)$$

where γ is the ratio of diffusional to rotational mobilities; advection enters through the substantial derivative $\mathcal{D}_t = \partial_t + \mathbf{v}\cdot\nabla$.

Giomi and Marchetti (2012) investigated stability of a polarized state to spontaneous advection and concentration perturbations affecting polarity. We will treat this problem differently, in the same way as in Sect. 6.2.4, and consider a 2D slice of unconstrained 3D space with circular symmetry relative to the unperturbed alignment. The analysis is more complicated due to the involvement of an additional variable and appearance of complex entries stemming from the linear splay term. The initial modulus at $c > 1$ is $\varrho_0 = \sqrt{c-1}$, and the base state is $p_1 = \varrho_0$, $p_2 = 0$, $c = c_0$ with $1 \leqslant c_0 \leqslant 2$. As usual, perturbations $\epsilon\widetilde{p}_i$, $\epsilon\widetilde{c}$ are expanded in the Fourier series as $\epsilon\widehat{\mathbf{p}}e^{i\mathbf{k}\cdot\mathbf{x}}$, $\epsilon\widehat{c}e^{i\mathbf{k}\cdot\mathbf{x}}$. The Fourier-transformed $O(\epsilon)$ molecular field is

$$\widehat{\mathbf{h}} = [-2\varrho_0^2\widehat{\mathbf{p}} + \varrho_0\widehat{c})]\delta_{i1} - k^2\mathbf{H}\cdot\widehat{\mathbf{p}} + \varsigma\widehat{c}, \qquad (6.4.15)$$

where $\varsigma_1 = ik\beta\cos\varphi$, $\varsigma_2 = ik\beta\sin\varphi$, and the tensor $\mathbf{H}$ is defined by (6.2.45).

The passive stress is linearized and Fourier-transformed as

$$\widehat{\sigma}_{ij}^{\mathrm{p}} = \tfrac{1}{2}\chi(p_i^0\widehat{h}_j + \widehat{h}_i p_j^0) - \tfrac{1}{2}(p_i^0\widehat{h}_j - \widehat{h}_i p_j^0), \qquad (6.4.16)$$

and the Navier–Stokes equation (6.2.16) is expressed after taking the curl as

$$\rho\mathcal{D}_t + \eta k^2\widehat{\Psi} = f_c\widehat{c} - \mathbf{f}\cdot\widehat{\mathbf{p}}, \quad f_c = \tfrac{1}{2}\chi\varrho_0^2\sin 2\varphi + \mathrm{i}k\beta\varrho_0(\chi-1)\sin\varphi,$$
$$\mathbf{f} = \begin{pmatrix}\varrho_0^3\sin 2\varphi\\ 0\end{pmatrix} - \frac{k^2}{2}\varrho_0\begin{pmatrix}\tfrac{1}{2}[1+\chi-\kappa\varrho_0^2]\sin 2\varphi\\ (1+2\chi)\sin^2\varphi + \kappa(1-2\chi\cos^2\varphi)\end{pmatrix}. \tag{6.4.17}$$

The transformed dynamic equation of $\widehat{p}$, similar to (6.2.39), is

$$\mathcal{D}_t\widehat{\mathbf{p}} = \widehat{\mathbf{h}} + \tfrac{1}{2}k^2\mathbf{g}\widehat{\Psi}, \quad \mathbf{g} = \begin{pmatrix}\chi\sin 2\varphi\\ -(1+\chi)\cos 2\varphi\end{pmatrix}. \tag{6.4.18}$$

The new element of the problem is concentration dynamics defined by (6.4.14):

$$\mathcal{D}_t\widehat{c} = -k^2\left(\gamma\widehat{p}_1 + D\widehat{c} + \varsigma\cdot\widehat{\mathbf{p}}\right). \tag{6.4.19}$$

Stability requires the eigenvalues λ of the following matrix to have negative real parts:

$$\mathbf{J} = \begin{pmatrix} -k^2\eta/\rho, & -f_1/\rho, & -f_2/\rho, & f_c/\rho\\ \tfrac{1}{2}k^2 g_1, & -2-k^2H_{11}, & -k^2H_{12}, & \varrho_0+\zeta_1\\ \tfrac{1}{2}k^2 g_2, & -k^2H_{21}, & -k^2H_{22}, & \zeta_2\\ 0, & -k^2(\gamma+\zeta_1), & -k^2(\gamma+\zeta_2), & -k^2D \end{pmatrix}. \tag{6.4.20}$$

The problem looks too complicated: the matrix is complex, so that the Routh–Hurwitz stability criterion does not apply. We start with the reduced problem with the constant concentration $c_0 = 2$ and $\varrho_0 = 1$ to see whether instabilities are possible also in this case. The simplest way is to check perturbations parallel or normal to the original alignment. At $\varphi = 0, \pi$, the term with the lowest power of k of the determinant of $\mathbf{J}$ is independent of the splay elasticity, and a monotonic instability takes place at $|\chi| > 2\sqrt{\eta}+1$. At $\varphi = \pm\pi/2$, the leading term is $\frac{1}{2}\rho^{-1}(4\eta-(1+\chi)^2$, independently of the bend elasticity, and there is an instability at $\frac{1}{2}\rho^{-1}\kappa(1+4\eta-\chi^2)$. The former limit is more restrictive. The leading term of the criterion of the oscillatory instability is positive definite.

Restoring the concentration variable, we find that the terms with the lowest power of k of all Hurwitz determinants are real, so that this stability criterion is applicable. The instability condition to parallel perturbations stemming from the leading term of the respective determinant is $\eta < \varrho_0(\chi^2-1)$. The similar condition for perpendicular perturbations is more complicated:

$$\eta < \frac{(\chi+1)[4\beta(\chi-1)\varrho_0 + (\chi+1)(\gamma\varrho^2+2D\varrho)]}{4(2D+\gamma\varrho-2\beta^2)}. \tag{6.4.21}$$

This limit may even grow indefinitely at large interaction coefficient β. Oscillatory instability is also possible, with the threshold conditions dependent on all parameters of the problem, and we may not forget of other perturbation angles. These instabilities encroach on the parametric domain where the system is stable in the absence of concentration inhomogeneities, and they take place with no added active stress.

6.4.4 Chemo-elastic Interactions

In elastic media, the mechanism of interaction between polarization and concentration inhomogeneities is different: through chemical modification rather than advection. The minimal scheme of interactions includes production of a signaling species c induced by deformation, polarization $\mathbf{p}$ directed along the gradient of this species, and deformation caused by polarization. Köpf and Pismen (2013b), adopting this general scheme, assumed the elastic layer to be incompressible but capable to stretch in the horizontal plane $\mathbf{x}$, while modifying its thickness $l = l_0 J^{-1}$, where l_0 is the unperturbed thickness and J is the determinant of the 2D deformation tensor $F_{ij} = \partial x_i/\partial X_j$, defined, as in Sect. 6.3.1, through the derivative of current $\mathbf{x}(t)$ with respect to reference $\mathbf{X}$ coordinates.

The directionality of the polar order parameter allows us to set up the problem in a way different from Sect. 6.3.1: coupling between deformation and polarization will be caused here not through the growth factor as in (6.3.2) but through the active force directed along $\mathbf{p}$, which formally appears as an external force, as it cannot be defined as a stress gradient. On the contrary, stress can accumulate within the layer under the action of a part of the active force that is not dissipated frictionally. In the absence of a growth factor, the strain tensor u_{ij} is identical to the deformation tensor, and the elastic Lagrangian (6.3.1) is modified by allowing for a a variable thickness in leu of the 2D incompressibility condition:

$$\mathcal{L}_{\text{el}} = \nu l \mathcal{L}_0, \quad \mathcal{L}_0 = \tfrac{1}{2}|u_{ij}|^2. \tag{6.4.22}$$

We neglect here the surface energy due to weak undulations of the free surface of the layer, since surface tension effects in solids are commonly negligible compared to elasticity. Varying (6.4.22) and adding an active force per unit volume $\mathbf{F} = f\mathbf{p}$, we obtain the dynamic equation

$$\eta \mathcal{D}_t x_i = \partial_j \left(J^{-1} u_{jk} \partial_k x_i - J^{-2} \mathcal{L}_0 \varepsilon_{ik} \varepsilon_{jl} \partial_l x_k \right) + J^{-1} f p_i, \tag{6.4.23}$$

where η is the effective friction coefficient. With the energy, time, and length units adopted before, the parameters of (6.4.23) are rescaled as $\eta \to \eta \Gamma K_1/(\nu l_0)$, $\gamma \to \gamma/\nu$ to transform this equation to the dimensionless form.

The polar Lagrangian retains the dimensionless form (6.4.12). Since, unlike Sect. 6.3.1, polarization is not incorporated in the elastic Lagrangian (6.4.22), we do not consider here possible effects of strain on the polarization direction and elastic anisotropy of a polarized medium. We also assume that rotation of the polarization vector induced by deformations is ineffective. With these simplifications, the only effect of deformation on evolution of the polar order parameter is mediated by the extension factor J through the replacement $\partial_j \to J^{-1}\partial_j$, and the dimensionless dynamic equation takes the form

$$\begin{aligned} \mathcal{D}_t p_i = {} & \left(c - 1 - |\mathbf{p}|^2\right) p_i + \partial_i (J^{-1} \partial_j p_j) \\ & + \kappa \partial_j \left[J^{-1} \left(\partial_j p_i - \partial_i p_j \right) \right] - \beta \partial_i (J^{-1} c). \end{aligned} \tag{6.4.24}$$

The reader is encouraged to try and enhance the model by incorporating the cross-interaction terms in (6.3.6) and (6.3.7).

The feedback loop is closed by making the scalar order parameter c strain-dependent. By symmetry, the simplest choice is to make its generation rate proportional to $J-1$. This coupling may be caused, for example, by increased adsorption of an active chemical in a stretched layer or by temperature rise due to increased opacity of thicker regions in an irradiated layer. In the biological context, this dependence may stem from modifying catalytic action of proteins due to strain-induced conformational changes. Allowing for the production rate proportional to the local strain with a coefficient $\varsigma > 0$ and a linear decay with a constant k, and including also, as in (6.4.14), the action of chemical potential derived from the polar Lagrangian, we write the concentration equation in the dimensionless form

$$\mathcal{D}_t c = J\partial_i \left[J^{-1}D\partial_i c + \gamma\partial_i(J^{-1}\beta\partial_j p_j - \tfrac{1}{2}\varrho^2)\right] - \chi c + \varsigma(J-1), \qquad (6.4.25)$$

where, to comply with the adopted time and length units, diffusivity is rescaled as $D \to D\Gamma K_1$, and the decay constant, as $\chi \to \chi\Gamma\alpha_0$. We have taken here into account that, due to incompressibility, concentration is not affected by deformation but the diffusional flux is proportional to the local thickness.

The above equation system becomes extremely convoluted if applied in a fixed coordinate frame where, besides computing the substantial derivative that includes the velocity $\mathbf{v} = \partial_t \mathbf{x}$, one would need to take care of transformations between the current and reference frames, as stipulated in Sect. 6.3.1. However, these complications do not affect the following linear stability analysis, and can be avoided at large deformations by carrying out numerical computations in the Lagrangian frame.

We explore now stability of the homogeneous unpolarized state $\mathbf{p} = 0$, $c = 0$, $\mathbf{x} = \mathbf{X}$ to infinitesimal perturbations. The linearized system determining bifurcation thresholds can be expressed through the 2D strain $\psi = J - 1 = \nabla_i(x_i - X_i)$ and splay $\phi = \partial_i p_i$ by taking the divergence of the linearized equations (6.4.23), (6.4.24):

$$\eta\,\partial_t\psi = \nabla^2\psi + f\phi, \quad \partial_t\phi = \nabla^2\phi - \phi - \beta\nabla^2 c, \qquad (6.4.26)$$

$$\partial_t c = D\nabla^2 c - \chi c + \varsigma\psi + \beta\gamma\nabla^2\phi. \qquad (6.4.27)$$

In order to reduce the number of parameters, we simplify the problem by dropping the inessential last term in (6.4.27) and rescale $c \to \varsigma c$, $\beta \to \beta/\varsigma$, so that (6.4.27) reduces to

$$\partial_t c = D\nabla^2 c - \chi c + \psi. \qquad (6.4.28)$$

The condition of monotonic instability stemming from the determinant of the Jacobi matrix of the system (6.4.26) (6.4.28) is

$$\beta f > k^2(\chi + Dk^2)(1 + k^2) \equiv g(k^2). \qquad (6.4.29)$$

The lowest force level at the instability limit is attained when the determinant vanishes at a wavenumber k corresponding to the maximum of the function $g(k^2)$:

$$k^2_{\max} = \frac{1}{3}\left[\sqrt{1 + \chi^2 + (3\beta f - \chi)/D} - (1 + \chi/D)\right]. \qquad (6.4.30)$$

This value is positive at $\beta f > \chi$. Using this in (6.4.29) leads to the relation between parameters of the problem at the instability of a Turing type. It is shown in Fig. 6.35a as a parametric relation between χ and βf at $D = 1$.

The wavenumber at the oscillatory instability is determined in a similar way by finding a wavenumber k corresponding to maximum of the second Routh–Hurwitz stability criterion at the point where it vanishes. This value, unlike that of the monotonic instability, depends on the effective viscosity η, and is quite convoluted; it is given below for the particular value $D = 1$:

$$k_{\rm max}^2 = \frac{1}{6}(1+\eta)^{-1}\left[\sqrt{(1+\chi)^2 + 3\eta^2(1-\chi)^2 - 6\eta\beta f} - \eta^2(1+4\eta+3\eta^2)\right]. \tag{6.4.31}$$

The relation between βf and χ at the instability threshold at $D = 1$ and some chosen values of η is shown in Fig. 6.35b. The oscillatory instability, unlike the monotonic one, takes place at negative values of the splay interaction parameter β.

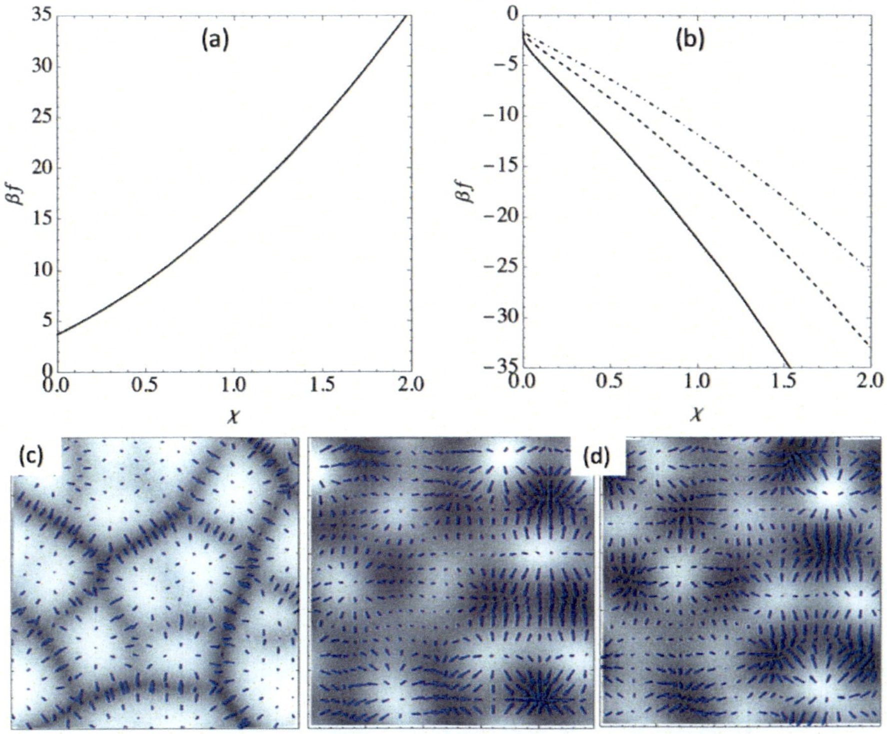

Fig. 6.35. (a,b) The dependence of βf on χ at the monotonic (a) and oscillatory (b) instability threshold at $D = 1$. The solid, dashed, and dot-dashed lines in (b) correspond to $\eta = 1, 1/2, 1/4$, respectively. (c) A stationary pattern midway through a coarsening sequence. (c,d) Antiphase snapshots of the oscillatory pattern. Lighter shades correspond to higher concentration levels; scaled arrows indicate the polarization vector $\mathbf{p}$ (Köpf and Pismen, 2013b, ©Elsevier)

The wavelengths are also quite different: k^2 is in the range between 1.5 and 4.5 in Fig. 6.35a and below 0.4 in Fig. 6.35b. The determinant also vanishes at $k = 0$, and numerical computations (Fig. 6.35c) show that the emerging patterns, starting from an initial perturbation, coarsen with time.

The biologically motivated application of this model (Köpf and Pismen, 2013c) addressed a spreading epithelial layer. The involvement of a signaling species is motivated here by experiments demonstrating the role of a particular enzyme in spreading (Nikolić et al, 2006). The substantial new element of this problem is the existence of a moving boundary. It is treated as a sharp interface by adding a wetting energy $\mathcal{F}_w$ to the overall free energy of a layer with the area S. The motion of the interface is described in terms of the spreading coefficient σ and the extension factor J:

$$\mathcal{F}_\mathrm{w} = \sigma S = \sigma \int_S J\, \mathrm{d}^2\mathbf{x}, \quad \sigma = \gamma_\mathrm{sf} - (\gamma_\mathrm{st} + \gamma_\mathrm{tf}), \tag{6.4.32}$$

where γ's are interfacial energies of the substrate-fluid (sf), substrate-tissue (st), and tissue-fluid (tf) interfaces. The variation of $\mathcal{F}_\mathrm{w}$ vanishes in the bulk where local movements do not change the covered area. This leaves only a boundary contribution of the form

$$\left.\frac{\delta \mathcal{F}_\mathrm{w}}{\delta x_i(\mathbf{x})}\right|_{\mathbf{x}\in\partial S} = \sigma \epsilon_{il}\epsilon_{jk} N_j \partial_k x_l = -f^{(\mathrm{w})}(\mathbf{x}) n_i, \tag{6.4.33}$$

where $\mathbf{N}$ and $\mathbf{n}$ denote the unit vectors normal to the boundary in the reference and current frame, respectively.

The computations carried out in the Lagrangian frame are in accord with the prominent feature of unconstrained spreading revealed in the experiment (Poujade et el, 2007): a nonuniform expansion of a monolayer resulting in the formation of sizable protrusions, commonly referred to as fingers. A typical pattern with stronger polarization in the vicinity of the boundary is shown in Fig. 6.36a. The histograms in Fig. 6.36b,c showing the distribution of angles between the local polarization and the velocity in the spreading direction demonstrate, in accordance with the experiment, far stronger correlation in fingers than in the bulk of the tissue.

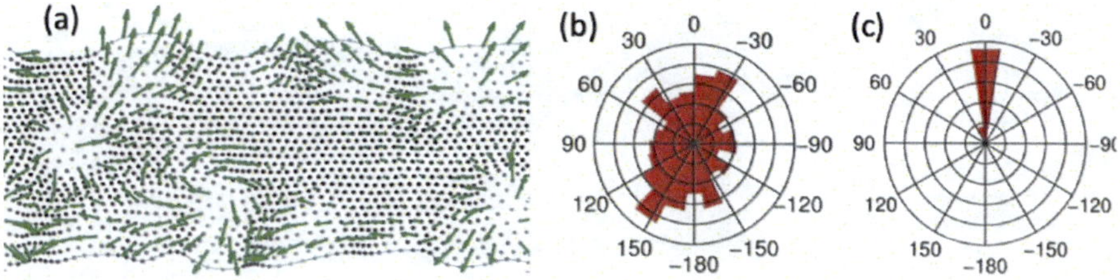

Fig. 6.36. (a) An example of swirling motion in a spreading epithelial layer initiated by noisy initial conditions. The arrows indicate the polarization amplitude and directions. (b,c) The distribution of the angle between the velocity and the unit vector in the spreading direction in the bulk region and in the finger at the wound edge, respectively (Köpf and Pismen, 2013c, ©The Royal Society of Chemistry)

6.4.5 Crawling Cells

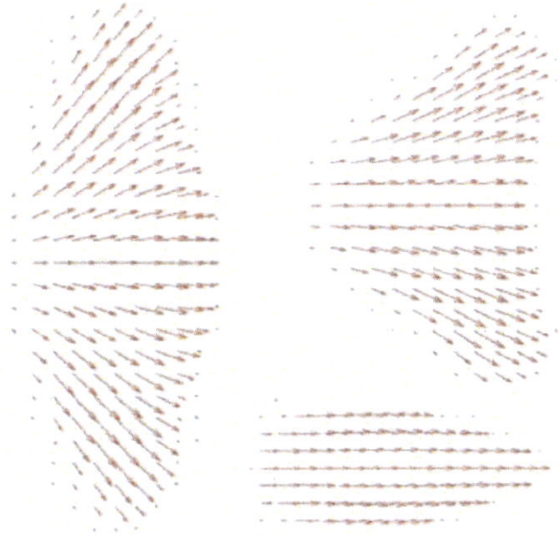

Fig. 6.37. Different shapes and the polarization fields in a moving droplet from the simulations by Tjhung et al (2015), ©Springer Nature

Instabilities in soft polar media have been viewed as a source of motion of cells crawling on a substrate, as well as spreading epithelial layers. The interior of a cell or a tissue is described by the same kind of models as expounded earlier in this section, with the motion of the boundary as the essential ingredient. The simple device enabling to avoid discontinuities while tracking an interface is the phase field model (Sect. 2.3.5) that includes a formal variable governed by a nonlinear diffusion equation having two stable uniform solutions. In order to comply with notation of Sect. 2.3.5, we denote the phase variable as u and its range as $|u| \leqslant 1$. The associated physical variable is density $\rho = \frac{1}{2}(u+1)$, and the two minima of the respective double-well potential (see Fig. 2.1), $\rho = 1$ and $\rho = 0$, are assigned, respectively, to the interior and the exterior of the cell, separated, similar to fronts in Sect. 2.1, by a narrow border region replacing a sharp boundary.

Tjhung et al (2015) incorporated the phase field in the polar Lagrangian, written here in a modified form

$$\mathcal{L} = V(u) + \tfrac{1}{2}D|\nabla u|^2 - \tfrac{1}{2}\alpha(u)\varrho^2(1 - \tfrac{1}{2}\varrho^2) + \tfrac{1}{2}K(\nabla \cdot \mathbf{p})^2 + \beta u \operatorname{div} \mathbf{p}. \quad (6.4.34)$$

The first term is a double-well potential (see Fig. 2.1) with the two minima assigned to the interior and the exterior of the cell. The second term containing the phase diffusivity D is responsible for the effective surface tension of the diffuse interface with the thickness proportional to $\sqrt{D}$. The third term is responsible for the influence of the phase field on the modulus of the polar vector. The next term is elasticity in the one-constant approximation. The last term is equivalent to the interaction term in (6.4.12), but here its primary role is in anchoring of $\mathbf{p}$ to the droplet surface: the energy is minimized if it points perpendicularly outwards at $\beta > 0$ and in the opposite direction at $\beta < 0$. The dynamics of the polar vector is derived from the Lagrangian in the usual way, with the hydrodynamic equation supplemented by the active stress $\sigma_{ij}^{\mathrm{a}} = -\zeta u p_i p_j$. The dynamics of the phase field is derived by varying the energy integral as in Sect. 2.3.5.

This model was the basis of extensive 3D simulations, which reproduced variegated shapes of moving cells. The distribution of polar orientation near the solid support in some simulation runs is shown in Fig. 6.37. Remarkably, earlier biologically motivated 3D simulations (Herant and Dembo, 2010), based on interplay among properties of the cell cytoplasm, its membrane, and the substrate, and not mentioning the polar order parameter at all, arrived at the same shapes of crawling cells as this, physically well motivated, model. This coincidence is quite understandable, as all sophisticated models contain a number of parameters, which can be adjusted to arrive at a result resembling crawling cells observed experimentally.

An earlier work (Ziebert et al, 2012) based on the phase field approach, though physically motivated, approached the problem in an informal way, not complying with the standard formalism. Their basic model includes, besides density (identified with the phase field), a single physical variable – the polarization vector $\mathbf{p}$. The dynamic equations, written in our notation, are

$$\partial_t \mathbf{p} = -(\alpha - \rho^2)\mathbf{p} + K\nabla^2\mathbf{p} - \beta\nabla\rho, \tag{6.4.35}$$

$$\partial_t \rho = D\nabla^2\rho - (1-\rho)(a-\rho)\rho - \alpha\mathbf{p}\cdot\nabla\rho. \tag{6.4.36}$$

Although these equations may look similar to (6.4.24), (6.4.25) with J set to unity, they are built on a different principle, aimed at transparency and efficiency of simulations, and apt to be amended and adjusted to experimental situations, as in later publications of the same group (Ziebert et al, 2016). In the first term, density plays the role of the modulus, but it cannot be defined as $|\mathbf{p}|^2$, and the linear term is meant to specify decay of polarization outside the cell. In the second term, K is denoted as a single elastic constant, but it is also viewed as the diffusivity of the polar vector. The polarization field $\mathbf{p}$ is not conserved, as it can be created by polymerization, which, as stipulated by the β-term, is proportional to the density gradient. Although it is

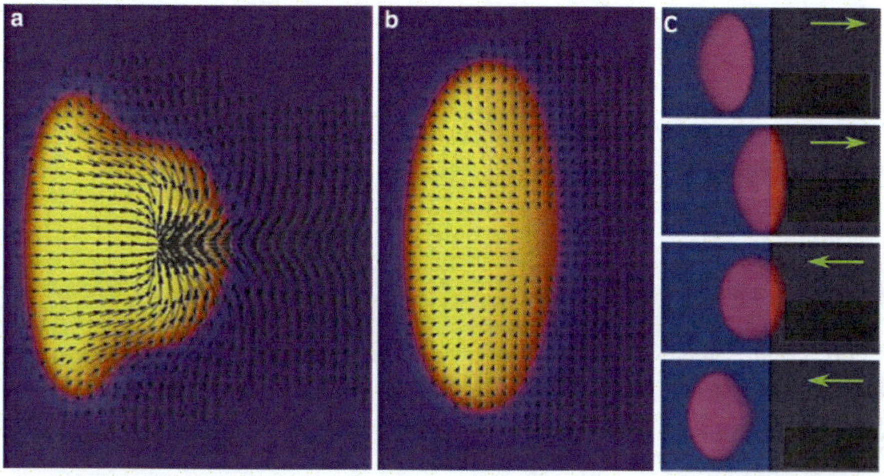

Fig. 6.38. (a,b) Diverse shapes and traction patterns in cells steadily moving to the left, dependent on parameters of the model. The phase field is color-coded, and the traction force exerted on the substrate is shown by the arrows. (c) Motion of a cell on a substrate where the adhesive strength is modulated by a step in the rate of adhesion formation. The cell is reflected from the low-adhesion region (Ziebert et al, 2016, ©Springer Nature)

identical to the interaction term in (6.4.24), it is not derived here from a Lagrangian, and is not matched by a respective term of chemical potential. Instead, the last term in (6.4.36) describes the essential effect of propulsion of the interface due to actin polymerization in the polarization direction. The parameter a biases the prevalence of "empty" and "filled" phases at $a > 1/2$ and $a < 1/2$, respectively, thereby allowing the cell to shrink or grow.

The orientational and mechanical elasticities are merged here in the same way as orientation and propagation directions are merged in the Vicsek model (Sect. 6.1.1), making the momentum conservation equation superfluous. Cell motion is caused by propagation of the diffuse boundary, which is of the same nature as propagation of fronts between alternative stationary states in Sect. 2.1.2. The authors observe that (6.4.36) can be cast as an advection equation

$$\partial_t \rho = \mathbf{v} \cdot \nabla^2 \rho, \qquad \mathbf{v} = \alpha \mathbf{p} - \nabla P \tag{6.4.37}$$

with the effective pressure P expressed as an integral of density. However, they take note (Ziebert et al, 2016) that the incompressibility condition for the flow velocity is not satisfied, which is the price to pay for the simplicity of the model.

The essential element of the problem is interaction of the cell with the substrate. In order to take it into account, Ziebert and Aranson (2013) amended the model by the dynamic equation of the density of adhesive contacts A:

$$\partial_t A = D_A \nabla^2 A + \rho(a_0 |\mathbf{p}|^2) + a_1 A^2) - A(b_0 + b_1 A^2), \tag{6.4.38}$$

containing a weak diffusivity of adhesive contacts D_A. Their formation is proportional to ρ and is enhanced by the intensity or polarization. Interactions between adhesive

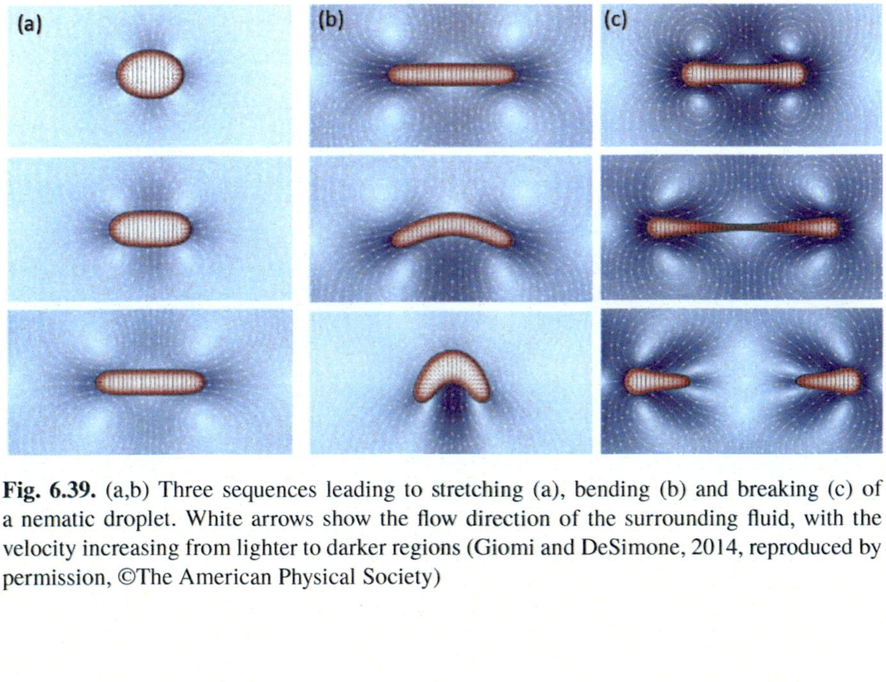

Fig. 6.39. (a,b) Three sequences leading to stretching (a), bending (b) and breaking (c) of a nematic droplet. White arrows show the flow direction of the surrounding fluid, with the velocity increasing from lighter to darker regions (Giomi and DeSimone, 2014, reproduced by permission, ©The American Physical Society)

sites is governed by the last term containing linear and cubic decay terms. Further modifications of the model (Ziebert et al, 2016) include the effect of the traction force exerted by the cell on the distribution of adhesive sites. An example of different shapes and traction patterns is shown in Fig. 6.38a,b. The sequence in Fig. 6.38c demonstrates how a cell, preferring to stay on a highly adhesive substrate, deforms and retracts when encountering a step in the in the rate of adhesion formation a_0. Cells crawling on the same substrate interact through their traction fields. As computational models absorb more biologically motivated detail, they further deviate from theoretical purity of the model by Tjhung et al (2015). More recently, a detailed sharp-interface description (see Sect. 6.4.4) applicable to the motion of both single and interacting cells has been proposed as the alternative to the phase field model (Bresler et al, 2019).

The phase field model has been also applied to an active nematic droplet immersed in a passive isotropic fluid (Giomi and DeSimone, 2014), in conjunction with the 2D Lagrangian based on the Q-tensor (Sect. 6.2.3). The simulations showed different kinds of evolution of an originally round droplet, shown in Fig. 6.39: stretching under the influence of the quadrupolar straining flow generated by the pair of +1/2 defects (a), bending due to splay instability (b), and breaking at a very large activity when surface tension fails to balance the straining flow. However, these simulations do not take into account the solid substrate, which brings in complex 3D effects at the contact line studied in the context of spreading passive nematic fluids (e.g., Lin et al, 2013).

References

1. Abou B, Wesfreid JE, and Roux S (2000), J. Fluid Mech. **416**, 217
2. Aegerter CM and Surko CM (2002), Physica D **166**, 155
3. Agashe SD (1985), IEEE Trans. Automatic Control **30**, 406
4. Aharoni H, Sharon E, and Kupferman R (2014) Phys. Rev. Lett. **113**, 257801
5. Ahlers G (1974), Phys. Rev. Lett. **33**, 1185
6. Alexander S and McTague J (1978), Phys. Rev. Lett. **41**, 702
7. Alonso S, Kähler R, Mikhailov AS, and Sagués F (2004), Phys. Rev. E **70**, 056201
8. Amengual A, Hernández-García E, Montagne R, and San Miguel M (1997), Phys. Rev. Lett. **78**, 4379
9. Angheluta L, Chen Z, Marchetti MC, and Bowick MJ (2021), New J. Phys. **23**, 033009
10. Aranson IS and Bishop A (1997), Phys. Rev. Lett. **79**, 4174
11. Aranson IS and Kramer L (2002), Rev. Mod Phys. **74**, 99
12. Aranson IS and Pismen LM (2000), Phys. Rev. Lett. **84**, 634
13. Aranson IS and Tsimring LS, (2003a), Phys. Rev. E **67**, 021305
14. Aranson IS and Tsimring LS, (2003b), Phys. Rev. E **74**, 031915
15. Aranson IS, Kramer L, and Weber A (1991), Physica D **33**, 376
16. Aranson IS, Aranson L, Kramer L, and Weber A (1992), Phys. Rev. A **46**, R2992
17. Aranson IS, Kramer L, and Weber A (1993), Phys. Rev. E **47**, 3231
18. Aranson IS, Kramer L, and Weber A (1994), Phys. Rev. Lett. **72**, 2316
19. Aranson IS, Bishop A, and Kramer L (1998), Phys. Rev. E **57**, 5276
20. Aranson IS, Malomed BA, Pismen LM, and Tsimring LS (2000), Phys. Rev. E **62**, R5
21. Arnold VI (1984), *Catastrophe Theory*, Springer-Verlag, Berlin
22. Aronson DG and Weinberger HF (1978), Adv Math **30**, 33
23. Ballerini M, . . . , Zdravkovic V (12 coauthors) (2008), Proc. Natl. Acad. Sci USA **105**, 1232
24. Barkley D (1992), Phys. Rev. Lett. **68**, 2090
25. Barkley D (1994), Phys. Rev. Lett. **72**, 164
26. Barré J, Chétrite R, Muratori M, and Peruani F (2015), J. Stat. Phys, **158**, 589
27. Bär M and Eiswirth M (1993), Phys. Rev. E **48**, 1635

L. Pismen, *Patterns and Interfaces in Dissipative Dynamics*, Springer Series in Synergetics,
https://doi.org/10.1007/978-3-031-29579-9

28. Bär M and Or-Guil M (1999), Phys. Rev. Lett. **82**, 1160
29. Bär M, Großmann R, Heidenreich S, and Peruani F (2020), Annu. Rev. Condens. Mat. Phys. **11**, 441
30. Batchelor GK (1967), *An Introduction to Fluid Dynamics*, Cambridge University Press, Cambridge
31. Bec J, Frisch U, and Khanin K (2000), J. Fluid Mech. **416**, 239
32. Bekki N and Nozaki K (1985), Phys. Lett. A **110**, 133
33. Ben-Jacob E, Brand HR, Dee G, Kramer L, and Langer JS (1985), Physica D **14**, 348
34. Bénard H (1900), Ann. Chim. Phys. **7** (Ser 23), 62
35. Bensimon D, Shraiman B, and Croquette V (1988), Phys. Rev. A **38**, 5461
36. Bergé P, Y Pomeau, and Vidal C (1984), *Order Within Chaos: towards a Deterministic Approach to Turbulence*, Wiley, New York
37. Beris AN and Edwards BJ, *Thermodynamics of Flowing Systems, With Internal Microstructure*, Oxford University Press, New York, 1994
38. Bertin E, Droz M, and Grégoire G (2006), Phys. Rev. E **74**, 022101
39. Bertin E, Droz M, and Grégoire G, (2009), J. Phys. A, Math. Theor. **42** 445001
40. Bertin E, Baskaran A, Chaté H, and Marchetti MC (2015), Phys. Rev. E **92** 042141
41. Bialek W, Cavagna A, Giardina I, Mora T, Silvestri E, Viale M, and Walczak AM (2012), Proc. Natl. Acad. Sci. USA **109**, 4786
42. Biot MA (1941), J. Appl. Phys. **12**, 155
43. Blair DL, Neicu T, and Kudrolli A (2003), Phys. Rev. E **67**, 031303
44. Blankschtein D and Hornreich RM (1985), Phys. Rev. B **32**, 3214
45. Bodenschatz E, Pesch W, and Kramer L (1988), Physica D **32**, 135
46. Bodenschatz E, Cannell DS, de Bruin JR, Ecke R, Hu YC, Lerman K, and Ahlers G (1992), Physica D **61**, 77
47. Boghosian BM, Chow CC, and Hwa T (1999), Phys. Rev. Lett. **83**, 5262
48. Bohr T, Huber G, and Ott E (1997), Physica D **106**, 95
49. Bohr T, Jensen MH, Paladin G, and Vulpiani A (1998), *Dynamical Systems Approach to Turbulence*, Cambridge University Press, New York
50. Bowman C and Newell AC (1998), Rev. Mod. Phys. **70**, 289
51. Boyer D and Viñals J (2001), Phys. Rev. E **63**, 061704
52. Boyer D and Viñals J (2002), Phys. Rev. E **65**, 046119
53. Brand HR (1989), Progr. Theor .Phys. Suppl. **99**, 442
54. Brand HR, Pleiner H, and Svenšek D (2014), Eur. Phys. J. E **37**, 83
55. Brand HR, Pleiner H, and Svenšek D (2018), Rheol. Acta **57**, 773
56. Braun E and Steinberg V (1991), EuroPhys. Lett. **15**, 167
57. Bray AJ (1995) in *Formation and Interaction of Topological Defects*, Davis and Brandenberger (eds), NATO ASI Series vol. 349, Plenum Press, New York, p. 105
58. Bresler Y, Palmieri B, and Grant M (2019), Eur. Phys. J. E **42**, 52
59. Brito C, Aranson IS, and Chaté H (2003), Phys. Rev. Lett. **90**, 068301
60. Brusch L, van Hecke M, Zimmermann MG, Bär M, and Torcini A (2000), Phys. Rev. Lett. **85**, 86

61. Brusch L, Torcini A, van Hecke M, Zimmermann MG, and Bär M (2001), Physica D **160**, 127
62. Brusch L, Torcini A, and Bär M (2003), Physica D **174**, 152
63. Buckmaster JD and Ludford GSS (1983), *Lectures on Mathematical Combustion*, Society for Industrial and Applied Mathematics, Philadelphia PA
64. Burgers JM (1948), Adv. Appl. Mech. **1**, 171
65. Busse FH (1978), Rep. Prog. Phys. **41**, 1929
66. Caprini L, Marconi UMB, and Puglisi A (2020), Phys. Rev. Lett. **124**, 078001
67. Chandrasekhar S (1961), *Hydrodynamic and Hydromagnetic Stability*, Oxford University Press, London
68. Chang HC, Demekhin EA, and Kopelevich DI (1993), J. Fluid Mech. **250**, 433
69. Chapman CJ and Proctor MRE (1980), J. Fluid Mech. **101**, 759
70. Chaté H (1994), Nonlinearity **7**, 185
71. Chaté H (2020), Annu. Rev. Condens. Matter Phys. **11** 189
72. Chaté H and Manneville P (1987), Phys. Rev. Lett. **58**, 112
73. Chaté H, Lemaître A, Marcq P, and Manneville P (1996), Physica A **224**, 348
74. Chaté H, Pikovsky A, and Rudzick O (1999), Physica D **131**, 17
75. Ciliberto S, Coullet P, Lega J, Pampaloni E, and Kortan AR (1990), Phys. Rev. Lett. **65** 2370
76. Clerc MG, Escaff D, and Rojas RG (2008), Europhys. Lett. **83** 28002.
77. Clerc MG, Escaff D, and Rojas RG (2022), Chaos Solitons Fractals **165** 112775
78. Cogan NG and Guy RD (2010), HFSP J. **4**, 11
79. Colinet P, Nepomnyashchy AA, and Legros JC (2002), EuroPhys. Lett. **57**, 480
80. Conte R and Musette M (1993), Physica D **69**, 1
81. Conte R, Musette M, Ng TW, and Wu C (2022), Phys. Rev. E **106** L042201
82. Cortese T, Eggers J, and Liverpool TB (2016), EuroPhys. Lett. **115** 28002
83. Coullet P and Emilsson K (1992), Physica D **61**, 119
84. Coullet P, Elphick C, Gil L, and Lega J (1987), Phys. Rev. Lett. **59**, 884
85. Coullet P, Frisch T, and Plaza T (1993), Physica D **59**, 75
86. Coullet P, Lega J, Houchmanzaden B, and Lajzerowicz J (1990), Phys. Rev. Lett. **65**, 1352
87. Coullet P, Riera C, and Tresser C (2000), Phys. Rev. Lett. **84**, 3069
88. Croquette V, Mory M, and Schosseler E (1983), J. Phys. (Paris) **44**, 293
89. Cross MC (1980), Phys. Fluids **23**, 1727
90. Cross MC and Hohenberg P (1993), Rev. Mod. Phys. **65**, 851
91. Cross MC and Newell AC (1984), Physica D **10**, 299
92. Csahók Z and Misbah C (1999), EuroPhys. Lett. **47**, 331
93. Davidsen J, Erichsen R, Kapral, R and Chaté H (2004), Phys. Rev. Lett. **93**, 018305
94. Davis SH (1969), J. Fluid Mech. **39**, 347
95. Dee GT and Langer JS (1983), Phys. Rev. Lett. **50**, 383
96. de Gennes PG, (1975) C. R. Acad. Sci., Ser. B **281**, 101
97. de Gennes PG and Prost J, (1993). *The Physics of Liquid Crystals*, Clarendon Press, Oxford

98. de Groot SR and Masur P (1984) Non-Equilibrium Thermodynamics, Dover, New York
99. De Wit A, Lima D, Dewel G, and Borckmans P (1996), Phys. Rev. E **54**, 261
100. Dewel G, Metens S, Hilali M, Borckmans P, and Price CB (1995), Phys. Rev. Lett. **74**, 4647
101. Doelman A (1995), Physica D **97**, 398
102. Doi M (1981) Polym. Sci., Polym. Phys. Ed. **19**, 229
103. Donnelly RJ (1991), *Quantized Vortices in Helium II*, Cambridge University Press, Cambridge
104. Doostmohammadi A, Ignés-Mullol J, Yeomans JM, and Sagués F (2018), Nat. Comm. **9**, 3246
105. Dubois-Violette E, Guazelli E, and Prost J (1983), Phil. Mag. A **48** 727
106. Echebarria B and Pérez-García C (1998), EuroPhys. Lett. **43**, 35
107. Echebarria B and Riecke H (2001), Physica D **158**, 45
108. Edwards BJ and Beris AN (1989), J. Rheol. **33**, 1189
109. Edwards S and Yeomans J (2009), Europhys. Lett. **85**, 18008
110. Efrati E, Sharon E, and Kupferman R (2013), Soft Matter **9**, 8187
111. Egolf DA and Greenside HS (1995), Phys. Rev. Lett. **74**, 1751
112. Elder KR, Gunton JD, and Goldenfeld N (1997), Phys. Rev. Lett. **56**, 1631
113. Elder KR, Katakowski M, Haataja M, and Grant M (2002), Phys. Rev. Lett. **88**, 245701
114. Elgeti J, Cates ME, and Marenduzzo D (2011), Soft Matter **7**, 3177
115. Elphick C, Meron E, and Spiegel EA (1988), Phys. Rev. Lett. **61**, 496
116. Elphick C, Hagberg A, and Meron E (1999), Phys. Rev. E **59**, 5285
117. Elphick C, Hagberg A, Malomed BA, and Meron E (1997), Phys. Lett. A **230**, 33
118. Ericksen JL (1961), Trans. Soc. Rheol **5**, 22
119. Evans R (1979), Adv. Phys. **28** 143
120. Faraday M (1831), Philos. Trans. R. Soc. Lond. **121**, 299
121. Fausti G, Tjhung E, Cates ME, and Nardini C (2021), Phys. Rev. Lett. **127**, 068001
122. Fife PC (1979), *Mathematical Aspects of Reacting and Diffusing Systems*, Springer-Verlag, Berlin
123. Fife PC (1985), J. Stat. Phys. **39**, 687
124. Fife PC (1988),*Dynamics of Internal Layers and Diffusive Interfaces*, SIAM, Philadelphia, PA
125. Fily Y and Marchetti MC (2012), Phys. Rev. Lett. **108**, 235702
126. Fily Y, Kafri Y, Solon AP, Tailleur J, and Turner A (2018), Physica A. **51** 044003
127. Finkelmann H, Ringsdorf H, and Wendorff JH (1978), Makromol. Chem. **179**, 273ÃŘ 276
128. FitzHugh R (1961), BioPhys. J. **1**, 445
129. Frank E (1946) Bull Amer. Math. Soc. **52**, 144
130. Frisch T and Rica S (1992), Physica D **61**, 155
131. Frisch U, She ZS, and Thual O (1986), J. Fluid Mech. **168**, 221
132. Frohoff-Hülsmann T, Wrembel J, and Thiele U, Phys. Rev. E **103**, 042602 (2021)

133. Frohoff-Hülsmann T, Thiele U, and Pismen LM (2023), Phil. Trans. R. Soc. A **381** 20220087
134. Gabbay M, Ott E, and Guzdar PN (1997), Phys. Rev. Lett. **78**, 2012
135. Gabbay M, Ott E, and Guzdar PN (1998), Physica D **118**, 371
136. Gallego R, Walgraef D, San Miguel M, and Toral R (2001), Phys. Rev. E **64**, 056218
137. Gershuni GZ and Zhukhovitsky EM (1976), *Convective Stability of Incompressible Fluids*, Keter, Jerusalem
138. Geyer D, Martin D, Tailleur J, and Bartolo D (2019), Phys. Rev. X **9**, 031043
139. Ginelli F (2016), Eur. Phys. J. Spec. Top. **225** 2099
140. Ginelli F and Chaté H (2010), Phys. Rev. Lett. **105** 168103
141. Ginzburg VL and Pitaevskii LP (1958), Sov. Phys. – JETP **34**, 858
142. Giomi L and DeSimone A (2014), Phys. Rev. Lett. **112** 147802
143. Giomi L and Marchetti MC (2012), Soft Matter **8**, 129
144. Giomi L, Marchetti MC, and Liverpool TB (2008), Phys. Rev. Lett. **101**, 198101
145. Giron B, Meerson B, and Sasorov PV (1998), Phys. Rev. E **58**, 4213
146. Glasner K and R Almgren (2000), Physica D **146** 328
147. Golovin AA, Nepomnyashchy AA, and Pismen LM (1994), Phys. Fluids **6**, 34
148. Golubitsky M and Schaeffer DG (1985), *Singularities and Groups in Bifurcation Theory*, Vols I, II (with I Stewart, 1988), Springer Verlag, New York
149. Gompper G et al (37 coauthors) (2020), J. Phys. Condens. Matter **32**, 193001
150. Goriely A (2006), J. Elasticity **84**, 281
151. Goryachev, A, Chaté H, and Kapral R (1999), Phys. Rev. Lett. **83**, 1878
152. Greenside HS and Coughran WM, Jr (1984), Phys. Rev. A **30**, 398
153. Grégoire G, Chaté H, and Tu Y (2003), Physica D **181**, 157
154. Griniasty I, Aharoni H, and Efrati E (2019), Phys. Rev. Lett. **123**, 127801
155. Griniasty I, Mostajeran C, and Cohen I (2021), Phys. Rev. Lett. **127**, 128001
156. Groh B and Mulder B (1999), Phys. Rev. E **59**, 5613
157. Großmann R, Romanczuk P, Bär M, and Schimansky-Geier L (2015), Eur. Phys. J. Spec. Top. **224**, 1325
158. Großmann R, Aranson IS, and Peruani F (2020), Nat. Comm. **11**, 5365
159. Hagan PS (1982), SIAM J. Appl. Math. **42**, 762
160. Hagberg A and Meron E (1994), Phys. Rev. Lett. **72**, 2494
161. Hagberg A, Yochelis A, Yizhak H, Elphick C, Pismen LM, and Meron E (2006), Physica D **217** 186
162. Haim D, Li G, Ouyang Q, McCormick WD, Swinney HL, Hagberg A, and Meron E (1996), Phys. Rev. Lett. **77**, 190
163. Haken, H (1983), *Synergetics* (1987), *Advanced Synergetics* (2004), *Synergetics, Introduction and Advanced Topics*, Springer Verlag, Berlin
164. Hakim V and Karma A (1999), Phys. Rev. E **60**, 5073
165. Harrington MJ, Razghandi K, Ditsch F, Guiducci L, Rueggeberg M, Dunlop JWC, Fratzl P, Neinhuis C, and Burgert I (2011), Nat. Commun. **2**, 337
166. Hébert M, Kant R, and de Gennes PG (1997), J. Phys. I **7**, 909
167. Helbing D and Molnár P (1995), Phys. Rev. E **51**, 4282
168. Helbing D, Farkas I, and Vicsek T (2000), Nature **407**, 487

169. Henry, H and Hakim V (2002), Phys. Rev. E **65**, 046235
170. Herant M and Dembo M (2010), Biophys. J. **98**, 1408
171. Hilali MF, Métens S, Borckmans P, and Dewel G (1995), Phys. Rev. E **51**, 2046
172. Hildebrand M, Kuperman M, Wio H, Mikhailov AS, and Ertl G (1999), Phys. Rev. Lett. **83**, 1475
173. Hoyle RB (1994), Phys. Rev. E **49**, 2875
174. Hoyle RB (2000), Phys. Rev. E **61**, 2506
175. Huber G, Alstrøm P, and Bohr T (1992), Phys. Rev. Lett. **69**, 2380
176. Ihle T (2014) Eur. Phys. J. Spec. Top., **223**, 1293
177. Ionov L, Stoychev G, Jehnichen D, and Sommer JU (2017), Appl. Mater. Interfaces **9**, 4873
178. Ipsen M, Kramer L, and Sørensen PG (2000), Phys. Rep. **337** 193
179. Ipsen M and van Hecke M (2001), Physica D **160**, 103
180. Ivantsov GP (1947), Dokl. Akad. Nauk SSSR **58**, 567
181. Jamgotchian H, Trivedi R, and Billia B (1993), Phys. Rev. E **47**, 4313
182. Janiaud B, Pumir A, Bensimon D, Croquette V, Richter H, and Kramer L (1992), Physica D **55**, 269
183. Joseph DD (1965), Phys. Fluids **8**, 2195
184. Kardar M, Parisi G, and Zhang Y (1986), Phys. Rev. Lett. **56**, 889
185. Karma A and Rappel W-J (1998), Phys. Rev. E **57**, 4323
186. Kawasaki K and Ohta T (1982), Physica A **116**, 573
187. Keener JP (1988), Physica D **31**, 269
188. Keener JP (1994), Physica D **70**, 61
189. Klein R and Majda A (1991), Physica D **53**, 267
190. Kléman M and Lavrentovich OD (2003) *Soft Matter Physics, An Introduction*, Springer-Verlag, New York
191. Knobloch E (2015), Annu. Rev. Cond. Matter Phys. **6**, 325
192. Knobloch E and de Luca J (1990), Nonlinearity **3**, 975
193. Kolmogorov A, Petrovsky I, and Piskunov N (1937), Bull. Moscow Univ., Math. Mech. **1**, 1
194. Köpf MH and Pismen LM (2013a), Eur. Phys. J. E **36**, 121
195. Köpf MH and Pismen LM (2013b), Physica D **259**, 48
196. Köpf MH and Pismen LM (2013c), Soft Matter **9**, 3727
197. Kramer L and Zimmermann W (1985), Physica D **16**, 221
198. Krischer K and Mikhailov AS (1994), Phys. Rev. Lett. **73**, 3165
199. Kruse K, Joanny J-F, Jülicher F, Prost J, and Sekimoto K, 2005 Eur. Phys. J. E **16**, 5
200. Kuramoto Y and Tsuzuki T (1976), Prog. Theor. Phys. **55**, 356
201. Kurtze, DA (1987), Phys. Rev. A **36**, 232
202. Kuznetsov EA, Nepomnyashchy AA and Pismen LM (1995), Phys. Lett. A **205**, 261
203. Landau (1944), Akad. Nauk Dokl. **44**, 339
204. Landau LD and Lifshitz EM (1970), *Theory of Elasticity*, Pergamon Press, Oxford,
205. Langer JS (1980), Rev. Mod. Phys. **52**, 1

206. Langer JS (1986), in *Directions in Condensed Matter*, p. 164, World Scientific, Singapore
207. La Porta A and Surko CM (2000), Physica D **139**, 177
208. Lauzeral J, Metens S, and Walgraef D (1993), EuroPhys. Lett. **24**, 707
209. Ledesma-Durán A and Aragón JL (2019), Sci. Rep. **9**, 11287
210. Lefever R, Nicolis G, and Prigogine I (1967), J. Chem. Phys. **47**, 1045
211. Lega J, Moloney JV, and Newell AC (1994), Phys. Rev. Lett. **73**, 2978
212. Leslie FM, Arch. Ration. Mech. Analysis **28**, 265 (1968)
213. Lestringant C and Audoly B (2017), J. Mech. Phys. Solids **103** 40
214. Li G, Qi Q, and Swinney HL (1996), J. Chem. Phys. **105**, 10830
215. Lifshitz IM and Slyozov VV (1958), Zh. Eksp. Teor. Fiz. **35**, 479 [Sov. Phys. JETP **8**, 331 (1959)]
216. Lin AL, Hagberg A, Meron E, and Swinney HL (2004), Phys. Rev. E **69**, 066217
217. Lin TS, Kondic L, Thiele U, and Cummings LJ, J. Fluid Mech. **729**, 214
218. Lorenz EN (1963), J. Atmos. Sci. **20**, 130
219. Mahault B, Patelli A and Chaté H (2018), J. Stat. Mech. 093202
220. Malomed BA (1984), Z. Phys. B, **55**, 241
221. Malomed BA, Nepomnyashchy AA, and Tribelsky MI (1989), Zh. Eksp. Teor. Fiz. **96**, 684
222. Malomed BA, Nepomnyashchy AA , and Tribelsky MI (1990), Phys. Rev. A **42**, 7244
223. Manneville P (1990), *Dissipative Structures and Weak Turbulence*, Academic Press, San Diego
224. Manneville P and Chaté H (1996), Physica D **96**, 30
225. Manneville P and Pomeau Y (1983), Philos. Mag. A **48**, 607
226. Marchetti MC, Joanny J-F, Ramaswamy S, Liverpool TB, Prost J, Rao M, and Simha RA (2013), Rev. Mod. Phys. **85**, 1143
227. Matthews PC and Cox S (2000), Nonlinearity, **13**, 1293
228. Meerson B and Sasorov PV (1996), Phys. Rev. E **53**, 3491
229. Meiron D and Newell AC (1985), Phys. Lett. A **113**, 289
230. Meixner M, De Wit A, Bose S, and Schöll E (1997), Phys. Rev. E **55**, 6690
231. Mermin ND (1979) Rev. Mod Phys. **51**, 591
232. Mikhailov AS (1991), *Foundations of Synergetics* I, *Distributed Active Systems* II (with AY Loskutov), *Complex Patterns*, Springer-Verlag, Berlin
233. Mikhailov AS, Davydov VA, and Zykov VS (1994), Physica D **70**, 1
234. Mineev-Weinstein M (1998), Phys. Rev. Lett. **80**, 2113
235. Modes CD and Warner M (2015), Phys. Rev. E **92**, 010401(R)
236. Mogilner A and Manhart A (2018), Annu. Rev. Fluid Mech. **50**, 347
237. Mostajeran C, Warner M, Ware TH, and White TJ (2016), Proc. R. Soc. A **472**, 20160112
238. Mullins WW and Sekerka RF (1963), J. Appl. Phys. **34**, 323
239. Murray JD (1989), *Mathematical Biology*, Springer, Berlin (2nd edn, 1993; 3rd edn, 2002/2003)
240. Nagumo J, Arimoto S, and Yoshizawa S (1962) Proc. IRE. **50**, 2061
241. Nédélec F, Surrey T, Maggs AC, and Leibler S (1997), Nature **389**, 305

242. Nepomnyashchy AA and Pismen LM (1991), Phys. Lett. A **153**, 427
243. Neu JC (1990), Physica D **43**, 385
244. Newell AC and Moloney JV (1992), *Nonlinear Optics*, Addison-Wesley, Redwood City, CA
245. Newell AC and Pomeau Y (1993), J. Phys. A **26**, L429
246. Newell AC and Whitehead JA (1969), J. Fluid Mech. **38**, 279
247. Newell AC, Passot T, and Lega J (1993), Ann. Rev. Fluid Mech. **25**, 184
248. Newell AC, Passot T, Bowman C, Ercolani N, and Indik R (1996), Physica D **97**, 185
249. Nicolis G (1995), *Introduction to Nonlinear Science*, Cambridge University Press, Cambridge
250. Nicolis G and Prigogine I (1977), *Self-Organization in Nonequilibrium Systems, From Dissipative Structures to Order through Fluctuations*, John Wiley, New York
251. Niethammer B and Pego RL (1999), J. Stat. Phys. **95**, 867
252. Normand C, Pomeau Y, and Velarde MG (1977), Rev. Mod Phys. **49**, 581
253. Nozaki K and Bekki N (1984), J. Phys. Soc. Jpn **53**, 1581
254. Nuz AE, Nepomnyashchy AA , and Pismen LM (1998), Physica A **299**, 179
255. Nuz AE, Nepomnyashchy AA , Golovin AA, Hari AA, and Pismen LM (2000), Physica D **135**, 233
256. Ohm C, Brehmer M, and Zentel R (2010), Adv. Mater. **22**, 3366
257. Ohta T, Mimura M, and Kobayashi R (1989), Physica D **34**, 115
258. Or-Guil M, Bode M, Schenk CP, and Purwins H-G (1998), Phys. Rev. E **57**, 6432
259. Ortoleva P and Ross J (1974), J. Chem. Phys. **60**, 5090
260. Ostwald W (1900), Z. Phys. Chem. **35**, 33
261. Ott E (1993) *Chaos in Dynamical Systems* (2nd edn 2002), Cambridge University Press, Cambridge
262. Ouyang Q and Swinney HL (1991), Nature **352**, 610
263. Palm E (1960), J. Fluid Mech. **8**, 183
264. Pampaloni, E, Residori S, Soria S, and Arecchi FT (1997), Phys. Rev. Lett. **78**, 1042
265. Park J-S and Lee KJ (1999), Phys. Rev. Lett. **83**, 5393
266. Park J-S and Lee KJ (2002), Phys. Rev. Lett. **88**, 224501
267. Park J-S, Woo S-J, and Lee KJ (2004), Phys. Rev. Lett. **93**, 098302
268. Passot T and Newell AC (1994), Physica D **74**, 301
269. Pelcé P and Sun J (1991), Physica D **48**, 353
270. Peña B and Pérez-García C (2000), EuroPhys. Lett. **51**, 300
271. Peruani F (2017), J. Phys. Soc. Jpn. **86**, 101010
272. Peruani F, Deutsch A, and Bär M (2006), Phys. Rev. E, **74**, 030904(R)
273. Peruani F, Deutsch A, and Bär M (2008), Eur. Phys. J. Spec. Top. **157**, 111
274. Peshkov A, Aranson IS, Bertin E, ChatÂńe H, and Ginelli F (2012), Phys. Rev. Lett. **109**, 268701
275. Peshkov A, Bertin E, Ginelli F, and Chaté H (2014), Eur. Phys. J. Spec. Top. **223**, 1315
276. Petrich DM and Goldstein RE (1994), Phys. Rev. Lett. **72**, 1120

277. Petrov V, Ouyang Q, and Swinney HL (1997), Nature **388**, 655
278. Pismen LM (1986), Dynamics and Stability of Systems **1**, 97
279. Pismen LM (1989), Physica D **36**, 1
280. Pismen LM (1994), J. Chem. Phys. **101**, 3135
281. Pismen LM (1999), *Vortices in Nonlinear Fields*, Clarendon Press, Oxford
282. Pismen LM (2001), Phys. Rev. E **64**, 021603
283. Pismen LM (2003), Physica D **184** 141
284. Pismen LM (2013), Phys. Rev. E **88**, 050502
285. Pismen LM (2021a) *Working with Dynamical Systems*, Taylor & Francis, Boca Raton FL
286. Pismen LM (2021b), *Active Matter Within and Around Us*, Springer Nature, Cham, Switzerland
287. Pismen LM (2022), Phys. Rev. E **106**, 034701
288. Pismen LM and Köpf MH (2014) Eur. Phys. J. Spec. Top. **223**, 1247
289. Pismen LM and Nepomnyashchy AA (1991), Phys. Rev. A **44**, R2243
290. Pismen LM and Nepomnyashchy AA (1992), Physica D **54**, 183
291. Pismen LM and Nepomnyashchy AA (1993), EuroPhys. Lett. **24**, 461
292. Pismen LM and Nepomnyashchy AA (1994), EuroPhys. Lett. **27**, 433
293. Pismen LM and Rodriguez JD (1990), Phys. Rev. A **42**, 2471
294. Pismen LM and Rubinstein BY (1999a), Int. J. Bifurcations Chaos **9**, 983
295. Pismen LM and Rubinstein BY (1999b), Chaos Solitons Fractals **10**, 761
296. Pleiner H, Svenšek D, and Brand HR (2016), Rheol. Acta **55**, 857.
297. Pomeau Y (1984), Lect. Notes Phys. **210**, 207
298. Pomeau Y (1986), Physica D **23**, 3
299. Pomeau Y and Manneville P (1979), J. Physique Lett. **40**, L609
300. Pomeau Y, Zaleski S, and Manneville P (1983), Phys. Rev. A **27**, 2710
301. Popp S, Stiller O, Aranson I, and Kramer L (1995), Physica D **84**, 398
302. Poujade M, Grasland-Mongrain E, Hertzog A, Jouanneau J, Chavrier P, Ladoux B, Buguin A, and Silberzan P (2007), Proc. Natl. Acad. Sci. USA **104**, 15988
303. Powell JA, Newell AC, and Jones CKRT (1991), Phys. Rev. A **44**, 3636
304. Proctor MRE and Jones CA (1988), J. Fluid Mech. **188**, 301
305. Purwins HG, Bodeker HU and Amiranashvili S (2010) Adv. Phys. **59** 485
306. Qian H and Mazenko GF (2004), Phys. Rev. E **69**, 011104
307. Qian T and Sheng P (1998), Phys. Rev. E **58** 7475
308. Rabinovich MI and Tsimring LS (1994), Phys. Rev. E **49**, R35
309. Ramaswamy S (2010), Ann. Rev. Cond. Matter Phys. **1**, 323
310. Rayleigh, Lord (1916), Proc. R. Soc. Lond. A **93**, 148
311. Rey AD (2004), J. Chem. Phys. **120**, 2010
312. Rey AD (2010), Soft Matter **6**, 3402
313. Reynolds WN, Pearson JE, and Ponce-Dawson S (1994), Phys. Rev. Lett. **72**, 2797
314. Ricca RL (1996), Fluid Dyn. Res. **18**, 245
315. Riecke H and Kramer L (2000), Physica D **137** 124
316. Rodriguez EK, Hoger A, and McCulloch AD (1994), J. Biomech. **27**, 455
317. Rodriguez JD, Pismen LM, and Sirovich L (1991), Phys. Rev. A **44**, 7980

318. Rønning J, Marchetti CM, Bowick MJ, and Angheluta L (2022), Proc. R. Soc. A **478**, 20210879
319. Rosenthal SB, Twomey CR, Hartnett AT, Wu HS, and Couzin ID (2015), Proc. Natl. Acad. Sci. USA **112**, 4690
320. Rössler OE (1976), Phys. Lett. A **57**, 397
321. Rousseau G, Chaté H, and Kapral R (1998), Phys. Rev. Lett. **80**, 5671
322. Sakaguchi H and Brand HR (1996), Physica D **117** 95
323. Sánchez-Garduño F and Castellanos V (2022), Lobachevskii J. Math. **43**, 141
324. Sandstede, B and Scheel A (2001), Phys. Rev. Lett. **86**, 171
325. Schenk, CP, Or-Guil M, Bode M, and Purwins H-G (1997), Phys. Rev. Lett. **78**, 3781
326. Schöpf W and Zimmermann W (1993), Phys. Rev. E **47**, 1739
327. Schwarz KW (1985), Phys. Rev. B **31**, 5782
328. Segel LA (1969), J. Fluid Mech. **38**, 203
329. Shankar S (2022), Phys. Today (8) 54
330. Shechtman, D, Blech I, Gratias D, and Cahn J (1984), Phys. Rev. Lett. **53**, 1951
331. Shilnikov LP (1965) Sov. Math. Dokl. **6**, 163
332. Shin H, Bowick MJ, and Xing X (2008), Phys. Rev. Lett. **101**, 037802
333. Shraiman BI, Pumir A, van Saarloos W, Hohenberg PC, Chaté H, and Holen M (1992), Physica D **57**, 241
334. Siggia ED and Zippelius A (1981), Phys. Rev. A **24**, 1036
335. Simha RA and Ramaswamy S (2002), Phys. Rev. Lett. **89**, 058101
336. Sirovich L and Everson R (1992), J. Supercomp. Appl. High Perf. Comp. **6**, 50
337. Sivashinsky GI (1977), Acta Astronaut. **4**, 1177
338. Sivashinsky GI (1983), Physica D **8**, 243
339. Solon AP, Caussin JB, Bartolo D, Chat/'e H, and Tailleur J (2015), Phys. Rev. E **92**, 062111
340. Solon AP, Fily Y, Baskaran A, Cates ME, Kafri Y, Kadar M, and Tailleur J (2015), Nat. Phys. **11**, 673
341. Solon AP, Stenhammar J, Cates ME, Kafri Y, and Tailleur J (2018), Phys. Rev. E **97**, 020602
342. Stark H and Lubensky TC (2005) Phys. Rev. E **72**, 051714
343. Strömbom D, Siljestam M, Park J, and Sumpter DJT (2015), Eur. Phys. J. Spec. Top. **224**, 3311
344. Swift J and Hohenberg P (1977), Phys. Rev. A **15**, 319
345. Takashima M (1981), J. Phys. Soc. Jpn. **50**, 2745
346. Tesauro G and Cross MC (1986), Phys. Rev. A **34**, 1363
347. Tesauro G and Cross MC (1987), Philos. Mag. A **56**, 703
348. Thampi SP, Golestanian R, and Yeomans JM (2014) Phil. Trans. R. Soc. A **372**, 20130366
349. Thiele U, Archer AJ, Robbins MJ, Gomez H, and Knobloch E (2013), Phys. Rev. E **87**, 042915
350. Thom R (1975), *Structural Stability and Morphogenesis, an Outline of a General Theory of Models*, Benjamin, Reading, MA
351. Timoshenko S (1925), J. Opt. Soc. Am. 11.3

352. Tjhung E, Tiribocchi A, Marenduzzo D, and Cates ME (2015). Nat. Commun. **6**, 5420
353. Tjhung E, Nardini C, and Cates ME (2018), Phys. Rev. X **8**, 031080
354. Toner J and Tu Y (1998), Phys. Rev. E **58**, 4828
355. Toner J, Tu Y, and Ramaswamy S (2005), Ann. Phys. **318**, 170
356. Tsimring LS (1996) Physica D **89**, 368
357. Turing AM (1952), Philos. Trans. R. Soc. London B **237**, 37
358. van der Waals JD (1894), Z. Phys. Chem. **13** 657 [English translation]
359. van Hecke M (1998), Phys. Rev. Lett. **80**, 1896
360. van Hecke M, Storm C, and van Saarloos W (1999), Physica D **134**, 1
361. van Saarloos W (1988), Phys. Rev. A **37**, 211
362. van Saarloos W (1989), Phys. Rev. A **39**, 6367
363. van Saarloos W (2003), Phys. Rep. **386**, 29
364. van Saarloos W and Hohenberg PC (1992), Physica D **56**, 303
365. Vicsek T, Czirok A, Ben-Jacob E, Cohen I, and Shochet O (1995), Phys. Rev. Lett. **75**, 1226
366. Vitelli V and Nelson DR (2006), Phys. Rev. E **74**, 021711
367. Voituriez R, Joanny J-F, and Prost J (2005), Europhys. Lett. **70** 404
368. Walgraef D (1997), *Spatio-Temporal Pattern Formation*, Springer, New York
369. Wang K, Steyn-Ross ML, Steyn-Ross DA, and Wilson, MT (2014), Int. J. Bifurc. Chaos **24** 1450101
370. Warner M and Terentjev EM (2003), *Liquid Crystal Elastomers*, Clarendon Press, Oxford
371. Wiener N and Rosenblueth A (1946), Arch. Inst. Cardiol. Mexico **205**, 3
372. Winfree AT (1987), *When Time Breaks Down*, Princeton University Press, Princeton, NJ
373. Winfree AT (1991), Chaos **1**, 303
374. Wittkowski R, Tiribocchi A, Stenhammar J, Allen RJ, Marenduzzo D, and Cates ME (2014), Nat. Commun. **5**, 4351
375. Wollkind DJ and Segel LA (1970), Philos Mag A **268**, 351
376. Wolpert L (2007), *Principles of development*, 3rd edition, Oxford University Press.
377. Yang L, Dolnik M, Zhabotinsky AM, and Epstein IR (2002), Phys. Rev. Lett. **88**, 208303
378. Yochelis A, Elphick C, Hagberg A, and Meron E (2004), Physica D **199**, 201
379. Yochelis A and Pismen LM (2006), Coll. Surf. A **274** 170
380. Zakharov AP, Leshansky AM, and Pismen LM (2016), Eur. Phys. J. E **39** 87
381. Zakharov AP and Pismen LM (2017), Soft Matter **13**, 2886
382. Zakharov AP and Pismen LM (2018), Phys. Rev. E **97**, 062705
383. Zakharov AP and Pismen LM (2019) Phys. Rev. Mater. **3**, 055603
384. Zaleski S, Pomeau Y, and Pumir A (1984), Phys. Rev. A **29**, 366
385. Zeldovich YB (1985), *The Mathematical Theory of Combustion and Explosions*, Consultants Bureau, New York
386. Zhabotinskii AM (1974), *Concentration Auto-oscillations* (in Russian), Nauka, Moscow

387. Ziebert F and Aranson IS, 2013, PLoS One **8**, e64511
388. Ziebert F, Swaminathan S, and Aranson IS (2012), J. R. Soc. Interface **9**, 1084
389. Ziebert F, Löber J, and Aranson IS (2016), in *Physical Models of Cell Motility*, ed. Aranson IS, p.1, Springer, Cham

Index

L. Pismen, *Patterns and Interfaces in Dissipative Dynamics*, Springer Series in Synergetics,
https://doi.org/10.1007/978-3-031-29579-9